AF790150

Operator Theory: Advances and Applications
Vol. 122

Editor:
I. Gohberg

Editorial Office:
School of Mathematical Sciences
Tel Aviv University
Ramat Aviv, Israel

Editorial Board:
J. Arazy (Haifa)
A. Atzmon (Tel Aviv)
J. A. Ball (Blacksburg)
A. Ben-Artzi (Tel Aviv)
H. Bercovici (Bloomington)
A. Böttcher (Chemnitz)
K. Clancey (Athens, USA)
L. A. Coburn (Buffalo)
K. R. Davidson (Waterloo, Ontario)
R. G. Douglas (Stony Brook)
H. Dym (Rehovot)
A. Dynin (Columbus)
P. A. Fillmore (Halifax)
P. A. Fuhrmann (Beer Sheva)
S. Goldberg (College Park)
B. Gramsch (Mainz)
G. Heinig (Chemnitz)
J. A. Helton (La Jolla)
M.A. Kaashoek (Amsterdam)
H.G. Kaper (Argonne)
S.T. Kuroda (Tokyo)
P. Lancaster (Calgary)
L.E. Lerer (Haifa)
E. Meister (Darmstadt)
B. Mityagin (Columbus)
V. V. Peller (Manhattan, Kansas)
J. D. Pincus (Stony Brook)
M. Rosenblum (Charlottesville)
J. Rovnyak (Charlottesville)
D. E. Sarason (Berkeley)
H. Upmeier (Marburg)
D. Voiculescu (Berkeley)
H. Widom (Santa Cruz)
D. Xia (Nashville)
D. Yafaev (Rennes)

Honorary and Advisory Editorial Board:
C. Foias (Bloomington)
P. R. Halmos (Santa Clara)
T. Kailath (Stanford)
P. D. Lax (New York)
M. S. Livsic (Beer Sheva)

Operator Theory and Analysis

The M.A. Kaashoek Anniversary Volume

Workshop in Amsterdam, November 12–14, 1997

H. Bart
I. Gohberg
A.C.M. Ran
Editors

Springer Basel AG

Editors:

H. Bart
Econometrisch Instituut
Erasmus Universiteit Rotterdam
Postbus 1738
3000 DR Rotterdam
The Netherlands
e-mail: bart@few.eur.nl

A.C.M. Ran
Divisie Wiskunde en Informatica
Faculteit der Exacte Wetenschappen
Vrije Universiteit
De Boelelaan 1081a
1081 HV Amsterdam
The Netherlands
e-mail: ran@cs.vu.nl

I. Gohberg
Department of Mathematical Sciences
Raymond and Beverly Sackler
Faculty of Exact Sciences
Tel Aviv University
69978 Ramat Aviv
Israel
e-mail: gohberg@math.tau.ac.il

2000 Mathematics Subject Classification 47-06

A CIP catalogue record for this book is available from the
Library of Congress, Washington D.C., USA

Deutsche Bibliothek Cataloging-in-Publication Data
Operator theory and analysis : the M. A. Kaashoek anniversary volume ;
workshop in Amsterdam, November 12 - 14, 1997 / H. Bart ... ed..
- Basel ; Boston ; Berlin : Birkhäuser, 2001
(Operator theory ; Vol. 122)

ISBN 978-3-0348-9502-6 ISBN 978-3-0348-8283-5 (eBook)
DOI 10.1007/978-3-0348-8283-5

This work is subject to copyright. All rights are reserved, whether the whole or part of the material is concerned, specifically the rights of translation, reprinting, re-use of illustrations, recitation, broadcasting, reproduction on microfilms or in other ways, and storage in data banks. For any kind of use permission of the copyright owner must be obtained.

© 2001 Springer Basel AG
Originally published by Birkhäuser Verlag Basel,Switzerland in 2001
Softcover reprint of the hardcover 1st edition 2001
Printed on acid-free paper produced from chlorine-free pulp. TCF ∞
Cover design: Heinz Hiltbrunner, Basel

ISBN 978-3-0348-9502-6

9 8 7 6 5 4 3 2 1

Preface

On November 12–14, 1997 a workshop was held at the Vrije Universiteit Amsterdam on the occasion of the sixtieth birthday of M.A. Kaashoek. The present volume contains the proceedings of this workshop.

The workshop was attended by 44 participants from all over the world: participants came from Austria, Belgium, Canada, Germany, Ireland, Israel, Italy, The Netherlands, South Africa, Switzerland, Ukraine and the USA. The atmosphere at the workshop was very warm and friendly. There where 21 plenary lectures, and each lecture was followed by a lively discussion.

The workshop was supported by: the Vakgroep Wiskunde of the Vrije Universiteit, the department of Mathematics and Computer Science of the Vrije Universiteit, the Stichting VU Computer Science & Mathematics Research Centre, the Thomas Stieltjes Institute for Mathematics, and the department of Economics of the Erasmus University Rotterdam.

The organizers would like to take this opportunity to express their gratitude for the support. Without it the workshop would not have been so successful as it was.

Table of Contents

M.A. Kaashoek, 2000

Curriculum Vitae of M.A. Kaashoek

Personal data:

Name: Marinus (Rien) A. Kaashoek
Born: Ridderkerk, The Netherlands; November 10, 1937
Degree: Ph. D., University of Leiden, The Netherlands, 1964; supervisor: A.C. Zaanen
Married: with Wilhelmina (Wies) E.J. Bakema; three children

Research interests:

Analysis and Operator Theory, and various connections between Operator Theory, Matrix Theory and Mathematical Systems Theory. In particular, Wiener-Hopf integral equations and Toeplitz operators and their nonstationary variants. State space methods for problems in Analysis. Also metric constrained interpolation problems, and various extension and completion problems for partially defined matrices or operators.

Positions:

- Assistant, University of Leiden, 1959–1962
- Junior staff member, University of Leiden, 1962–1965
- Senior Lecturer, Vrije Universiteit, Amsterdam, 1966–1969
- Professor, Vrije Universiteit, Amsterdam, 1969–

Visiting professorships:

- University of Maryland at College Park, January–July, 1975
- University of Calgary, Calgary, Alberta, Canada, August–September, 1981
- Ben Gurion University of the Negev, Beer Sheva, Israel, April, 1987
- Tel-Aviv University, Tel-Aviv, Israel, various periods

Fellowships:

- Z.W.O. Fellowship for postdoctoral studies at the University of California at Los Angeles, 1965–1966
- Royal Irish Academy Senior Visiting Fellowship, Spring 1970
- British Science Research Council Senior Visiting Fellowship, May–June, 1976

Award: Toeplitz Lecturer, Tel-Aviv, April 1991

Publications:

Author or co-author of more than 140 papers, many of which are written together with I. Gohberg. Co-author of six books. See the separate list of publications.

Ph. D. students:

1. N.P. Dekker, 1969 (supervisor: P.C. Baayen); subject: Joint numerical range and joint spectrum of Hilbert space operators.
2. H. Bart, 1973; subject: Meromorphic operator valued functions.
3. G.Ph.A. Thijsse, 1978; subject: Decomposition theorems for finite meromorphic operator functions.
4. H. den Boer, 1981; subject: Block diagonalization of matrix functions.
5. C.V.M. van der Mee, 1981 (co-supervisor: I. Gohberg); subject: Semigroup and factorization methods in transport theory.
6. A.C.M. Ran, 1984; subject: Semidefinite invariant subspaces, stability and applications.
7. L. Roozemond, 1987 (co-supervisor: I. Gohberg); subject: Systems of non-normal and first kind Wiener-Hopf equations.
8. H.J. Woerdeman, 1989 (co-supervisor: I. Gohberg); subject: Matrix and operator extensions.
9. R. Vreugdenhil, 1990 (co-supervisors: I. Gohberg, A.C.M. Ran); subject: Spectral theory of selfadjoint Wiener-Hopf and Toeplitz operators with rational symbols.
10. A.B. Kuijper, 1992 (co-supervisor: I. Gohberg); subject: The state space method for integro-differential equations of convolution type with rational matrix symbols.
11. G.J. Groenewald, 1993; subject: Wiener-Hopf factorization of rational matrix functions in terms of realization: an alternative approach.
12. J. Kos, 1995; (co-supervisor: I. Gohberg); subject: Time-dependent problems in linear operator theory.
13. D. Temme, 1996; (co-supervisors: H. Langer and A.C.M. Ran); subject: Dissipative operators in indefinite scalar product spaces.
14. D.R. Pik, 1999; (co-supervisor: D.Z. Arov); subject: Block lower triangular operators and optimal contractive systems.

Editorial work:

Member of the editorial board of

- Integral Equations and Operator Theory
- Operator Theory: Advances and Applications
- Asymptotic Analysis
- Mathematische Nachrichten
- Russian Mathematical Encyclopaedia

Co-editor of

- From A to Z, MC Tract 149, Mathematisch Centrum, 1982.
- Operator Theory and Systems, OT 19, Birkhäuser Verlag, Basel, 1986.
- Constructive Methods of Wiener-Hopf Factorization, OT 21, Birkhäuser Verlag, Basel, 1986.

- Linear Algebra in Signals, Systems, and Control, SIAM, 1988.
- The Gohberg Anniversary Collection, Vol. I, II, Birkhäuser Verlag, Basel, 1989.
- Proceedings of the International Symposium MTNS-89, Vol. I–III, Birkhäuser Verlag, Boston, 1990.
- Topics in Matrix and Operator Theory, OT 50, Birkhäuser Verlag, Basel, 1991.
- Challenges of a Generalized System Theory, Royal Netherlands Academy of Arts and Sciences, Verhandelingen, Afd. Natuurkunde, Eerste Reeks, deel 40, 1993.
- Papers dedicated to Jaap Korevaar, Indagationes Mathematicae, New Series 4(4), December 1993.
- Operator Theory in Function Spaces and Banach Lattices. The A.C. Zaanen Anniversary Volume, OT 75, Birkhäuser Verlag, Basel, 1995.
- Contributions to Operator Theory in Spaces with an Indefinite Metric. The Heinz Langer Anniversary Volume, OT 106, Birkhäuser Verlag, Basel, 1998.

Services:

- Member of various committees of the Vrije Universiteit and of the Dutch mathematical community.
- Member Board Faculty of Mathematics and Natural Sciences, 1973–1974.
- Chairman research committee of the Department of Mathematics and Computer Science of the Vrije Universiteit, 1978–1995
- Chairman Board Dutch Mathematical Society, 1980–1981.
- Co-organizer of the Amsterdam workshop on "Operator Theory and its Applications", 1986.
- Member and lecturer Dutch Network "System and Control Theory", 1987–1993.
- Member Tagungsleitung "Wiener-Hopf-Probleme", Oberwolfach 1986 and 1989.
- Co-organizer of the international conference on "Operator Theory: Advances and Applications", Calgary, 1988.
- Co-chairman of the 1989 International Symposium on the "Mathematical Theory of Networks and Systems" (MTNS-89) at Amsterdam.
- Member Organizing Committee of the Rotterdam workshop on "Matrix and Operator Theory", 1989.
- Member Steering Committee MTNS since 1989. Chairman of the committee 1989–1991.
- Member Steering Committee Workshops "Operator Theory and its Applications", since 1991.

- Member Organizing Committee of the "International Symposium on the Mathematical Theory of Networks and Systems" at Kobe (MTNS-91), at Regensburg (MTNS-93), at Padua (MTNS-98), at Perpignan (MTNS-2000).
- Chairman Board Thomas Stieltjes Institute for Mathematics, since 1993.
- Member Advisory Board of the Mathematical Institute of Leiden University, since 1996.
- Co-ordinator INTAS projects 93-0249, 93-0249-EXT and 97-M016.
- Dean Faculty of Mathematics and Computer Science of the Vrije Universiteit, October 1996–February 1998.
- Dean Faculty of Sciences of the Vrije Universiteit, since March 1998.

Publications of M.A. Kaashoek

Dissertation

Closed linear operators on Banach spaces, University of Leiden, Amsterdam, 1964.

Books

1. M.A. Kaashoek and T.T. West, *Locally compact semi-algebras with applications to spectral theory of positive operators*, North-Holland Mathematics Studies, vol. 9, North-Holland, Amsterdam, 1974, 102.
2. H. Bart, I. Gohberg and M.A. Kaashoek, *Minimal factorization of matrix and operator functions*, Birkhäuser Verlag, Basel, 1979, 277.
3. I. Gohberg, S. Goldberg and M.A. Kaashoek, *Classes of Linear Operators*, volume I, Birkhäuser Verlag, Basel, 1990, 468.
4. I. Gohberg, S. Goldberg and M.A. Kaashoek, *Classes of Linear Operators*, volume II, Birkhäuser Verlag, Basel, 1993, 552.
5. I. Gohberg, M.A. Kaashoek and F. van Schagen, *Partially specified matrices and operators: classification, completion, applications*, Birkhäuser Verlag, Basel, 1995, 333.
6. C. Foias, A. Frazho, I. Gohberg, and M.A. Kaashoek, *Metric constrained interpolation, commutant lifting and systems*, Birkhäuser Verlag, Basel, 1998, 587.

Papers

1. M.A. Kaashoek, Closed linear operators on Banach spaces, *Proc. Acad. Sci. Amsterdam* A **68** (1965), 405–414.
2. M.A. Kaashoek, Stability theorems for closed linear operators, *Proc. Acad. Sci. Amsterdam* A **68** (1965), 452–466.
3. M.A. Kaashoek, Ascent, descent, nullity and defect, a note on a paper by A.E. Taylor, *Math. Ann.* **172** (1967), 105–115.
4. M.A. Kaashoek and D.C Lay, On operators whose Fredholm set is the complex plane, *Pacific J. Math.* **21** (1967), 275–278.
5. M.A. Kaashoek and T.T. West, Locally compact monothetic semi-algebras, *Proc. London Math. Soc.* **18** (1968), 428–438.
6. M.A. Kaashoek, On the Riesz set of a linear operator, *Proc. Acad. Sci. Amsterdam* A **71** (1968), 46–53.
7. M.A. Kaashoek and T.T. West, Semi-simple locally compact monothetic semi-algebras, *Proc. Edinburgh Math. Soc.* **16** (1969), 215–219.

8. M.A. Kaashoek and T.T. West, Compact semigroups in commutative Banach algebras, *Proc. Camb. Phil. Soc.* **66** (1969), 265–274.
9. M.A. Kaashoek, Locally compact semi-algebras and spectral theory, *Nieuw Archief voor Wiskunde* **17** (1969), 8–16.
10. M.A. Kaashoek, On the peripheral spectrum of an element in a strict closed semi-algebra, in: *Hilbert space operators*, Colloquia Mathematica Societatis Janos Bolyai 5, 1970, Amsterdam, 1971, 319–332.
11. M.A. Kaashoek and M.R. Smyth, On operators T such that $f(T)$ is Riesz or meromorphic, *Proc. Royal Irish Acad.* **72** (section A) (1972), 81–87.
12. M.A. Kaashoek and D.C Lay, Ascent, descent, and commuting perturbations, *Trans. Amer. Math. Soc.* **169** (1972), 35–47.
13. M.A. Kaashoek, A note on the spectral properties of linear operators leaving invariant a convex set, *Proc. Acad. Sci. Amsterdam* A **76** (1973), 254–262.
14. H. Bart, M.A. Kaashoek and D.C. Lay, Stability properties of finite meromorphic operator functions, *Proc. Acad. Sci. Amsterdam* A **77** (1974), 217–259.
15. K.H. Förster and M.A. Kaashoek, The asymptotic behavior of the reduced minimum modulus of a Fredholm operator, *Proc. Amer. Math. Soc.* **49** (1975), 123–131.
16. H. Bart, M.A. Kaashoek and D.C. Lay, Relative inverses of meromorphic operator functions and associated holomorphic projection functions, *Math. Ann.* **218** (1975), 199–210.
17. I. Gohberg, M.A. Kaashoek and D.C. Lay, Spectral classification of operators and operator functions, *Bull. Amer. Math. Soc.* **82** (1976), 587–589.
18. M.A. Kaashoek, Recent developments in the spectral analysis of matrix and operator polynomials, in: *Proceedings Bicentennial Congress of the Wiskundig Genootschap*, part II, Math. Centre, Tract 101, Mathematical Centre, Amsterdam, 1978, 233–247.
19. M.A. Kaashoek, Commentary on a paper by C. Visser, in *Two decades of mathematics in the Netherlands*, Mathematical Centre, Amsterdam, 1978, 316–320.
20. I. Gohberg, M.A. Kaashoek and D.C. Lay, Equivalence, Linearization and Decomposition of holomorphic operator functions, *J. Funct. Analysis* **28** (1978), 102–144.
21. I. Gohberg, M.A. Kaashoek and F. van Schagen, Common multiples of operator polynomials with analytic coefficients, *Manuscripta Math.* **25** (1978), 279–314.
22. H. Bart, I. Gohberg and M.A. Kaashoek, Operator polynomials as inverses of characteristic functions, *Integral Equations and Operator Theory* **1** (1978), 1–8.
23. H. Bart, M.A. Kaashoek and D.C. Lay, The integral formula for the reduced algebraic multiplicity of meromorphic operator functions, *Proc. Edinb. Math. Soc.* **21** (1978), 65–72.

24. I. Gohberg and M.A. Kaashoek, Unsolved problems in matrix and operator theory, I. Partial multiplicities and additive perturbation, *Integral Equations and Operator Theory* **1** (1978), 278–283.
25. I. Gohberg, M.A. Kaashoek and L. Rodman, Spectral analysis of families of operator polynomials and a generalized Vandermonde matrix, I. The finite dimensional case, in: *Topics in Functional Analysis*. Advances in Mathematics, Supplementary Studies, vol. 3, Academic Press, London 1978, 91–128.
26. H. Bart, I. Gohberg and M.A. Kaashoek, Stable factorizations of monic matrix polynomials and stable invariant subspaces, *Integral Equations and Operator Theory* **1** (1978), 496–517.
27. I. Gohberg, M.A. Kaashoek and L. Rodman, Spectral analysis of families of operator polynomials and a generalized Vandermonde matrix, II. The infinite dimensional case, *J. Funct. Analysis* **30** (1978), 358–389.
28. I. Gohberg and M.A. Kaashoek, Unsolved problems in matrix and operator theory, II. Partial multiplicities for products, *Integral Equations and Operator Theory* **2** (1979), 116–120.
29. I. Gohberg, M.A. Kaashoek and F. van Schagen, Similarity of operator blocks and canonical forms. I. General results. Feedback equivalence and Kronecker indices, *Integral Equations and Operator Theory* **3** (1980), 350–396.
30. M.A. Kaashoek and M.P.A. van de Ven, A linearization for operator polynomials with coefficients in certain operator ideals, *Annali Mat. pura appl.* (IV) **15** (1980), 329–336.
31. H. Bart, I. Gohberg, M.A. Kaashoek and P. Van Dooren, Factorizations of transfer functions, *SIAM J. Control Opt.* **18** (1980), 675–696.
32. I. Gohberg, M.A. Kaashoek, L. Lerer and L. Rodman, Common multiples and common divisors of matrix polynomials, I. Spectral method, *Indiana University Mathematics Journal* **30** (1981), 321–356.
33. I. Gohberg, M.A. Kaashoek and F. van Schagen, Similarity of operator blocks and canonical forms, II. Infinite dimensional case and Wiener-Hopf factorization, in: *Topics in Modern Operator Theory*, OT 2, Birkhäuser Verlag, 1981, 121–170.
34. M.A. Kaashoek, C.V.M. van der Mee and L. Rodman, Analytic operator functions with compact spectrum, I. Spectral nodes, linearization and equivalence, *Integral Equations and Operator Theory* **4** (1981), 504–547.
35. M.A. Kaashoek, C.V.M. van der Mee and L. Rodman, Spectral analysis of analytic operator functions with compact spectrum, in: *Proc. Int. Symposium on Mathematical Theory of Networks and Systems* (Ed. N. Levan), Santa Monica, 1981, 134–138.
36. H. Bart, I. Gohberg and M.A. Kaashoek, Wiener-Hopf integral equations, Toeplitz matrices and linear systems, in: *Toeplitz Centennial* (Ed. I. Gohberg), OT 4, Birkhäuser Verlag, 1982, 85–135.

37. H. Bart, I. Gohberg and M.A. Kaashoek, Convolution equations and linear systems, *Integral Equations and Operator Theory* **5** (1982), 283–340.
38. M.A. Kaashoek, C.V.M. van der Mee and L. Rodman, Analytic operator functions with compact spectrum, II. Spectral pairs and factorization, *Integral Equations and Operator Theory* **5** (1982), 791–827.
39. I. Gohberg, M.A. Kaashoek, L. Lerer and L. Rodman, Common multiples and common divisors of matrix polynomials, II. Vandermonde and resultant matrices, *Linear and Multilinear Algebra* **12** (1982), 159–203.
40. I. Gohberg, M.A. Kaashoek and F. van Schagen, Rational matrix and operator functions with prescribed singularities, *Integral Equations and Operator Theory* **5** (1982), 673–717.
41. M.A. Kaashoek, C.V.M. van der Mee and L. Rodman, Analytic operator functions with compact spectrum, III. Hilbert space case: inverse problem and applications, *J. Operator Theory* **10** (1983), 219–250.
42. H. Bart, I. Gohberg and M.A. Kaashoek, The coupling method for solving integral equations, in: *Topics in Operator Theory, Systems and Networks, The Rehovot Workshop* (Eds. H. Dym, I. Gohberg), OT 12, Birkhäuser Verlag, Basel, 1984, 39–73. Addendum, *Integral Equations and Operator Theory* **8** (1985), 890–891.
43. H. Bart, I. Gohberg and M.A. Kaashoek, Wiener-Hopf factorization and realization, in: *Mathematical Theory of Networks and Systems, Proceedings of the MTNS-83 International Symposium*, Beer Sheva, Israel (Ed. P. Fuhrmann), Lecture notes in Control and Information Sciences, nr. 58, Springer Verlag, Berlin, 1984, 42–62.
44. I. Gohberg and M.A. Kaashoek, Time varying linear systems with boundary conditions and integral operators, I. The transfer operator and its properties, *Integral Equations and Operator Theory* **7** (1984), 325–391.
45. I. Gohberg, M.A. Kaashoek, L. Lerer and L. Rodman, Minimal divisors of rational matrix functions with prescribed zero and pole structure, in: *Topics in Operator Theory, Systems and Networks, The Rehovot Workshop* (Eds. H. Dym, I. Gohberg), OT 12, Birkhäuser Verlag, Basel, 1984, 241–275.
46. I. Gohberg, M.A. Kaashoek and F. van Schagen, Non-compact integral operators with semi-separable kernels and the discrete analogues: inversion and Fredholm properties, *Integral Equations and Operator Theory* **7** (1984), 642–703.
47. M.A. Kaashoek, Minimal factorization, linear systems and integral operators, in: *Operator and Function Theory* (Ed. S.C. Power), Reidel Publ. Co., 1985, 41–86.
48. H. Bart, I. Gohberg and M.A. Kaashoek, Fredholm theory of Wiener-Hopf equations in terms of realization of their symbols, *Integral Equations and Operator Theory* **8** (1985), 590–613.
49. M.A. Kaashoek, Analytic equivalence of the boundary eigenvalue operator function and its characteristic matrix function, *Integral Equations and Operator Theory* **9** (1986), 275–285.

50. I. Gohberg and M.A. Kaashoek, On minimality and stable minimality of time-varying linear systems with well-posed boundary conditions, *Int. J. Control* **43** (1986), 1401–1411.
51. I. Gohberg, M.A. Kaashoek and L. Lerer, Minimality and irreducibility of time-invariant linear boundary-value systems, *Int. J. Control* **44** (1986), 363–379.
52. H. Bart, I. Gohberg and M.A. Kaashoek, Wiener-Hopf factorization, inverse Fourier transforms and exponentially dichotomous operators, *J. Funct. Analysis* **68** (1986), 1–42.
53. I. Gohberg and M.A. Kaashoek, Various minimalities for systems with boundary conditions and integral operators, in: *Modelling, Identification and Robust Control* (Eds. C.I. Byrnes and A. Lindquist), North-Holland, 1986, 181–196.
54. I. Gohberg and M.A. Kaashoek, Similarity and reduction for time varying linear systems with well-posed boundary conditions, *SIAM J.Control Opt.* **24** (1986), 961–978.
55. H. Bart, I. Gohberg and M.A. Kaashoek, Wiener-Hopf equations with symbols analytic in a strip, in: *Constructive methods of Wiener-Hopf factorization* (Eds. I. Gohberg and M.A. Kaashoek), OT 21, Birkhäuser Verlag, Basel, 1986, 39–74.
56. I. Gohberg, M.A. Kaashoek, L. Lerer and L. Rodman, On Toeplitz and Wiener-Hopf operators with contourwise rational matrix and operator symbols, in: *Constructive methods of Wiener-Hopf factorization* (Eds. I. Gohberg and M.A. Kaashoek), OT 21, Birkhäuser Verlag, Basel, 1986, 75–126.
57. I. Gohberg and M.A. Kaashoek, Minimal factorization of integral operators and cascade decompositions, in: *Constructive methods of Wiener-Hopf factorization* (Eds. I. Gohberg and M.A. Kaashoek), OT 21, Birkhäuser Verlag, Basel, 1986, 157–230.
58. H. Bart, I. Gohberg and M.A. Kaashoek, Explicit Wiener-Hopf factorization and realization, in: *Constructive methods of Wiener-Hopf factorization* (Eds. I. Gohberg and M.A. Kaashoek), OT 21, Birkhäuser Verlag, Basel, 1986, 235–316.
59. H. Bart, I. Gohberg and M.A. Kaashoek, Invariants for Wiener-Hopf equivalence of analytic operator functions, in: *Constructive methods of Wiener-Hopf factorization* (Eds. I. Gohberg and M.A. Kaashoek), OT 21, Birkhäuser Verlag, Basel, 1986, 317–355.
60. H. Bart, I. Gohberg and M.A. Kaashoek, Multiplication by diagonals and reduction to canonical factorization, in: *Constructive methods of Wiener-Hopf factorization* (Eds. I. Gohberg and M.A. Kaashoek), OT 21, Birkhäuser Verlag, Basel, 1986, 357–372.
61. M.A. Kaashoek and A.C.M. Ran, Symmetric Wiener-Hopf factorization of self-adjoint rational matrix functions and realization, in: *Constructive methods of Wiener-Hopf factorization* (Eds. I. Gohberg and M.A. Kaashoek), OT 21, Birkhäuser Verlag, Basel, 1986, 373–409.

62. I. Gohberg and M.A. Kaashoek, Minimal representations of semi-separable kernels and systems with separable boundary conditions, *J. Math. Anal. Appl.* **124 (2)** (1987), 436–458.
63. I. Gohberg, M.A. Kaashoek and F. van Schagen, Szegö-Kac-Achiezer formulas in terms of realization of the symbol, *J. Funct. Analysis* **74** (1987), 24–51.
64. H. Bart, I. Gohberg and M.A. Kaashoek, The state space method in problems of analysis, in: *Proceedings First International Conference on Industrial and Applied Mathematics, Contributions from the Netherlands*, CWI 1987, 1–16.
65. I. Gohberg and M.A. Kaashoek, An inverse spectral problem for rational matrix functions, *Integral Equations and Operator Theory* **10** (1987), 437–465.
66. I. Gohberg, M.A. Kaashoek and L. Lerer, On minimality in the partial realization problem, *Systems and Control Letters* **9** (1987), 97–104.
67. I. Gohberg, M.A. Kaashoek and F. van Schagen, Rational contractive and unitary interpolants in realized form, *Integral Equations and Operator Theory* **11** (1988), 105–127.
68. M.A. Kaashoek and H.J. Woerdeman, Unique minimal rank extensions of triangular operators, *J. Math. Anal. Appl.* **131** (1988), 501–516.
69. I. Gohberg, M.A. Kaashoek and L. Lerer, Nodes and realization of rational matrix functions: minimality and applications, in: *Topics in Operator Theory and Interpolation*, OT 29, Birkhäuser Verlag, Basel, 1988, 181–232.
70. I. Gohberg, M.A. Kaashoek and A.C.M. Ran, Interpolation problems for rational matrix functions with incomplete data and Wiener-Hopf factorization, in: *Topics in interpolation theory of rational matrix-valued functions*, OT 33, Birkhäuser Verlag, Basel, 1988, 73–108.
71. I. Gohberg and M.A. Kaashoek, Regular rational matrix functions with prescribed pole and zero structure, in: *Topics in interpolation theory of rational matrix-valued functions* (Ed. I. Gohberg), OT 33, Birkhäuser Verlag, Basel, 1988, 109–122.
72. I. Gohberg and M.A. Kaashoek, Block Toeplitz operators with rational symbols, in: *Contributions to Operator Theory and its Applications* (Eds. I. Gohberg, J.W. Helton and L. Rodman), OT 35, Birkhäuser Verlag, Basel, 1988, 385–440.
73. I. Gohberg, M.A. Kaashoek and P. Lancaster, General theory of regular matrix polynomials and band Toeplitz operators, *Integral Equations and Operator Theory* **11** (1988), 776–882.
74. I. Gohberg, M.A. Kaashoek and H.J. Woerdeman, The band method for positive and contractive extension problems: An alternative version and new applications, *Integral Equations and Operator Theory* **12** (1989), 343–382.
75. M.A. Kaashoek and H.J. Woerdeman, Minimal lower separable representations: characterization and construction, in: *The Gohberg Anniversary*

Collection, vol. II (Eds. H. Dym, S. Goldberg, M.A. Kaashoek, P. Lancaster), OT 41, Birkhäuser Verlag, Basel, 329–344.

76. I. Gohberg, M.A. Kaashoek and F. van Schagen, Eigenvalues of completions of submatrices, *Lin. Multilinear Alg.* **25** (1989), 55–70.
77. I. Gohberg, M.A. Kaashoek and A.C.M. Ran, Partial pole and zero displacement by cascade connection, *SIAM Matrix Anal. Appl.* **10** (1989), 316–325.
78. I. Gohberg, M.A. Kaashoek and H.J. Woerdeman, The band method for positive and contractive extension problems, *J. Operator Theory* **22** (1989), 109–155.
79. M.A. Kaashoek and J.N.M.Schermer, Inversion of convolution equations on a finite interval and realization triples, *Integral Equations and Operator Theory* **13** (1990), 76–103.
80. I. Gohberg, M.A. Kaashoek and H.J. Woerdeman, The band method for extension problems and maximum entropy, in: *Signal Processing*, Part I (Eds. L. Auslander, T. Kailath, S. Mitter). The IMA Volumes in Mathematics and its Application, vol. 22, Springer Verlag, 1990, 75–94.
81. I. Gohberg, M.A. Kaashoek and A.C.M. Ran, Regular rational matrix functions with prescribed null and pole data except at infinity, *Lin. Alg. Appl.* **137/138** (1990), 387–412.
82. A. Ben-Artzi, I. Gohberg and M.A. Kaashoek, Invertibility and dichotomy of singular difference equations, in: *Topics in Operator Theory. Ernst D. Hellinger Memorial Volume* (Eds. L. de Branges, I. Gohberg and J. Rovnyak), OT 48, Birkhäuser Verlag, Basel, 1990, 157–184.
83. I. Gohberg, M.A. Kaashoek and A.C.M. Ran, Matrix polynomials with prescribed zero structure in the finite complex plane, in: *Topics in Matrix and Operator Theory* (Eds. H. Bart, I. Gohberg and M.A. Kaashoek), OT 50, Birkhäuser Verlag, Basel, 1991, 241–266.
84. I. Gohberg and M.A. Kaashoek, The Wiener-Hopf method for the transport equation: a finite dimensional version, in: *Modern Mathematical Methods in Transport Theory* (Eds. W. Greenberg and J. Polewczak), OT 51, Birkhäuser Verlag, Basel, 1991, 20–33.
85. I. Gohberg, M.A. Kaashoek and H.J. Woerdeman, A maximum entropy principle in the general framework of the band method. *J. Funct. Analysis* **95** (1991), 231–254.
86. I. Gohberg, M.A. Kaashoek and H.J. Woerdeman, A note on extensions of band matrices with maximal and submaximal invertible blocks, *Lin. Alg. App.* **150** (1991), 157–166.
87. I. Gohberg, M.A. Kaashoek and L. Lerer, A directional partial realization problem, *Systems and Control Letters* **17** (1991), 305–314.
88. I. Gohberg and M.A. Kaashoek, The state space method for solving singular integral equations. In: *Mathematical System Theory. The influence of Kalman* (Ed. A.C. Antoulas), Springer-Verlag, 1991, 509–523.
89. I. Gohberg, M.A. Kaashoek and H.J. Woerdeman, The time variant extension problems of Nehari type and the band method. In: H_∞*-Control Theory*

(Eds. E. Mosca, L. Pandolphi), Lecture Notes in Mathematics 1496, Springer-Verlag, 1991, 309–323.

90. I. Gohberg, M.A. Kaashoek and H.J. Woerdeman, The band method for several positive extension problems of non-band type, *J. Operator Theory* **26** (1991), 191–218.
91. I. Gohberg and M.A. Kaashoek, Asymptotic formulas of Szegö-Kac-Achiezer type, *Asymptotic Analysis* **5** (1992), 187–220.
92. J.A. Ball, I. Gohberg and M.A. Kaashoek, Nevanlinna-Pick interpolation for time-varying input-output maps: The discrete case, in: *Time-variant systems and interpolation* (Ed. I. Gohberg), OT 56, Birkhäuser Verlag, Basel, 1992, 1–51.
93. J.A. Ball, I. Gohberg and M.A. Kaashoek, Nevanlinna-Pick interpolation for time-varying input-output maps: The continuous time case, in: *Time-variant systems and interpolation* (Ed. I. Gohberg), OT 56, Birkhäuser Verlag, Basel, 1992, 52–89.
94. J.A. Ball, I. Gohberg and M.A. Kaashoek, Time-varying systems: Nevanlinna-Pick interpolation and sensitivity minimization, in: *Recent Advances in Mathematical Theory of Systems, Control, Networks and Signal Processing I, Proceedings MTNS-91* (Eds. H. Kimura, S. Kodama) Mita Press, Tokyo, 1992, 53–58.
95. I. Gohberg, M.A. Kaashoek and L. Lerer, Minimality and realization of discrete time-varying systems, in: *Time-Variant Systems and Interpolation* (Ed. I. Gohberg), Birkhäuser Verlag, Basel, 1992, 261–296.
96. I. Gohberg, M.A. Kaashoek and A.C.M. Ran, Factorizations of and extensions to J-unitary matrix functions on the unit circle, *Integral Equations and Operator Theory* **15** (1992), 262–300.
97. I. Gohberg, M.A. Kaashoek and L. Lerer, Minimal rank completion problems and partial realization, in: *Recent Advances in Mathematical Theory of Systems, Control, Networks and Signal Processing I, Proceedings MTNS-91* (Eds. H. Kimura, S. Kodama) Mita Press, Tokyo, 1992, 65–70.
98. J.A. Ball, I. Gohberg and M.A. Kaashoek, Reduction of the abstract four block problem to a Nehari problem, in: *Continuous and discrete Fourier transforms, extension problems and Wiener-Hopf equations* (Ed. I. Gohberg), OT 58, Birkhäuser Verlag, Basel, 1992, 121–141.
99. I. Gohberg and M.A. Kaashoek, The band extension on the real line as a limit of discrete band extensions, I. The main limit theorem, in: *Operator Theory and Complex Analysis* (Eds. T. Ando and I. Gohberg), OT 59, Birkhäuser Verlag, Basel, 1992, 191–220.
100. I. Gohberg and M.A. Kaashoek, The band extension on the real line as a limit of discrete band extensions, II. The entropy principle, in: *Continuous and discrete Fourier transforms, extension problems and Wiener-Hopf equations* (Ed. I. Gohberg), OT 58, Birkhäuser Verlag, Basel, 1992, 71–92.

101. M.A. Kaashoek, A.C.M. Ran and L. Rodman, Local minimal factorizations of rational matrix functions in terms of null and pole data: formulas for the factors, *Integral Equations and Operator Theory* **16** (1993), 98–130.
102. A. Ben-Artzi, I. Gohberg and M.A. Kaashoek, Invertibility and dichotomy of differential operators on a half line, *J. Dynamics and Differential Equations* **5 (1)** (1993), 1–36.
103. M.A. Kaashoek and S.M. Verduyn Lunel, Characteristic matrices and spectral properties of evolutionary systems, *Trans. Amer. Math. Soc.* **334 (2)** (1992), 479–517.
104. I. Gohberg, M.A. Kaashoek and F. van Schagen, On the local theory of regular analytic matrix functions, *Lin. Alg. Appl.* **182** (1993), 9–25.
105. J.A. Ball, I. Gohberg and M.A. Kaashoek, Bitangential interpolation for input-output operators of time-varying systems: The discrete time case, in *New aspects in Interpolation and Completion Theories* (Ed. I. Gohberg), OT 64, Birkhäuser Verlag, Basel, 1993, 33–72.
106. A. Ben-Artzi, I. Gohberg and M.A. Kaashoek, Time-varying generalizations of invertibility and Fredholm theorems for Toeplitz operators, *Indag. Math.* **4** (1993), 385–405.
107. I. Gohberg, M.A. Kaashoek and H.J. Woerdeman, The band method for bordered algebras, in: *Contributions to Operator Theory and its Applications. The Ando Anniversary Volume* (Eds. T. Furuta, I. Gohberg, T. Nakazi), OT 62, Birkhäuser Verlag, Basel, 1993, 85–97.
108. A. Ben-Artzi, I. Gohberg and M.A. Kaashoek, Exponentially dominated infinite block matrices of finite Kronecker rank, *Integral Equations and Operator Theory* **18** (1994), 30–77.
109. M.A. Kaashoek and J. Kos, The Nehari-Takagi problem for input-output operators of time-varying continuous time systems, *Integral Equations and Operator Theory* **18** (1994), 435–467.
110. J.A. Ball, G. Groenewald, M.A. Kaashoek and J. Kim, Column reduced rational matrix functions with given null-pole data in the complex plane, *Linear Alg. Appl.* **203/204** (1994), 67–110.
111. I. Gohberg and M.A. Kaashoek, Projection method for block Toeplitz operators with operator-valued symbols, in: *Toeplitz operators and related topics. The Harold Widom Anniversary Volume* (Eds. E.L. Basor and I. Gohberg), OT 71, Birkhäuser Verlag, Basel, 1994, 79–104.
112. J.A. Ball, I. Gohberg and M.A. Kaashoek, H_∞-control and interpolation for time-varying systems, in: *Systems and Networks: Mathematical Theory and Applications*, vol. I (Eds. U. Helmke, R. Mennicken, J. Saurer), Akademie Verlag, Berlin, 1994, 33–48.
113. J.A. Ball, I. Gohberg and M.A. Kaashoek, Bitangential interpolation for input-output maps of time-varying systems: the continuous time case, *Integral Equations and Operator Theory* **20** (1994), 1–43.

114. M.A. Kaashoek and S.M. Verduyn Lunel, An integrability condition on the resolvent for hyperbolicity of the semigroup, *J. Differential Equations* **112** (1994), 374–406.
115. I. Gohberg, M.A. Kaashoek and L. Rodman, Local and global analytic equivalence of analytic operator functions, in: *Linear and Complex Analysis Problem Book 3*, Part I, (Eds. V. Havin and N.K. Nikolski), Springer Verlag, Berlin, 1994, 205–206.
116. J.A. Ball, I. Gohberg and M.A. Kaashoek, Input-output operators of J-unitary time-varying continuous time systems, in: *Operator theory in functions spaces and Banach lattices*, The A.C. Zaanen anniversary volume (Eds. C.B. Huijsmans, M.A. Kaashoek and B. de Pagter), OT 75, Birkhäuser Verlag, Basel, 1995, 75–94.
117. J.A. Ball, I. Gohberg and M.A. Kaashoek, Two-sided Nudelman interpolation for input-output operators of discrete time-varying system, *Integral Equations and Operator Theory* **21** (1995), 174–211.
118. A. Ben-Artzi, I. Gohberg and M.A. Kaashoek, Discrete nonstationary bounded real lemma in indefinite metrics, the strict contractive case, in: *Operator Theory and boundary eigenvalue problems* (Eds. I. Gohberg and H. Langer), OT 80, Birkhäuser Verlag, 1995, 49–78.
119. J.A. Ball, I. Gohberg and M.A. Kaashoek, A frequency response function for linear time-varying systems, *Math Control Signals Systems* **8** (1995), 334–351.
120. M.A. Kaashoek, State space theory of rational matrix functions and applications, Lecture Series 4 in: *Lectures on Operator Theory and its Applications* (Ed. P. Lancaster). Fields Institute Monographs 3, Amer. Math. Soc. 1995, 233–333.
121. I. Gohberg, M.A. Kaashoek and J. Kos, The asymptotic behaviour of the singular values of matrix powers and applications, *Lin. Alg. Appl.* **245** (1996), 55–76.
122. I. Gohberg, M.A. Kaashoek and J. Kos, Classification of linear time-varying difference equations under kinematic similarity, *Integral Equations and Operator Theory* **25** (1996), 55–76.
123. J.A. Ball, I. Gohberg and M.A. Kaashoek, The band method and Grassmannian approach for completion and extension problems, in: *Recent developments in operator theory and its applications* (Eds. I. Gohberg, P. Lancaster and P.N. Shivakumar), OT 87, Birkhäuser Verlag, Basel, 1996, 17–60.
124. I. Gohberg, M.A. Kaashoek and L. Lerer, Factorization of banded lower triangular infinite matrices, *Lin. Alg. Appl.* **247** (1996), 347–357.
125. H. Bart, I. Gohberg and M.A. Kaashoek, Wiener-Hopf equations and linear systems, in: *Proc. Symposia Appl. Math.* 52, Amer. Math. Soc. Providence RI, 1996, 115–128.
126. C. Foias, A. Frazho, I. Gohberg and M.A. Kaashoek, Discrete time-variant interpolation as classical interpolation with an operator argument, *Integral Equations and Operator Theory* **26** (1996), 371–403.

127. M.A. Kaashoek, C.V.M. van der Mee and A.C.M. Ran, Weighting operator patterns of Pritchard-Salamon realizations, *Integral Equations and Operator Theory* **27** (1997), 48–70.
128. J.A. Ball, I. Gohberg and M.A. Kaashoek, Nudelman interpolation and the band method, *Integral Equations and Operator Theory* **27** (1997), 253–284.
129. C. Foias, A. Frazho, I. Gohberg and M.A. Kaashoek, A time-variant version of the commutant lifting theorem and nonstationary interpolation, *Integral Equations and Operator Theory* **28** (1997), 158–190.
130. I. Gohberg, M.A. Kaashoek and F. van Schagen, Operator blocks and quadruples of subspaces: classification and the eigenvalue completion problem, *Lin. Alg. Appl.* **269** (1998), 65–89.
131. M.A. Kaashoek and D.R. Pik, Factorization of lower triangular unitary operators with finite Kronecker index into elementary factors, Proceedings IWOTA-95, OT 103, Birkhäuser Verlag, Basel, 1998, 183–217.
132. C. Foias, A. Frazho, I. Gohberg and M.A. Kaashoek, Parametrization of all solutions of the three chains completion problem, *Integral Equations and Operator Theory* **29** (1997), 455–490.
133. D.Z. Arov, M.A. Kaashoek and D.R. Pik, Minimal and optimal linear discrete time-invariant dissipative scattering systems, *Integral Equations and Operator Theory* **29** (1997), 127–154.
134. I. Gohberg, M.A. Kaashoek and A.L. Sakhnovich, Canonical systems with rational spectral densities: explicit formulas and applications, *Math. Nachr.* **194** (1998), 93–125.
135. I. Gohberg, M.A. Kaashoek and A.L. Sakhnovich, Pseudo-canonical systems with rational Weyl functions: explicit formulas and applications, *J. Differential Eqs.* **146** (1998), 375–398.
136. I. Gohberg, M.A. Kaashoek and A.L. Sakhnovich, Sturm-Liouville systems with rational Weyl functions: explicit formulas and applications, *Integral Equations and Operator Theory* **30** (1998), 338–377.
137. C. Foias, A. Frazho, I. Gohberg and M.A. Kaashoek, The maximum principle for the three chains completion problem, *Integral Equations and Operator Theory* **30** (1998), 67–82.
138. D.Z. Arov, M.A. Kaashoek and D.R. Pik, Optimal time-variant systems and factorization of operators, I: minimal and optimal systems, *Integral Equations and Operator Theory* **31** (1998), 389–420.
139. M.A. Kaashoek and A.C.M. Ran, Norm bounds for Volterra integral operators and time varying linear systems with finite horizon, in: *Contributions to operator theory in spaces with an indefinite metric. The Heinz Langer volume*, OT 106, Birkhäuser Verlag, Basel, 1998, 275–290.
140. M.A. Kaashoek, C.V.M. van der Mee and A.C.M. Ran, Wiener-Hopf factorization of transfer functions of extended Pritchard-Salamon realizations, *Math. Nachr.* **196** (1998), 71–102.

141. M.A. Kaashoek and C.G. Zeinstra, The band method and generalized Carathédory-Toeplitz interpolation at operator points, *Integral Equations and Operator Theory* **33** (1999), 175–210.
142. I. Gohberg and M.A. Kaashoek, State space methods for analysis problems involving rational matrix functions, in: *Dynamical systems, control, coding, computer vision*, (Eds. G. Picci and D.S. Gilliam), Birkhäuser Verlag, 1999, 93–110.
143. D.Z. Arov, M.A. Kaashoek and D.R. Pik, Optimal time-variant systems and factorization of operators, II: Factorization, *J. Operator Theory* **43** (2000), 263–294.
144. I. Gohberg, M.A. Kaashoek and A.L. Sakhnovich, Canonical systems on the line with rational spectral densities: explicit formulas, in: *Operator Theory and Related Topics, Volume I of the Proceedings of the Mark Krein International Conference, Odessa, Ukraine, August 1–22, 1997*, OT 117, Birkhäuser Verlag, Basel, 2000, 127–139.
145. I. Gohberg, M.A. Kaashoek and J. Kos, Classification of linear time-varying periodic difference equations under periodic or kinematic similarity, *SIAM Matrix Analysis* **21** (2000), 481–507.
146. I. Gohberg, M.A. Kaashoek and F. van Schagen, Finite section method for linear ordinary differential equations, *J. Differential Eqs.* **163** (2000), 312–334.

Opening Address

Israel Gohberg

Dear Colleagues and Dear Friends
Ladies and Gentlemen

I am thankful for this opportunity to open the conference on the occasion of the sixtieth birthday of M.A. Kaashoek, the founder and head of the Analysis and Operator Theory group in Amsterdam at the Vrije Universiteit. He and his group have made important contributions to traditional operator theory as well as developing new areas. I was closely related to this group and helped to build and nurture it. Together we established and developed connections between operator theory, integral equations and systems theory. Those connections allowed us to enrich considerably all theories involved and to build a method, based on state space techniques, for solving explicitly problems in Analysis.

Over a period of 20 years the group developed into an important world renowned School of Operator Theory and its Applications. Here were organized many summer schools, and national and international conferences. The members of this School published numerous monographs and important papers, and several outstanding young experts were born and bred in this School.

I have had the pleasure to work together with Rien for more than twenty years. Five books (with H. Bart, S. Goldberg, F. van Schagen, C. Foias and A. Frazho) were written as well as many papers. One book with Harm Bart and Andre Ran is in preparation. Together we educated many graduate students. I will always remember with joy and satisfaction those weeks, and weekends, when we worked together. Usually the weekends included, as well as attractive mathematics, also interesting excursions and bicycle tours in Netherlands and enjoyable dinners with the family. The parties organized by the hospitable Kaashoek family were always remarkable for the excellent food and very warm and friendly atmosphere. I hope that this wonderful time is not over and we will be able to continue to enjoy this warm hospitality for many years.

By the way we have very different mentalities by descent and education, but during the twenty years of joint work there were never any arguments between us. As can be expected, sometimes our opinions differed, but always the difference was resolved before reaching the dangerous level of an argument. Each time this required convincing discussions, flexibility and compromise. As a result this led to a better understanding and decisions that satisfied both of us.

I would like to mention that Rien is a wonderful teacher, friend and coauthor. Rien can listen to others and can express precisely and persistently his wishes. Discussions with him are always pleasant. He is not only very talented and capable, but also very reliable, organized, kind and generous. I learned many good things from him. Today, on the occasion of his sixtieth birthday, I would like to congratulate Rien on his outstanding achievements and to thank him heartily for this wonderful lasting collaboration and friendship. I would also like to congratulate today the faculteit of Wiskunde and Informatica on its excellent group in Analysis. In my opinion it is certainly a day of celebration for the entire Vrije Universiteit.

Rien Kaashoek obtained an excellent mathematical education in Leiden University in the school of one of the famous pioneers of functional analysis, Professor A.C. Zaanen. Rien's assignment was in new territory, unbounded operator theory. He successfully started his research in a new area for this school. At the end of the sixties he started to build up his group at the Vrije Universiteit.

I first met Rien at a conference in Tihany, Hungary, in 1970. We could hardly communicate; at that time I did not speak English and he did not speak Russian. The next time we met was after I immigrated to Israel. Our meeting was in College Park, Maryland, where he was spending his sabbatical and I was visiting as a guest of Seymour Goldberg. This was at the beginning of 1975. Soon we were already looking for an area of joint interest. We continued our work in Amsterdam at the end of 1975 and in the spring and autumn of 1976, and finished this first period with three important things: two large papers and a nice bicycle tour to the Keukenhof. The first paper, jointly with David Lay, was in a sense in the area of our old interests, and the second, with Harm Bart, was in a new area. It was mostly on matrix polynomials and used the approach of characteristic functions. This was the beginning of connections with the interests of the Odessa School of M.G. Krein and with mathematical systems theory and control from Electrical Engineering. Our first paper was soon published in the American Journal of Functional Analysis; the second was rejected by a respected journal. Unfortunately the reviewer did not make the slightest effort to understand the paper or to understand the framework in which it was written. This paper was a new beginning, and we decided that it would be fitting to publish it in the first issue of the new journal, Integral Equations and Operator Theory. Finally, the Keukenhof astonished me by its rich beauty of tulips.

After this beginning we met many times for extended periods in Amsterdam and Tel-Aviv. I was visiting Amsterdam on a regular basis more than three months per year and Rien spent a large part of his sabbaticals in Tel-Aviv. We published many papers in many different areas, all the time extending our fields of interest. Let me mention some of them: factorization of matrix-functions and operator-functions and its applications to integral and differential equations, homogeneous and nonhomogeneous interpolation problems for matrix-functions and applications to problems in control, completion problems for matrices and applications, applications of the state space method to different problems in analysis, inverse spectral problems and applications to nonlinear equations, different time dependent problems in analysis, and the band method.

One of the important features of our work was its closeness to electrical engineering (especially systems theory). We were able to speak to the theoretically oriented engineers and to understand their mathematical problems. We listened to many talks by engineers and participated in their conferences. We were active in the steering committee of MTNS, and one of the big meetings of MTNS was organized at the VU (co-chaired by M.A. Kaashoek, A.C.M. Ran and J.H. van Schuppen). We certainly did not become engineers, we remained mathematicians, but without any doubt all of this influenced our work and gave us a beautiful feeling of having next to us the motivations and applications.

I was planning to have a second part to my talk. The second part was to have been devoted entirely to the paper which was rejected, and I had planned to present in detail the ideas contained in this paper and their developments. Unfortunately I had to undergo an urgent and difficult operation, and I have had to postpone my plans for another time. For the same reason I am unable to participate in this conference in which I was so keen to participate. I regret very much having to miss this conference and meetings with close friends.

I am very happy to open the conference dedicated to the sixtieth birthday of my close friend and coauthor Marinus Adriaan Kaashoek. Dear guests, welcome to Amsterdam and have a nice time in this lovely city. Special thanks to my friend Seymour Goldberg for presenting this address to the conference.

Marinus (Rien) Adrianus Kaashoek: Memories at the Occasion of His Sixtieth Birthday by Three of His PhD Students

Reminiscences by Harm Bart

The first time I heard about Rien Kaashoek was as follows. I was a student at the Free University in Amsterdam and was unsuccesfully struggling with a question about Stieltjes integrals. When I asked one of the assistant professors, he also was not able to come up with an answer. At the end of our conversation, he said: we have a new faculty member by the name of Kaashoek. His field is Analysis, so maybe you should ask him. By the way, the name sounded to me like Kazuc (or something like that), so I had the impression that the person in question was a foreigner from Eastern Europe. He wasn't and turned out to be a genuine Dutchman with, as a matter of fact, a very Dutch name. I approached him and he solved the problem. So he lived up to the expectations. Probably he does not remember, but this was my first encounter with Rien.

In the final stages of my study in Mathematics, all examinations were oral. If I remember correctly, I did one such examination with Rien, and this on the basis of notes taken by a fellow student (Rikus Kok). I got a high mark, so certainly the notes were all right. It should also be said that they reflected the crisp and clear style in which the course (Operator Theory) was given. The style that, as we all know, is Rien's trademark.

From what was said above, it may seem that as a student I had little contact with Rien. This, however, is not true. My first scientific production is a research memorandum (report), written while I was a student, containing an application of the Banach fixed point theorem to differential equations. Rien was my main sparring partner in this project. Also, I rounded off my studies by doing a project on Invariant Subspaces under Rien's guidance. In all these activitities, I came to appreciate his penetrating insight, precision and drive for clarity in formulation.

After finishing my studies, I decided to pursue an academic career and was looking for a position as a PhD student. For some time, I felt inclined to go into General Topology, a field that was covered at the Free University by the charismatic and inspiring Maarten Maurice. But, under the strong influence of Rien, I changed my mind and, when a position became available, I asked Rien to be my supervisor. He gave me the thesis of Lothrop Mittenthal to read. The topic was operator valued functions and generalization of spectral theory, and Rien told me that in his opinion there was progress to be made in this area. He was right and in 1973 I defended my thesis "Meromorphic Operator Valued Functions". I was Rien's first (official) PhD student. (Unofficially he had an earlier one, but I was the first he guided from the beginning to the very end.)

After obtaining my PhD degree, I got a permanent position at the Free University. In this way, Rien and I had an excellent opportunity to continue our cooperation. As a matter of fact, this happened in the somewhat wider framework of a cooperation with David Lay from the University of Maryland. The work we did with the three of us was very much stimulated by the study of some papers by Israel Gohberg and his co-authors. We received this material from Bernhard Gramsch in a rather peculiar translation from Russian into German made by a non-mathematician. Nevertheless the papers did their work. In 1975 personal contact with Israel Gohberg, who had by then left the Soviet Union, took place and Rien sensed very quickly that there were tremendous possibilities here for broadening the scope of the research done in Amsterdam. And indeed the semi-periodic presence of Israel Gohberg at the Free University that Rien was able to arrange had an enormous impact on the group there. As an example, I mention that one of his questions at the weekly seminar ("What about stability?") induced as many papers as it contained words. It is worth noting that the seminar started in the mid seventies and is, after about 24 years, still going on! It is one of the signs that Rien made a lasting impact.

Until a couple of years ago, I was still involved in all these activities. But then I became entangled in administrative duties at the Erasmus University where I was appointed in 1984. Rien himself is now also a dean and our recent conversations are more about "managing" a faculty than about mathematics. I hope that somewhere in the future things will swing back to mathematics, so that we will be able to work again on the problems that, after all, interest us more than (almost) anything else.

Memories of Rien can never go without memories of his wife Wies. Already for a long time Rien and Wies are good friends of my wife and me. Long enough to see the children of both families grow up. Long enough also to know that the friendship will last forever.

Let me close these reminiscenses with a small anecdote. In 1974 (or 1975) Rien and I were invited to a conference in Dublin organised by Rien's good friend Trevor West. It so happened that we were travelling together. On the way from Amsterdam to Dublin we had to change planes at Heathrow, London. While we were waiting suddenly the sign "cancelled" flashed up on the message board. It turned out that due to a strike the pilots from British Airways did not fly on Dublin. (Lothrop Mittenthal found out that those from Air Lingus did; they probably wanted to be home before dark.) So we had to stay overnight in London with the extra complication that our luggage was lost (and in fact already forwarded to Dublin). The next day we could catch a flight as stand by passengers. After about an hour the plane began to descend. Then, almost at the moment of touch down, its nose went up and with roaring engines it started to climb again. There was still something wrong due to the strike. About half an hour later, we landed in Shannon at the West coast of Ireland. After we got a meal, we were stowed into a train which, to say the least, was extremely uncomfortable. I then found out that Rien is not very much in favour of travelling in general and certainly does not like travelling in this way. I flatter myself with the thought that at this occasion I was of some support to him.

Reminiscences by André Ran

As a third year student the choice of a specialization for the final part of the studies in mathematics was a hard one for me. At that time at the Vrije Universiteit in Amsterdam there where many good opportunities. One could do set theoretic topology in a very active group, one could go into probability theory or statistics, computer science was a possibility, and finally there was a small but very active group in operator theory and analysis. For me the deciding factor was a course in functional analysis that was thaught by Rien Kaashoek, with some of the lectures being given by Harm Bart. This course was not only very appealing because of the nice mix of deep theory and concrete applications, but also a joy to follow because of the enthusiasm of the teachers, and their skill in presenting the material. Only when preparing for the examination did I discover that the path they had led us on through the material was one with many turn-offs, many of which where dead-end streets, others opening up into endless mazes. I learned later, from several courses and many lectures that I heard from Rien, that he has the amazing ability to make the most difficult material sound easy, and to present it in a very smooth way, so that the audience comes away with an understanding of the essentials, without being burdened by the details. These details, when presented by him, are always painstakingly precise. I do not recall an ε-δ reasoning by Rien that did not end up precisely with the desired quantity being less than or equal to ε. Never 2ε, or $M\varepsilon$ or something like that, always precisely ε. The same precision and attention to detail can be observed in his writing.

A few years later, I had the pleasure of having Rien as advisor for my masters thesis and later my PhD thesis. Many mornings where filled with discussions in his office, regularly once a week for at least one hour. I learned a lot of mathematics in these discussions, and from Rien I also learned how to present material, both in writing and in oral presentations. Rien was always encouraging and supportive, and his criticism was constructive and delivered in a positive manner.

It was at that time that Rien's PhD students started calling him "de baas" (the boss). But that title was used with respect and admiration by all of us. Indeed, Rien was a good boss: always more coach and mentor, friend even, always interested in the personal affairs as well as the mathematical affairs. There is only one occassion that I remember him being an old-fashioned boss to one of the masters students, for good reason. It seems it worked on that one occassion, but I don't think Rien liked the experience, as it does not fit with his character.

Obviously, when I was offered the chance to work in his group on a permanent basis after spending one year in the United States, I was very happy. We worked on a number of projects over the last fourteen years, and the experience is still enjoyable (and I hope he can say: likewise). I consider it a pleasure and a privilige to work with him, and hope that we may continue for many years to come.

Reminiscences by Hugo Woerdeman

The first time I approached Professor Kaashoek I was an undergraduate student in his Functional Analysis class, feeling pretty much like a nobody. Naturally I introduced myself, but it turned out to be unnecessary. “Oh, I know who you are”, Rien said. And so a relationship of mutual respect started between us. Indeed, Rien is one of the most respectful people that I know, one of his many qualities that I want to highlight here. But let me not get ahead of myself, and return to our early days. It was not long after our first meeting that I started an honors thesis (afstudeerproject) in Rien's group. My advisor was Leonid Lerer, who was visiting at the time, and under his guidance I apparently did well enough to be asked by Rien to stay on as his Ph.D. student. With a warning though, that the job market might be tough by the time I would finish. I don't think any warning could have stopped me, but I certainly was not struck by this one. As Rien's student, surely I would find a position!

A stimulating environment with lots of interaction marked my time as a graduate student. First of all there were several fellow students, of which my office mate Leen Roozemond kept me the most amused. Then there was a multitude of visitors who we were encouraged to talk to and maybe collaborate with. Also, we had our weekly seminar consisting of detailed talks. And last but certainly not least, we had an abundance of one-on-one interaction with Rien. Clearly, though, all the respect my fellow students and I felt for Rien did not stop us from being part of the bonding experience of making fun of our ‘baas’ (=boss). The most rewarding topic was usually his, oh shall we say, attention for details, making us sometimes boast “I think I got the new record of corrected comma's!”. Usually we got the comma placements down after a while, and we ultimately graduated.

My graduation was a special occasion for me and must have cost Rien ample preparation time. Not only did he give me a very eloquent and warm address during the official ceremony, but he also chose to prove my only non-mathematical theorem of my thesis. Before I continue I must explain an already abandoned Dutch tradition. It used to be required that a Ph.D. thesis was accompanied by at least six statements that “are not on the main topic of the thesis”. In this way, one's well roundedness was tested. Over the years it had grown to be a custom that the last statement should be one that everybody could understand, and sometimes even smile about. Resisting any pressure of trying to be funny, I decided to make a connection between my Montessori elementary education and my poor memory for factoids. When during the official ceremony I was jokingly asked by Israel Gohberg for a proof, I was tongue-tied. But as a good advisor, Rien took care of my embarrassment and provided a proof later that evening. By quizzing both Andre Ran, who had had a classical education (and a little bit of help), and myself on several factoids (“what is M.G. Krein's date of birth?”), the theorem was easily proven. When I think of my graduation I always think of Rien's humorous and lively performance as quizmaster. That, and the memorable whisper “now you can call me Rien”.

Rien, let me take this opportunity to thank you for everything you have done for me. You have given me a great education, have always been respectful to me, and have always been very supportive. You have introduced me to exciting mathematics, shown how it is communicated and taught me to be generous with recognizing others. I view you as a great mathematician, an excellent and dedicated teacher, and a very good friend. Happy 60th, and I look forward to a future with a lot of continued interaction!

Operator Theory:
Advances and Applications, Vol. 122
© 2001 Birkhäuser Verlag Basel/Switzerland

On the Separation of Certain Spectral Components of Selfadjoint Operator Matrices

Vadim Adamyan and Reinhard Mennicken

Dedicated to Professor Marinus A. Kaashoek on the occasion of his 60th birthday

The papers [AL], [AdLMSau] were devoted to the characterization of certain spectral subspaces of selfadjoint operators $\mathbf{L}$, which are generated by symmetric 2×2 operator matrices of the form

$$\mathbf{L}_0 = \begin{pmatrix} A & B \\ B^* & D \end{pmatrix}$$

in the product of Hilbert spaces $\mathcal{H}_1$ and $\mathcal{H}_2$; the entries A, B and D are not necessarily bounded operators in or between these spaces. One of the main assumptions in those investigations was the weak separation of the spectra of the selfadjoint operators A and D, i.e., $\max \sigma(D) \leq \min \sigma(A)$.

The present article deals with the case when the spectra of the diagonal operators A and D overlap, i.e., it is assumed that A is bounded from below, D is bounded from above and $\max \sigma(D) > \min \sigma(A)$.

The results are applied to the study of basis properties of certain subsequences of the eigenfunctions of some Sturm-Liouville boundary value problems with λ-dependent potentials.

1 Introduction

Investigations of small ideal plasma oscillations on the basis of linearized equations of magnetohydrodynamics lead to spectral problems for selfadjoint operators formally given as operator matrices

$$\mathbf{L}_0 = \begin{pmatrix} A & B \\ B^* & D \end{pmatrix}$$

in the Hilbert space product $\mathcal{H} = \mathcal{H}_1 \times \mathcal{H}_2$ (endowed with the scalar product $\langle \cdot, \cdot \rangle_{\mathcal{H}} = \langle \cdot, \cdot \rangle_{\mathcal{H}_1} + \langle \cdot, \cdot \rangle_{\mathcal{H}_2}$). For many important plasma equilibrium configurations the entries A, B and D of $\mathbf{L}_0$ are unbounded operators in $\mathcal{H}_1$, $\mathcal{H}_2$ or between them, respectively, satisfying the following assumptions:

(a) A and D are selfadjoint operators in the Hilbert spaces $\mathcal{H}_1$ and $\mathcal{H}_2$, respectively,
(b) A is bounded from below and D is bounded from above,
(c) B is a densely defined closed operator such that $\mathcal{D}(|A|^{\frac{1}{2}}) \subset \mathcal{D}(B^*)$ and $\mathcal{D}(B) \subset \mathcal{D}(D)$.

The conditions (a)–(c) ensure the existence of a selfadjoint operator $\mathbf{L}$ associated with the symmetric operator $\mathbf{L}_0$ [AdLMSau].

The assumption (b) implies that either

(d$_1$) $\alpha := \sup \sigma(D) \leq \beta := \inf \sigma(A)$

or

(d$_2$) $\beta < \alpha$.

The case (d$_1$) is considered in [AdL], [AdLMSau]. These papers are primarily concerned with the description of the spectral subspaces of $\mathbf{L}$ corresponding to the intervals $(-\infty, \alpha]$ and (β, ∞) and the study of basis properties of the eigenfunctions related to the spectrum of $\mathbf{L}$ in these intervals. The fact that $\sigma(A)$ and $\sigma(D)$ are actually separated in the case (d$_1$) plays a decisive part in the proofs and constructions of [AdL], [AdLMSau].

The main result of [AdL], [AdLMSau] may be summarized as follows.

Let us denote by $\mathbf{E}(t)$, $-\infty < t < \infty$, the spectral function of $\mathbf{L}$ and let

$$\mathcal{L}_- := \mathcal{R}(\mathbf{E}(\alpha+)), \qquad \mathcal{L}_+ := \mathcal{R}(I - \mathbf{E}(\alpha+)) = \mathcal{L}_-^\perp$$

be the spectral subspaces of $\mathbf{L}$ corresponding to the intervals $(-\infty, \alpha]$ and $(\alpha, +\infty)$.

Theorem 1.1 *Let the assumptions* (a), (b), (c) *and* (d$_1$) *be fulfilled and suppose that* $\mathcal{N}(A-\alpha) \cap \mathcal{N}(B^*) = \{0\}$, *where* $\mathcal{N}(\ldots)$ *are the null-spaces of the corresponding operators. Then there exists a contraction* K *from* $\mathcal{H}_1$ *into* $\mathcal{H}_2$ *such that the following statements hold:*

(i) *The spectral subspaces* $\mathcal{L}_+$ *and* $\mathcal{L}_-$ *admit graph representations with the angular operator* K, *i.e.,*

$$\mathcal{L}_+ = \left\{ \begin{pmatrix} x \\ Kx \end{pmatrix} \middle| \, x \in \mathcal{H}_1 \right\}, \quad \mathcal{L}_- = \left\{ \begin{pmatrix} -K^*x \\ x \end{pmatrix} \middle| \, x \in \mathcal{H}_2 \right\}. \tag{1.1}$$

(ii) *If* B *and* D *are bounded operator, then* K *satisfies the Riccati equation*

$$KBK + KA - DK - B^* = 0.$$

An equivalent form of this equation, namely

$$K(A-z)^{-1} - (I + KG(\overline{z})^*)W(z)^{-1}(G(z) - K) = 0, z \in \rho(A) \cap \rho(L)$$

where

$$G(z) = B^*(A-z)^{-1}, \; W(z) = z + B^*(A-z)^{-1}B - D, \; z \in \rho(A),$$

remains true, if B *and* D *are unbounded operators with the change that in* $W(z)$ *the operator* $B^*(A-z)^{-1}B$ *has to be substituted by its bounded extension* $\overline{B^*(A-z)^{-1}B}$.

(iii) *The spectral subspace $\mathcal{L}_+$ is unitarily equivalent to the Hilbert space $(\mathcal{H}_1, \langle (I+K^*K)\cdot, \cdot\rangle)$. The corresponding unitary operator U is the restriction of the operator P_1,*

$$P_1\begin{pmatrix} x_1 \\ x_2 \end{pmatrix} = x_1, \begin{pmatrix} x_1 \\ x_2 \end{pmatrix} \in \mathcal{H},$$

to $\mathcal{L}_+$. The operator $\mathrm{L}_+ := U\mathbf{L}_{|\mathcal{L}_+} U^*$ is selfadjoint and explicitly given by

$$\begin{aligned} \mathcal{D}(\mathrm{L}_+) &:= \{x \in \mathcal{H}_1 \mid x + G(0)^*Kx \in \mathcal{D}(A),\ Kx \in \mathcal{D}(D)\}, \\ \mathrm{L}_+x &:= A(x + G(0)^*Kx) - \beta G(0)^*Kx,\ x \in \mathcal{D}(\mathrm{L}_+). \end{aligned}$$

Let us assume in addition to the conditions (a)–(d_1) that

(e) A^{-1} is compact.

Since by (d_1) the inequality $D \le \alpha \le \beta$ holds, it follows from [ALMS], [AL], [AMS] that

- $\sigma_{ess}(\mathbf{L}) \subset (-\infty, \alpha)$;
- $(\alpha, \beta) \subset \rho(\mathbf{L})$ *if* $\alpha < \beta$;
- *the spectrum of* $\mathbf{L}$ *in* (β, ∞) *consists only of isolated eigenvalues of finite multiplicities*;
- β *is not an accumulation point of* $\sigma(\mathbf{L}) \cap (\beta, \infty)$.

Let $(\lambda_j)_1^\infty$ be the increasing sequence of eigenvalues of $\mathbf{L}$ in (β, ∞) counted with respect to their multiplicities and

$$\mathbf{e}_j = \begin{pmatrix} e_j \\ Ke_j \end{pmatrix}, \ j = 1, 2, \ldots,$$

a corresponding orthonormal system $(\mathbf{e}_j)_1^\infty$ of eigenvectors. By Theorem 1.1 the system $(e_j)_1^\infty$ forms an orthonormal basis of eigenvectors of L_+ in $(\mathcal{H}_1, \langle (I + K^*K)\cdot, \cdot\rangle)$ corresponding to the sequence of eigenvalues $(\lambda_j)_1^\infty$.

Theorem 1.2 *Under the assumptions* (a)–(e) *the system* $(e_j)_1^\infty$ *forms a Riesz basis in* $\mathcal{H}_1$.

The purpose of this note is to find analogues of the Theorems 1.1, 1.2 for the case (d_2). The basic idea consists in reducing this case to (d_1) by some extension of $\mathcal{H}_2$ on the account of $\mathcal{H}_1$ in the representation of $\mathcal{H}$ with a corresponding change of the block operator representation of $\mathbf{L}_0$.

From now on we assume that (d_2) holds, i.e., $\beta < \alpha$. Without loss of generality we will assume that $A \gg 0$.

Let $E_A(t)$, $-\infty < t < \infty$, denote the spectral function of A. For $\gamma \ge \alpha$ write

(1.2) $$\mathcal{H}_1(\gamma) := \mathcal{R}(I - E_A(\gamma-)), \qquad \mathcal{H}_2(\gamma) := \mathcal{R}(E_A(\gamma-)).$$

and consider the operators

$$\begin{cases} A_\gamma := A_{|\mathcal{H}_1(\gamma)}, & A_{\gamma 2} := A_{|\mathcal{H}_2(\gamma)}; \\ B_{\gamma_1} := \overline{(I - \mathbf{E}_A(\gamma -))B}, & B_{\gamma_2} := \overline{\mathbf{E}_A(\gamma -)B} \end{cases} \tag{1.3}$$

Observe that the mappings A_{γ_2} and B_{γ_2} are bounded operators on $\mathcal{H}_2(\gamma)$ and $\mathcal{H}_2$, respectively. The adjoint operators $B^*_{\gamma_1}$ and $B^*_{\gamma_2}$ are given by $B^*(I - E_A(\gamma -))$ and $B^*(I - E_A(\gamma -))$. According to the representation $\mathcal{H} = \mathcal{H}_1(\gamma) \times \mathcal{H}_2(\gamma) \times \mathcal{H}_2$ the operator matrix $\mathbf{L}_0$ can be decomposed into the following 3×3 matrix operator:

$$\mathbf{L}_0 = \begin{pmatrix} A_\gamma & 0 & B_{\gamma_1} \\ 0 & A_{\gamma_2} & B^*_{\gamma_2} \\ B^*_{\gamma_2} & B^*_{\gamma_2} & D \end{pmatrix}. \tag{1.4}$$

We define $\hat{\mathcal{H}}_1 := \mathcal{H}_1(\gamma)$, $\hat{\mathcal{H}}_2 := \mathcal{H}_2(\gamma) \times \mathcal{H}_2$ and correspondingly

$$\begin{cases} B_\gamma := (0, B_{\gamma_1}) \\ D_\gamma := \begin{pmatrix} A_{\gamma_2} & B\gamma_2 \\ B^*_{\gamma_2} & D \end{pmatrix}. \end{cases} \tag{1.5}$$

B_γ is a densely defined closed linear operator from $\hat{\mathcal{H}}_2$ into $\hat{\mathcal{H}}_1$, D_γ is a self-adjoint operator in $\hat{\mathcal{H}}_2$. Since $\mathcal{H} = \hat{\mathcal{H}}_1 \times \hat{\mathcal{H}}_2$ we see that $\mathbf{L}_0$ has the following representation:

$$\mathbf{L}_0 = \begin{pmatrix} A_\gamma & B_\gamma \\ B^*_\gamma & D_\gamma \end{pmatrix}. \tag{1.6}$$

The operators A_γ, B_γ and D_γ fulfill the assumptions (a), (b) and (c). By construction $\sigma(A_\gamma) \geq \gamma$ $(\geq \alpha)$, however, the condition (d_1) may not hold for A_γ and D_γ. The matter is that the "perturbation" B_{γ_2} may shift the spectrum of the diagonal matrix diag (A_{γ_2}, D) to the right. In the next section a criterion for the validity of the inequality $\max \sigma(D_\gamma) \leq \gamma$ is proved, which means that the condition (d_1) is fulfilled for A_γ and D_γ. It follows that analogues of the Theorems 1.1 and 1.2 hold with respect to the spectral subspace

$$\mathcal{L}_+(\gamma) := \mathcal{R}(I - \mathbf{E}(\gamma +)) = \mathcal{L}_-(\gamma)^\perp.$$

In the last section of this work the obtained results are applied to study basis properties of certain system of eigenfunctions for a Sturm-Liouville boundary problem with a λ-dependent potential.

2 Conditions for the Separation of Spectral Components

We start with the following general statement.

Proposition 2.1 *Let R and S be selfadjoint operators acting in Hilbert spaces $\mathcal{K}_1$ and $\mathcal{K}_2$, respectively, and suppose that they are bounded from above. Further, let Q be a bounded operator from $\mathcal{K}_1$ into $\mathcal{K}_2$. Assume that*

$$\max \sigma(R) \cup \sup \sigma(S) \leq \gamma. \tag{2.1}$$

Then for the selfadjoint operator

$$M := \begin{pmatrix} R & Q \\ Q^* & S \end{pmatrix} \tag{2.2}$$

in the product space $\mathcal{K}_1 \times \mathcal{K}_2$ the following two conditions are equivalent:

i) $\max \sigma(M) \leq \gamma$;

ii) *for each $\gamma' > \gamma$*

$$\|(\gamma' - S)^{-\frac{1}{2}} Q^* (\gamma' - R)^{-1} Q (\gamma' - S)^{-\frac{1}{2}}\| < 1. \tag{2.3}$$

Proof: For $z \in \rho(R)$ the Schur-Frobenius factorization

$$z - M = \begin{pmatrix} I & 0 \\ Q^*(R-z)^{-1} & I \end{pmatrix} \begin{pmatrix} z - R & 0 \\ 0 & W(z) \end{pmatrix} \begin{pmatrix} I & (R-z)^{-1} Q \\ 0 & I \end{pmatrix} \tag{2.4}$$

holds, where

$$W(z) := z - S - Q^*(z - R)^{-1} Q. \tag{2.5}$$

Now suppose that γ fulfills the condition i) and let $\gamma' > \gamma$. Consider the formula (2.4) for $z = \gamma'$ and notice that $\gamma' - R \gg 0$ and $\gamma' - S \gg 0$. Therefore $\gamma' - M \gg 0$ implies $W(\gamma') \gg 0$. From the representation

$$W(\gamma) = (\gamma' - S)^{\frac{1}{2}} [I - (\gamma' - S)^{-\frac{1}{2}} Q^* (\gamma' - R)^{-1} Q (\gamma' - S)^{-\frac{1}{2}}) (\gamma' - S)^{\frac{1}{2}}. \tag{2.6}$$

we conclude that

$$(\gamma' - S)^{-\frac{1}{2}} Q^* (\gamma' - R)^{-1} Q (\gamma' - S)^{-\frac{1}{2}} \ll I \tag{2.7}$$

which is equivalent to the inequality (2.3) so that the statement ii) holds.

Vice versa, suppose that the condition ii) is fulfilled. Then $W(\gamma') \gg 0$ and thus $\gamma' - M \gg 0$ for all $\gamma' > \gamma$. This means that $(\gamma, \infty) \subset \rho(M)$ whence the condition i) holds. □

Remark 2.2 *If the inequality* (2.3) *is fulfilled for some* $\gamma > \max \sigma(R) \cup \sigma(S)$, *then for all* $\gamma' \geq \gamma$.

Proof: For $z \in \rho(R)$ we have

$$W'(z) = I + Q^*(z - R)^{-2}Q \tag{2.8}$$

and hence $W'(\gamma) \gg 0$ for all $\gamma > \sigma(R)$. Thus $W(\gamma)$ is a nondecreasing operator function. This proves Remark 2.2. □

Applying Proposition 2.1 to the operator D_γ defined in 1.5 yields

Proposition 2.3 *Let the assumptions* (a), (b), (c) *and* (d_2) *hold. Suppose that* $\gamma > \alpha = \max \sigma(D)$ *and the inequality*

$$\|(\gamma' - D)^{-\frac{1}{2}} B^*_{\gamma_2}(\gamma' - A_{\gamma_2})^{-1} B_{\gamma_2}(\gamma' - D)^{-\frac{1}{2}}\| < 1. \tag{2.9}$$

holds for all $\gamma' > \gamma$. *Then*

$$\sigma(D_\gamma) \leq \gamma \leq \min \sigma(A_\gamma). \tag{2.10}$$

The next propositions contain sufficient conditions on the spectrum of the operator A which provide the existence of some $\gamma \geq \alpha$ such that (2.9) holds for all $\gamma' > \gamma$.

Proposition 2.4 *Let the assumptions* (a), (b), (c) *and* (d_2) *be fulfilled. Suppose that there is a double sequence of real numbers*

$$\beta = \min \sigma(A) = \nu_1^- \leq \nu_1^+ < \nu_2^- \leq \nu_2^+ < \nu_3^- \leq \nu_3^+ < \cdots \tag{2.11}$$

such that

$$\lim_{j\to\infty} \sup\, (\nu_{j+1}^- - \nu_j^+) = +\infty, \tag{2.12}$$

and

$$\sigma(A) \subset \bigcup_j [\nu_j^-, \nu_j^+]. \tag{2.13}$$

Then there is a number $\gamma > \alpha = \max \sigma(D)$ *such that inequality* (2.9) *holds for all* $\gamma' \geq \gamma$.

Proof: Notice that by our assumptions on A and B the operator $J := B^* A^{-\frac{1}{2}}$ is defined and bounded on $\mathcal{H}_1$. For $\gamma > \beta = \min \sigma(A)$ we set

$$m(\gamma) := \max \sigma(A) \cap (-\infty, \gamma).$$

and obtain the estimate

$$\|B^*_{\gamma_2}\| = \|B^* A^{-\frac{1}{2}} \cdot A^{\frac{1}{2}} E_A(\gamma-)\| \leq \|J\| \cdot m(\gamma)^{\frac{1}{2}}.$$

Choose a natural number j such that $\nu_{j+1}^- > \alpha$ and

$$\frac{\nu_j^+}{\nu_{j+1}^- - \alpha} \|J\|^2 \frac{1}{\nu_{j+1}^- - \nu_j^+} < 1.$$

Take $\gamma = \nu_{j+1}^-$ and let $\gamma' \geq \gamma$. Then

$$\begin{aligned} &\|(\gamma' - D)^{-\frac{1}{2}} B_{\gamma_2}^* (\gamma - A_{\gamma_2'})^{-1} B_{\gamma_2} (\gamma' - D)^{-\frac{1}{2}}\| \\ &\leq \frac{m(\gamma)}{\gamma' - \alpha} \|J\|^2 \frac{1}{\gamma' - m(\gamma)} \\ &\leq \frac{\nu_j^+}{\nu_{j+1}^- - \alpha} \|J\|^2 \frac{1}{\nu_{j+1}^- - \nu_j^+} < 1, \end{aligned}$$

which completes the Proof of Proposition 2.4. □

Remark 2.5 *If B is a bounded operator on $\mathcal{H}$, the assumption* (2.12) *can be weakened. The claim of Proposition* 2.4 *remains true if*

$$\limsup_{j \to \infty} \nu_{j+1}^- (\nu_{j+1}^- - \nu_j^+) = +\infty. \tag{2.14}$$

Proof: Similar as above we choose a natural number j such that $\nu_{j+1}^- > \alpha$ and

$$\frac{\nu_j^+}{\nu_{j+1}^- - \alpha} \|B\|^2 \frac{1}{\nu_{j+1}^- (\nu_{j+1}^- - \nu_j^+)} < 1.$$

Again take $\gamma = \nu_{j+1}^-$ and suppose that $\gamma > \gamma'$. Then

$$\begin{aligned} &\|(\gamma' - D)^{-\frac{1}{2}} B_{\gamma_2}^* (\gamma - A_{\gamma_2'})^{-1} B_{\gamma_2} (\gamma' - D)^{-\frac{1}{2}}\| \\ &\leq \frac{1}{\gamma' - \alpha} \|B\|^2 \frac{1}{\gamma' - m(\gamma)} \\ &\leq \frac{\nu_j^+}{\nu_{j+1}^- - \alpha} \|B\|^2 \frac{1}{\nu_{j+1}^- (\nu_{j+1}^- - \nu_j^+)} < 1, \end{aligned}$$

□

Lemma 2.6 *Let R be a selfadjoint operator in an infinite-dimensional Hilbert space. Suppose that it is bounded from below and that it has a compact resolvent. By $(\mu_j)_1^\infty$ denote the non-decreasing sequence of eigenvalues of R counted according to their multiplicities. Then the following statements hold:*

i) *If the resolvent of R is nuclear, then*

$$\limsup_{j\to\infty} (\mu_{j+1} - \mu_j) = +\infty. \tag{2.15}$$

ii) *If the resolvent of R belongs to the Hilbert-Schmidt class, then*

$$\limsup_{j\to\infty} \mu_{j+1}(\mu_{j+1} - \mu_j) = +\infty. \tag{2.16}$$

Proof: Suppose without loss of generality that $R \gg 0$. If the relation (2.15) would not hold, then

$$M := \sup_j(\mu_{j+1} - \mu_j) < \infty.$$

This would imply that

$$\mu_{j+1} \le \mu_1 + M \cdot j$$

and hence the contradiction

$$\operatorname{Tr} A^{-1} = \sum_1^\infty \frac{1}{\mu_j} \ge \sum_1^\infty \frac{1}{\mu_1 + M \cdot j} = +\infty.$$

To prove the second assertion we assume that $A \gg 0$. By assumption A^{-1} is a Hilbert-Schmidt operator and hence A^{-2} a trace class operator. Since for $j \in \mathbb{N}$

$$2\mu_{j+1}(\mu_{j+1} - \mu_j) \ge (\mu_{j+1} + \mu_j)(\mu_{j+1} - \mu_j) = \mu_{j+1}^2 - \mu_j^2$$

and by the first statement

$$\limsup_{j\to\infty} (\mu_{j+1}^2 - \mu_j^2) = +\infty,$$

the proof of the lemma is complete. □

As a direct consequence of Theorem 1.1 and Proposition 2.3 we obtain

Theorem 2.7 *Let the assumptions* (a), (b), (c) *and* (d_2) *be fulfilled. Suppose that there is a number* $\gamma \le \alpha = \max \sigma(D)$ *such that inequality* (2.9) *holds for all* $\gamma' > \gamma$. *Assume that* $\mathcal{N}(A_\gamma - \gamma) \cap \mathcal{N}(B^*) = \{0\}$. *Then all the statements of* Theorem 1.1 *hold with respect to the representation* $\mathcal{H} = \hat{\mathcal{H}}_1 \times \hat{\mathcal{H}}_2$, *the entries* A_γ, B_γ *and* D_γ *of* $\mathbf{L}$, *and the spectral subspaces* $\mathcal{L}_-(\gamma)$, $\mathcal{L}_+(\gamma)$.

The following statement is an immediate consequence of Theorem 1.2.

Theorem 2.8 *Suppose in addition to the assumptions of Theorem* 2.7 *that the resolvent of* A_α *is compact. Then the spectrum of* $\mathbf{L}$ *in* $(\gamma, +\infty)$ *consists of isolated eigenvalues of finite multiplicity with no finite limit point. By* $\{\lambda_j\}_1^\infty$ *denote the* *non-decreasing sequence of eigenvalues of* $\mathbf{L}$ *in* $(\gamma, +\infty)$ *counted according to their multiplicities. Let*

$$\mathbf{e}_j = \begin{pmatrix} e_j \\ K e_j \end{pmatrix}, \quad j = 1, 2, \ldots,$$

be a corresponding orthonormal sequence of eigenvectors of $\mathbf{L}$. *Then the system* $(e_j)_1^\infty$ *is a Riesz basis in* $\mathcal{H}_1(\gamma)$.

Remark 2.9 i) *Under assumptions of Theorem* 2.8 *one can choose* γ *such that* $\mathcal{N}(A_\gamma - \gamma) = \{0\}$ *without referring to the assumption* $\mathcal{N}(A_\gamma - \gamma) \cap \mathcal{N}(B^*) = \{0\}$.

ii) *If* A *(or* A_α*) has a nuclear resolvent or* B *is a bounded operator and the resolvent of* A *(or* A_α*) belongs to the Hilbert-Schmidt class, then the existence of a number* γ *in Theorem* 2.7 *follows and thus the assertions of Theorem* 2.8 *are fulfilled.*

3 Example

In this section we consider a simple example from ordinary differential equation.

Let u, v be bounded function on the interval $I := [0, 1]$,

$$\|v\|_\infty = \operatorname*{ess\,sup}_{x \in I} | v(x) | \neq 0.$$

The function u is assumed to be real valued, v is allowed to take complex values. We consider the Hilbert spaces $\mathcal{H}_1 = \mathcal{H}_2 = \mathrm{L}^2(I)$, $\mathcal{H} = \mathcal{H}_1 \times \mathcal{H}_2$. Let $\mathrm{H}_2(I)$ denote the Sobolev space of order 2, i.e.,

$$\mathrm{H}_2(I) := (y \in \mathrm{L}^2(I) \mid y, y' \text{absolutely continuous}, y'' \in \mathrm{L}^2(I)).$$

Define the operator A in $L^2(I)$ as follows:

$$\begin{cases} \mathcal{D}(A) := \{y \in \mathrm{H}_2^2(I) \mid y(0) = y(1) = 0\}, \\ Ay = -y'' \qquad (y \in \mathcal{D}(A)). \end{cases}$$

Further, let B and D denote the bounded multiplication operators

$$By = uy, \qquad Dy = vy \qquad (y \in L^2(I)).$$

A and D are selfadjoint operators; $A \gg 0$, $\sigma(A)$ consists of the simple eigenvalues $\mu_n = n^2\pi^2$, $n = 1, 2, \ldots$, whence A^{-1} is a nuclear operator,

$$\alpha := \max \sigma(D) = \operatorname*{ess\,sup}_{x \in I} u(x), \quad \beta = \min \sigma(A) = \pi^2.$$

The functions $\sqrt{2}\sin \pi n x, n = 1, 2, \dots$ form an orthonormal system of eigenvectors of A related to eigenvalues μ_n.

Let us consider the selfadjoint operator $\mathbf{L}$ in $\mathcal{H} = \mathrm{L}^2(I) \times \mathrm{L}^2(I)$ given as the block operator matrix

$$\mathbf{L} = \begin{pmatrix} A & B \\ B^* & D \end{pmatrix} = \begin{pmatrix} -\frac{d^2}{dx^2} & v \\ \overline{v} & u \end{pmatrix}$$

on the domain $\mathcal{D}(A) \times \mathrm{L}^2(I)$. It is easy to see that $\sigma(\mathbf{L}) \cap (\alpha, \infty)$ consists only of isolated simple eigenvalues λ_n, which coincide with the eigenvalues of the Sturm-Liouville boundary problem

$$\text{(3.1)} \qquad \begin{cases} -y''(x) - \frac{|v(x)|^2}{u(x)-\lambda} y(x) = \lambda y(x), & x \in I, \\ y(0) = y(1) = 0. \end{cases}$$

The associated eigenvectors $\mathbf{e}_n$ of $\mathbf{L}$ are connected with the corresponding eigenfunctions e_n of the boundary problem (3.1) by the relation

$$\mathbf{e}_n = \begin{pmatrix} e_n, \\ -\frac{\overline{v} e_n}{u-\lambda_n} \end{pmatrix}.$$

Set $n_0 = 0$ if $\alpha < \pi^2 = \beta$. Otherwise, choose the minimal $n_0 \in \mathbb{N}$ such that the conditions

$$\text{(3.2)} \qquad \begin{cases} \alpha < (n_0 + 1)^2 \pi^2, \\ \operatorname*{ess\,sup}_{x \in I} \dfrac{1}{\pi^2(2n_0 + 1)} \dfrac{|\, v(x)\,|^2}{\pi^2(n_0 + 1)^2 - u(x)} < 1 \end{cases}$$

hold. Take $\gamma = \mu_{n_0+1} = (n_0 + 1)^2\pi^2$ and let $\gamma' > \gamma$. Similarly as in the Proof of Remark 2.5 we obtain the estimate

$$\begin{aligned} &\|(\gamma' - D)^{-\frac{1}{2}} B_\gamma^*(\gamma' - A_\gamma)^{-1} B_\gamma (\gamma' - D)^{-\frac{1}{2}}\| \\ &\leq \operatorname*{ess\,sup}_{x \in I} \frac{|\, v(x)\,|^2}{(\mu_{n_0+1} - u(x))\,(\mu_{n_0+1} - \mu_{n_0})} < 1. \end{aligned}$$

Notice that $\mathcal{N}(A_\gamma - \gamma) \cap \mathcal{N}(B_\gamma^*) = \{0\}$ since $\|v\|_\infty \neq 0$. Thus the assumptions of Theorem 2.7 are fulfilled. They also hold for any $\gamma = \mu_{n+1}, n \geq n_0$. By P_n, $n \in \mathbb{N}$, we denote the orthogonal projection of $\mathbf{L}^2(I)$ onto the subspace

$$\text{(3.3)} \qquad \mathcal{H}_1(\mu_n) := \left\{ f \in \mathbf{L}^2(I) : \int_0^1 f(x) \sin k\pi x dx = 0, k = 1, 2, \dots n \right\}$$

Applying Theorem 2.7, Remark 2.9 and Theorem 2.8 yields

Theorem 3.1 *Let* $n_0 \in \mathbb{N}$ *be minimal such that the conditions* (3.2) *hold. By* $(\lambda_j)_1^\infty$ *denote the sequence of the eigenvalues of the Sturm-Liouville boundary problem* (3.1) *in the interval* $[\pi^2 n^2, +\infty)$, $n \geq n_0$ *and let* y_j, $j \in \mathbb{N}$ *be the corresponding normalized eigenfunctions. Then the system* $\{P_n y_j : j \in \mathbb{N}\}$ *is a Riesz basis in the Hilbert space* (3.3).

Acknowledgements

Some part of this work was done while the first author was visiting the University of Regensburg with support of the Deutsche Forschungsgemainschaft within the frameworks of the joint German-Ukrainian project "Spectral Theory of Systems of Differential Operators of Mixed Order and Applications". Vadim Adamyan wishes to express his thanks to the Deutsche Forshungsgemainschaft for its financial support and to the University of Regensburg for its hospitality.

References

[AL] V. Adamyan and H. Langer, Spectral properties of a class of operator-valued functions, *J. Operator Theory* **33** (1995), 259–277.

[AdLMSau] V. Adamyan, H. Langer, R. Mennicken and J. Saurer, Spectral components of selfadjoint block operator matrices with unbounded entries, *Math. Nachr.* **178** (1996), 43–80.

[AMSau] V. Adamyan, R. Mennicken and J. Saurer, On the discrete spectrum of some selfadjoint operator matrices, *J. Operator Theory* **39** (1998), 3–41.

Odessa State University
Department of Theoretical Physics
Dvorjanska 2
270026 Odessa
Ukraine

Universität Regensburg
NWF I – Mathematik
93040 Regensburg
Germany

Operator Theory:
Advances and Applications, Vol. 122
© 2001 Birkhäuser Verlag Basel/Switzerland

A Coisometric Realization for Triangular Integral Operators

D. Alpay, V. Bolotnikov, A. Dijksma and Y. Peretz

Dedicated to Rien Kaashoek on the occasion of his 60-th birthday

In this paper we introduce and develop analogs of de Branges-Rovnyak spaces in the setting of lower triangular integral operators and the corresponding coisometric realization theorem.

1 Introduction

Time-invariant discrete time dissipative systems are characterized by a $\mathbb{C}^{m\times n}$-valued Schur (that is, analytic and contractive) function $S(z)$ in the open unit disk $\mathbb{D}$. This function is called the transfer function of the system and the system is called the realization of the function; see [16]. The kernel

$$\frac{I_m - S(z)S(\omega)^*}{1 - z\overline{\omega}} \tag{1.1}$$

is nonnegative in $\mathbb{D}$, and determines a reproducing kernel Hilbert space which was introduced and thoroughly studied by L. de Branges and J. Rovnyak; see [12]. These spaces play an important role in for example system theory, interpolation theory and operator models; see [15], [1] for more on these links and for references.

A time-invariant dissipative system is characterized by a $\mathbb{C}^{m\times n}$-valued Schur function $S(z)$ in the open right half plane $i\mathbb{C}_-$; see [18, p. 128–129]. Now the kernel

$$\frac{I_m - S(z)S(\omega)^*}{-(z + \overline{\omega})} \tag{1.2}$$

is nonnegative in $i\mathbb{C}_-$, and determines a reproducing kernel Hilbert space $\mathcal{H}_L(S)$. It is the state space of a coisometric realization of S in the following sense. Let $\alpha \in i\mathbb{C}_-$ be such that $\mathrm{Re}(\alpha) \geq 1$. Then the formulas

$$\begin{aligned} A(f) &= \frac{f(z) - f(\alpha)}{z - \alpha} \\ B(\xi) &= \frac{S(z) - S(\alpha)}{z - \alpha}\xi \\ C(f) &= f(\alpha) \\ D(\xi) &= S(\alpha)\xi \end{aligned}$$

define the colligation

$$\begin{pmatrix} A & B \\ C & D \end{pmatrix} : \begin{pmatrix} \mathcal{H}_L(S) \\ \mathbb{C}^n \end{pmatrix} \longrightarrow \begin{pmatrix} \mathcal{H}_L(S) \\ \mathbb{C}^m \end{pmatrix}$$

with the property that for every $z \in i\mathbb{C}_-$ with $|z - \alpha| < 1$ and any $\xi \in \mathbb{C}^n$,

$$S(z)\xi = (D + (z-\alpha)C(I_{\mathcal{H}_L(S)} - (z-\alpha)A)^{-1}B)(\xi). \tag{1.3}$$

The colligation is coisometric in the sense that

$$\begin{pmatrix} T & 0 \\ (\frac{1}{2}+\alpha)C & I_m \end{pmatrix} \begin{pmatrix} T^* & (\frac{1}{2}+\overline{\alpha})C^* \\ 0 & I_m \end{pmatrix}$$

$$= \begin{pmatrix} N & B \\ (\frac{1}{2}-\alpha)C & D \end{pmatrix} \begin{pmatrix} N^* & (\frac{1}{2}-\overline{\alpha})C^* \\ B^* & D^* \end{pmatrix}$$

where $T = I_{\mathcal{H}_L(S)} + (\frac{1}{2}+\alpha)A$ and $N = -I_{\mathcal{H}_L(S)} + (\frac{1}{2}-\alpha)A$, and it is closely outer-connected which means that

$$\bigcap_{|z-\alpha|<1} \operatorname{Ker} C(I_{\mathcal{H}_L(S)} - (z-\alpha)A)^{-1} = \{0\}.$$

This new definition of a coisometry was introduced in [2] and [3]. There the denominator of the reproducing kernel is of the form $a(z)a(\omega)^* - b(z)b(\omega)^*$ (where $a(z)$ and $b(z)$ are analytic in some region of $\mathbb{C}$) instead of just $-(z+\omega^*)$, which corresponds to the choice $a(z) = \frac{1}{2} - z$ and $b(z) = \frac{1}{2} + z$.

In this paper we present a nonstationary version of this coisometric realization. When one considers nonstationary discrete time systems, the Schur functions are replaced by upper triangular contractions from ℓ_2 to ℓ_2; see [4], [5]. For Hankel-norm approximation, interpolation problems, and completion problems in this setting we mention the works [14], [10] and [13], and [17]. Here we shall consider the coisometric realization in the nonstationary continuous time setting. Now Schur functions in $i\mathbb{C}_-$ are replaced by lower triangular integral contraction operators. This was first done by J. Ball, I. Gohberg, and M.A. Kaashoek in [9], [8], [7], and we refer to these papers for a further discussion about this topic and its application to interpolation problems.

We describe the contents of this paper: In Section 2 we review the nonstationary continuous time setting needed in the sequel. In Section 3 we define and study resolvent operators; a resolvent operator appears as the main operator in the coisometric colligation. The state space of this colligation is considered in Section 4. For the convenience of the reader at the end of Sections 2 and 4 we summarize the analogies between the continuous stationary and continuous nonstationary notions. Finally, in Section 5 we prove the existence of a coisometric colligation whose "characteristic function" coincides with the nonstationary analog of a Schur function.

The isometric and unitary realizations will be considered elsewhere.

2 Preliminaries and Notations

In this section we review and further develop the continuous nonstationary setting introduced first by the authors of [8]. The main results of this section are: Lemma 2.4, Lemma 2.5 and Lemma 2.6 in which we list properties of the generalized point evaluation map. At the end of the section we give a table which shows the analogy between the stationary and nonstationary settings for the continuous time case.

We first set some notation. By $L_2^{m\times n}(\mathbb{R})$ we denote the Hilbert space of all $\mathbb{C}^{m\times n}$-valued measurable functions on $\mathbb{R}$ which are square integrable with norm

$$\|f\|^2_{L_2^{m\times n}(\mathbb{R})} = \int_{-\infty}^{\infty} \mathrm{Tr}\,(f(t)^* f(t))\, dt$$

and by $L_\infty^{m\times n}(\mathbb{R})$ the set of all $\mathbb{C}^{m\times n}$-valued measurable functions on $\mathbb{R}$ which are essentially bounded. We write $L_2^m(\mathbb{R})$ and $L_\infty^m(\mathbb{R})$ for the spaces $L_2^{m\times 1}(\mathbb{R})$ and $L_\infty^{m\times 1}(\mathbb{R})$. For $W \in L_\infty^{m\times n}(\mathbb{R})$ we denote by $\mathcal{M}_W^\ell$ and $\mathcal{M}_W^r$ the operator of multiplication by W from the left and from the right respectively; when no confusion is expected, we simply write W instead of $\mathcal{M}_W^\ell$ or $\mathcal{M}_W^r$. If $\|W\|$ is the norm of the operator $W : L_2^n(\mathbb{R}) \longrightarrow L_2^m(\mathbb{R})$ of multiplication by the function $W(t)$, that is,

$$(W\psi)(t) = W(t)\psi(t), \quad t \in \mathbb{R},$$

then

$$\|W\| = \|W\|_{L_\infty^{m\times n}} := ess\sup_{t\in\mathbb{R}} \|W(t)\|_{\mathbb{C}^{m\times n}}.$$

By Δ we denote the operator of differentiation acting in $L_2^n(\mathbb{R})$ with domain $\mathrm{Dom}(\Delta)$, which is the subspace of all locally absolutely continuous functions ψ in $L_2^n(\mathbb{R})$ such that ψ' still belongs to $L_2^n(\mathbb{R})$. For given $s \in \mathbb{R}$, $x_s \in \mathbb{C}^n$ and $W(t) \in L_\infty^{n\times n}(\mathbb{R})$ the differential equation

$$\frac{dx}{dt} = W(t)x(t), \quad t \in \mathbb{R}, \tag{2.1}$$

has a unique solution

$$x(t) = U_W(t)U_W(s)^{-1}x_s, \quad t \in \mathbb{R},$$

satisfying $x(s) = x_s$. In this expression $U_W(t)$ is the unique $n \times n$ matrix-valued function which is locally absolutely continuous and satisfies

$$\frac{dU_W(t)}{dt} = W(t)U_W(t), \quad U_W(0) = I_n.$$

The matrix function $U_W(t)$ is called the *fundamental matrix* for the differential equation (2.1) and is invertible for every $t \in \mathbb{R}$. The matrix function $U_W(t)U_W(s)^{-1}$ is called the *transition matrix* for the differential equation (2.1).

By definition, the differential equation (2.1) admits a *dichotomy* if there exist a projection P on $\mathbb{C}^n$ (called the dichotomy) and a pair of constants $M > 0$ and $0 < a < 1$ such that

$$\|U_W(t)PU_W(s)^{-1}\|_{\mathbb{C}^{n\times n}} \leq Ma^{t-s}, \quad \text{for} \quad t \geq s,$$

and

$$\|U_W(t)(I_n - P)U_W(s)^{-1}\|_{\mathbb{C}^{n\times n}} \leq Ma^{s-t}, \quad \text{for} \quad t \leq s.$$

If a dichotomy exists, it is unique. The connection between the existence of a dichotomy and the invertibility of the operator $T = \Delta - W$ is established in the next theorem; for the proof, see [11].

Theorem 2.1 *Let $W \in L_\infty^{n\times n}(\mathbb{R})$. The operator $T = \Delta - W$ with* $\mathrm{Dom}(T) = \mathrm{Dom}(\Delta)$ *is invertible if and only if the ordinary differential equation* (2.1) *admits a dichotomy. If T is invertible, its inverse is given by the formula*

$$\begin{aligned}(T^{-1}(\psi))(t) &= \int_{-\infty}^{t} U_W(t)PU_W(s)^{-1}\psi(s)ds \\ &\quad - \int_t^{\infty} U_W(t)(I_n - P)U_W(s)^{-1}\psi(s)ds\end{aligned}$$

for $t \in \mathbb{R}$ and $\psi \in L_2^{n\times p}(\mathbb{R})$, where P and $U_W(t)$ are the dichotomy and the fundamental matrix of (2.1)*, respectively, and*

$$\|T^{-1}\| \leq \frac{2M}{\ln a^{-1}}. \tag{2.2}$$

In this paper we only consider the case where the dichotomy $P = 0$: A function W is called an $n \times n$ *anti-stable time-varying point* (or *astv-point* for short) if $W \in L_\infty^{n\times n}(\mathbb{R})$ and there exist two constants $M > 0$ and $0 < a < 1$ such that

$$\|U_W(t)U_W(s)^{-1}\| \leq Ma^{s-t}, \quad s \geq t.$$

In this case $T = \Delta - W$ is invertible and its inverse

$$(T^{-1}(\psi))(t) = -\int_t^{\infty} U_W(t)U_W(s)^{-1}\psi(s)ds, \quad t \in \mathbb{R}, \tag{2.3}$$

is a bounded upper triangular integral operator (see definition below) with $\|T^{-1}\| \leq \frac{M}{\ln a^{-1}}$.

If $W(t) = \omega I_n$ with $\mathrm{Re}\,\omega > 0$, then W is an astv-point. Indeed, the fundamental matrix is $U_W(t) = e^{\omega t}I_n$, and thus,

$$\|U_W(t)U_W(s)^{-1}\| = \|e^{\omega(t-s)}I_n\| \leq e^{-\mathrm{Re}(\omega)(s-t)},$$

that is (2.1) admits a dichotomy $P = 0$ with the pair of constants $M = 1$ and $a = e^{-\mathrm{Re}(\omega)}$. Hence the open right half plane $i\mathbb{C}_-$ is included in the set of astv-points.

By $\mathcal{W}^{m\times n}$ we denote the set of all operators $T : L_2^n(\mathbb{R}) \longrightarrow L_2^m(\mathbb{R})$ of the form

$$(T(\psi))(t) = k_0(t)\psi(t) + \int_{-\infty}^{\infty} k(t,s)\psi(s)ds, \quad t \in \mathbb{R}, \tag{2.4}$$

where $k_0 \in L_\infty^{m\times n}(\mathbb{R})$ and the kernel function $k(t,s)$ is a $\mathbb{C}^{m\times n}$-valued function on $\mathbb{R} \times \mathbb{R}$ whose entries are measurable and satisfy

$$\sup_{t-s=\tau} \|k(t,s)\| \leq \ell_k(\tau)$$

for some $\ell_k \in L_1(\mathbb{R})$. An element in $\mathcal{W}^{m\times n}$ can be viewed as a matrix whose entries belong to $\mathcal{W} := \mathcal{W}^{1\times 1}$. The set $\mathcal{W}$ is closed under sums and products of operators and is called the *nonstationary Wiener algebra on the real line*. It has been shown in [8, Lemma 1.2] that every $T \in \mathcal{W}^{m\times n}$ defines via (2.4) a bounded linear operator from $L_2^n(\mathbb{R})$ into $L_2^m(\mathbb{R})$ and the representation (2.4) is unique up to perturbations in $k_0(t)$ and $k(t,s)$ on subsets of measure zero of $\mathbb{R}$ and $\mathbb{R} \times \mathbb{R}$, respectively.

An operator $T \in \mathcal{W}^{m\times n}$ is called *lower triangular* if the kernel function in (2.4) is zero a.e. on $s \geq t$. In this case

$$(T(\psi))(t) = k_0(t)\psi(t) + \int_{-\infty}^{t} k(t,s)\psi(s)ds, \quad t \in \mathbb{R}, \tag{2.5}$$

and we denote the set of such operators by $\mathcal{LW}^{m\times n}$. By $\mathcal{UW}^{m\times n}$ we denote the set of all *upper triangular* operators $T \in \mathcal{W}^{m\times n}$ for which the kernel function k is zero a.e. on $s \leq t$.

Let W be an $m \times m$ astv-point and let $T \in \mathcal{LW}^{m\times n}$ be given by (2.5). Then by (2.3), $(\Delta - W)^{-1} \in \mathcal{UW}^{m\times m}$ and the kernel of the integral operator $(\Delta - W)^{-1}T \in \mathcal{W}^{m\times n}$ is given by

$$\begin{cases} -U_W(t)U_W(s)^{-1}k_0(s) - \displaystyle\int_s^{\infty} U_W(t)U_W(\tau)^{-1}k(\tau,s)d\tau & t \leq s, \\ -\displaystyle\int_t^{\infty} U_W(t)U_W(\tau)^{-1}k(\tau,s)d\tau & t > s. \end{cases} \tag{2.6}$$

Note that if $D : L_2^m(\mathbb{R}) \to L_2^m(\mathbb{R})$ is the operator of multiplication from the left by the function $D \in L_\infty^{m\times m}(\mathbb{R})$, then $(\Delta - W)^{-1}D \in \mathcal{UW}^{m\times n}$. The following theorem shows that for suitable D the difference

$$(\Delta - W)^{-1}T - (\Delta - W)^{-1}D$$

is lower triangular. For the proof we refer to [8, Proposition 1.4].

Theorem 2.2 *Let $T \in \mathcal{LW}^{m\times n}$ and let W be an $m\times m$ astv-point. Then there is a unique operator $D : L_2^m(\mathbb{R}) \to L_2^m(\mathbb{R})$ of multiplication from the left by the function $D \in L_\infty^{m\times m}(\mathbb{R})$ such that*

$$(\Delta - W)^{-1}(T - D) \in \mathcal{LW}^{m\times n}. \tag{2.7}$$

It is given by

$$\begin{aligned} D(t) &= T^\wedge(W)(t) := k_0(t) + \int_0^\infty \\ & U_W(t)U_W(t+\tau)^{-1}k(t+\tau,t)d\tau,\ t\in\mathbb{R}, \end{aligned} \tag{2.8}$$

where $k_0(t)$ and $k(t,s)$ are the singular part and the kernel function, respectively, in the representation (2.5) *of T. The kernel of the integral operator* (2.7) *is given by*

$$\begin{cases} -\displaystyle\int_t^\infty U_W(t)U_W(\tau)^{-1}k(\tau,s)d\tau & t>s, \\ 0 & t\le s, \end{cases} \tag{2.9}$$

where T is of the form (2.5).

The map $W \mapsto T^\wedge(W)$ is called the *generalized left point evaluation map* on $\mathcal{LW}^{m\times n}$. For the time-invariant case this map reduces to the usual point evaluation in the frequency domain, that is, to the Laplace transform. Indeed, if $W(t) = \omega I_m$ with $\mathrm{Re}(\omega) > 0$ and

$$(T(\psi))(t) = \int_{-\infty}^t k(t-s)\psi(s)ds,\ \ t\in\mathbb{R},$$

where $k \in L_1^{m\times m}(\mathbb{R})$ with $\mathrm{supp}(k) \subset [0,\infty)$, then

$$T^\wedge(W)(t) = \int_0^\infty e^{-\omega\tau}k(\tau)d\tau =: k^\wedge(\omega), \quad t\in\mathbb{R}.$$

It is easy to check that if $T \in \mathcal{LW}^{m\times n}$ and $F \in \mathcal{LW}^{p\times m}$, then for every $p\times p$ astv-point W

$$(FT)^\wedge(W) = (F^\wedge(W)T)^\wedge(W). \tag{2.10}$$

Let $T \in \mathcal{LW}^{m\times n}$ and let V be an $n\times n$ astv-point. Then $T(\Delta - V)^{-1} \in \mathcal{W}^{m\times n}$ and its kernel function is given by

$$\begin{cases} -k_0(t)U_V(t)U_V(s)^{-1} - \displaystyle\int_{-\infty}^s k(t,\tau)U_V(\tau)U_V(s)^{-1}d\tau & t\le s, \\ -\displaystyle\int_{-\infty}^s k(t,\tau)U_V(\tau)U_V(s)^{-1}d\tau & t>s. \end{cases} \tag{2.11}$$

The following analog of Theorem 2.2 shows that for suitable D the operator

$$T(\Delta - W)^{-1} - D(\Delta - W)^{-1}$$

is lower triangular. For the proof we refer to [8, Proposition 1.4].

Theorem 2.3 *Let $T \in \mathcal{LW}^{m\times n}$ and let V be an $n \times n$ astv-point. Then there is a unique operator $D : L_2^m(\mathbb{R}) \to L_2^m(\mathbb{R})$ of multiplication from the left by the function $D \in L_\infty^{m\times m}(\mathbb{R})$ such that*

$$(T - D)(\Delta - V)^{-1} \in \mathcal{LW}^{m\times n}. \tag{2.12}$$

It is given by

$$\begin{aligned} D(t) &= T^{\triangle}(V)(t) := k_0(t) \\ &\quad + \int_0^\infty k(t, t-\tau)U_V(t-\tau)U_V(t)^{-1}d\tau, \quad t \in \mathbb{R}. \end{aligned} \tag{2.13}$$

The kernel of the integral operator (2.12) *is given by*

$$\begin{cases} 0 & t \le s, \\ -\displaystyle\int_{-\infty}^{s} k(t,\tau)U_V(\tau)U_V(s)^{-1}d\tau & t > s. \end{cases} \tag{2.14}$$

We refer to the map $V \mapsto T^{\triangle}(V)$ as the *generalized right point evaluation map* on $\mathcal{LW}^{m\times n}$. It has the following property: If $T \in \mathcal{LW}^{m\times n}$, $F \in \mathcal{LW}^{p\times m}$ and V is an $n \times n$ astv-point, then

$$(FT)^{\triangle}(V) = (FT^{\triangle}(V))^{\triangle}(V). \tag{2.15}$$

Lemma 2.4 *If $T \in \mathcal{LW}^{m\times n}$, W is an $m \times m$ astv-point and V is an $n \times n$ astv-point, then*

$$((T - T^{\triangle}(V))(\Delta - V)^{-1})^{\wedge}(W) = ((\Delta - W)^{-1}(T - T^{\wedge}(W)))^{\triangle}(V).$$

Proof: From (2.9) and (2.13) we obtain

$$\begin{aligned} &((\Delta - W)^{-1}(T - T^{\wedge}(W)))^{\triangle}(V)(t) \\ &= -\int_0^\infty \int_t^\infty U_W(t)U_W(\tau)^{-1}k(\tau, t-\alpha)U_V(t-\alpha)U_V(t)^{-1}d\tau d\alpha. \end{aligned}$$

Similarly, from (2.14) and (2.8) we get

$$\begin{aligned} &((T - T^{\triangle}(V))(\Delta - V)^{-1})^{\wedge}(W)(t) \\ &= -\int_0^\infty \int_{-\infty}^t U_W(t)U_W(t+\alpha)^{-1}k(t+\alpha, \tau)U_V(\tau)U_V(t)^{-1}d\tau d\alpha \\ &= -\int_0^\infty \int_t^\infty U_W(t)U_W(\tau)^{-1}k(\tau, t-\alpha)U_V(t-\alpha)U_V(t)^{-1}d\tau d\alpha. \end{aligned}$$

□

By $\mathcal{S}_2^{m\times n}$ we denote the set of all Hilbert-Schmidt integral operators $T : L_2^{n\times p}(\mathbb{R}) \longrightarrow L_2^{m\times p}(\mathbb{R})$. It is well known that $T \in \mathcal{S}_2^{m\times n}$ if and only if it admits a representation of the form

$$(T(\psi))(t) = \int_{-\infty}^{\infty} k(t,s)\psi(s)ds, \quad t \in \mathbb{R},$$

where the kernel function k belongs to $L_2^{m\times n}(\mathbb{R}\times\mathbb{R})$. $\mathcal{S}_2^{m\times n}$ is a Hilbert space under the inner product

$$\langle F, G\rangle_{\mathcal{S}_2^{m\times n}} := \int_{-\infty}^{\infty}\int_{-\infty}^{\infty} \operatorname{Tr}\left(g(t,s)^* f(t,s)\right) dtds = \langle f, g\rangle_{L_2^{m\times n}(\mathbb{R}\times\mathbb{R})},$$

where f and g are the kernels of the operators F and G, respectively.

The closed subspaces of lower triangular and upper triangular operators in $\mathcal{S}_2^{m\times n}$ will be denoted by $\mathcal{LS}_2^{m\times n}$ and $\mathcal{US}_2^{m\times n}$. We denote by p and q the orthogonal projections $p : \mathcal{S}_2^{m\times n} \longrightarrow \mathcal{LS}_2^{m\times n}$ and $q : \mathcal{S}_2^{m\times n} \longrightarrow \mathcal{US}_2^{m\times n}$.

Lemma 2.5 *If $F \in \mathcal{LS}_2^{m\times n}$ and W is an $m\times m$ astv-point, then $F^\wedge(W) \in L_2^{m\times n}(\mathbb{R}) \cap L_\infty^{m\times n}(\mathbb{R})$.*

Proof: By Theorem 2.2, the function

$$F^\wedge(W)(t) = \int_0^\infty U_W(t)U_W(t+\tau)^{-1} f(t+\tau, t)d\tau, \quad t \in \mathbb{R}, \tag{2.16}$$

belongs to $L_\infty^{m\times n}(\mathbb{R})$. We show that it also belongs to $L_2^{m\times n}(\mathbb{R})$. Indeed,

$$\begin{aligned}
\|F^\wedge(W)(t)\|^2_{\mathbb{C}^{m\times n}} &= \left\| \int_t^\infty U_W(t)U_W(s)^{-1} f(s,t)ds \right\|^2_{\mathbb{C}^{m\times n}} \\
&\le \left(\int_t^\infty \|U_W(t)U_W(s)^{-1} f(s,t)\|_{\mathbb{C}^{m\times n}} ds \right)^2 \\
&\le \left(\int_t^\infty \|U_W(t)U_W(s)^{-1}\|_{\mathbb{C}^{m\times m}} \|f(s,t)\|_{\mathbb{C}^{m\times n}} ds \right)^2 \\
&\le \left(\int_t^\infty \|U_W(t)U_W(s)^{-1}\|^2_{\mathbb{C}^{m\times m}} ds \right) \left(\int_t^\infty \|f(s,t)\|^2_{\mathbb{C}^{m\times n}} ds \right) \\
&\le \left(\int_t^\infty M^2 a^{2(s-t)} ds \right) \left(\int_t^\infty \|f(s,t)\|^2_{\mathbb{C}^{m\times n}} ds \right) \\
&= \left(\int_0^\infty M^2 a^{2\tau} d\tau \right) \left(\int_t^\infty \|f(s,t)\|^2_{\mathbb{C}^{m\times n}} ds \right) \\
&= \frac{M^2}{2\ln a^{-1}} \int_t^\infty \|f(s,t)\|^2_{\mathbb{C}^{m\times n}} ds,
\end{aligned}$$

and therefore,

$$\begin{aligned}\int_{-\infty}^{\infty} \|F^{\wedge}(W)(t)\|^2_{\mathbb{C}^{m\times n}}dt &\leq \frac{M^2}{2\ln a^{-1}}\int_{-\infty}^{\infty}\int_t^{\infty} \|f(s,t)\|^2_{\mathbb{C}^{m\times n}}dsdt \\ &\leq \frac{M^2}{2\ln a^{-1}}\|f\|^2_{L_2^{m\times n}(\mathbb{R}\times\mathbb{R})}.\end{aligned}$$

Therefore, the function $F^{\wedge}(W)$ belongs to $L_2^{m\times n}(\mathbb{R})$ and

$$\|F^{\wedge}(W)\|_{L_2^{m\times n}(\mathbb{R})} \leq \frac{M}{(2\ln a^{-1})^{1/2}}\|f\|_{L_2^{m\times n}(\mathbb{R}\times\mathbb{R})}. \tag{2.17}$$

□

The subspace $\mathcal{LS}_2^{m\times n}$ is a reproducing kernel Hilbert space with reproducing kernel $(W-\Delta)^{-*}$ in the sense of formula (2.18) below.

Lemma 2.6 *Let W be an $m\times m$ astv-point and let G be the operator of multiplication from the left by the function $G \in L_2^{m\times n}(\mathbb{R}) \cap L_\infty^{m\times n}(\mathbb{R})$. Then the set of all operators of the form $(W-\Delta)^{-*}G$ is dense in $\mathcal{LS}_2^{m\times n}$ and for all $F \in \mathcal{LS}_2^{m\times n}$,*

$$\langle F, (W-\Delta)^{-*}G\rangle_{\mathcal{LS}_2^{m\times n}} = \langle F^{\wedge}(W), G\rangle_{L_2^{m\times n}(\mathbb{R})}. \tag{2.18}$$

Proof: By (2.3), the kernel of the operator $(W-\Delta)^{-*}G$ is given by

$$k(t,s) = \begin{cases} U_W(t)^{-*}U_W(s)^*G(s) & t \geq s, \\ 0 & t < s. \end{cases}$$

Thus

$$\begin{aligned}&\int_{-\infty}^{\infty}\int_{-\infty}^{\infty} \|k(t,s)\|^2_{\mathbb{C}^{m\times n}}dtds \\ &= \int_{-\infty}^{\infty}\int_s^{\infty} \|U_W(t)^{-*}U_W(s)^*G(s)\|^2_{\mathbb{C}^{m\times n}}dtds \\ &\leq \int_{-\infty}^{\infty}\int_s^{\infty} \|U_W(t)^{-*}U_W(s)^*\|^2_{\mathbb{C}^{m\times m}}\|G(s)\|^2_{\mathbb{C}^{m\times n}}dtds \\ &= \int_{-\infty}^{\infty} \|G(s)\|^2_{\mathbb{C}^{m\times n}}\left(\int_s^{\infty} \|U_W(t)^{-*}U_W(s)^*\|^{\mathbb{C}^{m\times m}}dt\right)ds \\ &\leq \int_{-\infty}^{\infty} \|G(s)\|^2_{\mathbb{C}^{m\times n}}\left(\int_s^{\infty} M^2a^{2(t-s)}dt\right)ds \\ &= \frac{M^2}{2\ln a^{-1}}\|G\|^2_{L_2^{m\times n}(\mathbb{R})},\end{aligned}$$

which shows that $k(t,s)$ belongs to $L_2^{m\times n}(\mathbb{R}\times\mathbb{R})$. Hence $(W-\Delta)^{-*}G \in \mathcal{LS}_2^{m\times n}$ and

$$\|(W-\Delta)^{-*}G\|_{\mathcal{LS}_2^{m\times n}} \leq \frac{M}{(2\ln a^{-1})^{1/2}}\|G\|_{L_2^{m\times n}(\mathbb{R})}. \tag{2.19}$$

If $F \in \mathcal{LS}_2^{m\times n}$ has kernel $f(t,s)$, then

$$\begin{aligned}
&\langle F, (W-\Delta)^{-*}G\rangle_{\mathcal{LS}_2^{m\times n}}\\
&= \langle f, k\rangle_{L_2^{m\times n}(\mathbb{R}\times\mathbb{R})} = \int_{-\infty}^{\infty}\int_{-\infty}^{\infty} \operatorname{Tr} k(t,s)^* f(t,s)dtds\\
&= \int_{-\infty}^{\infty}\int_{s}^{\infty} \operatorname{Tr} G(s)^* U_W(s)U_W(t)^{-1} f(t,s)dtds\\
&= \int_{-\infty}^{\infty} \operatorname{Tr} G(s)^* \left(\int_{s}^{\infty} U_W(s)U_W(t)^{-1} f(t,s)dt\right)ds\\
&= \int_{-\infty}^{\infty} \operatorname{Tr} G(s)^* F^{\wedge}(W)(s)ds\\
&= \langle F^{\wedge}(W), G\rangle_{L_2^{m\times n}(\mathbb{R})},
\end{aligned}$$

which proves (2.18). To prove the density claim, assume that

$$\langle F, (W-\Delta)^{-*}G\rangle_{\mathcal{LS}_2^{m\times n}} = 0$$

for every choice of $G \in L_2^{m\times n}(\mathbb{R}) \cap L_\infty^{m\times n}(\mathbb{R})$ and of an $m\times m$ astv-point W. Then $F^{\wedge}(W)(t) = 0$ for all $t \in \mathbb{R}$ and all $m\times m$ astv-points W, since the set $L_2^{m\times n}(\mathbb{R}) \cap L_\infty^{m\times n}(\mathbb{R})$ is dense in $L_2^{m\times n}(\mathbb{R})$. Choose $W(t) = \omega I$ with $\operatorname{Re}\omega > 0$. Then,

$$\begin{aligned}
F^{\wedge}(W)(t) &= \int_t^{\infty} U_W(t)U_W(\tau)^{-1} f(\tau,t)d\tau\\
&= \int_t^{\infty} e^{\omega(t-\tau)} f(\tau,t)d\tau = 0, \quad t\in\mathbb{R}.
\end{aligned}$$

Setting $f_t(\tau) = f(t+\tau,t)$ we get

$$\int_t^{\infty} e^{\omega(t-\tau)} f(\tau,t)d\tau = \int_0^{\infty} e^{-\omega\tau} f_t(\tau)d\tau = f_t^{\wedge}(\omega),$$

where $f_t^{\wedge}(\omega)$ is the Laplace transform of f_t. Thus $f(t+\tau,t) = f_t(\tau) = 0$ for any $t,\tau\in\mathbb{R}$, and therefore $f=0$ and $F=0$. □

Similarly, the set $\mathcal{LS}_2^{m\times n}$ is a reproducing kernel space in the following sense: The set of all operators of the form $G(V-\Delta)^{-*}$ where V is an $n\times n$ astv-point and G is

Stationary Setting	Nonstationary Setting
The complex space $\mathbb{C}^{m\times n}$.	The sets $L_2^{m\times n}(\mathbb{R}) \cap L_\infty^{m\times n}(\mathbb{R})$.
The open right half plane $i\mathbb{C}_-$.	The sets of $m \times m$ and $n \times n$ astv-points.
The multiplication by the variable z.	The differentiation operator Δ acting in $L_2^{n\times n}(\mathbb{R})$.
The Banach space $\mathbf{H}_\infty^{m\times n}(i\mathbb{C}_-)$ of analytic functions in the open right half plane.	The nonstationary Wiener algebra $\mathcal{LW}^{m\times n}$ of lower triangular integral operators.
Schur functions $S \in \mathbf{H}_\infty^{m\times n}(i\mathbb{C}_-)$ with $\|S\|_\infty \leq 1$.	Lower triangular contractions $S \in \mathcal{LW}^{m\times n}$ with $\|S\| \leq 1$.
The point evaluation of $f \in \mathbf{H}_\infty^{p\times q}(i\mathbb{C}_-)$ at $\omega \in \mathbb{D}$, that is, the unique vector $f(\omega) \in \mathbb{C}^{p\times q}$ such that the function $\frac{f(z)-f(\omega)}{z-\omega}$ is analytic.	The unique multiplication operators $T^\wedge(W)$, $T^\Delta(V)$ such that $(\Delta - W)^{-1}(T - T^\wedge(W))$ and $(T - T^\Delta(V))(\Delta - V)^{-1}$ are lower triangular integral operators.
The resolvent operators and equation $\mathcal{R}_\alpha - \mathcal{R}_\beta = (\alpha - \beta)\mathcal{R}_\alpha\mathcal{R}_\beta$.	The resolvent operators and equations $\mathcal{R}_V^\ell - \mathcal{R}_W^\ell = \mathcal{R}_V^\ell(\mathcal{M}_V^\ell - \mathcal{M}_W^\ell)\mathcal{R}_W^\ell$ and $\mathcal{R}_V^r - \mathcal{R}_W^r = \mathcal{R}_V^r(\mathcal{M}_V^r - \mathcal{M}_W^r)\mathcal{R}_W^r$.
The Hardy space $\mathbf{H}_2^{n\times m}(i\mathbb{C}_-)$ of the open right half plane $i\mathbb{C}_-$.	The space of lower triangular Hilbert-Schmidt operators $\mathcal{LS}_2^{n\times m}$.
Cauchy's formula for $\mathbf{H}_2(i\mathbb{C}_-)$ functions.	Formulas (2.18) and (2.20)

the operator of multiplication from the left by a function $G \in L_2^{m\times n}(\mathbb{R}) \cap L_\infty^{m\times n}(\mathbb{R})$ is dense in $\mathcal{LS}_2^{m\times n}$ and the reproducing kernel property

$$\langle F, G(V-\Delta)^{-*}\rangle_{\mathcal{LS}_2^{m\times n}} = \langle F^\Delta(W), G\rangle_{L_2^{m\times n}(\mathbb{R})} \tag{2.20}$$

holds.

In the next table we display the analogy between stationary and nonstationary cases, for the continuous time setting. The last item pertains to the next section. The operator $\mathcal{R}_\alpha$ on the left side is the backwards shift operator A defined in the Introduction.

3 Resolvent Operators and Resolvent Equations

In this section we give a generalization of the resolvent equation in the continuous time nonstationary setting. With an $m \times m$ astv-point V, we associate the resolvent operators $\mathcal{R}_V^\ell, \mathcal{R}_V^r : \mathcal{LS}_2^{m\times n} \longrightarrow \mathcal{LS}_2^{m\times n}$ defined by

$$\mathcal{R}_V^\ell(F) = (\Delta - V)^{-1}(F - F^\wedge(V)), \quad \mathcal{R}_V^r(F) = (F - F^\Delta(V))(\Delta - V)^{-1},$$

and the multiplication operators $\mathcal{M}_V^\ell, \mathcal{M}_V^r : \mathcal{LS}_2^{m\times n} \longrightarrow \mathcal{LS}_2^{m\times n}$ defined by

$$\mathcal{M}_V^\ell(F) = VF, \quad \mathcal{M}_V^r(F) = FV.$$

By (2.6) and (2.9),

$$\mathcal{R}_V^\ell(F) = (\Delta - V)^{-1}(F - F^\wedge(V)) = p((\Delta - V)^{-1}F)$$

and hence, $\|\mathcal{R}_V^\ell\| \le \|(\Delta - V)^{-1}\|$. The same norm estimate holds for $\mathcal{R}_V^r$ on account of (2.11) and (2.14). Obviously, $\|\mathcal{M}_V^\ell\| \le \|V\|$ and $\|\mathcal{M}_V^r\| \le \|V\|$.

Lemma 3.1 *If W and V are $n \times n$ astv-points, then*

$$(\Delta - W)^{-1}(W - V)(\Delta - V)^{-1} = (\Delta - W)^{-1} - (\Delta - V)^{-1}. \tag{3.1}$$

Proof: Let $\psi \in L_2^n(\mathbb{R})$. Then

$$\begin{aligned}
&(\Delta - W)^{-1}(W - V)(\Delta - V)^{-1}(\psi)(t)\\
&= \int_t^\infty U_W(t)U_W(s)^{-1}\left(\int_s^\infty (W(s) - V(s))U_V(s)U_V(\tau)^{-1}\psi(\tau)d\tau\right)ds\\
&= \int_t^\infty U_W(t)\left(\int_t^\tau U_W(s)^{-1}(W(s) - V(s))U_V(s)ds\right)U_V(\tau)^{-1}\psi(\tau)d\tau\\
&= -\int_t^\infty U_W(t)(U_W(\tau)^{-1}U_V(\tau) - U_W(t)^{-1}U_V(t))U_V(\tau)^{-1}\psi(\tau)d\tau\\
&= (\Delta - W)^{-1}(\psi)(t) - (\Delta - V)^{-1}(\psi)(t).
\end{aligned}$$

□

Lemma 3.2 *Let $F \in \mathcal{LS}_2^{m\times n}$. Then for all $m \times m$ astv-points W and V,*

$$((W - V)(\Delta - V)^{-1}(F - F^\wedge(V)))^\wedge(W) = F^\wedge(W) - F^\wedge(V), \tag{3.2}$$

and for all $n \times n$ astv-points W and V

$$((F - F^\Delta(W))(\Delta - W)^{-1}(W - V))^\Delta(V) = F^\Delta(W) - F^\Delta(V). \tag{3.3}$$

Proof: The kernel function of the operator $(W - V)(\Delta - V)^{-1}(F - F^\wedge(V))$ is given by

$$\begin{cases} -\displaystyle\int_t^\infty (W(t) - V(t))U_V(t)U_V(\tau)^{-1}f(\tau, s)d\tau & t \ge s,\\ 0 & t < s.\end{cases}$$

Therefore

$$
\begin{aligned}
&((W-V)(\Delta-V)^{-1}(F-F^{\wedge}(V)))^{\wedge}(W)(t)\\
&=-\int_t^{\infty} U_W(t)U_W(s)^{-1}\left(\int_s^{\infty}(W(s)-V(s))U_V(s)U_V(\tau)^{-1}f(\tau,t)d\tau\right)ds\\
&=-\int_t^{\infty} U_W(t)\left(\int_t^{\tau} U_W(s)^{-1}(W(s)-V(s))U_V(s)ds\right)U_V(\tau)^{-1}f(\tau,t)d\tau\\
&=\int_t^{\infty} U_W(t)(U_W(\tau)^{-1}U_V(\tau)-U_W(t)^{-1}U_V(t))U_V(\tau)^{-1}f(\tau,t)d\tau\\
&=F^{\wedge}(W)(t)-F^{\wedge}(V)(t),
\end{aligned}
$$

which proves (3.2); (3.3) is proved quite similarly. □

Corollary 3.3 *Let $F \in \mathcal{LS}_2^{m\times n}$ and let V be an $n \times n$ astv-point. Then*

$$(F(V-\Delta)^{-*}(V+V^*))^{\Delta}(V)=F^{\Delta}(V). \tag{3.4}$$

Proof: Since $\Delta^* = -\Delta$, putting $W = -V^*$ in (3.1) we get

$$(V-\Delta)^{-*}(V+V^*)(\Delta-V)^{-1}=(\Delta-V)^{-1}-(V-\Delta)^{-*}, \tag{3.5}$$

which implies, in particular, that

$$
\begin{aligned}
&q(F(V-\Delta)^{-*}(V+V^*)(\Delta-V)^{-1})\\
&\quad=q(F(\Delta-V)^{-1}-F(V-\Delta)^{-*})\\
&\quad=q(F(\Delta-V)^{-1})\\
&\quad=q((F-F^{\Delta}(V))(\Delta-V)^{-1})+q(F^{\Delta}(V)(\Delta-V)^{-1})\\
&\quad=F^{\Delta}(V)(\Delta-V)^{-1}.
\end{aligned}
$$

On the other hand, for any $G \in \mathcal{LS}_2^{m\times n}$ it holds that $q(G(\Delta-V)^{-1}) = G^{\Delta}(V)$ $(\Delta-V)^{-1}$. Thus

$$
\begin{aligned}
&q(F(V-\Delta)^{-*}(V+V^*)(\Delta-V)^{-1})\\
&\quad=(F(V-\Delta)^{-*}(V+V^*))^{\Delta}(V)(\Delta-V)^{-1}
\end{aligned}
$$

and therefore $(F(V-\Delta)^{-*}(V+V^*))^{\Delta}(V)=F^{\Delta}(V)$. □

Corollary 3.4 *If V and W are $m \times m$ astv-points, then*

$$\mathcal{R}^{\ell}_W-\mathcal{R}^{\ell}_V=\mathcal{R}^{\ell}_W(\mathcal{M}^{\ell}_W-\mathcal{M}^{\ell}_V)\mathcal{R}^{\ell}_V \tag{3.6}$$

and

$$\mathcal{R}^{r}_W-\mathcal{R}^{r}_V=\mathcal{R}^{r}_W(\mathcal{M}^{r}_W-\mathcal{M}^{r}_V)\mathcal{R}^{r}_V. \tag{3.7}$$

Proof: Let $F \in \mathcal{LS}_2^{m\times n}$. Using the last two lemmas we get

$$\begin{aligned}
&\mathcal{R}_W^\ell(\mathcal{M}_W^\ell - \mathcal{M}_V^\ell)\mathcal{R}_V^\ell(F)\\
&\quad = (\Delta - W)^{-1}((W-V)(\Delta - V)^{-1}(F - F^\wedge(V))\\
&\qquad - ((W-V)(\Delta-V)^{-1}(F - F^\wedge(V)))^\wedge(W))\\
&\quad = ((\Delta - W)^{-1} - (\Delta - V)^{-1})(F - F^\wedge(V))\\
&\qquad - (\Delta - W)^{-1}(F^\wedge(W) - F^\wedge(V))\\
&\quad = (\Delta - W)^{-1}(F - F^\wedge(W))\\
&\qquad - (\Delta - V)^{-1}(F - F^\wedge(V))\\
&\quad = \mathcal{R}_W^\ell(F) - \mathcal{R}_V^\ell(F).
\end{aligned}$$

The proof of the second part goes in much the same way. □

Corollary 3.5 *Let V and W be $m\times m$ astv-points such that $\|W - V\|\|(\Delta - V)^{-1}\| < 1$. Then*

$$\begin{aligned}
\mathcal{R}_W^\ell &= \mathcal{R}_V^\ell(I - (\mathcal{M}_W^\ell - \mathcal{M}_V^\ell)\mathcal{R}_V^\ell)^{-1}\\
\mathcal{R}_W^r &= \mathcal{R}_V^r(I - (\mathcal{M}_W^r - \mathcal{M}_V^r)\mathcal{R}_V^r)^{-1}\\
I + (\mathcal{M}_W^\ell - \mathcal{M}_V^\ell)\mathcal{R}_W^\ell &= (I - (\mathcal{M}_W^\ell - \mathcal{M}_V^\ell)\mathcal{R}_V^\ell)^{-1}\\
I + (\mathcal{M}_W^r - \mathcal{M}_V^r)\mathcal{R}_W^r &= (I - (\mathcal{M}_W^r - \mathcal{M}_V^r)\mathcal{R}_V^r)^{-1}\\
I + \mathcal{R}_W^\ell(\mathcal{M}_W^\ell - \mathcal{M}_V^\ell) &= (I - \mathcal{R}_V^\ell(\mathcal{M}_W^\ell - \mathcal{M}_V^\ell))^{-1}\\
I + \mathcal{R}_W^r(\mathcal{M}_W^r - \mathcal{M}_V^r) &= (I - \mathcal{R}_V^r(\mathcal{M}_W^r - \mathcal{M}_V^r))^{-1}.
\end{aligned}$$

Proof: The equalities follow from the resolvent equation (3.6) provided the indicated inverses exist. But this follows from the assumption that $\|W - V\|\|(\Delta - V)^{-1}\| < 1$, which implies that

$$\|(\mathcal{M}_W^\ell - \mathcal{M}_V^\ell)\,\mathcal{R}_V^\ell\| \le \|W - V\|\|(\Delta - V)^{-1}\| < 1.$$

□

4 The State Spaces $\mathcal{H}_L(S)$ and $\mathcal{H}_R(S)$

Assume that $S \in \mathcal{LW}^{m\times n}$ is a contraction from $L_2^n(\mathbb{R})$ into $L_2^m(\mathbb{R})$: $\|S\| \le 1$. Then the multiplication operators $\mathcal{M}_S^\ell : \mathcal{LS}_2^{n\times p} \longrightarrow \mathcal{LS}_2^{m\times p}$ and $\mathcal{M}_S^r : \mathcal{LS}_2^{q\times m} \longrightarrow \mathcal{LS}_2^{q\times n}$ defined by $\mathcal{M}_S^\ell(F) = SF$ and $\mathcal{M}_S^r(F) = FS$ are also contractions. So the operators $I - \mathcal{M}_S^r\mathcal{M}_S^{r*}$ and $I - \mathcal{M}_S^\ell\mathcal{M}_S^{\ell*}$ are positive and the ranges of the square roots of these operators provided with the lifted norm (see below) are Hilbert spaces which we denote by $\mathcal{H}_L(S)$ and $\mathcal{H}_R(S)$. As in the stationary case (as recalled in the Introduction) they are state spaces for a left and right coisometric realizations of S. The left case will be discussed in Section 5. The right case can be done similarly; we will give the formulas without proofs. In this section we give various characterizations for S to be a contraction, we discuss the Hilbert spaces $\mathcal{H}_L(S)$ and $\mathcal{H}_R(S)$, and prove that multiplication operators by bounded functions define bounded operators on these spaces.

Lemma 4.1 *Let S be a lower triangular operator in $\mathcal{LW}^{m\times n}$.*

(a) *The adjoint of $\mathcal{M}_S^\ell$ is given by*

$$\mathcal{M}_S^{\ell *}(F) = p(S^*F)$$

and acts on the left kernel of $\mathcal{LS}_2^{m\times p}$ as the operator of multiplication from the left:

$$\mathcal{M}_S^{\ell *}((W-\Delta)^{-*}G) = S^{\wedge}(W)^*(W-\Delta)^{-*}G, \tag{4.1}$$

where W is an $m \times m$ astv-point and G is the operator of multiplication from the left by a function $G \in L_2^{m\times p}(\mathbb{R}) \cap L_\infty^{m\times p}(\mathbb{R})$.

(b) *The adjoint of $\mathcal{M}_S^r$ is given by*

$$\mathcal{M}_S^{r*}(F) = p(FS^*)$$

and acts on the right kernel of $\mathcal{LS}_2^{q\times n}$ as the operator of multiplication from the right:

$$\mathcal{M}_S^{r*}(H(V-\Delta)^{-*}) = H(V-\Delta)^{-*}S^{\Delta}(V)^*, \tag{4.2}$$

where V is an $n \times n$ astv-point and H is the operator of multiplication from the left by a function $H \in L_2^{q\times n}(\mathbb{R}) \cap L_\infty^{q\times n}(\mathbb{R})$.

In the case of reproducing kernel Hilbert spaces, (4.1) and (4.2) show that the kernels are the eigenfunctions of the adjoints of the multiplication operators.

Proof: For $F \in \mathcal{LS}_2^{m\times p}$ and $N \in \mathcal{LS}_2^{n\times p}$ we have

$$\begin{aligned}
\langle N, \mathcal{M}_S^{\ell *}(F)\rangle_{\mathcal{LS}_2^{n\times p}} &= \langle SN, F\rangle_{\mathcal{LS}_2^{m\times p}} = \mathrm{Tr} F^*SN \\
&= \mathrm{Tr}(S^*F)^*N = \langle N, S^*F\rangle_{\mathcal{LS}_2^{n\times p}} \\
&= \langle N, p(S^*F)\rangle_{\mathcal{LS}_2^{n\times p}}.
\end{aligned}$$

Let W be an $m \times m$ astv-point, let G be the operator of multiplication from the left by a function $G \in L_2^{m\times p}(\mathbb{R}) \cap L_\infty^{m\times p}(\mathbb{R})$ and let $N \in \mathcal{LS}_2^{n\times p}$. Then by (2.10),

$$\begin{aligned}\langle N, \mathcal{M}_S^{\ell *}((W-\Delta)^{-*}G)\rangle_{\mathcal{LS}_2^{n\times p}} &= \langle SN, (W-\Delta)^{-*}G\rangle_{\mathcal{LS}_2^{m\times p}} \\ &= \langle (SN)^\wedge(W), G\rangle_{L_2^{m\times p}} \\ &= \langle (S^\wedge(W)N)^\wedge(W), G\rangle_{L_2^{m\times p}} \\ &= \langle S^\wedge(W)N, (W-\Delta)^{-*}G\rangle_{\mathcal{LS}_2^{m\times p}} \\ &= \langle N, S^\wedge(W)^*(W-\Delta)^{-*}G\rangle_{\mathcal{LS}_2^{n\times p}}\end{aligned}$$

and thus $\mathcal{M}_S^{\ell *}((W-\Delta)^{-*}G) = S^\wedge(W)^*(W-\Delta)^{-*}G$. The proof of (b) is similar and will be omitted. □

For $S \in \mathcal{LW}^{m\times n}$, W an $m \times m$ astv-point and V an $n \times n$ astv-point we define the lower triangular operators

$$K_S^L(W) = (I_{L_2^{m\times m}(\mathbb{R})} - SS^\wedge(W)^*)(W-\Delta)^{-*} \tag{4.3}$$

$$K_S^R(V) = (V-\Delta)^{-*}(I_{L_2^{n\times n}(\mathbb{R})} - S^\Delta(V)^*S). \tag{4.4}$$

Theorem 4.2 *Let S be a lower triangular operator in $\mathcal{LW}^{m\times n}$. The following are equivalent:*

1. *The operator S is a contraction from $L_2^{n\times p}(\mathbb{R})$ to $L_2^{m\times p}(\mathbb{R})$.*
2. *The operator $\mathcal{M}_S^\ell : \mathcal{LS}_2^{n\times p} \longrightarrow \mathcal{LS}_2^{m\times p}$ is a contraction operator.*
3. *The kernel $K_S^L(V, W) := K_S^L(W)^\wedge(V)$ is nonnegative in the following sense: For every choice of $\ell \in \mathbb{N}$, of $m \times m$ astv-points W_i and of multiplication operators from the left by functions $G_i \in L_2^{m\times p}(\mathbb{R}) \cap L_\infty^{m\times p}(\mathbb{R})$, $i = 1, \ldots, \ell$, we have*

$$\sum_{j,i=1}^{\ell} \langle K_S^L(W_i, W_j)G_j, G_i\rangle_{L_2^{m\times p}(\mathbb{R})} \geq 0. \tag{4.5}$$

4. *The operator $\mathcal{M}_S^r : \mathcal{LS}_2^{q\times m} \longrightarrow \mathcal{LS}_2^{q\times n}$ is a contraction operator.*
5. *The kernel $K_S^R(V, W) := K_S^R(W)^\Delta(V)$ is nonnegative in the following sense: For every choice of $\ell \in \mathbb{N}$, of $n \times n$ astv-points W_i and of multiplication operators from the right by functions $G_i \in L_2^{n\times q}(\mathbb{R}) \cap L_\infty^{n\times q}(\mathbb{R})$, $i = 1, \ldots, \ell$, we have*

$$\sum_{i,j=1}^{\ell} \langle G_j K_S^L(W_i, W_j), G_i\rangle_{L_2^{n\times q}}(\mathbb{R}) \geq 0.$$

Proof: We first prove the equivalence of **1** and **2**. If S is a contraction then,

$$\|\mathcal{M}_S^\ell(F)\|_{\mathcal{LS}_2^{m\times p}} = \|SF\|_{\mathcal{LS}_2^{m\times p}} \leq \|S\|\|F\|_{\mathcal{LS}_2^{n\times p}} \leq \|F\|_{\mathcal{LS}_2^{n\times p}}.$$

Conversely, let $S \in \mathcal{LW}^{m\times n}$ be given by

$$S(\psi)(t) = k_0(t)\psi(t) + \int_{-\infty}^{t} k_S(t,s)\psi(s)ds$$

and assume that $\mathcal{M}_S^\ell$ is a contraction. Then the operator $I_{\mathcal{LS}_2^{n\times p}} - \mathcal{M}_S^{\ell*}\mathcal{M}_S^\ell$ is nonnegative and therefore for every $F \in \mathcal{LS}_2^{n\times p}$

$$\begin{aligned}
0 &\leq \langle (I_{\mathcal{LS}_2^{n\times p}} - \mathcal{M}_S^{\ell*}\mathcal{M}_S^\ell)(F), F\rangle_{\mathcal{LS}_2^{n\times p}} \\
&= \langle F, F\rangle_{\mathcal{LS}_2^{n\times p}} - \langle SF, SF\rangle_{\mathcal{LS}_2^{m\times p}} \\
&= \langle f, f\rangle_{L_2^{n\times p}(\mathbb{R}\times\mathbb{R})} - \langle k_{SF}, k_{SF}\rangle_{L_2^{m\times p}(\mathbb{R}\times\mathbb{R})}
\end{aligned}$$

where $f(t,s) \in L_2^{n\times p}(\mathbb{R}\times\mathbb{R})$ and $k_{SF}(t,s) \in L_2^{m\times p}(\mathbb{R}\times\mathbb{R})$ are the kernel functions of the operators F and SF, respectively. For $\xi \in L_2^{n\times p}(\mathbb{R})$ define

$$\xi_s(t) = \begin{cases} \xi(t) & t \geq s, \\ 0 & t < s, \end{cases}$$

and let $\eta \in L_2^{p\times p}(\mathbb{R})$ be the function given by

$$\eta(t) = \begin{cases} I_p & \text{if } c \leq t \leq d, \\ 0 & \text{otherwise.} \end{cases}$$

Then the function $f(t,s) := \xi_s(t)\eta(s) \in L_2^{n\times p}(\mathbb{R}\times\mathbb{R})$ is the kernel function of a lower triangular operator F. Since

$$\begin{aligned}
SF(\psi)(t) &= k_0(t)\int_{-\infty}^{t} f(t,s)\psi(s)ds + \int_{-\infty}^{t} k_S(t,\tau) \\
&\quad \int_{-\infty}^{\tau} f(\tau,s)\psi(s)dsd\tau \\
&= \int_{-\infty}^{t} k_0(t)f(t,s)\psi(s)ds + \int_{-\infty}^{t} \\
&\quad \int_{s}^{t} k_S(t,\tau)f(\tau,s)\psi(s)d\tau ds \\
&= \int_{-\infty}^{t}\left(k_0(t)f(t,s) + \int_{s}^{t} k_S(t,\tau)f(\tau,s)d\tau\right)\psi(s)ds,
\end{aligned}$$

we have

$$k_{SF}(t,s) = \begin{cases} k_O(t)f(t,s) + \int_s^t k_S(t,\tau)f(\tau,s)d\tau & t \geq s, \\ 0 & t < s. \end{cases}$$

Moreover,

$$\begin{aligned} k_O(t)f(t,s) &+ \int_s^t k_S(t,\tau)f(\tau,s)d\tau \\ &= k_O(t)\xi_s(t)\eta(s) + \int_s^t k_S(t,\tau)\xi_s(\tau)d\tau\eta(s) \\ &= k_O(t)\xi_s(t)\eta(s) + \int_{-\infty}^t k_S(t,\tau)\xi_s(\tau)d\tau\eta(s) \\ &= \left(k_O(t)\xi_s(t) + \int_{-\infty}^t k_S(t,\tau)\xi_s(\tau)d\tau\right)\eta(s) \\ &= S(\xi_s)(t)\eta(s), \end{aligned}$$

thus

$$\begin{aligned} \langle f, f\rangle_{L_2^{n\times p}(\mathbb{R}\times\mathbb{R})} &- \langle k_{SF}, k_{SF}\rangle_{L_2^{m\times p}(\mathbb{R}\times\mathbb{R})} \\ &= \int_c^d (\langle \xi_s, \xi_s\rangle_{L_2^{n\times p}(\mathbb{R})} - \langle S(\xi_s), S(\xi_s)\rangle_{L_2^{m\times p}(\mathbb{R})})\, ds \\ &= \int_c^d \langle (I - S^*S)(\xi_s), \xi_s\rangle_{L_2^{n\times p}(\mathbb{R})}\, ds \end{aligned}$$

(where S^* is the adjoint of S considered as an integral operator from $L_2^{n\times p}(\mathbb{R})$ into $L_2^{m\times p}(\mathbb{R})$) is nonnegative in every interval $[c, d]$, and therefore

$$\langle (I - S^*S)(\xi_s), \xi_s\rangle_{L_2^{n\times p}(\mathbb{R})} \geq 0$$

for every truncated function ξ_s as above. Since the truncated functions form a dense set in $L_2^{n\times p}(\mathbb{R})$, $I - S^*S$ is nonnegative on $L_2^{n\times p}(\mathbb{R})$ and S is a contractive integral operator.

We now prove the equivalence of **2** and **3**. If the operator $\mathcal{M}_S^\ell$ is a contraction then,

$$\langle (I_{\mathcal{L}\mathcal{S}_2^{m\times p}} - \mathcal{M}_S^\ell \mathcal{M}_S^{\ell*})(F), F\rangle_{\mathcal{L}\mathcal{S}_2^{m\times p}} \geq 0$$

for any $F \in \mathcal{L}\mathcal{S}_2^{m\times p}$. Setting $F = \sum_{j=1}^\ell (W_j - \Delta)^{-*}G_j$, where the W_j are $m\times m$ astv-points and the G_j are multiplication operators from the left by functions

$G_j \in L_2^{m\times p}(\mathbb{R}) \cap L_\infty^{m\times p}(\mathbb{R})$, $j = 1, \ldots, \ell$, we get

$$\begin{aligned} 0 &\le \langle (I - \mathcal{M}_S^\ell \mathcal{M}_S^{\ell *})(F), F\rangle_{\mathcal{LS}_2^{m\times p}} \\ &= \left\langle \sum_{j=1}^{\ell} (I - SS^\wedge(W_j)^*)(W_j - \Delta)^{-*}G_j, \sum_{i=1}^{n}(W_i - \Delta)^{-*}G_i \right\rangle_{\mathcal{LS}_2^{m\times p}} \\ &= \left\langle \sum_{j=1}^{\ell} K_S^L(W_j)G_j, \sum_{i=1}^{\ell}(W_i - \Delta)^{-*}G_i \right\rangle_{\mathcal{LS}_2^{m\times p}}. \end{aligned}$$

Therefore (4.5) holds and the kernel $K_S^L(V, W)$ is nonnegative. Conversely, if the kernel $K_S^L(V, W)$ is nonnegative, the same computation shows that $I_{\mathcal{LS}_2^{m\times p}} - \mathcal{M}_S^\ell \mathcal{M}_S^{\ell *}$ is nonnegative on the set of all operators of the form $\sum_{j=1}^n (W_j - \Delta)^{-*}G_j$. Since this set is dense in $\mathcal{LS}_2^{m\times p}$ we conclude that $I_{\mathcal{LS}_2^{m\times p}} - \mathcal{M}_S^\ell \mathcal{M}_S^{\ell *}$ is nonnegative on $\mathcal{LS}_2^{m\times p}$, which implies that the operator $\mathcal{M}_S^\ell$ is a contraction.

The proof of the equivalence with **4** and **5** can be given in a similar way and is omitted. □

Assume S satisfies one of the equivalent conditions of the theorem. Then we associate with S two Hilbert spaces $\mathcal{H}_L(S)$ and $\mathcal{H}_R(S)$. Following de Branges we define for $F \in \mathcal{LS}_2^{n\times p}$,

$$\kappa_L(F) = \sup_{H \in \mathcal{LS}_2^{n\times p}} \{\|F + SH\|^2_{\mathcal{LS}_2^{m\times p}} - \|H\|^2_{\mathcal{LS}_2^{n\times p}}\}.$$

Then the space

$$\mathcal{H}_L(S) := \{F \in \mathcal{LS}_2^{m\times p} | \kappa_L(F) < \infty\} \tag{4.6}$$

is a Hilbert space with norm $\|F\|_{\mathcal{H}_L(S)} = \sqrt{\kappa_L(F)}$. This is a particular case of a result by Ando; see [6]. The definition of the norm implies that the inclusion map $\mathcal{H}_L(S) \hookrightarrow \mathcal{LS}_2^{n\times p}$ is a contraction. From [6, Theorem 3.9, p. 27] it also follows that $\mathcal{H}_L(S)$ is equal to the operator range $\mathrm{Ran}(I - \mathcal{M}_S^\ell \mathcal{M}_S^{\ell *})^{\frac{1}{2}}$ with the lifted norm

$$\|(I - \mathcal{M}_S^\ell \mathcal{M}_S^{\ell *})^{\frac{1}{2}} u\|_{\mathrm{Ran}(I - \mathcal{M}_S^\ell \mathcal{M}_S^{\ell *})^{\frac{1}{2}}} = \|(I - \pi)u\|_{\mathcal{LS}_2^{m\times p}},$$

where π is the orthogonal projection onto the kernel of $(I - \mathcal{M}_S^\ell \mathcal{M}_S^{\ell *})$. Moreover, the space $\mathcal{H}_L(S)$ is also a left reproducing kernel Hilbert space with reproducing kernel $K_S^L(W)$ in the sense that for all $m\times m$ astv-points W and all $H\in L_2^{m\times p}(\mathbb{R})\cap L_\infty^{m\times p}(\mathbb{R})$, the function

$$\begin{aligned} K_S^L(W)H &= (I-\mathcal{M}_S^\ell\mathcal{M}_S^{\ell*})(W-\Delta)^{-*}H \\ &= (I-\mathcal{M}_S^\ell\mathcal{M}_S^{\ell*})^{\frac{1}{2}}(I-\mathcal{M}_S^\ell\mathcal{M}_S^{\ell*})^{\frac{1}{2}}(W-\Delta)^{-*}H \end{aligned}$$

belongs to $\mathcal{H}_L(S)$ and then

$$\langle F, K_S^L(W)H\rangle_{\mathcal{H}_L(S)} = \operatorname{Tr} H^*F^\wedge(W). \tag{4.7}$$

Indeed, if $F=(I-\mathcal{M}_S^\ell\mathcal{M}_S^{\ell*})^{\frac{1}{2}}u\in\mathcal{H}_L(S)$ with $u\in\mathcal{LS}_2^{p\times m}$,

$$\begin{aligned} \langle F, K_S^L(W)H\rangle_{\mathcal{H}_L(S)} &= \langle u, (I-\mathcal{M}_S^\ell\mathcal{M}_S^{\ell*})^{\frac{1}{2}}(W-\Delta)^{-*}H\rangle_{\mathcal{LS}_2^{p\times m}} \\ &= \langle F, (W-\Delta)^{-*}H\rangle, \end{aligned}$$

which on account of (2.18), implies (4.7).
Similarly, the space

$$\mathcal{H}_R(S) := \{F\in\mathcal{LS}_2^{q\times n} | \kappa_R(F)<\infty\}$$

with norm $\|F\|_{\mathcal{H}_R(S)}=\sqrt{\kappa_R(F)}$, where

$$\kappa_R(F) = \sup_{H\in\mathcal{LS}_2^{q\times m}} \{\|F+hS\|^2_{\mathcal{LS}_2^{q\times n}} - \|H\|^2_{\mathcal{LS}_2^{q\times m}}\},$$

is a Hilbert space which is contractively included in $\mathcal{LS}_2^{q\times n}$. It is equal to the operator range $\operatorname{Ran}(I-\mathcal{M}_S^r\mathcal{M}_S^{r*})^{\frac{1}{2}}$ with the lifted norm and a right reproducing kernel Hilbert space with reproducing kernel $K_S^R(W)$ in the sense that

$$\langle F, GK_S^r(W)\rangle_{\mathcal{H}_R(S)} = \operatorname{Tr} G^*F^\Delta(W).$$

Note that for $S=0$, the last inequality coincides with formula (2.20).
The following lemma will be used in Section 5.

Lemma 4.3 *For $W\in L_\infty^{p\times p}(\mathbb{R})$, the multiplication operator $\mathcal{M}_W^r:\mathcal{H}_L(S)\longrightarrow\mathcal{H}_L(S)$ defined by $\mathcal{M}_W^r(F)=FW$ is bounded. If $\|W\|_{L_\infty^{p\times p}}<1$, then $\mathcal{M}_W^r$ is a strict contraction.*

Proof: Let $F=\sum_{j=1}^n K_S^L(V_j)G_j$ where the V_j are $m\times m$ astv-points and the G_j are multiplication operators from the left by functions $G_j\in L_2^{m\times p}(\mathbb{R})\cap L_\infty^{m\times p}(\mathbb{R})$, $j=1,\dots,n$. Then $\mathcal{M}_W^r(F)=\sum_{j=1}^n K_S^L(V_j)G_jW$, where now

$G_jW \in L_2^{m\times p}(\mathbb{R}) \cap L_\infty^{m\times p}(\mathbb{R})$ for $j = 1, \ldots, n$ and therefore $\mathcal{M}_W^r(F)$ belongs to $\mathcal{H}_L(S)$. The matrix-valued function

$$X(t) := \sum_{i,j=1}^{n} G_i(t)^* K_S^L(V_i, V_j)(t) G_j(t), \quad t \in \mathbb{R},$$

is nonnegative and $\|F\|^2_{\mathcal{H}_L(S)} = \int_{-\infty}^{\infty} \operatorname{Tr} X(t)dt = \|X^{1/2}\|^2_{L_2^{p\times p}(\mathbb{R})}$.

Thus

$$\begin{aligned}
\|\mathcal{M}_W^r(F)\|^2_{\mathcal{H}_L(S)} &= \int_{-\infty}^{\infty} \operatorname{Tr} W(t)^* X(t) W(t) dt \\
&= \int_{-\infty}^{\infty} \operatorname{Tr} W(t)^* X(t)^{1/2} X(t)^{1/2} W(t) dt \\
&= \|X^{1/2} W\|^2_{L_2^{p\times p}(\mathbb{R})} \\
&\leq \|W\|^2_{L_\infty^{p\times p}(\mathbb{R})} \|X^{1/2}\|^2_{L_2^{p\times p}(\mathbb{R})} \\
&= \|W\|^2_{L_\infty^{p\times p}(\mathbb{R})} \|F\|^2_{\mathcal{H}_L(S)}
\end{aligned}$$

and therefore the operator $\mathcal{M}_W^r$ is a strict contraction. □

Similarly, if $W \in L_\infty^{q\times q}(\mathbb{R})$ and $\|W\|_{L_\infty^{q\times q}(\mathbb{R})} < 1$, then $\mathcal{M}_W^\ell : \mathcal{H}_R(S) \longrightarrow \mathcal{H}_R(S)$ defined by $\mathcal{M}_W^\ell(F) = WF$ is a strict contraction operator.

We now complete the table began in Section 2.

In the last item, the symbols $S^{*\nabla}(W)$ and $S^{*\vee}(W)$ denote the analog of the left and right point evaluations for upper triangular operators. Let $S \in \mathcal{LW}^{m\times n}$ and let W be an $n \times n$ astv-point. Then $(\Delta - W)^{-*}S^* \in \mathcal{W}^{n\times m}$ and its kernel function is given by

$$\text{(4.8)} \qquad \begin{cases} -\displaystyle\int_{-\infty}^{t} U_W(t)^{-*} U_W(\tau)^* k_S(s,\tau)^* d\tau & t < s, \\[2ex] -U_W(t)^{-*} U_W(s)^* k_0(s)^* \\ \quad -\displaystyle\int_{-\infty}^{s} U_W(t)^{-*} U_W(\tau)^* k_S(s,\tau)^* d\tau & t \geq s. \end{cases}$$

Taking $t = s$ in (4.8) and changing the sign we get the function

$$S^{*\nabla}(W)(t) = k_0(t)^* + \int_0^{\infty} U_W(t)^{-*} U_W(t-\tau)^* k_S(t, t-\tau)^* d\tau, \quad t \in \mathbb{R}.$$

Note that if $S \in \mathcal{LW}^{m\times n}$, then $S^* \in \mathcal{UW}^{n\times m}$ and

$$S^{\Delta}(W)(t)^* = S^{*\nabla}(W)(t).$$

Stationary Setting	Nonstationary Setting
The reproducing kernels $K_S^L(z,\omega)=\frac{I_p-S(z)S(\omega)^*}{-(z+\overline{\omega})}$ and $K_S^R(z,\omega)=\frac{I_q-S(\omega)^*S(z)}{-(z+\overline{\omega})}$ of $\mathcal{H}_L(S)$ and $\mathcal{H}_R(S)$	The reproducing kernels $K_S^L(W)=(I-SS^\wedge(W)^*)(W-\Delta)^{-*}$ and $K_S^R(W)=(W-\Delta)^{-*}$ $(I-S^\Delta(W)^*S)$ of $\mathcal{H}_L(S)$ and $\mathcal{H}_R(S)$
For a Schur function $S\in\mathbf{H}_\infty^{p\times q}$ $(i\mathbb{C}_-)$ the Schur conjugate function $\tilde{S}(z)=S(z^*)^*\in\mathbf{H}_\infty^{q\times p}(i\mathbb{C}_-)$	For a Schur upper triangular operator $S\in\mathcal{LW}^{m\times n}$ the Schur lower triangular operator $S^*\in\mathcal{UW}^{n\times m}$.
The reproducing kernels $K_{\tilde{S}}^L(z,\omega)=\frac{I_q-\tilde{S}(z)\tilde{S}(\omega)^*}{-(z+\overline{\omega})}$ and $K_{\tilde{S}}^R(z,\omega)=\frac{I_p-\tilde{S}(\omega)^*\tilde{S}(z)}{-(z+\overline{\omega})}$ of the spaces associated to $\tilde{S}$.	The reproducing kernels $K_{S^*}^L(W)=(I-S^*S^{*\nabla}(W)^*)$ $(W-\Delta)^{-1}$ and $K_{S^*}^R(W)$ $=(W-\Delta)^{-1}(I-S^{*\vee}(W)^*S^*)$ of the spaces associated to S^*.

If now W is an $m\times m$ astv-point, then $S^*(\Delta-W)^{-*}\in\mathcal{W}^{n\times m}$ and its kernel function is given by

$$\begin{cases} -k_0(t)^*U_W(t)^{-*}U_W(s)^* \\ \quad -\int_t^\infty k_S(\tau,t)^*U_W(\tau)^{-*}U_W(s)^*d\tau & t\geq s, \\ -\int_s^\infty k_S(\tau,t)^*U_W(\tau)^{-*}U_W(s)^*d\tau & t<s. \end{cases} \tag{4.9}$$

Taking $t=s$ in (4.9) and changing the sign we get

$$S^\vee(W)(t)=k_0(t)^*+\int_0^\infty k_S(t+\tau,t)^*U_W(t+\tau)^{-*}U_W(t)^*d\tau,\quad t\in\mathbb{R}.$$

Note that $S^*\in\mathcal{UW}^{n\times m}$ and

$$S^\wedge(W)(t)^*=S^{*\vee}(W)(t).$$

These notions will not be used in the sequel. We have added them to the table for the sake of completeness.

5 The Coisometric Realization

In this section we prove the existence of a coisometric realization of a contraction $S\in\mathcal{LW}^{m\times n}$ in which $\mathcal{H}_L(S)$ is a state space. The coisometric realization with state space $\mathcal{H}_R(S)$ will be delineated without proofs (see Theorem 5.6). To make our method work, we need the technical condition that S is a Hilbert-Schmidt operator, that is,

$$S(\psi)(t) = \int_{-\infty}^{t} k_S(t,s)\psi(s)ds, \quad t \in \mathbb{R},$$

where $k_S \in L_2^{m\times n}(\mathbb{R}\times\mathbb{R})$ with $\|S\| \leq 1$. It allows us to define bounded operators in the realization (see (5.1) below). There is a motivation for this assumption which originates from the theory of second order stationary stochastic processes. There one considers (say scalar) stationary processes with covariance of the form $\delta(t-s)+k(t-s)$, where δ is the Dirac distribution and k is in $L_2(\mathbb{R})$; thus this covariance is the kernel of a nonnegative integral operator of the form

$$\psi(t) \mapsto \psi(t) + \int_{-\infty}^{t} k(t-s)\psi(s)ds.$$

The nonstationary analog of this operator is the operator of the form

$$\Upsilon(\psi)(t) = \psi(t) + \int_{-\infty}^{t} k_\Upsilon(t,s)\psi(s)ds,$$

where $k_\Upsilon(t,s)$ is a function from $L_2(\mathbb{R}\times\mathbb{R})$ with a nonnegative real part. Then the operator $S = (I-\Upsilon)(I+\Upsilon)^{-1}$ is a Hilbert-Schmidt integral operator with operator norm $\|S\| \leq 1$.

Let $S \in \mathcal{L}\mathcal{S}_2^{m\times n}$ be defined by

$$S(\psi)(t) = \int_{-\infty}^{t} k_S(t,s)\psi(s)ds, \quad t \in \mathbb{R}, \quad \psi \in L_2^{n\times p}(\mathbb{R}),$$

and assume $\|S\| \leq 1$. Fix an integer p and a $p\times p$ astv-point V such that $\|(\Delta - V)^{-1}\| \leq 1$ (for example, take $V = \alpha I_p$ with $\mathrm{Re}(\alpha) \geq 1$). We will study the following colligation associated with S and centered at V:

$$\mathcal{V} = \begin{pmatrix} \mathbf{A} & \mathbf{B} \\ \mathbf{C} & \mathbf{D} \end{pmatrix} : \begin{pmatrix} \mathcal{H}_L(S) \\ L_2^{n\times p}(\mathbb{R}) \end{pmatrix} \longrightarrow \begin{pmatrix} \mathcal{H}_L(S) \\ L_2^{m\times p}(\mathbb{R}) \end{pmatrix}$$

where

$$\begin{array}{rcl} \mathbf{A}(F) &=& (F - F^\Delta(V))(\Delta - V)^{-1} \\ \mathbf{B}(G) &=& (SG - (SG)^\Delta(V))(\Delta - V)^{-1} \\ \mathbf{C}(F) &=& F^\Delta(V) \\ \mathbf{D}(G) &=& (SG)^\Delta(V). \end{array} \tag{5.1}$$

Here the operators $\mathbf{B}$ and $\mathbf{D}$ are defined on functions $G \in L_2^{n\times p}(\mathbb{R}) \cap L_\infty^{n\times p}(\mathbb{R})$. In the next theorem we will show that they can be extended to bounded operators on $L_2^{n\times p}(\mathbb{R})$ and also that $\mathbf{A}$ and $\mathbf{C}$ are well defined bounded operators. We prove that $\mathcal{V}$ is a coisometric colligation in a sense to be defined, closely outer connected and that it provides a realization for S.

Theorem 5.1 *The operators* **A**, **B**, **C**, *and* **D** *defined in* (5.1) *are bounded and their adjoints are given by*

$$\begin{aligned}
(5.2)\qquad \mathbf{A}^*(K_S^L(W)G) &= (-K_S^L(W)G - S\Psi^\wedge(V)^*)(V-\Delta)^{-*} \\
(5.3)\qquad \mathbf{B}^*(K_S^L(W)G)(t) &= (G^*(\Delta - W)^{-1}(S - S^\wedge(W)))^\wedge(V)(t)^* \\
(5.4)\qquad \mathbf{C}^*(G) &= (G - S(G^*S)^\wedge(V)^*)(V-\Delta)^{-*} \\
(5.5)\qquad \mathbf{D}^*(G) &= (G^*S)^\wedge(V)^*,
\end{aligned}$$

where $G \in L_2^{m\times p}(\mathbb{R}) \cap L_\infty^{m\times p}(\mathbb{R})$, W *is an* $m \times m$ *astv-point and*

$$(5.6)\qquad \Psi = G^*(\Delta - W)^{-1}(S - S^\wedge(W)).$$

Proof: To show that the operator **A** is well defined and bounded, we will use the characterization (4.6) of $\mathcal{H}_L(S)$. For $F \in \mathcal{H}_L(S)$ we shall show that

$$\sup_{H\in\mathcal{L}\mathcal{S}_2^{n\times p}} \{\|(F - F^\Delta(V))(\Delta - V)^{-1} + SH\|^2_{\mathcal{L}\mathcal{S}_2^{m\times p}} - \|H\|^2_{\mathcal{L}\mathcal{S}_2^{n\times p}}\} < \infty.$$

Since $\|(\Delta - V)^{-1}\| \le 1$, it follows that

$$\begin{aligned}
&\|H(\Delta - V)^{-*}\|^2_{\mathcal{S}_2^{n\times p}} - \|H\|^2_{\mathcal{S}_2^{n\times p}} + \|SH\|^2_{\mathcal{S}_2^{m\times p}} - \|SH(\Delta - V)^{-*}\|^2_{\mathcal{S}_2^{m\times p}} \\
&= \operatorname{Tr}((\Delta - V)^{-1}H^*(I - S^*S)H(\Delta - V)^{-*} - H^*(I - S^*S)H) \\
&= \operatorname{Tr}((I - S^*S)^{1/2}H(\Delta - V)^{-*}(\Delta - V)^{-1}H^*(I - S^*S)^{1/2} \\
&\quad -(I - S^*S)^{1/2}HH^*(I - S^*S)^{1/2}) \\
&= -\operatorname{Tr}((I - S^*S)^{1/2}H(I - (\Delta - V)^{-*} \\
&\quad (\Delta - V)^{-1})H^*(I - S^*S)^{1/2}) \le 0,
\end{aligned}$$

and therefore

$$\begin{aligned}
&\|(F - F^\Delta(V))(\Delta - V)^{-1} + SH\|^2_{\mathcal{L}\mathcal{S}_2^{m\times p}} - \|H\|^2_{\mathcal{L}\mathcal{S}_2^{n\times p}} \\
&= \|F(\Delta - V)^{-1} - F^\Delta(V)(\Delta - V)^{-1} + SH\|^2_{\mathcal{S}_2^{m\times p}} - \|H\|^2_{\mathcal{S}_2^{n\times p}} \\
&= \|F^\Delta(V)(\Delta - V)^{-1}\|^2_{\mathcal{S}_2^{m\times p}} + \|F(\Delta - V)^{-1} + SH\|^2_{\mathcal{S}_2^{m\times p}} \\
&\quad -2\operatorname{Re}\langle F^\Delta(V)(\Delta - V)^{-1}, F(\Delta - V)^{-1} + SH\rangle_{\mathcal{S}_2^{m\times p}} - \|H\|^2_{\mathcal{S}_2^{n\times p}} \\
&= \|F(\Delta - V)^{-1} + SH\|^2_{\mathcal{S}_2^{m\times p}} - \|F^\Delta(V)(\Delta - V)^{-1}\|^2_{\mathcal{S}_2^{m\times p}} - \|H\|^2_{\mathcal{S}_2^{n\times p}} \\
&= \|F(\Delta - V)^{-1}\|^2_{\mathcal{S}_2^{m\times p}} + \|SH\|^2_{\mathcal{S}_2^{m\times p}} + 2\operatorname{Re}\langle F(\Delta - V)^{-1}, SH\rangle_{\mathcal{S}_2^{m\times p}} \\
&\quad -\|F^\Delta(V)(\Delta - V)^{-1}\|^2_{\mathcal{S}_2^{m\times p}} - \|H\|^2_{\mathcal{S}_2^{n\times p}}
\end{aligned}$$

$$
\begin{aligned}
&= \|F\|^2_{\mathcal{S}_2^{m\times p}} + 2\mathrm{Re}\langle F, SH(\Delta - V)^{-*}\rangle_{\mathcal{S}_2^{m\times p}} \\
&\quad + \|SH(\Delta - V)^{-*}\|^2_{\mathcal{S}_2^{m\times p}} - \|H(\Delta - V)^{-*}\|^2_{\mathcal{S}_2^{n\times p}} \\
&\quad + \|H(\Delta - V)^{-*}\|^2_{\mathcal{S}_2^{n\times p}} - \|H\|^2_{\mathcal{S}_2^{n\times p}} \\
&\quad + \|SH\|^2_{\mathcal{S}_2^{m\times p}} - \|SH(\Delta - V)^{-*}\|^2_{\mathcal{S}_2^{m\times p}} \\
&\quad + \|F(\Delta - V)^{-1}\|^2_{\mathcal{S}_2^{m\times p}} - \|F\|^2_{\mathcal{S}_2^{m\times p}} - \|F^{\Delta}(V)(\Delta - V)^{-1}\|^2_{\mathcal{S}_2^{m\times p}} \\
&\le \sup_{H\in\mathcal{L}\mathcal{S}_2^{n\times p}} (\|F + SH(\Delta - V)^{-*}\|^2_{\mathcal{S}_2^{m\times p}} - \|H(\Delta - V)^{-*}\|^2_{\mathcal{S}_2^{n\times p}}) \\
&\quad + \sup_{H\in\mathcal{L}\mathcal{S}_2^{n\times p}} (\|H(\Delta - V)^{-*}\|^2_{\mathcal{S}_2^{n\times p}} - \|H\|^2_{\mathcal{S}_2^{n\times p}} \\
&\quad + \|SH\|^2_{\mathcal{S}_2^{m\times p}} - \|SH(\Delta - V)^{-*}\|^2_{\mathcal{S}_2^{m\times p}}) \\
&\quad + \|F(\Delta - V)^{-1}\|^2_{\mathcal{S}_2^{m\times p}} - \|F\|^2_{\mathcal{S}_2^{m\times p}} - \|F^{\Delta}(V)(\Delta - V)^{-1}\|^2_{\mathcal{S}_2^{m\times p}} \\
&\le \|F\|^2_{\mathcal{H}_L(S)} + \|F(\Delta - V)^{-1}\|^2_{\mathcal{S}_2^{m\times p}} - \|F\|^2_{\mathcal{S}_2^{m\times p}} \\
&\quad - \|F^{\Delta}(V)(\Delta - V)^{-1}\|^2_{\mathcal{S}_2^{m\times p}} \\
&\quad + \sup_{H\in\mathcal{L}\mathcal{S}_2^{n\times p}} (\|H(\Delta - V)^{-*}\|^2_{\mathcal{S}_2^{n\times p}} - \|H\|^2_{\mathcal{S}_2^{n\times p}} \\
&\quad + \|SH\|^2_{\mathcal{S}_2^{m\times p}} - \|SH(\Delta - V)^{-*}\|^2_{\mathcal{S}_2^{m\times p}}) \\
&\le \|F\|^2_{\mathcal{H}_L(S)} + \|F(\Delta - V)^{-1}\|^2_{\mathcal{S}_2^{m\times p}} - \|F\|^2_{\mathcal{S}_2^{m\times p}} \\
&\quad - \|F^{\Delta}(V)(\Delta - V)^{-1}\|^2_{\mathcal{S}_2^{m\times p}},
\end{aligned}
$$

which shows that the operator **A** is everywhere defined. Moreover, **A** is a contraction:

$$
\begin{aligned}
&\|A(F)\|^2_{\mathcal{H}_L(S)} \\
&\quad = \sup_{H\in\mathcal{L}\mathcal{S}_2^{n\times p}} \{\|(F - F^{\Delta}(V))(\Delta - V)^{-1} + SH\|^2_{\mathcal{L}\mathcal{S}_2^{m\times p}} - \|H\|^2_{\mathcal{L}\mathcal{S}_2^{n\times p}}\} \\
&\quad \le \|F\|^2_{\mathcal{H}_L(S)} + \|F(\Delta - V)^{-1}\|^2_{\mathcal{S}_2^{m\times p}} - \|F\|^2_{\mathcal{S}_2^{m\times p}} \\
&\qquad - \|F^{\Delta}(V)(\Delta - V)^{-1}\|^2_{\mathcal{S}_2^{m\times p}} \\
&\quad \le \|F\|^2_{\mathcal{H}_L(S)} + \|(\Delta - V)^{-1}\| \|F\|^2_{\mathcal{S}_2^{m\times p}} - \|F\|^2_{\mathcal{S}_2^{m\times p}} \le \|F\|^2_{\mathcal{H}_L(S)}.
\end{aligned}
$$

The operator **C** is bounded as follows from

$$\begin{aligned}\|F^{\triangle}(V)\|_{L_2^{m\times p}(\mathbb{R})} &\leq \frac{M}{(2\ln a^{-1})^{1/2}}\|f\|_{L_2^{m\times p}(\mathbb{R}\times\mathbb{R})}\\ &= \frac{M}{(2\ln a^{-1})^{1/2}}\|F\|_{\mathcal{L}\mathcal{S}_2^{m\times p}}\\ &\leq \frac{M}{(2\ln a^{-1})^{1/2}}\|F\|_{\mathcal{H}_L(S)}.\end{aligned}$$

Here we used an estimate similar to the inequality (2.17) and the fact that $\mathcal{H}_L(S)$ is contractively included in $\mathcal{L}\mathcal{S}_2^{m\times p}$.

Next we show that for any $G \in L_2^{n\times p}(\mathbb{R}) \cap L_\infty^{n\times p}(\mathbb{R})$ the operator $(SG - (SG)^{\triangle}(V))(\Delta - V)^{-1}$ belong to $\mathcal{H}_L(S)$. Since S is an Hilbert-Schmidt operator, we have $SG \in \mathcal{L}\mathcal{S}_2^{m\times p}$, and on account of Lemma 2.5, $(SG)^{\triangle}(V) \in L_2^{m\times n}(\mathbb{R}) \cap L_\infty^{m\times n}(\mathbb{R})$. By inequality (2.19), the operator $(SG)^{\triangle}(V)(\Delta - V)^{-1}$ and then also $(SG - (SG)^{\triangle}(V))(\Delta - V)^{-1}$ belongs to $\mathcal{L}\mathcal{S}_2^{m\times n}$. To show that it belongs to $\mathcal{H}_L(S)$, we shall prove that

$$\sup_{H\in\mathcal{L}\mathcal{S}_2^{n\times p}} \{\|(SG - (SG)^{\triangle}(V))(\Delta - V)^{-1} + SH\|^2_{\mathcal{L}\mathcal{S}_2^{m\times p}} - \|H\|^2_{\mathcal{L}\mathcal{S}_2^{n\times p}}\} < \infty.$$

Indeed,

$$\begin{aligned}&\|(SG - (SG)^{\triangle}(V))(\Delta - V)^{-1} + SH\|^2_{\mathcal{L}\mathcal{S}_2^{m\times p}} - \|H\|^2_{\mathcal{L}\mathcal{S}_2^{n\times p}}\\ &= \|SG(\Delta - V)^{-1} - (SG)^{\triangle}(V)(\Delta - V)^{-1}\\ &\quad + SH\|^2_{\mathcal{S}_2^{m\times p}} - \|H\|^2_{\mathcal{L}\mathcal{S}_2^{n\times p}}\\ &= \|SG(\Delta - V)^{-1} + SG\|^2_{\mathcal{S}_2^{m\times p}} + \|(SG)^{\triangle}(V)(\Delta - V)^{-1}\|^2_{\mathcal{S}_2^{m\times p}}\\ &\quad -2\mathrm{Re}\langle (SG)^{\triangle}(V)(\Delta - V)^{-1}, SG(\Delta - V)^{-1}\\ &\quad + SH\rangle^2_{\mathcal{S}_2^{m\times p}} - \|H\|^2_{\mathcal{L}\mathcal{S}_2^{n\times p}}\\ &= \|SG(\Delta - V)^{-1} + SH\|^2_{\mathcal{S}_2^{m\times p}}\\ &\quad -\|(SG)^{\triangle}(V)(\Delta - V)^{-1}\|^2_{\mathcal{S}_2^{m\times p}} - \|H\|^2_{\mathcal{L}\mathcal{S}_2^{n\times p}}\\ &\leq \|G(\Delta - V)^{-1} + H\|^2_{\mathcal{S}_2^{n\times p}}\\ &\quad -\|(SG)^{\triangle}(V)(\Delta - V)^{-1}\|^2_{\mathcal{S}_2^{m\times p}} - \|H\|^2_{\mathcal{L}\mathcal{S}_2^{n\times p}}\\ &= \|G(\Delta - V)^{-1}\|^2_{\mathcal{S}_2^{n\times p}} + \|H\|^2_{\mathcal{S}_2^{n\times p}}\end{aligned}$$

$$-\|(SG)^{\triangle}(V)(\Delta - V)^{-1}\|^2_{\mathcal{S}_2^{m\times p}} - \|H\|^2_{\mathcal{LS}_2^{n\times p}}$$
$$= \|G(\Delta - V)^{-1}\|^2_{\mathcal{S}_2^{n\times p}} - \|(SG)^{\triangle}(V)(\Delta - V)^{-1}\|^2_{\mathcal{S}_2^{m\times p}},$$

and since the last term does not depend on H, the indicated supremum is finite. By the previous estimate and inequality (2.19), we have

$$\begin{aligned}
\|\mathbf{B}(G)\|^2_{\mathcal{H}_L(S)} &= \kappa_L(BG) \\
&= \sup_{H\in\mathcal{LS}_2^{n\times p}} \{\|(SG - (SG)^{\triangle}(V))(\Delta - V)^{-1} \\
&\quad + SH\|^2_{\mathcal{LS}_2^{m\times p}} - \|H\|^2_{\mathcal{LS}_2^{n\times p}}\} \\
&\leq \|G(\Delta - V)^{-1}\|^2_{\mathcal{S}_2^{n\times p}} \\
&\leq \frac{M^2}{2\ln a^{-1}}\|G\|^2_{L_2^{n\times p}(\mathbb{R})},
\end{aligned}$$

and therefore, $\mathbf{B}$ has a unique bounded extension to $L_2^{n\times p}(\mathbb{R})$.

For $G \in L_2^{n\times p}(\mathbb{R}) \cap L_\infty^{n\times p}(\mathbb{R})$ and $H \in L_2^{m\times p}(\mathbb{R}) \cap L_\infty^{m\times p}(\mathbb{R})$, we have

$$\begin{aligned}
|\langle \mathbf{D}(G), H\rangle_{L_2^{m\times p}(\mathbb{R})}| &= |\langle (SG)^{\triangle}(V), H\rangle_{L_2^{m\times p}(\mathbb{R})}| \\
&= |\langle SG, H(V - \Delta)^{-*}\rangle_{\mathcal{LS}_2^{m\times p}}| \\
&= |\langle S, H(V - \Delta)^{-*}G^*\rangle_{\mathcal{LS}_2^{m\times n}}| \\
&\leq \|S\|_{\mathcal{LS}_2^{m\times n}}\|H(V - \Delta)^{-*}G^*\|_{\mathcal{LS}_2^{m\times n}}.
\end{aligned}$$

Form the arguments in the Proof of Lemma 2.6 leading up to the inequality (2.19), we get

$$\begin{aligned}
&\|H(V - \Delta)^{-*}G^*\|^2_{\mathcal{LS}_2^{m\times n}} \\
&= \int_{-\infty}^{\infty}\int_s^{\infty} \|H(t)U_V(t)^{-*}U_V(s)^*G(s)^*\|^2_{\mathbb{C}^{p\times n}}dtds \\
&\leq \|H\|^2_{L_2^{m\times p}(\mathbb{R})}\int_{-\infty}^{\infty}\int_s^{\infty} \|U_V(t)^{-*}U_V(s)^*G(s)^*\|^2_{\mathbb{C}^{p\times n}}dtds \\
&\leq \|H\|^2_{L_2^{m\times p}(\mathbb{R})}\|(V - \Delta)^{-*}G^*\|^2_{\mathcal{LS}_2^{p\times n}} \\
&\leq \frac{M^2}{2\ln a^{-1}}\|H\|^2_{L_2^{m\times p}(\mathbb{R})}\|G\|^2_{L_2^{n\times p}(\mathbb{R})}.
\end{aligned}$$

Hence

$$|\langle \mathbf{D}(G), H\rangle_{L_2^{m\times p}(\mathbb{R})}| \leq \frac{M}{(2\ln a^{-1})^{1/2}}\|S\|_{\mathcal{LS}_2^{m\times n}}\|H\|_{L_2^{m\times p}(\mathbb{R})}\|G\|_{L_2^{n\times p}(\mathbb{R})}.$$

Thus $\mathbf{D}$ has a unique bounded extension to $L_2^{n\times p}(\mathbb{R})$. Note that we used the fact that $S \in \mathcal{LS}^{m\times n}$ to show that $\mathbf{B}$ and $\mathbf{D}$ are bounded.

We now compute the adjoints of the operators of the colligation $\mathcal{V}$. For $G \in L_2^{n\times p}(\mathbb{R}) \cap L_\infty^{n\times p}(\mathbb{R})$ and $H \in L_2^{m\times p}(\mathbb{R}) \cap L_\infty^{m\times p}(\mathbb{R})$, we have

$$\begin{aligned}
\langle G, \mathbf{D}^*(H)\rangle_{L_2^{n\times p}(\mathbb{R})} &= \langle \mathbf{D}(G), H\rangle_{L_2^{m\times p}(\mathbb{R})} \\
&= \langle (SG)^{\Delta}(V), H\rangle_{L_2^{m\times p}(\mathbb{R})} \\
&= \langle SG, H(V-\Delta)^{-*}\rangle_{\mathcal{LS}_2^{m\times p}} \\
&= \langle H^*S, (V-\Delta)^{-*}G^*\rangle_{\mathcal{LS}_2^{p\times n}} \\
&= \langle (H^*S)^{\wedge}(V), G^*\rangle_{L_2^{p\times n}(\mathbb{R})} \\
&= \langle G, (H^*S)^{\wedge}(V)^*\rangle_{L_2^{n\times p}(\mathbb{R})},
\end{aligned}$$

and thus (5.5) holds. Furthermore, for $m \times m$ astv-points W,

$$\begin{aligned}
&\langle G, \mathbf{B}^*(K_S^L(W)H)\rangle_{L_2^{n\times p}(\mathbb{R})} \\
&= \langle (SG-(SG)^{\Delta}(V))(\Delta-V)^{-1}, K_S^L(W)H\rangle_{\mathcal{H}_L(S)} \\
&= \langle ((SG-(SG)^{\Delta}(V))(\Delta-V)^{-1})^{\wedge}(W), H\rangle_{L_2^{m\times p}(\mathbb{R})} \\
&= \langle ((\Delta-W)^{-1}(SG-(SG)^{\wedge}(W)))^{\Delta}(V), H\rangle_{L_2^{m\times p}(\mathbb{R})} \\
&= \langle (\Delta-W)^{-1}(SG-(SG)^{\wedge}(W)), H(V-\Delta)^{-*}\rangle_{\mathcal{LS}_2^{m\times p}} \\
&= \langle (\Delta-W)^{-1}(S-S^{\wedge}(W))G, H(V-\Delta)^{-*}\rangle_{\mathcal{LS}_2^{m\times p}} \\
&= \langle H^*(\Delta-W)^{-1}(S-S^{\wedge}(W)), (V-\Delta)^{-*}G^*\rangle_{\mathcal{LS}_2^{p\times n}} \\
&= \langle (H^*(\Delta-W)^{-1}(S-S^{\wedge}(W)))^{\wedge}(V), G^*\rangle_{L_2^{p\times n}(\mathbb{R})} \\
&= \langle G, (H^*(\Delta-W)^{-1}(S-S^{\wedge}(W)))^{\wedge}(V)^*\rangle_{L_2^{n\times p}(\mathbb{R})},
\end{aligned}$$

which proves (5.3). Now we calculate $\mathbf{C}^*$. For $G, H \in L_2^{m\times p}(\mathbb{R}) \cap L_\infty^{m\times p}(\mathbb{R})$ and $m \times m$ astv-points W,

$$\begin{aligned}
&\langle (\mathbf{C}^*(G))^{\wedge}(W), H\rangle_{L_2^{m\times p}(\mathbb{R})} \\
&= \langle \mathbf{C}^*(G), K_S^L(W)H\rangle_{\mathcal{H}_L(S)} \\
&= \langle G, (K_S^L(W)H)^{\Delta}(V)\rangle_{L_2^{m\times p}(\mathbb{R})} \\
&= \langle G(V-\Delta)^{-*}, K_S^L(W)H\rangle_{\mathcal{LS}_2^{m\times p}}
\end{aligned}$$

$$= \langle G(V-\Delta)^{-*}, (I - SS^{\wedge}(W)^{*})(W-\Delta)^{-*}H\rangle_{\mathcal{L}\mathcal{S}_2^{m\times p}}$$

$$= \langle G(V-\Delta)^{-*}, (W-\Delta)^{-*}H\rangle_{\mathcal{L}\mathcal{S}_2^{m\times p}}$$

$$-\langle G(V-\Delta)^{-*}, SS^{\wedge}(W)^{*}(W-\Delta)^{-*}H\rangle_{\mathcal{L}\mathcal{S}_2^{m\times p}}.$$

The last summand can be written as

$$\langle G(V-\Delta)^{-*}, SS^{\wedge}(W)^{*}(W-\Delta)^{-*}H\rangle_{\mathcal{L}\mathcal{S}_2^{m\times p}}$$

$$= \langle (V-\Delta)^{-*}H^{*}, G^{*}SS^{\wedge}(W)^{*}(W-\Delta)^{-*}\rangle_{\mathcal{L}\mathcal{S}_2^{p\times m}}$$

$$= \langle H^{*}, (G^{*}SS^{\wedge}(W)^{*}(W-\Delta)^{-*})^{\wedge}(V)\rangle_{L_2^{p\times m}(\mathbb{R})}$$

$$= \langle H^{*}, ((G^{*}S)^{\wedge}(V)S^{\wedge}(W)^{*}(W-\Delta)^{-*})^{\wedge}(V)\rangle_{L_2^{p\times m}(\mathbb{R})}$$

$$= \langle (V-\Delta)^{-*}H^{*}, (G^{*}S)^{\wedge}(V)S^{*}(W-\Delta)^{-*}\rangle_{\mathcal{S}_2^{p\times m}}$$

$$= \langle S(G^{*}S)^{\wedge}(V)^{*}(V-\Delta)^{-*}, (W-\Delta)^{-*}H\rangle_{\mathcal{L}\mathcal{S}_2^{m\times p}}.$$

Therefore

$$\langle (\mathbf{C}^{*}(G))^{\wedge}(W), H\rangle_{L_2^{m\times p}(\mathbb{R})}$$

$$= \langle G(V-\Delta)^{-*} - S(G^{*}S)^{\wedge}(V)^{*}(V-\Delta)^{-*}, (W-\Delta)^{-*}H\rangle_{\mathcal{L}\mathcal{S}_2^{m\times p}}$$

$$= \langle ((G - S(G^{*}S)^{\wedge}(V)^{*})(V-\Delta)^{-*})^{\wedge}(W), H\rangle_{L_2^{m\times p}(\mathbb{R})}$$

from which we conclude that (5.4) holds.

We now compute $\mathbf{A}^{*}$. For $G, H \in L_2^{m\times p}(\mathbb{R}) \cap L_\infty^{m\times p}(\mathbb{R})$, $m\times m$ astv-points U and W, and $F := K_S^L(U)G$.

$$\langle (\mathbf{A}^{*}(K_S^L(W)H))^{\wedge}(U), G\rangle_{L_2^{m\times p}(\mathbb{R})}$$

$$= \langle \mathbf{A}^{*}(K_S^L(W)H), F\rangle_{\mathcal{H}_L(S)}$$

$$= \langle K_S^L(W)H, (F - F^{\Delta}(V))(\Delta - V)^{-1}\rangle_{\mathcal{H}_L(S)}$$

$$= \langle \mathbf{A}^{*}(K_S^L(W)H), K_S^L(U)G\rangle_{\mathcal{H}_L(S)}$$

$$= \langle H, ((F - F^{\Delta}(V))(\Delta - V)^{-1})^{\wedge}(W)\rangle_{L_2^{m\times p}(\mathbb{R})}$$

$$= \langle H, ((\Delta - W)^{-1}(F - F^{\wedge}(W)))^{\Delta}(V)\rangle_{L_2^{m\times p}(\mathbb{R})}$$

$$= \langle H(V-\Delta)^{-*}, (\Delta - W)^{-1}(F - F^{\wedge}(W))\rangle_{\mathcal{L}\mathcal{S}_2^{m\times p}}$$

$$= \langle H(V-\Delta)^{-*}, (\Delta - W)^{-1}F\rangle_{\mathcal{S}_2^{m\times p}}$$

$$= \langle (\Delta - W)^{-*} H(V - \Delta)^{-*}, F \rangle_{\mathcal{L}\mathcal{S}_2^{m\times p}}$$
$$= \langle (\Delta - W)^{-*} H(V - \Delta)^{-*}, (I - SS^\wedge(U)^*)(U - \Delta)^{-*} G \rangle_{\mathcal{L}\mathcal{S}_2^{m\times p}}$$
$$= \langle (\Delta - W)^{-*} H(V - \Delta)^{-*}, (U - \Delta)^{-*} G \rangle_{\mathcal{L}\mathcal{S}_2^{m\times p}}$$
$$- \langle (\Delta - W)^{-*} H(V - \Delta)^{-*}, SS^\wedge(U)^*(U - \Delta)^{-*} G \rangle_{\mathcal{L}\mathcal{S}_2^{m\times p}}.$$

With Ψ defined by (5.6) the last term can be written as

$$\langle (\Delta - W)^{-*} H(V - \Delta)^{-*}, SS^\wedge(U)^*(U - \Delta)^{-*} G \rangle_{\mathcal{L}\mathcal{S}_2^{m\times p}}$$
$$= \langle S^*(\Delta - W)^{-*} H(V - \Delta)^{-*}, S^\wedge(U)^*(U - \Delta)^{-*} G \rangle_{\mathcal{S}_2^{m\times p}}$$
$$= \langle (S^* - S^\wedge(W)^*)(\Delta - W)^{-*} H(V - \Delta)^{-*},$$
$$S^\wedge(U)^*(U - \Delta)^{-*} G \rangle_{\mathcal{S}_2^{m\times p}}$$
$$+ \langle S^\wedge(W)^*(\Delta - W)^{-*} H(V - \Delta)^{-*}, S^\wedge(U)^*(U - \Delta)^{-*} G \rangle_{\mathcal{S}_2^{m\times p}}$$
$$= \langle (\Psi^* - \Psi^\wedge(V)^*)(V - \Delta)^{-*}, S^\wedge(U)^*(U - \Delta)^{-*} G \rangle_{\mathcal{S}_2^{m\times p}}$$
$$+ \langle \Psi^\wedge(V)^*(V - \Delta)^{-*}, S^\wedge(U)^*(U - \Delta)^{-*} G \rangle_{\mathcal{S}_2^{m\times p}}$$
$$+ \langle S^\wedge(W)^*(\Delta - W)^{-*} H(V - \Delta)^{-*}, S^\wedge(U)^*(U - \Delta)^{-*} G \rangle_{\mathcal{S}_2^{m\times p}}$$
$$= \langle \Psi^\wedge(V)^*(V - \Delta)^{-*}, S^\wedge(U)^*(U - \Delta)^{-*} G \rangle_{\mathcal{S}_2^{m\times p}}$$
$$+ \langle S^\wedge(W)^*(\Delta - W)^{-*} H(V - \Delta)^{-*}, S^\wedge(U)^*(U - \Delta)^{-*} G \rangle_{\mathcal{S}_2^{m\times p}}$$
$$= \langle \Psi^\wedge(V)^*(V - \Delta)^{-*} + S^\wedge(W)^*(\Delta - W)^{-*} H(V - \Delta)^{-*},$$
$$S^\wedge(U)^*(U - \Delta)^{-*} G \rangle_{\mathcal{S}_2^{m\times p}}$$
$$= \langle \Psi^\wedge(V)^*(V - \Delta)^{-*} + S^\wedge(W)^*(\Delta - W)^{-*} H(V - \Delta)^{-*},$$
$$S^*(U - \Delta)^{-*} G \rangle_{\mathcal{S}_2^{m\times p}}$$
$$- \langle \Psi^\wedge(V)^*(V - \Delta)^{-*} + S^\wedge(W)^*(\Delta - W)^{-*} H(V - \Delta)^{-*},$$
$$(S^* - S^\wedge(U)^*)(U - \Delta)^{-*} G \rangle_{\mathcal{S}_2^{m\times p}}$$
$$= \langle \Psi^\wedge(V)^*(V - \Delta)^{-*} + S^\wedge(W)^*(\Delta - W)^{-*} H(V - \Delta)^{-*},$$
$$S^*(U - \Delta)^{-*} G \rangle_{\mathcal{S}_2^{m\times p}}$$
$$= \langle S(\Psi^\wedge(V)^* + S^\wedge(W)^*(\Delta - W)^{-*} H)(V - \Delta)^{-*},$$
$$(U - \Delta)^{-*} G \rangle_{\mathcal{L}\mathcal{S}_2^{m\times p}}.$$

Combining these formulas we get

$$\begin{aligned}
&\langle (\mathbf{A}^*(K_S^L(W)H))^\wedge(U), G\rangle_{L_2^{m\times p}(\mathbb{R})} \\
&\quad = \langle (\Delta - W)^{-*}H(V-\Delta)^{-*}, (U-\Delta)^{-*}G\rangle_{\mathcal{L}\mathcal{S}_2^{m\times p}} \\
&\qquad -\langle S(\Psi^\wedge(V)^* + S^\wedge(W)^*(\Delta - W)^{-*}H)(V-\Delta)^{-*}, \\
&\qquad (U-\Delta)^{-*}G\rangle_{\mathcal{L}\mathcal{S}_2^{m\times p}} \\
&\quad = \langle (-K_S^L(W)H - S\Psi^\wedge(V)^*)(V-\Delta)^{-*}, (U-\Delta)^{-*}G\rangle_{\mathcal{L}\mathcal{S}_2^{m\times p}} \\
&\quad = \langle ((-K_S^L(W)H - S\Psi^\wedge(V)^*)(V-\Delta)^{-*})^\wedge(U), G\rangle_{L_2^{m\times p}(\mathbb{R})}
\end{aligned}$$

and therefore (5.2) holds. □

Theorem 5.2 *The colligation $\mathcal{V}$ is coisometric in the sense that if*

$$\mathbf{T} := -I_{\mathcal{H}_L(S)} - \mathbf{A}\left(\frac{1}{2}I_{\mathcal{H}_L(S)} - \mathcal{M}_V^r\right)$$

and

$$\mathbf{N} := I_{\mathcal{H}_L(S)} + \mathbf{A}\left(\frac{1}{2}I_{\mathcal{H}_L(S)} + \mathcal{M}_V^r\right),$$

then

$$\begin{aligned}
&\begin{pmatrix} \mathbf{T} & 0 \\ \mathbf{C}(\frac{1}{2}I_{\mathcal{H}_L(S)} - \mathcal{M}_V^r) & I_{L_2^{m\times p}(\mathbb{R})}\end{pmatrix}\begin{pmatrix} \mathbf{T}^* & (\frac{1}{2}I_{\mathcal{H}_L(S)} - \mathcal{M}_{V*}^r)\,\mathbf{C}^* \\ 0 & I_{L_2^{m\times p}(\mathbb{R})}\end{pmatrix} \\
&= \begin{pmatrix} \mathbf{N} & \mathbf{B} \\ \mathbf{C}(\frac{1}{2}I_{\mathcal{H}_L(S)} + \mathcal{M}_V^r) & \mathbf{D}\end{pmatrix}\begin{pmatrix} \mathbf{N}^* & (\frac{1}{2}I_{\mathcal{H}_L(S)} + \mathcal{M}_{V*}^r)\,\mathbf{C}^* \\ \mathbf{B}^* & \mathbf{D}^*\end{pmatrix}.
\end{aligned} \tag{5.7}$$

Proof: We have to show the next three equalities

$$\mathbf{T}\mathbf{T}^* = \mathbf{N}\mathbf{N}^* + \mathbf{B}\mathbf{B}^* \tag{5.8}$$

$$\begin{aligned}
&\mathbf{C}(\tfrac{1}{2}I_{\mathcal{H}_L(S)} - \mathcal{M}_V^r)(\tfrac{1}{2}I_{\mathcal{H}_L(S)} - \mathcal{M}_{V*}^r)\mathbf{C}^* + I_{L_2^{m\times p}(\mathbb{R})} \\
&\quad = \mathbf{C}(\tfrac{1}{2}I_{\mathcal{H}_L(S)} + \mathcal{M}_V^r)(\tfrac{1}{2}I_{\mathcal{H}_L(S)} + \mathcal{M}_{V*}^r)\mathbf{C}^* + \mathbf{D}\mathbf{D}^*
\end{aligned} \tag{5.9}$$

and

$$\mathbf{T}\left(\frac{1}{2}I_{\mathcal{H}_L(S)} - \mathcal{M}_{V*}^r\right)\mathbf{C}^* = \mathbf{N}\left(\frac{1}{2}I_{\mathcal{H}_L(S)} + \mathcal{M}_{V*}^r\right)\mathbf{C}^* + \mathbf{B}\mathbf{D}^*. \tag{5.10}$$

In the calculations below, G is an arbitrary element from $L_2^{m\times p}(\mathbb{R}) \cap L_\infty^{m\times p}(\mathbb{R})$. We first prove (5.9). Using (3.4) we have

$$\begin{aligned}
&(\mathbf{C}(\mathcal{M}_V^r + \mathcal{M}_{V*}^r)\,\mathbf{C}^*)(G) \\
&\quad = \mathbf{C}\,((G - S(G^*S)^\wedge(V)^*)(V-\Delta)^{-*}(V+V^*)) \\
&\quad = ((G - S(G^*S)^\wedge(V)^*)(V-\Delta)^{-*}(V+V^*))^\Delta(V) \\
&\quad = (G - S(G^*S)^\wedge(V)^*)^\Delta(V) = G - (S(G^*S)^\wedge(V)^*)^\Delta(V) \\
&\qquad (I_{L_2^{m\times p}(\mathbb{R})} - \mathbf{DD}^*)\,(G),
\end{aligned}$$

and this implies (5.9). Now we show (5.10):

$$\begin{aligned}
&(\mathbf{A}(\mathcal{M}_V^r + \mathcal{M}_{V*}^r)\,\mathbf{C}^*)(G) \\
&\quad = \mathbf{A}\,((G - S(G^*S)^\wedge(V)^*)(V-\Delta)^{-*}(V+V^*)) \\
&\quad = ((G - S(G^*S)^\wedge(V)^*)(V-\Delta)^{-*}(V+V^*) \\
&\qquad -((G - S(G^*S)^\wedge(V)^*)(V-\Delta)^{-*}(V+V^*))^\Delta(V))(\Delta - V)^{-1} \\
&\quad = ((G - S(G^*S)^\wedge(V)^*)(V-\Delta)^{-*}(V+V^*) \\
&\qquad -(G - (S(G^*S)^\wedge(V)^*))^\Delta(V))(\Delta - V)^{-1} \\
&\quad = (G - S(G^*S)^\wedge(V)^*)((\Delta - V)^{-1} - (V-\Delta)^{-*}) \\
&\qquad -(G - (S(G^*S)^\wedge(V)^*))^\Delta(V))(\Delta - V)^{-1} \\
&\quad = -(G - S(G^*S)^\wedge(V)^*)(V-\Delta)^{-*} \\
&\qquad -(S(G^*S)^\wedge(V)^* - (S(G^*S)^\wedge(V)^*)^\Delta(V))(\Delta - V)^{-1} \\
&\quad = -\mathbf{C}^*(G) - \mathbf{BD}^*(G).
\end{aligned}$$

Hence

$$\mathbf{BD}^* = -(I_{\mathcal{H}_L(S)} + \mathbf{A}(\mathcal{M}_V^r + \mathcal{M}_{V*}^r))\,\mathbf{C}^*,$$

and now (5.10) follows from

$$\frac{1}{2}(\mathbf{N} - \mathbf{T}) = I_{\mathcal{H}_L(S)} + \mathbf{A}\mathcal{M}_V^r.$$

To prove (5.8) we observe that

$$\mathbf{A}^*(K_S^L(W)H) = (-K_S^L(W)H - S\mathbf{B}^*(K_S^L(W)H))(V-\Delta)^{-*},$$

and hence

$$\begin{aligned}
&\mathbf{A}\,(\mathcal{M}_V^r + \mathcal{M}_{V*}^r)\,\mathbf{A}^*(K_S^L(W)G) \\
&\quad = ((-K_S^L(W)G - S\mathbf{B}^*(K_S^L(W)G))(V-\Delta)^{-*}(V+V^*)
\end{aligned}$$

$$-((-K_S^L(W)G)^{\triangle}(V) - (S\mathbf{B}^*(K_S^L(W)G))^{\triangle}(V)))(\Delta - V)^{-1}$$
$$= -(K_S^L(W)G - (K_S^L(W)G)^{\triangle}(V))(\Delta - V)^{-1} - \mathbf{BB}^*(K_S^L(W)G)$$
$$-(-K_S^L(W)G - S\mathbf{B}^*(K_S^L(W)G))(V - \Delta)^{-*}$$
$$= -\mathbf{A}\,(K_S^L(W)G) - \mathbf{BB}^*(K_S^L(W)G) - \mathbf{A}^*(K_S^L(W)G).$$

Thus

$$\mathbf{BB}^* = -(\mathbf{A} + \mathbf{A}\,(\mathcal{M}_V^r + \mathcal{M}_{V*}^r)\,\mathbf{A}^* + \mathbf{A}^*) = \mathbf{TT}^* - \mathbf{NN}^*.$$

The last equality follows directly from the definitions of **T** and **N**. □

In the next theorem we prove the realization formula. We denote by Ω_V the set of all $p \times p$ astv-points W such that $\|W - V\| < 1$.

Theorem 5.3 *For* $G \in L_2^{n\times p}(\mathbb{R}) \cap L_\infty^{n\times p}(\mathbb{R})$ *and* $W \in \Omega_V$, *we have*

$$(SG)^{\triangle}(W) = (\mathbf{D} + \mathbf{C}(\mathcal{M}_W^r - \mathcal{M}_V^r)(I_{\mathcal{H}_L(S)} \tag{5.11}$$
$$-\mathbf{A}\,(\mathcal{M}_W^r - \mathcal{M}_V^r))^{-1}\mathbf{B})(G).$$

Proof: Since $W \in \Omega_V$, $\|W - V\|\|(\Delta - V)^{-1}\| < 1$ (recall that we assumed $\|(\Delta - V)^{-1}\| \leq 1$) and hence we may apply Corollary 3.5. This gives that

$$\mathcal{R}_W^r = (I - \mathcal{R}_V^r(\mathcal{M}_W^r - \mathcal{M}_V^r))^{-1}\mathcal{R}_V^r$$

is a bounded operator in $\mathcal{H}_L(S)$. From $\mathbf{A} = \mathcal{R}_V^r|_{\mathcal{H}_L(S)}$ and $\mathbf{B}(G) = \mathcal{R}_V^r(SG)$, we have that

$$(I_{\mathcal{H}_L(S)} - \mathbf{A}(\mathcal{M}_W^r - \mathcal{M}_V^r))^{-1}\mathbf{B}(G)$$
$$= (I - \mathcal{R}_V^r(\mathcal{M}_W^r - \mathcal{M}_V^r))^{-1}\mathcal{R}_V^r(SG)$$
$$= (SG - (SG)^{\triangle}(W))(\Delta - W)^{-1} = \mathcal{R}_W^r(SG).$$

Using (3.3) we obtain

$$\mathbf{C}(\mathcal{M}_W^r - \mathcal{M}_V^r)(I_{\mathcal{H}_L(S)} - \mathbf{A}(\mathcal{M}_W^r - \mathcal{M}_V^r))^{-1}\mathbf{B}(G)$$
$$= ((SG - (SG)^{\triangle}(W))(\Delta - W)^{-1}(W - V))^{\triangle}(V)$$
$$= (SG)^{\triangle}(W) - (SG)^{\triangle}(V)$$
$$= (SG)^{\triangle}(W) - \mathbf{D}(G),$$

from which the theorem follows. □

Theorem 5.4 *The colligation* $\mathcal{V}$ *is closely outer-connected in the sense that*

$$\bigcap_{W\in\Omega_V} \operatorname{Ker} \mathbf{C}\,(I_{\mathcal{H}_L(S)} - (\mathcal{M}_W^r - \mathcal{M}_V^r)\,\mathbf{A})^{-1} = \{0\}. \tag{5.12}$$

Proof: By the first and last formulas in Corollary 3.5, we have for $F \in \mathcal{H}_L(S)$ and $W \in \Omega_V$,

$$\begin{aligned} &F + (F - F^{\Delta}(W))(\Delta - W)^{-1}(W - V) \\ &= (I + (\mathcal{M}^r_W - \mathcal{M}^r_V)\mathcal{R}^r_W)(F) \\ &= (I - (\mathcal{M}^r_W - \mathcal{M}^r_V)\mathcal{R}^r_V)^{-1}(F) \\ &= (I - (\mathcal{M}^r_W - \mathcal{M}^r_V)\,\mathbf{A})^{-1}(F). \end{aligned}$$

Thus

$$\begin{aligned} &\mathbf{C}(I - (\mathcal{M}^r_W - \mathcal{M}^r_V)\,\mathbf{A})^{-1}(F) \\ &= F^{\Delta}(V) + ((F - F^{\Delta}(W))(\Delta - W)^{-1}(W - V))^{\Delta}(V) \\ &= F^{\Delta}(V) + F^{\Delta}(W) - F^{\Delta}(V) = F^{\Delta}(W). \end{aligned}$$

Therefore if $F \in \bigcap_{W \in \Omega_V} \operatorname{Ker} \mathbf{C}\,(I_{\mathcal{H}_L(S)} - (\mathcal{M}^r_W - \mathcal{M}^r_V)\,\mathbf{A})^{-1}$ then $F^{\Delta}(W) = 0$ for all $W \in \Omega_V$, which implies $F = 0$. □

In what follows we show that the colligation (5.1) is unique up to a unitary equivalence.

Let $\mathcal{F}_W : L_2^{n\times p}(\mathbb{R}) \longrightarrow L_2^{m\times p}(\mathbb{R})$ be the operator defined by

$$\begin{aligned} \mathcal{F}_W(G) = (SG)^{\Delta}(W) &= (\mathbf{D} + \mathbf{C}(\mathcal{M}^r_W - \mathcal{M}^r_V) \\ &(I_{\mathcal{H}_L(S)} - \mathbf{A}(\mathcal{M}^r_W - \mathcal{M}^r_V))^{-1}\mathbf{B})(G). \end{aligned}$$

Then one can show that

$$\begin{aligned} &I_{L_2^{m\times p}(\mathbb{R})} - \mathcal{F}_W\mathcal{F}_U^* \\ &= \mathbf{C}\,(I_{\mathcal{H}_L(S)} - (\mathcal{M}^r_W - \mathcal{M}^r_V\mathbf{A}))^{-1}(\mathcal{M}^r_W + \mathcal{M}^r_{U^*}) \\ &\times(I_{\mathcal{H}_L(S)} - \mathbf{A}^*(\mathcal{M}^r_{U^*} - \mathcal{M}^r_{V^*}))^{-1}\mathbf{C}^*. \end{aligned} \tag{5.13}$$

We will say that the $p \times p$ astv-point V is *normal* if there exists a $\nu \in i\mathbb{C}_-$ such that νI_p belongs to Ω_V, that is, νI_p is an astv-point such that $\|\nu I_p - V\| < 1$.

Theorem 5.5 *Let*

$$\begin{pmatrix} \tilde{\mathbf{A}} & \tilde{\mathbf{B}} \\ \tilde{\mathbf{C}} & \tilde{\mathbf{D}} \end{pmatrix} : \begin{pmatrix} \mathcal{H} \\ L_2^{n\times p}(\mathbb{R}) \end{pmatrix} \longrightarrow \begin{pmatrix} \mathcal{H} \\ L_2^{m\times p}(\mathbb{R}) \end{pmatrix}$$

be a coisometric closely outer-connected colligation in the sense of (5.7) *and of* (5.12), *respectively, where $\mathcal{H}$ is a separable Hilbert space $L_\infty^{p\times p}(\mathbb{R})$ right module. Assume* (1) *that for every $G \in L_2^{n\times p}(\mathbb{R}) \cap L_\infty^{n\times p}(\mathbb{R})$ and any $W \in \Omega_V$,*

$$(SG)^{\Delta}(W) = (\tilde{\mathbf{D}} + \tilde{\mathbf{C}}\,(\mathcal{M}^r_W - \mathcal{M}^r_V)(I_{\mathcal{H}} - \tilde{\mathbf{A}}\,(\mathcal{M}^r_W - \mathcal{M}^r_V))^{-1}\tilde{\mathbf{B}})(G),$$

(2) *the moment-type condition holds: for every* $W, U \in \Omega_V$,

$$\begin{aligned}&\mathbf{C}\,(I_{\mathcal{H}_L(S)} - (\mathcal{M}^r_W - \mathcal{M}^r_V)\,\mathbf{A})^{-1}(I_{\mathcal{H}_L(S)} - \mathbf{A}^*(\mathcal{M}^r_{U*} - \mathcal{M}^r_{V*}))^{-1}\mathbf{C}^*\\ &= \tilde{\mathbf{C}}\,(I_{\mathcal{H}} - (\mathcal{M}^r_W - \mathcal{M}^r_V)\,\tilde{\mathbf{A}})^{-1}(I_{\mathcal{H}} - \tilde{\mathbf{A}}^*(\mathcal{M}^r_{U*} - \mathcal{M}^r_{V*}))^{-1}\tilde{\mathbf{C}}^*,\end{aligned} \tag{5.14}$$

and (3) *the astv-point V is normal. Then there exists a unitary map* $\sigma : \mathcal{H}_L(S) \longrightarrow \mathcal{H}$ *such that*

$$\begin{pmatrix}\tilde{\mathbf{A}} & \tilde{\mathbf{B}}\\ \tilde{\mathbf{C}} & \tilde{\mathbf{D}}\end{pmatrix} = \begin{pmatrix}\sigma & 0\\ 0 & I\end{pmatrix}\begin{pmatrix}\mathbf{A} & \mathbf{B}\\ \mathbf{C} & \mathbf{D}\end{pmatrix}\begin{pmatrix}\sigma^* & 0\\ 0 & I\end{pmatrix}. \tag{5.15}$$

Proof: Let $\mathbf{R}$ be the linear relation on $\mathcal{H}_L(S)$ spanned by all the pairs of the form:

$$\begin{aligned}&((I_{\mathcal{H}_L(S)} - \mathbf{A}^*(\mathcal{M}^r_{U*} - \mathcal{M}^r_{V*}))^{-1}\mathbf{C}^*(G),\\ &(I_{\mathcal{H}} - \tilde{\mathbf{A}}^*(\mathcal{M}^r_{U*} - \mathcal{M}^r_{V*}))^{-1}\tilde{\mathbf{C}}^*(G)),\end{aligned}$$

where $G \in L_2^{m\times p}(\mathbb{R}) \cap L_\infty^{m\times p}(\mathbb{R})$ and $U \in \Omega_V$. Using the moment-type condition (5.14) we have for every $G_1, G_2 \in L_2^{m\times p}(\mathbb{R}) \cap L_\infty^{m\times p}(\mathbb{R})$ and $W, U \in \Omega_V$ that

$$\begin{aligned}&\langle (I_{\mathcal{H}_L(S)} - \mathbf{A}^*(\mathcal{M}^r_{U*} - \mathcal{M}^r_{V*}))^{-1}\mathbf{C}^*(G_1),\\ &\quad (I_{\mathcal{H}_L(S)} - \mathbf{A}^*(\mathcal{M}^r_{W*} - \mathcal{M}^r_{V*}))^{-1}\mathbf{C}^*(G_2)\rangle_{\mathcal{H}_L(S)}\\ &= \langle \mathbf{C}\,(I_{\mathcal{H}_L(S)} - (\mathcal{M}^r_W - \mathcal{M}^r_V)\mathbf{A})^{-1}(I_{\mathcal{H}_L(S)} - \mathbf{A}^*\\ &\quad (\mathcal{M}^r_{U*} - \mathcal{M}^r_{V*}))^{-1}\mathbf{C}^*(G_1), G_2\rangle_{L_2^{m\times p}(\mathbb{R})}\\ &= \langle \tilde{\mathbf{C}}\,(I_{\mathcal{H}} - (\mathcal{M}^r_W - \mathcal{M}^r_V)\tilde{\mathbf{A}})^{-1}(I_{\mathcal{H}} - \tilde{\mathbf{A}}^*(\mathcal{M}^r_{U*} - \mathcal{M}^r_{V*}))^{-1}\\ &\quad \tilde{\mathbf{C}}^*(G_1), G_2\rangle_{L_2^{m\times p}(\mathbb{R})}\\ &= \langle (I_{\mathcal{H}} - \tilde{\mathbf{A}}^*(\mathcal{M}^r_{U*} - \mathcal{M}^r_{V*}))^{-1}\tilde{\mathbf{C}}^*(G_1), (I_{\mathcal{H}} - \tilde{\mathbf{A}}^*\\ &\quad (\mathcal{M}^r_{W*} - \mathcal{M}^r_{V*}))^{-1}\tilde{\mathbf{C}}^*(G_2)\rangle_{\mathcal{H}}.\end{aligned}$$

Therefore, the relation $\mathbf{R}$ is well defined and isometric. Since both colligations are closely outer-connected, the relation $\mathbf{R}$ has a dense domain and dense range. Thus the relation $\mathbf{R}$ is the graph of a unitary operator $\sigma : \mathcal{H}_L(S) \longrightarrow \mathcal{H}$ satisfying

$$\begin{aligned}&\sigma((I_{\mathcal{H}_L(S)} - \mathbf{A}^*(\mathcal{M}^r_{U*} - \mathcal{M}^r_{V*}))^{-1}\mathbf{C}^*(G))\\ &= (I_{\mathcal{H}} - \tilde{\mathbf{A}}^*(\mathcal{M}^r_{U*} - \mathcal{M}^r_{V*}))^{-1}\tilde{\mathbf{C}}^*(G),\end{aligned}$$

for every $G \in L_2^{m\times p}(\mathbb{R}) \cap L_\infty^{m\times p}(\mathbb{R})$ and $U \in \Omega_V$.
Similarly, $\tilde{\sigma} : \mathcal{H} \longrightarrow \mathcal{H}_L(S)$ with

$$\begin{aligned}&\tilde{\sigma}((I_{\mathcal{H}} - \tilde{\mathbf{A}}^*(\mathcal{M}^r_{U*} - \mathcal{M}^r_{V*}))^{-1}\tilde{\mathbf{C}}^*(G))\\ &= (I_{\mathcal{H}_L(S)} - \mathbf{A}^*(\mathcal{M}^r_{U*} - \mathcal{M}^r_{V*}))^{-1}\mathbf{C}^*(G),\end{aligned}$$

for every $G \in L_2^{m\times p}(\mathbb{R}) \cap L_\infty^{m\times p}(\mathbb{R})$ and $U \in \Omega_V$, also defines a unitary operator. These two operators satisfy the equations $\sigma\tilde{\sigma} = I_{\mathcal{H}}$ and $\tilde{\sigma}\sigma = I_{\mathcal{H}_L(S)}$ on dense sets, and therefore on the whole related space.
Since

$$\begin{aligned}
I_{L_2^{m\times p}(\mathbb{R})} &- \mathcal{F}_W\mathcal{F}_U^* \\
&= \mathbf{C}\,(I_{\mathcal{H}_L(S)} - (\mathcal{M}_W^r - \mathcal{M}_V^r)\mathbf{A})^{-1}(\mathcal{M}_W^r + \mathcal{M}_{U*}^r) \\
&\quad (I_{\mathcal{H}_L(S)} - \mathbf{A}^*(\mathcal{M}_{U*}^r - \mathcal{M}_{V*}^r))^{-1}\mathbf{C}^* \\
&= \mathbf{C}\,(I_{\mathcal{H}_L(S)} - (\mathcal{M}_W^r - \mathcal{M}_V^r)\mathbf{A})^{-1}(\mathcal{M}_W^r + \mathcal{M}_{U*}^r) \\
&\quad \tilde{\sigma}(I_{\mathcal{H}} - \tilde{\mathbf{A}}^*(\mathcal{M}_{U*}^r - \mathcal{M}_{V*}^r))^{-1}\tilde{\mathbf{C}}^* \\
&= \tilde{\mathbf{C}}\,(I_{\mathcal{H}} - (\mathcal{M}_W^r - \mathcal{M}_V^r)\tilde{\mathbf{A}})^{-1}(\mathcal{M}_W^r + \mathcal{M}_{U*}^r) \\
&\quad (I_{\mathcal{H}} - \tilde{\mathbf{A}}^*(\mathcal{M}_{U*}^r - \mathcal{M}_{V*}^r))^{-1}\tilde{\mathbf{C}}^* \\
&= \mathbf{C}\,(I_{\mathcal{H}_L(S)} - (\mathcal{M}_W^r - \mathcal{M}_V^r)\mathbf{A})^{-1}\sigma^*(\mathcal{M}_W^r + \mathcal{M}_{U*}^r) \\
&\quad (I_{\mathcal{H}} - \tilde{\mathbf{A}}^*(\mathcal{M}_{U*}^r - \mathcal{M}_{V*}^r))^{-1}\tilde{\mathbf{C}}^*,
\end{aligned}$$

and since both colligations are closely outer-connected, we have that

$$(\mathcal{M}_W^r + \mathcal{M}_{U*}^r)\tilde{\sigma} = \sigma^*(\mathcal{M}_W^r + \mathcal{M}_{U*}^r). \tag{5.16}$$

By the assumption that V is normal, and taking $W = \nu I_p = U$, then from (5.16) we have $(\nu + \overline{\nu})\tilde{\sigma} = (\nu + \overline{\nu})\sigma^*$, that is, $\tilde{\sigma} = \sigma^*$.

Let M and a be the pair of constants and P be the projection related to the dichotomous differential equation

$$\frac{dx}{dt} = V(t)x(t), \quad t \in \mathbb{R}.$$

Let $U_V(t)$ be the related fundamental matrix, and let $W = V + \lambda I_p$. Then W would be an astv-point in Ω_V if $\lambda \in \mathbb{D} \cap \{\omega \in \mathbb{C} | \mathrm{Re}(\omega) > \ln(a)\}$ (note that this set contains a neighborhood of 0) . Indeed, the fundamental matrix $U_W(t)$ related to the differential equation

$$\frac{dx}{dt} = W(t)x(t), \quad t \in \mathbb{R},$$

is $U_W(t) = e^{\lambda t}U_V(t)$. Thus

$$\begin{aligned}
\|U_W(t)U_W(s)^{-1}\| &= \|e^{\lambda(t-s)}U_V(t)U_V(s)^{-1}\| \\
&\leq Me^{-\mathrm{Re}(\lambda)(s-t)}e^{\ln(a)(s-t)} \\
&= Me^{(\ln(a)-\mathrm{Re}(\lambda))(s-t)} = M\tilde{a}^{s-t},
\end{aligned}$$

where $\tilde{a} = e^{\ln(a)-\mathrm{Re}(\lambda)}$. Therefore, $0 < \tilde{a} < 1$ if $\mathrm{Re}(\lambda) > \ln(a)$. On the other hand $W \in \Omega_V$ if $\|W - V\| < 1$ and this holds if $\lambda \in \mathbb{D}$. Since

$$\sigma(I_{\mathcal{H}_L(S)} - \mathbf{A}^*(\mathcal{M}^r_{U*} - \mathcal{M}^r_{V*}))^{-1}\mathbf{C}^* = (I_{\mathcal{H}} - \tilde{\mathbf{A}}^*(\mathcal{M}^r_{U*} - \mathcal{M}^r_{V*}))^{-1}\tilde{\mathbf{C}}^*,$$

and taking $U = V + \mu I_p$ where $\mu \in \mathbb{D} \cap \{\omega \in \mathbb{C} | \mathrm{Re}(\omega) > \ln(a)\}$ we get

$$\sigma(I_{\mathcal{H}_L(S)} - \overline{\mu}\mathbf{A}^*)^{-1}\mathbf{C}^* = (I_{\mathcal{H}} - \overline{\mu}\tilde{\mathbf{A}}^*)^{-1}\tilde{\mathbf{C}}^*. \tag{5.17}$$

Thus for $\mu = 0$ we have $\sigma\mathbf{C}^* = \tilde{\mathbf{C}}^*$ and therefore

$$\tilde{\mathbf{C}} = \mathbf{C}\sigma^*. \tag{5.18}$$

Still from (5.17) we have

$$\sigma\mathbf{C}^* + \overline{\mu}\sigma\mathbf{A}^*(I_{\mathcal{H}_L(S)} - \overline{\mu}\mathbf{A}^*)^{-1}\mathbf{C}^* = \tilde{\mathbf{C}}^* + \overline{\mu}\tilde{\mathbf{A}}^*(I_{\mathcal{H}} - \overline{\mu}\tilde{\mathbf{A}}^*)^{-1}\tilde{\mathbf{C}}^*,$$

and using (5.18) we get

$$\begin{aligned}\sigma\mathbf{A}^*(I_{\mathcal{H}_L(S)} - \overline{\mu}\mathbf{A}^*)^{-1}\mathbf{C}^* &= \tilde{\mathbf{A}}^*(I_{\mathcal{H}} - \overline{\mu}\tilde{\mathbf{A}}^*)^{-1}\tilde{\mathbf{C}}^* \\ &= \tilde{\mathbf{A}}^*\sigma(I_{\mathcal{H}_L(S)} - \overline{\mu}\mathbf{A}^*)^{-1}\mathbf{C}^*.\end{aligned}$$

Since the sets of all elements of the form $(I_{\mathcal{H}_L(S)} - \overline{\mu}\mathbf{A}^*)^{-1}\mathbf{C}^*(G)$ and $(I_{\mathcal{H}} - \overline{\mu}\tilde{\mathbf{A}}^*)^{-1}\tilde{\mathbf{C}}^*(G)$ are dense in $\mathcal{H}_L(S)$ and in $\mathcal{H}$ respectively, we have that $\sigma\mathbf{A}^* = \tilde{\mathbf{A}}^*\sigma$ from which we conclude

$$\tilde{\mathbf{A}} = \sigma\mathbf{A}\sigma^*. \tag{5.19}$$

Since $(SG)^{\triangle}(V) = \mathbf{D}(G) = \tilde{\mathbf{D}}(G)$ we have

$$\tilde{\mathbf{D}} = \mathbf{D}. \tag{5.20}$$

Now, using the realization formula with $W = V + \lambda I_p$ with $\lambda \in \mathbb{D} \cap \{\omega \in \mathbb{C} | \mathrm{Re}(\omega) > \ln(a)\}$, and (5.20) we get $\lambda\mathbf{C}(I_{\mathcal{H}_L(S)} - \lambda\mathbf{A})^{-1}\mathbf{B} = \lambda\tilde{\mathbf{C}}(I_{\mathcal{H}} - \lambda\tilde{\mathbf{A}})^{-1}\tilde{\mathbf{B}}$. Thus

$$\begin{aligned}\mathbf{B}^*(I_{\mathcal{H}_L(S)} - \overline{\lambda}\mathbf{A}^*)^{-1}\mathbf{C}^* &= \tilde{\mathbf{B}}^*(I_{\mathcal{H}} - \overline{\lambda}\tilde{\mathbf{A}}^*)^{-1}\tilde{\mathbf{C}}^* \\ &= \tilde{\mathbf{B}}^*\sigma(I_{\mathcal{H}_L(S)} - \overline{\lambda}\mathbf{A}^*)^{-1}\mathbf{C}^*,\end{aligned}$$

and therefore $\mathbf{B}^* = \tilde{\mathbf{B}}^*\sigma$, from which we conclude

$$\tilde{\mathbf{B}} = \sigma\mathbf{B}. \tag{5.21}$$

From (5.18)–(5.21) we have (5.15). □

To complete the presentation we briefly state the results related with the space $\mathcal{H}_R(S)$. Let

$$\mathcal{V} = \begin{pmatrix} \mathbf{A} & \mathbf{B} \\ \mathbf{C} & \mathbf{D} \end{pmatrix} : \begin{pmatrix} \mathcal{H}_R(S) \\ L_2^{q\times m}(\mathbb{R}) \end{pmatrix} \longrightarrow \begin{pmatrix} \mathcal{H}_R(S) \\ L_2^{q\times n}(\mathbb{R}) \end{pmatrix},$$

where

$$\begin{aligned} \mathbf{A}(F) &= (\Delta - V)^{-1}(F - F^\wedge(V)) \\ \mathbf{B}(G) &= (\Delta - V)^{-1}(GS - (GS)^\wedge(V)) \\ \mathbf{C}(F) &= F^\wedge(V) \\ \mathbf{D}(G) &= (GS)^\wedge(V). \end{aligned}$$

Here $G \in L_2^{q\times m}(\mathbb{R}) \cap L_\infty^{q\times m}(\mathbb{R})$, and V is a $q \times q$ astv-point such that $\|(\Delta - V)^{-1}\| < 1$. Then the colligation $\mathcal{V}$ is well defined, coisometric in the sense that

$$\begin{pmatrix} \mathbf{T} & 0 \\ \mathbf{C}\,(\frac{1}{2}I_{\mathcal{H}_R(S)} - \mathcal{M}_V^\ell) & I_{L_2^{q\times n}(\mathbb{R})} \end{pmatrix} \begin{pmatrix} \mathbf{T}^* & (\frac{1}{2}I_{\mathcal{H}_R(S)} - \mathcal{M}_{V^*}^\ell)\,\mathbf{C}^* \\ 0 & I_{L_2^{q\times n}(\mathbb{R})} \end{pmatrix}$$
$$= \begin{pmatrix} \mathbf{N} & \mathbf{B} \\ \mathbf{C}\,(\frac{1}{2}I_{\mathcal{H}_R(S)} + \mathcal{M}_V^\ell) & \mathbf{D} \end{pmatrix} \begin{pmatrix} \mathbf{N}^* & (\frac{1}{2}I_{\mathcal{H}_R(S)} + \mathcal{M}_{V^*}^\ell)\,\mathbf{C}^* \\ \mathbf{B}^* & \mathbf{D}^* \end{pmatrix}$$

where

$$\mathbf{T} := -I_{\mathcal{H}_R(S)} - \mathbf{A}\left(\frac{1}{2}I_{\mathcal{H}_R(S)} - \mathcal{M}_V^\ell\right)$$

and

$$\mathbf{N} := I_{\mathcal{H}_R(S)} + \mathbf{A}\left(\frac{1}{2}I_{\mathcal{H}_R(S)} + \mathcal{M}_V^\ell\right),$$

and $\mathcal{V}$ is closely outer-connected:

$$\bigcap_{W\in\Omega_V} \operatorname{Ker} \mathbf{C}(I_{\mathcal{H}_R(S)} - (\mathcal{M}_W^\ell - \mathcal{M}_V^\ell)\mathbf{A})^{-1} = \{0\}.$$

The next theorem summarizes the realization result:

Theorem 5.6 *For $G \in L_2^{q\times m}(\mathbb{R}) \cap L_\infty^{q\times m}(\mathbb{R})$ and $W \in \Omega_V$, we have*

$$(GS)^\wedge(W) = (\mathbf{D} + \mathbf{C}\,(\mathcal{M}_W^\ell - \mathcal{M}_V^\ell)(I_{\mathcal{H}_R(S)} - \mathbf{A}\,(\mathcal{M}_W^\ell - \mathcal{M}_V^\ell))^{-1}\mathbf{B})(G).$$

If moreover, the central point V is normal, that is, there exists a point $\nu \in i\mathbb{C}_-$ such that $\|\nu I_q - V\| < 1$, then the colligation $\mathcal{V}$ is unique up to a unitary equivalence in the sense of Theorem 5.5.

References

[1] D. Alpay, Algorithme de Schur, espaces à noyau reproduisant et théorie des systèmes, volume 6 of Panoramas et Synthèses. *Société Mathématique de France*, 1998.

[2] D. Alpay, V. Bolotnikov, A. Dijksma and H.S.V. de Snoo, On some operator colligations and associated reproducing kernel Hilbert spaces, volume 61 of *Operator theory: Advances and Applications*, pages 89–159. Birkhäuser Verlag, Basel, 1993.

[3] D. Alpay, V. Bolotnikov, A. Dijksma and H.S.V. de Snoo, On some operator colligations and associated reproducing kernel Pontryagin spaces. *J. Funct. Anal.* **136** (1996), 39–80.

[4] D. Alpay, P. Dewilde and H. Dym, On the existence and construction of solutions to the partial lossless inverse scattering problem with applications to estimation theory. *IEEE Trans. Inform. Theory* **35** (1989), 1184–1205.

[5] D. Alpay and Y. Peretz, Special realizations for Schur upper triangular operators. In *Contributions to operator theory in spaces with an indefinite metric. The Heinz Langer anniversary volume*, volume 106 of *Operator theory: Advances and Applications*, pages 37–90. Birkhäuser Verlag, Basel, 1998.

[6] T. Andô, *de Branges spaces and analytic operator functions.* Lecture notes, Hokkaido University, Sapporo, 1990.

[7] J. Ball, I. Gohberg and M. Kaashoek, Input-output operators of J-unitary time-varying continuous time systems. In *Essays dedicated to A.C. Zaanen*, volume 75 of *Operator Theory: Advances and Applications*, pages 57–94. Birkhäuser Verlag, Basel, 1995.

[8] J. Ball, I. Gohberg and M.A. Kaashoek, Nevanlinna-Pick interpolation for time-varying input-output maps: the continuous case. In I. Gohberg, editor, *Time-variant systems and interpolation*, volume 56 of *Operator Theory: Advances and Applications*, pages 52–89. Birkhäuser Verlag, Basel, 1992.

[9] J. Ball, I. Gohberg and M.A. Kaashoek, Bitangential interpolation for input-output maps of time-varying systems: the continuous time case. *Integral Equations Operator Theory* **20** (1994), 1–43.

[10] J. Ball, I. Gohberg and M.A. Kaashoek, Two sided Nudelman interpolation for input-output operators of discrete time-varying systems. *Integral Equations Operator Theory* **21** (1995), 174–211.

[11] A. Ben-Artzi and I. Gohberg, Dichotomy of systems and invertibility of linear ordinary differential operators. In I. Gohberg, editor, *Time-variant systems and interpolation*, volume 56 of *Operator Theory: Advances and Applications*, pages 90–119. Birkhäuser Verlag, Basel, 1992.

[12] L. de Branges and J. Rovnyak, *Square summable power series.* Holt, Rinehart and Winston, NewYork, 1966.

[13] P. Dewilde and H. Dym, Interpolation for upper triangular operators. In I. Gohberg, editor, *Time-variant systems and interpolation*, volume 56 of *Operator Theory: Advances and Applications*, pages 153–260. Birkhäuser Verlag, Basel, 1992.

[14] P. Dewilde and A. van der Veen, On the Hankel-norm approximation of upper-triangular operators and matrices. *Integral Equations Operator Theory* **17** (1993), 1–45.

[15] H. Dym, J contractive matrix functions, reproducing kernel spaces and interpolation, volume 71 of CBMS Lecture Notes. *Amer. Math. Soc.*, Rhodes Island, 1989.

[16] P.A. Fuhrmann, *Linear systems and operators in Hilbert space*. McGraw-Hill international book company, 1981.

[17] J. Kos, *Time-dependent problems in linear operator theory*. Ph.D.-thesis, Vrije Universiteit, Amsterdam, 1995.

[18] A.H. Zemanian, *Realizability theory for continuous linear systems*. Dover Publications, Inc., NewYork, 1995.

Daniel Alpay
Department of Mathematics
Ben-Gurion University of the Negev
POB 653. Beer-Sheva 84105, Israel
dany@cs.bgu.ac.il

Vladimir Bolotnikov
Department of Mathematics
Ben-Gurion University of the Negev
POB 653. Beer-Sheva 84105, Israel
vladi@cs.bgu.ac.il

Aad Dijksma
Department of Mathematics
University of Groningen, POB 800
9700 AV Groningen, The Netherlands
a.dijksma@math.rug.nl

Yossi Peretz
Department of Mathematics
Ben-Gurion University of the Negev
POB 653. Beer-Sheva 84105, Israel
peretzyo@cs.bgu.ac.il

MSC: 46E22, 47A48

Operator Theory:
Advances and Applications, Vol. 122
© 2001 Birkhäuser Verlag Basel/Switzerland

Some Remarks on the Inverse Monodromy Problem for 2×2 Canonical Differential Systems

Damir Z. Arov and Harry Dym

To Rien Kaashoek on his sixtieth birthday with best wishes

A deep theorem of de Branges establishes the uniqueness of real normalized solutions of the inverse monodromy problem for 2×2 canonical systems of differential equations with a symplectic monodromy matrix which is inner with respect to the signature matrix $\begin{bmatrix} 0 & i \\ -i & 0 \end{bmatrix}$. In this paper, this theorem is used to show that the inverse monodromy problem for such systems of equations has a unique normalized solution for monodromy matrices which are J-inner with respect to any 2×2 signature matrix which is not definite if and only if the monodromy matrix $U(\lambda)$ has zero exponential type in either the upper or lower half plane (or equivalently, if and only if the exponential type of $U(\lambda)$ is equal to the exponential type of its determinant). A complete description of the set of solutions is furnished in the opposite case. Enroute simple proofs are given for a number of well known type formulas.

1 Introduction

Let J be a 2×2 signature matrix with one eigenvalue equal to 1 and one eigenvalue equal to -1. Then J is unitarily equivalent to the signature matrix

$$\mathcal{J} = \begin{bmatrix} 0 & i \\ -i & 0 \end{bmatrix}. \tag{1.1}$$

Let $\mathcal{E}$ denote the set of 2×2 mvf's (matrix valued functions) with entire entries and let $\mathcal{U}(J)$ denote the set of J-inner mvf's with respect to the open upper half plane $\mathbb{C}_+$; $\mathcal{U}(J)$ is the set of 2×2 mvf's which are meromorphic in $\mathbb{C}_+$ and meet the following two conditions:

(1) $U(\lambda)$ is J-contractive in $\mathbb{C}_+$, i.e.,

$$U(\lambda) J U(\lambda)^* \leq J \tag{1.2}$$

at every point $\lambda \in \mathbb{C}_+$ at which $U(\lambda)$ is holomorphic.

D.Z. Arov wishes to thank the Weizmann Institute of Science for support and hospitality; H. Dym wishes to thank Renee and Jay Weiss for endowing the chair which supports his research.

(2) The boundary values $U(\mu)$, which exist as nontangential limits at a.e. point $\mu \in \mathbb{R}$ are J-unitary, i.e.,

$$U(\mu)JU(\mu)^* = J \tag{1.3}$$

for a.e. $\mu \in \mathbb{R}$.

Condition (1.3) guarantees that $U(\lambda)$ is invertible in $\mathbb{C}_+$ except for an isolated set of points. Thus, as usual, we shall define $U(\lambda)$ for $\lambda \in \mathbb{C}_-$, the open lower half plane, by the formula

$$U(\lambda) = J\{U^\#(\lambda)\}^{-1}J,$$

where

$$f^\#(\lambda) = f(\bar{\lambda})^* = \overline{f(\bar{\lambda})}^\tau$$

for any mvf $f(\lambda)$ and the superscript τ stands for the matrix transpose.

Let $U \in \mathcal{E} \cap \mathcal{U}(J)$ with $U(0) = I_2$ and let[1]

$$\ell = -i \operatorname{trace}\{U'(0)J\}. \tag{1.4}$$

Then, by a theorem of Potapov [Po], there exists a 2×2 mvf $H(x)$ that is measurable with respect to the Lebesgue measure such that

(1.5) (1) $H(x) \geq 0$ for a.e. $x \in [0, \ell]$.

(1.6) (2) trace $H(x) = 1$ for a.e. $x \in [0, \ell]$.

(3) The solution $U_x(\lambda) = U(x, \lambda)$ of the equation

$$U(x,\lambda) = I_2 + i\lambda \int_0^x U(s,\lambda)H(s)ds\,J, \quad 0 \leq x \leq \ell, \tag{1.7}$$

agrees with U at the point $x = \ell$:

$$U(\ell, \lambda) = U(\lambda). \tag{1.8}$$

The mvf $U_x(\lambda) = U(x, \lambda)$, $0 \leq x \leq \ell$, considered in (1.7) is the fundamental matrix of solutions of the canonical differential system

$$\frac{dy}{dx}(x,\lambda) = i\lambda y(x,\lambda)H(x)J, \quad 0 \leq x \leq \ell, \tag{1.9}$$

i.e., it is the solution of the Cauchy problem

$$\left.\begin{aligned} &\frac{dU}{dx}(x,\lambda) = i\lambda U(x,\lambda)H(x)J, \ 0 \leq x \leq \ell, \\ &U(0,\lambda) = I_2. \end{aligned}\right\}$$

[1] Formula (1.4) gives the correct value for the length of the interval only when the Hermitian $H(x)$ is subject to (1.6).

In this formulation, $H(x)$ is commonly referred to as the Hermitian, $U_x(\lambda) = U(x,\lambda)$, $0 \le x \le \ell$, as the matrizant and $U(\lambda) = U(\ell,\lambda)$ as the monodromy matrix of the canonical differential system (1.9). Note that the conditions (1.5) and (1.6) automatically insure that $H(x)$ is summable on $[0,\ell]$ since ℓ is finite.

Conversely, for any summable nonnegative mvf $H(x)$, the integral equation (1.7) has a unique solution $U_x(\lambda) = U(x,\lambda)$ which is both entire and J-inner (in the variable λ) with $U_x(0) = I_2$ for every $x \in [0,\ell]$.

The inverse monodromy problem for the canonical differential system (1.9) with monodromy matrix $U(\lambda)$ is to describe the set of summable mvf's $H(x)$ for which (1.5), (1.7) and (1.8) hold. Each such $H(x)$ is said to be a solution of this inverse monodromy problem. If also (1.6) holds, then $H(x)$ is said to be a normalized solution. In this case, the length ℓ of the interval is uniquely determined by the monodromy matrix $U(\lambda)$ by formula (1.4) because

$$\int_0^\ell H(s)ds = -iU'(0)J.$$

In general, even under the normalization condition (1.6), there may be many solutions $H(x)$ to the inverse monodromy problem for a given monodromy matrix $U \in \mathcal{E}\cap\mathcal{U}(J)$ with $U(0) = I_2$. Typically, uniqueness will prevail only if extra constraints are imposed on the monodromy matrix $U(\lambda)$ and/or the class of Hermitians $H(x)$. A deep theorem of L. de Branges [dB1]–[dB4] (especially pp. 46–47 of [dB4]) and [dB7] establishes uniqueness in the class of real normalized solutions $H(x)$ of the inverse monodromy problem when $J = \mathcal{J}$ and the monodromy matrix $U(\lambda)$ is subject to the symmetry constraint

$$\overline{U(\overline{\lambda})} = U(\lambda) \text{ for every } \lambda \in \mathbb{C}. \tag{1.10}$$

Theorem 1.1 *Let $U \in \mathcal{E}\cap\mathcal{U}(\mathcal{J})$ with $U(0) = I_2$ and suppose further that $U(\lambda)$ is subject to the symmetry constraint* (1.10). *Then there exists exactly one normalized solution $H(x)$ to the inverse monodromy problem for the canonical differential system* (1.9) *with $J = \mathcal{J}$ and monodromy matrix $U(\lambda)$ such that $H(x) = \overline{H(x)}$ for a.e. point $x \in [0,\ell]$.*

To be more precise, de Branges dealt with the signature matrix $-\mathcal{J}$. However, it is readily checked that this implies uniqueness for the inverse monodromy problem for canonical differential systems with $m = 2$ in the class of real normalized Hermitians $H(x)$, for any signature matrix J for which there exists a real unitary 2×2 matrix V such that

$$V^*JV = -\mathcal{J}. \tag{1.11}$$

In particular this includes the choice $J = \mathcal{J}$, since

$$\begin{bmatrix} 0 & 1 \\ 1 & 0 \end{bmatrix} \mathcal{J} \begin{bmatrix} 0 & 1 \\ 1 & 0 \end{bmatrix} = -\mathcal{J}.$$

To check the assertion (in the sentence preceding formula (1.11)), let H_1 and H_2 be real normalized solutions of the inverse monodromy for the canonical differential system (1.9) with monodromy matrix $U(\lambda)$ and a signature matrix J which satisfies (1.11). Then $V^*H_1(x)V$ and $V^*H_2(x)V$ are both real normalized solutions of the inverse monodromy problem for the canonical differential system (1.9) with monodromy matrix $V^*U(\lambda)V$ and signature matrix $V^*JV = -\mathcal{J}$. Therefore, by the cited theorem of de Branges,

$$V^*H_1(x)V = V^*H_2(x)V \quad \text{for a.e. } x \in [0, \ell].$$

We shall say that an entire mvf $U(\lambda)$ with transpose $U(\lambda)^\tau$ is symplectic if

$$U(\lambda)^\tau \mathcal{J} U(\lambda) = \mathcal{J} \tag{1.12}$$

for every point $\lambda \in \mathbb{C}$. The next two lemmas clarify the connection between real Hermitians $H(x)$, condition (1.10) and some classes of symplectic mvf's. For more information, see Section 5 of [ArD4].

Lemma 1.1 *Let $U \in \mathcal{E} \cap \mathcal{U}(J)$ with $U(0) = I_2$ and $\overline{J} = -J$. Let $H(x)$ be a solution of the inverse monodromy problem for the canonical differential system* (1.9) *with monodromy matrix $U(\lambda)$ and matrizant $U_x(\lambda) = U(x, \lambda)$, $0 \le x \le \ell$. Then the following two statements are equivalent:*

(1) $U_x(\lambda) = \overline{U_x(\overline{\lambda})}$ *for every point* $x \in [0, \ell]$.

(2) $H(x) = \overline{H(x)}$ *for a.e. point* $x \in [0, \ell]$.

Proof: Formula (1.7) and the assumption $\overline{J} = -J$ clearly imply that

$$\overline{U(x, \overline{\lambda})} = I_2 + i\lambda \int_0^x \overline{U(s, \overline{\lambda})}\, \overline{H(s)} ds J$$

for every point $\lambda \in \mathbb{C}$ and $x \in [0, \ell]$. The rest follows easily upon comparing this formula with (1.7). □

It is readily checked that for any 2×2 matrix A with transpose A^τ,

$$A^\tau \mathcal{J} A = \mathcal{J} \Longleftrightarrow \det A = 1. \tag{1.13}$$

This result and a simple calculation yields the following conclusion:

Lemma 1.2 *Let $U \in \mathcal{E} \cap \mathcal{U}(\pm\mathcal{J})$. Then the following three conditions are equivalent:*

(1) $\det U(\lambda) = 1$.

(2) $U(\lambda)^\tau \mathcal{J} U(\lambda) = \mathcal{J}$.

(3) $\overline{U(\overline{\lambda})} = U(\lambda)$.

In this paper we shall use Theorem 1.1 to establish the uniqueness of the set of normalized solutions to the inverse monodromy problem for two other classes of monodromy matrices and shall given elementary proofs of some important formulas connected with the exponential type of the solution $U_x(\lambda)$ of (1.9) which have been obtained elsewhere by difficult estimates. To define these classes, we need a couple of definitions and some information on the growth of entire J-inner mvf's.

Let

$$\tau_f = \limsup_{\nu\uparrow\infty} \frac{\ln\|f(i\nu)\|}{\nu} \quad \text{and} \quad \delta_f = \limsup_{\nu\uparrow\infty} \frac{\ln|\det f(i\nu)|}{\nu}$$

for $f \in \mathcal{E}$ and let

$$\mathcal{E}(a_1, a_2) = \{f \in \mathcal{E} : \ \tau_{f^\#} = a_1 \quad \text{and} \quad \tau_f = a_2\}.$$

We shall make use of the following facts which we formulate as a lemma for ease of future references.

Lemma 1.3 *Let $U \in \mathcal{E} \cap \mathcal{U}(J)$ (with $J \neq \pm I_2$, as always in this paper). Then:*

(1) *$U \in \mathcal{E}(a_1, a_2)$ with $0 \le a_1 < \infty$ and $0 \le a_2 < \infty$.*

(2) *$U(\lambda)$ is of exponential type with*

$$\text{type}(U) = \max\{a_1, a_2\}.$$

(3) $\det\{U(\lambda)\} = e^{i\lambda(a_1-a_2)}\det\{U(0)\}$.

(4) $\delta_{U^\#} = a_1 - a_2$ *and* $\delta_U = a_2 - a_1$.

(5) $a_2 = 0 \Longleftrightarrow \tau_{U^\#} = \delta_{U^\#}$.

(6) $a_1 = 0 \Longleftrightarrow \tau_U = \delta_U$.

(7) *If $U_1 \in \mathcal{U}(J)$ is a left J-inner divisor of U, i.e., if $U_1^{-1}U \in \mathcal{U}(J)$, then $U_1 \in \mathcal{E}(a_1^{(1)}, a_2^{(1)}) \cap \mathcal{U}(J)$, where*

$$0 \le a_1^{(1)} \le a_1 \quad \textit{and} \quad 0 \le a_2^{(1)} \le a_2.$$

(8) *$e^{i\lambda\alpha}U(\lambda) \in \mathcal{E}(a_1 + \alpha, a_2 - \alpha)) \cap \mathcal{U}(J)$ for every real number α in the interval $-a_1 \le \alpha \le a_2$. Moreover, for $\alpha \in \mathbb{R}$, $e^{i\lambda\alpha}U(\lambda) \notin \mathcal{U}(J)$ if $\alpha \notin [-a_1, a_2]$.*

Proof: Items (1), (2) and (7) are well known. In particular, it suffices to establish the result for $W \in \mathcal{E} \cap \mathcal{U}(j)$, where j is the diagonal signature matrix $1 \oplus (-1)$. But then (1) follows from Theorems 2.3 and 3.8 of [ArD1], upon applying the former to $w_{11}^\#$ and w_{22}, where $w_{ij}(\lambda)$ denotes the ij entry in the 2×2 mvf $W(\lambda)$. Subsequently, (7) follows from Theorem 3.4 of [ArD1], since $\tau_{w_{11}^\#} = \tau_{b_1^{-1}}$ and $\tau_{w_{22}} = \tau_{b_2^{-1}}$, where b_1 and b_2 are inner factors of $s_{11} = (w_{11}^\#)^{-1}$ and $s_{22} = (w_{22})^{-1}$, respectively (see Subsection 3.5 of [ArD1]; the order doesn't matter here because we are dealing with scalar valued functions). Finally, (2) follows from a theorem of M.G. Krein [Kr]; for a more accessible reference, see e.g., Section 4 (especially the discussion of formula (4.43)) in [GoGo] and the notes to that section.

Next, since the inner functions $b_1(\lambda)$ and $b_2(\lambda)$ referred to just above are both entire (see e.g., Lemma 3.9 of [ArD1]) it follows that

$$b_1(\lambda) = e^{i\lambda a_1} b_1(0) \quad \text{and} \quad b_2(\lambda) = e^{i\lambda a_2} b_2(0).$$

Therefore, (3) drops out by Lemma 3.5 of [ArD1], (4) is immediate from (3), whereas (5) and (6) are immediate from (4).

Finally, the cited growth conditions in (8) are easily obtained by calculating the indicated limits. To check that $e^{i\lambda\alpha}U(\lambda)$ is J inner, it is enough to check that its Potapov-Ginsburg transform is a 2×2 inner mvf, which follows from the maximum principle. (More information is available in the work of Simakova; see e.g., Theorem 3.2 of [ArD1], and the rest of that section for additional discussion and references.) □

We remark that items (1), (2) and (7) of Lemma 1.3 are valid for general entire $m \times m$ J inner functions $U(\lambda)$ with $J \neq \pm I_m$ and $m \geq 2$. Analogues of the remaining statements for $m \geq 2$ are discussed in [ArD3] and [ArD4]. In particular, if$J$ is unitarily equivalent to $I_p \oplus (-I_q)$, then it follows from Theorem 4.1 of [ArD4], that

$$a_2 = 0 \Longrightarrow \tau_{U^\#} \leq \delta_{U^\#} \leq p\tau_{U^\#} \tag{1.14}$$

and

$$a_1 = 0 \Longrightarrow \tau_U \leq \delta_U \leq q\tau_U. \tag{1.15}$$

The converse in (1.14) [resp. (1.15)] will hold if $p = 1$ [resp. $q = 1$].

The abstract and the Table of Contents give a pretty good idea of what this paper is about. In addition, the following table may serve as a useful road map to the main uniqueness theorems which are considered in this paper. In the table it is always assumed that $U \in \mathcal{E} \cap \mathcal{U}(J)$ with $U(0) = I_2$ and $J \neq I_2$ and that the Hermitian $H(x)$ is summable and positive semidefinite a.e. The table is then read as follows: If the monodromy matrix $U(\lambda)$ meets the extra condition in column one and the signature matrix J meets the extra condition in column two, then there exists one and only one solution $H(x)$ of the inverse monodromy problem for the canonical differential system (1.9) with monodromy matrix $U(\lambda)$ in the class of Hermitians delineated in column three. The relevant theorems and comments are indicated in column four.

The condition $U \in \mathcal{E}(a_1, a_2) \cap \mathcal{U}(J)$ with $a_1a_2 = 0$ holds if and only if

$$\text{type}\{U\} = \text{type}\{det\ U\};$$

see Lemma 4.1. (For yet other criteria, see Theorems 2.4 and 3.2.) In the opposite case, i.e., if $a_1a_2 > 0$, then there are infinitely many normalized solutions of the

Table 1 Classes of Uniqueness

	$U(\lambda)$	J	$H(x)$	Reference
1.	$U(\lambda) = \overline{U(\overline{\lambda})}$	$J = \mathcal{J}$	trace$\{H(x)\} = 1$ $H(x) = \overline{H(x)}$	Theorem 1.1 (de Branges)
2.	$U(\lambda) = \overline{U(\overline{\lambda})}$	$V^*JV = \mathcal{J}$ V real unitary	trace$\{H(x)\} = 1$ $H(x) = \overline{H(x)}$	Simple variant of Theorem 1.1
3.	$U \in \mathcal{E}(a, 0)$	–	trace$\{H(x)\} = 1$	Theorem 2.3
4.	$U \in \mathcal{E}(0, a)$	–	trace$\{H(x)\} = 1$	Theorem 3.1
5.	$U \in \mathcal{E}(a, a)$	–	trace $H(x) = 1$ det $U_x(\lambda) = 1$	Theorem 2.2 and (3) of Lemma 1.3

inverse monodromy problem under consideration (unless additional constraints are imposed, as in rows 1, 2 and 5). A description of these solutions is furnished in Theorem 4.2. In addition, in Theorem 7.2 we show that if $U \in \mathcal{E}(a, 0) \cap \mathcal{U}(J)$ [resp. $U \in \mathcal{E}(0, a) \cap \mathcal{U}(J)$] with $U(0) = I_2$, then there is at most one solution $H(x)$ of the inverse monodromy problem with trace$\{H(x)J\} = 1$ [resp. trace$\{H(x)J\} = -1$] a.e. A number of simple examples are considered in Section 6 to illustrate the three classes of J-inner mvf's which are introduced there. This section also contains an instructive calculation that serves to verify that the space $\mathcal{H}(U) \cap L_2^2(\mathbb{R})$ is indeed dense in $\mathcal{H}(U)$ for the right A-regular J-inner mvf $U(\lambda)$ considered in Example 3, as it should be.

Additional information on the inverse monodromy problem for canonical integral and differential systems of order $m \times m$ can be found in [ArD1]–[ArD4]. However, the analysis there is restricted primarily to the case of strongly regular monodromy matrices. For additional perspective and references, see the appendix of the monograph [GK] by Gohberg and Krein.

Conventions

The symbol J will always denote a 2×2 signature matrix with one eigenvalue equal to $+1$ and one eigenvalue equal to -1. The solution $H(x)$ of the inverse monodromy problem is always assumed to be summable on the given interval $[0, \ell]$ with $H(x) \geq 0$ a.e.

2 Monodromy Matrices with Zero Type in the Upper Half Plane

Lemma 2.1 *Let* $U \in \mathcal{E}(a, 0) \cap \mathcal{U}(J)$ *with* $U(0) = I_2$. *Let* $H(x)$ *be a solution of the inverse monodromy problem for the canonical differential system* (1.9) *with* *monodromy matrix* $U(\lambda)$, *let* $U_x(\lambda) = U(x, \lambda)$ *denote the corresponding matrizant and let*

$$\tau(x) = \tau_{U_x^\#} \quad for \quad x \in [0, \ell]. \tag{2.1}$$

Then:

(1) $U_x \in \mathcal{E}(\tau(x), 0) \cap \mathcal{U}(J)$ *for every* $x \in [0, \ell]$.

(2) $\text{trace}\{H(x)J\} \geq 0$ *for a.e.* $x \in [0, \ell]$.

(3) $\tau(x) = \int_0^x \text{trace}\{H(s)J\}ds \quad for\ every \quad x \in [0, \ell]$,

i.e., $\tau(x)$ *is an absolutely continuous nondecreasing function of* x *on* $[0, \ell]$ *with* $\tau(0) = 0$ *and* $\tau(\ell) = a$.

Proof: The assumption $U \in \mathcal{E} \cap \mathcal{U}(J)$ implies that det $U(\lambda)$ is an invertible entire function of exponential type and hence that

$$\det U(\lambda) = e^{i\alpha\lambda}$$

for some constant α. Therefore, since $U(0) = I_2$,

$$i\alpha = (\det U)'(0) = \text{trace}\{U'(0)\},$$

i.e.,

$$\det U(\lambda) = \exp\{\lambda \text{ trace } U'(0)\}.$$

Applying this formula to the solution $U_x(\lambda)$ of (1.7), we obtain[2]

$$\det U_x(\lambda) = \exp\left[i\lambda \int_0^x \text{trace}\{H(s)J\}ds\right].$$

Therefore, by item (3) of Lemma 1.3,

$$\tau(x) = \tau_{U_x^\#} = \int_0^x \text{trace}\{H(s)J\}ds.$$

Item (5) of the same lemma guarantees that (1) holds and also that $\tau(x)$ is nondecreasing, since

$$U_{x_1}^{-1}U_{x_2} \in \mathcal{E} \cap \mathcal{U}(J) \text{ for } 0 \leq x_1 \leq x_2 \leq \ell.$$

The remaining assertions of the lemma are now selfevident. □

---

[2]This formula is valid for general $m \times m$ canonical differential systems with $m \geq 2$; see e.g. Theorem 0.1 of [ArD3].

Corollary: *Let $U(\lambda)$, $H(x)$ and $\tau(x)$ be as in Lemma 2.1 and let $J = \mathcal{J}$ and*

$$H(x) = \begin{bmatrix} \alpha(x) & \beta(x) \\ \overline{\beta(x)} & \gamma(x) \end{bmatrix}. \tag{2.2}$$

Then

$$\tau'(x) = \frac{\beta(x) - \overline{\beta(x)}}{i}. \tag{2.3}$$

Proof: By Lemma 2.1,

$$\tau'(x) = \text{trace}\{H(x)\mathcal{J}\} \text{ for a.e. } x \in [0, \ell].$$

But that is readily seen to coincide with the formula (2.3). □

Lemma 2.2 *Let $U \in \mathcal{E}(a, 0) \cap \mathcal{U}(J)$ with $U(0) = I_2$. Let $H(x)$ be a normalized solution of the inverse monodromy problem for the canonical differential system* (1.9) *with monodromy matrix $U(\lambda)$. Then*

$$\det H(x) = 0 \textit{ for a.e. } x \in [0, \ell]. \tag{2.4}$$

Proof: It suffices to consider the case $J = \mathcal{J}$. Let

$$\tau(x) = \tau_{U_x^\#} = \text{type}(U_x),$$
$$\tau_1(x) = \tau(x)/2$$

and let

$$U_1(x, \lambda) = \exp\{-i\lambda\tau_1(x)\}U(x, \lambda).$$

Then, since $\tau_{U_x} = 0$, it is readily checked that

$$U_1(x, \cdot) \in \mathcal{E}(\tau_1(x), \tau_1(x)) \cap \mathcal{U}(\mathcal{J})$$

and

$$\frac{\partial U_1}{\partial x} = i\lambda U_1(x, \lambda)H_1(x)\mathcal{J},$$

where

$$H_1(x) = H(x) - \tau_1'(x)\mathcal{J}.$$

Therefore, by (2.3),

$$H_1(x) = \begin{bmatrix} \alpha(x) & \beta(x) \\ \overline{\beta}(x) & \gamma(x) \end{bmatrix} - \frac{1}{2}\begin{bmatrix} 0 & (\beta - \overline{\beta})(x) \\ (\overline{\beta} - \beta)(x) & 0 \end{bmatrix}$$
$$= \begin{bmatrix} \alpha(x) & (\beta + \overline{\beta})(x)/2 \\ (\beta + \overline{\beta})(x)/2 & \gamma(x) \end{bmatrix} = \overline{H_1(x)}$$

and

$$\det\ H_1(x) - (\tau_1'(x))^2 = \det\ H(x). \tag{2.5}$$

Thus, as $H(x) \geq 0$, it follows that

$$H_1(x) \geq 0 \quad \text{for a.e.} \quad x \in [0, \ell]$$

and

$$f(x) = \sqrt{\det\ H_1(x)} \geq \tau_1'(x) \quad \text{for a.e.} \quad x \in [0, \ell]. \tag{2.6}$$

The next objective is to prove the opposite inequality. To this end, let

$$\beta_1(x) = (\beta + \overline{\beta})(x)/2$$

and

$$H_2(x) = H_1(x) + f(x)\mathcal{J}.$$

Then

$$H_2(x) = \begin{bmatrix} \alpha(x) & (\beta_1 + if)(x) \\ (\beta_1 - if)(x) & \gamma(x) \end{bmatrix}$$

and thus, as

$$\begin{aligned} \det\ H_2(x) &= \alpha(x)\gamma(x) - \beta_1(x)^2 - f(x)^2 \\ &= \det\ H_1(x) - f(x)^2 = 0 \end{aligned} \tag{2.7}$$

for a.e. point $x \in [0, \ell]$, it follows that

$$H_2(x) \geq 0 \quad \text{for a.e.} \quad x \in [0, \ell].$$

Let $U_2(x, \lambda)$ denote the matrizant corresponding to $H_2(x)$ and let

$$\varphi(x) = \int_0^x f(s)ds.$$

Then it is readily checked that

$$\frac{\partial}{\partial x}[\exp\{-i\lambda\varphi(x)\}U_2(x,\lambda)] = i\lambda\ \exp\{-i\lambda\varphi(x)\}U_2(x)H_1(x)$$

and hence that

$$e^{-i\lambda\varphi(x)}U_2(x,\lambda) = U_1(x,\lambda) \quad \text{for every} \quad x \in [0, \ell].$$

Thus,

$$U_2(x, \cdot) \in \mathcal{E}(\tau_1(x) + \varphi(x), \tau_1(x) - \varphi(x))$$

for every $x \in [0, \ell]$. In particular, this implies that the absolutely continuous function $\tau_1(x) - \varphi(x)$ is nondecreasing on the interval $[0, \ell]$ and hence that

$$\tau_1'(x) \geq f(x) \quad \text{for a.e.} \quad x \in [0, \ell]. \tag{2.8}$$

Therefore, upon combining inequalities (2.6) and (2.8), we obtain first that

$$\tau_1'(x) = f(x) = \sqrt{\det\ H_1(x)}$$

and then that

$$H_2(x) = H_1(x) + f(x)\mathcal{J} = H(x),$$

which, in view of (2.7), serves to complete the proof of the lemma. □

Theorem 2.1 *Let $U \in \mathcal{E} \cap \mathcal{U}(\pm\mathcal{J})$ with $U(0) = I_2$. Let $H(x) = \overline{H(x)}$ be a real normalized solution of the inverse monodromy problem for the canonical differential system* (1.9) *with monodromy matrix $U(\lambda)$, let $U_x(\lambda) = U(x, \lambda)$ be the corresponding matrizant and let $\tau(x) =$* type *U_x. Then*

$$\tau(x) = \tau_{U_x^{\#}} = \tau_{U_x} \tag{2.9}$$

and

$$\tau(x) = \int_0^x \sqrt{\det\ H(s)}ds \tag{2.10}$$

for every $x \in [0, \ell]$.

Proof: Assertion (2.9) is immediate from Lemmas 1.1 and 1.3. To verify (2.10), let

$$\widetilde{U}(x, \lambda) = e^{i\lambda\tau(x)}U(x, \lambda).$$

Then $\widetilde{U}_x \in \mathcal{E}(2\tau(x), 0) \cap \mathcal{U}(\pm\mathcal{J})$ and, as

$$\left(\frac{\partial \widetilde{U}_x}{\partial\lambda}\right)(0) = i\tau(x)I_2 + \left(\frac{\partial U_x}{\partial\lambda}\right)(0),$$

the corresponding Hermitian $\widetilde{H}(x)$ satisfies the equation

$$\int_0^x \widetilde{H}(s)ds = \tau(x)(\pm\mathcal{J}) + \int_0^x H(s)ds,$$

for every $x \in [0, \ell]$. Therefore, since $\widetilde{H}(x)$ is a normalized solution of the inverse monodromy problem for a monodromy matrix $\widetilde{U} \in \mathcal{E}(a, 0)$, it follows from Lemma 2.2 that

$$\det\ \widetilde{H}(x) = 0 \quad \text{for every} \quad x \in [0, \ell]$$

and hence, by direct calculation, just as in the Proof of Lemma 2.2, that (2.10) holds. □

Theorem 2.2 *Let $U \in \mathcal{E}(a, a) \cap \mathcal{U}(J)$ with $U(0) = I_2$. Then there is exactly one normalized solution $H(x)$ of the inverse monodromy problem for the canonical differential system* (1.9) *with monodromy matrix $U(\lambda)$ such that the corresponding matrizant $U_x(\lambda)$ meets condition* (2.9) *for every $x \in [0, \ell]$. In this instance, formula* (2.10) *also holds for every $x \in [0, \ell]$.*

Proof: Let V be a unitary matrix such that $V^*JV = \mathcal{J}$. Then $\widetilde{U}_x(\lambda) = V^*U_x(\lambda)V$ belongs to $\mathcal{E}(\tau(x), \tau(x)) \cap \mathcal{U}(\mathcal{J})$ for every $x \in [0, \ell]$ and is the matrizant of the canonical differential system (1.9) with monodromy matrix $V^*U(\lambda)V$ and normalized Hermitian $\widetilde{H}(x) = V^*H(x)V$. Moreover, by Lemma 1.3, $\det\{\widetilde{U}_x(\lambda)\} = 1$

and hence, by Lemma 1.2, $\widetilde{U}_x(\lambda)$ meets condition (1.10). Thus $\widetilde{H}(x)$ is real and so, by Theorem 2.1,

$$\tau(x) = \int_0^x \sqrt{\det\ \widetilde{H}(s)ds} = \int_0^x \sqrt{\det\ H(s)ds}$$

for every point $x \in [0, \ell]$, as claimed. The uniqueness of $\widetilde{H}(x)$, and so too of $H(x)$, follows from Theorem 1.1. □

Theorem 2.3 *Let* $U \in \mathcal{E}(a, 0) \cap \mathcal{U}(J)$ *with* $U(0) = I_2$. *Then there exists exactly one normalized solution* $H(x)$ *of the inverse monodromy problem for the canonical differential system* (1.9) *with monodromy matrix* $U(\lambda)$. *Moreover, if* $U_x(\lambda) = U(x, \lambda)$, $0 \leq x \leq \ell$, *denotes the corresponding matrizant and* $\tau(x) =$ type(U_x), *then:*

(1) $U_x \in \mathcal{E}(\tau(x), 0) \cap \mathcal{U}(J)$ *for every* $x \in [0, \ell]$.

(2) $\tau(x) = \int_0^x \text{trace}\{H(s)J\}ds$ *for every* $x \in [0, \ell]$.

(3) $\text{trace}\{H(x)J\} \geq 0$ *for a.e.* $x \in [0, \ell]$.

(4) $\det\ H(x) = 0$ *for a.e.* $x \in [0, \ell]$.

Proof: Let us suppose first that $J = \mathcal{J}$, let $U_x(\lambda) = U(x, \lambda)$, $0 \leq x \leq \ell$, denote the matrizant corresponding to $H(x)$ and let

$$2\tau_1(x) = \tau(x) = \text{type}(U_x)$$

for every point $x \in [0, \ell]$. Then, by Lemma 2.1, $\tau(x)$ is absolutely continuous with

$$\tau'(x) = \text{trace}\{H(x)\mathcal{J}\}$$

and hence, upon setting

$$U_1(x, \lambda) = e^{-i\lambda\tau_1(x)} U(x, \lambda),$$

it follows from the analysis in Lemma 2.2 that $U_1(x, \lambda)$ is the matrizant of a canonical system with Hermitian

$$H_1(x) = H(x) - \tau_1'(x)\mathcal{J} \tag{2.11}$$

that is both real and normalized. By de Branges' theorem (Theorem 1.1), there is only one such Hermitian and hence only one matrizant $U_1(x, \lambda)$. Therefore, since

$$\tau_1(x) = \text{type}\ U_1(x, \cdot),$$

there is only one such function $\tau_1(x)$. Thus $H(x)$ is uniquely defined by formula (2.11).

To this point we have established the existence of a unique normalized solution $H(x)$ to the inverse monodromy problem when the monodromy matrix $U \in$

$\mathcal{E}(a, 0) \cap \mathcal{U}(\mathcal{J})$. Suppose next that $H_1(x)$ and $H_2(x)$ are normalized solutions to the inverse monodromy problem when the monodromy matrix $U \in \mathcal{E}(a, 0) \cap \mathcal{U}(J)$ and $V^*JV = \mathcal{J}$ for some unitary matrix V. But then $V^*H_1(x)V$ and $V^*H_2(x)V$ are normalized solutions of the inverse monodromy problem with monodromy matrix $V^*U(\lambda)V \in \mathcal{E}(a, 0) \cap \mathcal{U}(\mathcal{J})$. Therefore $V^*H_1(x)V = V^*H_2(x)V$ for a.e. $x \in [0, \ell]$ by the preceding discussion. This clearly establishes the asserted uniqueness. The remaining four items in the statement of the theorem are immediate from the preceding two lemmas. □

Remark 2.1 Conditions (3) and (4) in Theorem 2.3 imply that there exists a 2×1 measurable mvf $h(x)$ with square summable entries on $[0, \ell]$ such that

$$H(x) = h(x)h(x)^* \quad \text{and} \quad h(x)^*Jh(x) \geq 0 \tag{2.12}$$

for a.e. $x \in [0, \ell]$. Moreover, $H(x)$ is normalized if and only if $h(x)^*h(x) = 1$ for a.e. $x \in [0, \ell]$.

The next theorem is a partial converse to Theorem 2.3.

Theorem 2.4 *Let $U \in \mathcal{E} \cap \mathcal{U}(J)$ with $U(0) = I_2$ and let $H(x)$ be a normalized solution of the inverse monodromy problem for the canonical differential system* (1.9) *with monodromy matrix $U(\lambda)$ such that*

$$\text{trace}\{H(x)J\} \geq 0 \quad \textit{and} \quad \det\, H(x) = 0 \quad \textit{for a.e.} \quad x \in [0, \ell].$$

Then $U \in \mathcal{E}(a, 0)$, where

$$a = \text{type}(U) = \int_0^\ell \text{trace}\{H(s)J\}ds.$$

Proof: We may assume without loss of generality that $J = \mathcal{J}$. Then, upon writing

$$H(x) = \begin{bmatrix} \alpha(x) & \beta(x) \\ \overline{\beta(x)} & \gamma(x) \end{bmatrix} \quad \text{and} \quad H_1(x) = \begin{bmatrix} \alpha(x) & \beta_1(x) \\ \beta_1(x) & \gamma(x) \end{bmatrix}$$

with $\beta_1(x) = \frac{\beta(x)+\overline{\beta(x)}}{2}$, we obtain

$$\text{trace}\{H(x)\mathcal{J}\} = \frac{\beta(x) - \overline{\beta(x)}}{i}$$

and

$$\det\, H_1(x) = \det\, H(x) + \frac{1}{4}[\text{trace}\{H(x)\mathcal{J}\}]^2.$$

This proves that

$$\det\, H_1(x) \geq 0 \quad \text{for a.e.} \quad x \in [0, \ell]$$

and hence we see that

$$H_1(x) \geq 0, \quad H_1(x) = \overline{H_1(x)} \quad \text{and} \quad \text{trace}\, H_1(x) = 1$$

for a.e. $x \in [0, \ell]$. Let $U_1(x, \lambda)$, $0 \le x \le \ell$, denote the matrizant corresponding to $H_1(x)$. Then, by Theorem 2.1,

$$U_1(x, \cdot) \in \mathcal{E}(\varphi(x), \varphi(x)),$$

where

$$\varphi(x) = \int_0^x \sqrt{\det H_1(s)}ds = \int_0^x \frac{\beta(s) - \overline{\beta}(s)}{2i} ds.$$

Thus, the mvf

$$\Omega_x(\lambda) = \Omega(x, \lambda) = e^{i\lambda\varphi(x)} U_1(x, \lambda)$$

enjoys the following properties:

$$\begin{aligned} -i\left(\frac{\partial \Omega_x}{\partial \lambda}\right)(0)J &= \varphi(x)\mathcal{J} + \int_0^x H_1(s)ds \\ &= \int_0^x \{H_1(s) + \varphi'(s)\mathcal{J}\}ds \\ &= \int_0^x H(s)ds. \end{aligned}$$

But this serves to identify $\Omega_x(\lambda)$ as the matrizant corresponding to $H(x)$. Therefore, the given monodromy matrix

$$U(\lambda) = \Omega_{\varphi(\ell)}(\lambda)$$

and hence as

$$\Omega_{\varphi(x)} \in \mathcal{E}(2\varphi(x), 0)$$

it follows that

$$U(\lambda) \in \mathcal{E}(a, 0),$$

where

$$a = 2\varphi(\ell) = \int_0^\ell \operatorname{trace}\{H(s)\mathcal{J}\}ds.$$

□

3 Monodromy Matrices with Zero Type in the Lower Half Plane

In this section we restate a number of the main conclusions from Section 2 for monodromy matrices $U(\lambda)$ which belong to the class $\mathcal{E}(0, a) \cap \mathcal{U}(J)$ instead of to the class $\mathcal{E}(a, 0) \cap \mathcal{U}(J)$. We do not present proofs because all the stated results are easily obtained from their counterparts in Section 2 by a change of notation. The basic observation is that if

$$\widetilde{U}(\lambda) = U(-\lambda),$$

then

$$U \in \mathcal{E}(0,a) \cap \mathcal{U}(J) \Longleftrightarrow \widetilde{U} \in \mathcal{E}(a,0) \cap \mathcal{U}(-J).$$

Moreover, in the same vein,

$$U(x,\lambda) = I_p + i\lambda \int_0^x U(s,\lambda)H(s)ds\,J, \qquad 0 \le x \le \ell,$$

if and only if

$$U(x,-\lambda) = I_p + i\lambda \int_0^x U(s,-\lambda)H(s)ds(-J), \qquad 0 \le x \le \ell.$$

Lemma 3.1 *Let $U \in \mathcal{E}(0,a) \cap \mathcal{U}(J)$ with $U(0) = I_2$. Let $H(x)$ be a solution of the inverse monodromy problem for the canonical differential system* (1.9) *with monodromy matrix $U(\lambda)$, let $U_x(\lambda) = U(x,\lambda)$ denote the corresponding matrizant and let*

(3.1) $$\tau(x) = \tau_{U_x} \quad \text{for} \quad x \in [0,\ell].$$

Then:

(1) $U_x \in \mathcal{E}(0,\tau(x)) \cap \mathcal{U}(J)$ *for every* $x \in [0,\ell]$.

(2) $\operatorname{trace}\{H(x)J\} \le 0$ *for a.e.* $x \in [0,\ell]$.

(3) $$\tau(x) = -\int_0^x \operatorname{trace}\{H(s)J\}ds \quad \text{for every} \quad x \in [0,\ell],$$

i.e., $\tau(x)$ is an absolutely continuous nondecreasing function of x on $[0,\ell]$ *with $\tau(0) = 0$ and $\tau(\ell) = a$.*

Corollary: *Let $U(\lambda)$ and $H(x)$ be as in Lemma* 3.1 *and let $J = \mathcal{J}$ and*

(3.2) $$H(x) = \begin{bmatrix} \alpha(x) & \beta(x) \\ \overline{\beta(x)} & \gamma(x) \end{bmatrix}.$$

Then

(3.3) $$\tau'(x) = -\frac{\beta(x) - \overline{\beta(x)}}{i}.$$

Lemma 3.2 *Let $U \in \mathcal{E}(0,a) \cap \mathcal{U}(J)$ with $U(0) = I_2$. Let $H(x)$ be a normalized solution of the inverse monodromy problem for the canonical differential system* (1.9) *with monodromy matrix $U(\lambda)$. Then*

(3.4) $$\det H(x) = 0 \quad \text{for a.e.} \quad x \in [0,\ell].$$

Theorem 3.1 *Let $U \in \mathcal{E}(0,a) \cap \mathcal{U}(J)$ with $U(0) = I_2$. Then there exists exactly one normalized solution $H(x)$ of the inverse monodromy problem for the canonical differential system* (1.9) *with monodromy matrix $U(\lambda)$. Moreover,* *if* $U_x(\lambda) = U(x,\lambda)$, $0 \le x \le \ell$, *denotes the corresponding matrizant and* $\tau(x) = \mathrm{type}(U_x)$, *then:*

(1) $U_x \in \mathcal{E}(0,\tau(x)) \cap \mathcal{U}(J)$ *for every* $x \in [0,\ell]$.

(2) $\tau(x) = -\int_0^x \mathrm{trace}\{H(s)J\}ds$ *for every* $x \in [0,\ell]$.

(3) $\mathrm{trace}\{H(x)J\} \le 0$ *for a.e.* $x \in [0,\ell]$.

(4) $\det\ H(x) = 0$ *for a.e.* $x \in [0,\ell]$.

Remark 3.1 Conditions (3) and (4) in Theorem 3.1 imply that there exists a 2×1 measurable mvf $h(x)$ with square summable entries on $[0,\ell]$ such that

$$H(x) = h(x)h(x)^* \text{ and } h(x)^* J h(x) \le 0 \tag{3.5}$$

for a.e. $x \in [0,\ell]$. Moreover, $H(x)$ is normalized if and only if $h(x)^*h(x) = 1$ for a.e. $x \in [0,\ell]$.

The next theorem is a partial converse to Theorem 3.1.

Theorem 3.2 *Let* $U \in \mathcal{E} \cap \mathcal{U}(J)$ *with* $U(0) = I_2$ *and let* $H(x)$ *be a normalized solution of the inverse monodromy problem for the canonical differential system* (1.9) *with monodromy matrix* $U(\lambda)$ *such that*

$$\mathrm{trace}\{H(x)J\} \le 0 \quad \text{and} \quad \det\ H(x) = 0 \quad \text{for a.e.} \quad x \in [0,\ell].$$

Then $U \in \mathcal{E}(0,a)$, *where*

$$a = \mathrm{type}(U) = -\int_0^a \mathrm{trace}\{H(s)J\}ds.$$

4 Monodromy Matrices with Non Zero Type in Both Half Planes

In this section we shall obtain some conclusions for monodromy matrices $U \in \mathcal{E}(a_1,a_2) \cap \mathcal{U}(J)$ with $U(0) = I_2$ and $a_1a_2 > 0$. In this setting the exponential type of $U(\lambda)$ is no longer equal to the exponential type of $\det\{U(\lambda)\}$:

Lemma 4.1 *Let* $U \in \mathcal{E}(a_1,a_2) \cap \mathcal{U}(J)$ *with* $U(0) = I_2$. *Then*

$$\mathrm{type}\{U\} = \mathrm{type}\{\det\ U\} \tag{4.1}$$

if and only if $a_1a_2 = 0$, *i.e., if and only if either* $a_1 = 0$ *or* $a_2 = 0$.

Proof: By Lemma 1.3, the equality (4.1) will hold if and only if $\max\{a_1,a_2\} = |a_1 - a_2|$. But it is readily seen that this is only possible if either $a_1 = 0$ or $a_2 = 0$. □

Theorem 4.1 *Let $U \in \mathcal{E}(a_1, a_2) \cap \mathcal{U}(J)$ with $U(0) = I_2$ and $a_1 a_2 > 0$. Let $H(x)$, $0 \le x \le \ell$, be a normalized solution of the inverse monodromy problem for the canonical differential system* (1.9) *with monodromy matrix $U(\lambda)$. Let $U_x(\lambda) = U(x, \lambda)$ denote the corresponding matrizant and let*

$$\tau_1(x) = \tau_{U_x^\#} \quad \text{and} \quad \tau_2(x) = \tau_{U_x} \quad \text{for} \quad 0 \le x \le \ell.$$

Then:

(1) *$\tau_1(x)$ and $\tau_2(x)$ are absolutely continuous nondecreasing functions of x on the interval* $[0, \ell]$ *with $\tau_j(0) = 0$ and $\tau_j(\ell) = a_j$ for $j = 1, 2$.*

(2) *The mvf*

$$U_+(x, \lambda) = e^{i\lambda\tau_2(x)} U(x, \lambda) \qquad [\textit{resp.}\ U_-(x, \lambda) = e^{-i\lambda\tau_1(x)} U(x, \lambda)]$$

is the matrizant of the canonical differential system (1.9) *with normalized Hermitian*

$$H_+(x) = H(x) + \tau_2'(x) J \qquad [\textit{resp.}\ H_-(x) = H(x) - \tau_1'(x) J]$$

and monodromy matrix

$$U_+(\lambda) = e^{i\lambda a_2} U(\lambda) \qquad [\textit{resp.}\ U_-(\lambda) = e^{-i\lambda a_1} U(\lambda)].$$

(3) *The monodromy matrix*

$$U_+ \in \mathcal{E}(a, 0) \cap \mathcal{U}(J) \qquad [\textit{resp.}\ U_- \in \mathcal{E}(0, a) \cap \mathcal{U}(J)],$$

where $a = a_1 + a_2$.

(4) *The matrizant*

$$U_+(x, \cdot) \in \mathcal{E}(\tau_+(x), 0) \cap \mathcal{U}(J) \qquad [\textit{resp.}\ U_-(x, \cdot) \in \mathcal{E}(0, \tau_-(x)) \cap \mathcal{U}(J)],$$

where

$$\tau_+(x) = \tau_-(x) = \tau_1(x) + \tau_2(x).$$

(5) $\tau_1(x) - \tau_2(x) = \int_0^x \operatorname{trace}\{H(s)J\} ds$ *for* $0 \le x \le \ell$.

Proof: By (7) of Lemma 1.3, $\tau_1(x)$ and $\tau_2(x)$ are nondecreasing functions of x on the interval $[0, \ell]$. Moreover, by (8) of Lemma 1.3,

$$U_+(x, \cdot) \in \mathcal{E}(\tau_1(x) + \tau_2(x), 0) \cap \mathcal{U}(J).$$

Therefore, by Lemma 2.1,

$$\tau_+(x) = \tau_1(x) + \tau_2(x)$$

is absolutely continuous with respect to x on the interval $[0, \ell]$. Thus, the same holds true for $\tau_1(x)$ and $\tau_2(x)$, since they are both nondecreasing.

Next, it is readily checked by straightforward differentiation that $U_+(x, \lambda)$ is the matrizant corresponding to the canonical differential system (1.9) with normalized Hermitian $H_+(x)$. Consequently, by Theorem 2.2,

$$\begin{aligned} \tau_1(x) + \tau_2(x) &= \int_0^x \operatorname{trace}\{H_+(s)J\}ds \\ &= \int_0^x \operatorname{trace}\{H(s)J\}dx + 2\tau_2(x), \end{aligned}$$

which yields (5). The remaining assertions in (2), (3) and (4) that are connected with $U_-(\lambda)$, $U_-(x, \lambda)$ and $H_-(x)$ are verified in much the same way, to complete the proof. □

Notice that formula (3) of Lemma 2.1 and formula (3) of Lemma 3.1 are special cases of formula (4.2).

Theorem 4.2 *Let $U \in \mathcal{E}(a_1, a_2) \cap \mathcal{U}(J)$ with $U(0) = I_2$ and $a_1 a_2 > 0$. Let*

$$U_+(\lambda) = e^{i\lambda a_2}U(\lambda) \quad and \quad U_-(\lambda) = e^{-i\lambda a_1}U(\lambda).$$

Then $U_+ \in \mathcal{E}(a, 0) \cap \mathcal{U}(J)$, $U_+(0) = I_2$, $U_- \in \mathcal{E}(0, a) \cap \mathcal{U}(J)$, $U_-(0) = I_2$, where $a = a_1 + a_2$.

Let $H_+(x)$ and $H_-(x)$ be the unique normalized solutions of the inverse monodromy problem for canonical differential systems of the form (1.9) with monodromy matrices $U_+(\lambda)$ and $U_-(\lambda)$, respectively. Let $U_+(x, \lambda)$ and $U_-(x, \lambda)$ denote the matrizants of these systems and let $\tau(x) = \tau_+(x) = \operatorname{type}(U_+(x, \cdot))$ and $\tau_-(x) = \operatorname{type}(U_-(x, \cdot))$. Then:

(1) $\tau(x) = \tau_+(x) = \tau_-(x)$ *for* $0 \le x \le \ell$.

(2) *$\tau(x)$ is an absolutely continuous nondecreasing function of x on the interval $[0, \ell]$ with $\tau(0) = 0$ and $\tau(\ell) = a_1 + a_2$.*

(3) $U_+(x, \lambda) = e^{i\lambda\tau(x)}U_-(x, \lambda)$ *for* $0 \le x \le \ell$.

(4) $H_+(x) = H_-(x) + \tau'(x)J$ *for a.e.* $x \in [0, \ell]$.

Let $t_1(x)$ be an absolutely continuous function of x on the interval $[0, \ell]$ such that

$$\begin{aligned} &0 \le t_1'(x) \le \tau'(x) \quad a.e. \text{ on } \quad [0, \ell], \\ &t_1(0) = 0 \quad and \quad t_1(\ell) = a_1. \end{aligned}$$

Let $t_2(x) = \tau(x) - t_1(x)$. Then

$$H(x) = H_+(x) - t_2'(x)J = H_-(x) + t_1'(x)J$$

is a normalized solution of the inverse monodromy problem for the canonical differential system (1.9) with monodromy matrix $U(\lambda)$.

Proof: The first four assertions (1)–(4) are immediate from Theorem 4.1. Next, the assumption that $0 \leq t_1'(x) \leq \tau'(x)$ guarantees that

$$e^{i\lambda t_1(x)} U_-(x, \lambda) \in \mathcal{E}(t_1(x), \tau(x) - t_1(x)) \cap \mathcal{U}(J) \quad for\ every \quad x \in [0, \ell].$$

Then, by straightforward differentiation, it is readily checked that $e^{i\lambda t_1(x)} U_-(x, \lambda)$, [resp. $e^{-i\lambda t_2(x)} U_+(x, \lambda)$] is the matrizant of the canonical differential system (1.9) with normalized Hermitian $H_-(x) + t_1'(x)J$ [resp. $H_+(x) - t_2'(x)J$] and monodromy matrix $e^{i\lambda a_1} U_-(\lambda) = U(\lambda)$ [resp. $e^{-i\lambda a_2} U_+(\lambda) = U(\lambda)$]. □

Theorem 4.2 gives a description of the set of all normalized solutions of the inverse monodromy problem for the canonical differential system (1.9) when the monodromy matrix $U \in \mathcal{E}(a_1, a_2) \cap \mathcal{U}(J)$ with $a_1 a_2 > 0$. Some related type estimates (for the case $m = 2$) are discussed on pages 127–130 of de Branges' book [dB7]. A number of analogues of Theorem 4.2 for the case of $m \times m$ canonical systems with $m \geq 2$ are provided in [ArD4] under the added assumption that the monodromy matrix is strongly regular; see especially Theorem 4.9.

Theorem 4.3 *Let $U \in \mathcal{E} \cap \mathcal{U}(J)$ with $U(0) = I_2$. Then a normalized solution of the inverse monodromy problem for the canonical differential system* (1.9) *with monodromy matrix $U(\lambda)$ is unique if and only if condition* (4.1) *holds.*

Proof: By construction, the functions $t_j(x)$ are absolutely continuous nondecreasing functions of x on $[0, \ell]$ with $t_j(0) = 0$ and $t_j(\ell) = a_j$ for $j = 1, 2$ and

$$t_1(x) + t_2(x) = \tau(x).$$

In particular, we will have uniqueness if and only if either $a_1 = 0$ (which forces $t_1(x) \equiv 0$) or $a_2 = 0$ (which forces $t_2(x) \equiv 0$). But in view of Lemma 4.1, this is equivalent to condition (4.1). □

We remark that Theorem 4.3 yields an analogue of the condition for uniqueness in the class of normalized solutions of the inverse monodromy problem for $m \times m$ canonical differential systems in the definite case with $J = I_m$ that is presented in Theorem 0.2 of [ArD3]. The latter is a consequence of the characterization by Brodskii and Kisilevskii of unicellular dissipative Volterra operators; see Theorem 30.1 of Brodskii [Br].

5 Reparametrizations

In this section we shall establish analogues of a number of theorems from Sections 2 and 3 for nonnormalized solutions $H(x)$ of the inverse monodromy problem for the canonical differential system (1.9). In particular, we shall replace the assumption that trace $H(x) = 1$ a.e., by the less restrictive pair of constraints that

(5.1) $$H(x) \in L_1^{2\times 2}[0, \ell] \quad \text{and} \quad \text{trace} H(x) > 0 \quad \text{for a.e.} \quad x \in [0, \ell].$$

Lemma 5.1 *Let* $U \in \mathcal{E} \cap \mathcal{U}(J)$ *with* $U(0) = I_2$ *and let* $H(x)$ *be a solution of the inverse monodromy problem for the canonical differential system* (1.9) *with monodromy matrix* $U(\lambda)$. *Suppose further that* trace $H(x) > 0$ *for a.e.* $x \in [0, \ell]$ *and let*

$$\varphi_1(x) = \int_0^x \text{trace}\{H(s)\}ds \quad \textit{for} \quad x \in [0, \ell]. \tag{5.2}$$

Then:

(1) $\varphi_1(x)$ *is a strictly increasing absolutely continuous function of* x *on the interval* $[0, \ell]$.

(2) *The inverse function* $x = \psi_1(y)$ *to* $y = \varphi_1(x)$ *is a strictly increasing absolutely continuous function of* y *on the interval* $[0, \varphi_1(\ell)]$ *and*

$$1 = \varphi_1'(\psi_1(y))\psi_1'(y) = \text{trace}\{H(\psi_1(y))\}\psi_1'(y) \tag{5.3}$$

for a.e. point $y \in [0, \varphi_1(\ell)]$.

(3) *The mvf*

$$H_1(y) = H(\psi_1(y))\psi_1'(y) \quad \textit{for a.e.} \quad y \in [0, \varphi_1(\ell)] \tag{5.4}$$

is a normalized solution of the inverse monodromy problem for $U(\lambda)$.

Proof: The first two assertions are either selfevident or based on standard calculations.

To verify the third assertion, observe first that, by assumption, $U(\lambda)$ is equal to the solution $U(\ell, \lambda)$ at $x = \ell$ of the integral equation

$$\begin{aligned} U(x, \lambda) &= I_2 + i\lambda \int_0^x U(s, \lambda)H(s)ds J \\ &= I_2 + i\lambda \int_0^{\varphi_1(x)} U(\psi_1(u), \lambda)H(\psi_1(u))\psi_1'(u)du J, \\ & \quad 0 \leq x \leq \ell. \end{aligned}$$

Hence,

$$U_1(y, \lambda) = U(\psi_1(y), \lambda)$$

is a solution of the integral equation

$$U_1(y, \lambda) = I_2 + i\lambda \int_0^y U_1(y, \lambda)H_1(y)dy, \quad 0 \leq y \leq \varphi_1(\ell),$$

where $H_1(y)$ is given by (5.4) and is therefore seen to be a normalized solution of the inverse monodromy problem for $U(\lambda)$ by formula (5.3). □

Lemma 5.2 *Let $U \in \mathcal{E}(a,0) \cap \mathcal{U}(J)$ with $U(0) = I_2$, let $H_1(y)$, $0 \le y \le \ell_1$, be the unique normalized solution of the inverse monodromy problem for the canonical differential system* (1.9) *with monodromy matrix $U(\lambda)$, let $H(x)$, $0 \le x \le \ell$, be any solution of this inverse monodromy problem with* trace $H(x) > 0$ *for a.e. $x \in [0,\ell]$ and let $\varphi_1(x)$ be defined by formula* (5.2). *Then*

$$H(x) = \varphi_1'(x) H_1(\varphi_1(x)) \quad \textit{for a.e. } x \in [0,\ell]. \tag{5.5}$$

Proof: This is immediate from formula (5.4) and the relations

$$y = \varphi_1(x) \quad \text{for} \quad x \in [0,\ell] \quad \text{and} \quad x = \psi_1(y) \quad \text{for} \quad y \in [0, \varphi_1(\ell)].$$

□

Remark 5.1 Let $U(\lambda)$ and $H_1(y)$, $0 \le y \le \ell_1$, be as in Lemma 5.2 and let $\varphi_1(x)$ be any absolutely continuous function of x on the interval $[0,\ell]$ with $\varphi_1'(x) > 0$ for a.e. $x \in [0,\ell]$, $\varphi_1(0) = 0$ and $\varphi_1(\ell) = \ell_1$. Then the mvf $H(x)$ given by formula (5.5) is a solution of the inverse monodromy problem considered in Lemma 5.2 that meets the condition (5.1).

The next result strengthens Theorems 2.3 and 2.4.

Theorem 5.2 *Let $U \in \mathcal{E} \cap \mathcal{U}(J)$ with $U(0) = I_2$ and let $H(x)$ be a solution of the inverse monodromy problem for the canonical differential system* (1.9) *with monodromy matrix $U(\lambda)$ such that* trace $H(x) > 0$ *for a.e. $x \in [0,\ell]$. Then:*

$$U \in \mathcal{E}(a,0) \Longleftrightarrow \text{trace } H(x)J \ge 0 \ \textit{and} \ \det H(x) = 0 \quad \textit{for a.e. } x \in [0,\ell]. \tag{5.6}$$

Moreover, in this setting,

$$a = \int_0^\ell \text{trace}\{H(s)J\}ds. \tag{5.7}$$

Proof: Let $U \in \mathcal{E}(a,0)$ and let $H_1(y)$, $0 \le y \le \ell_1$, denote the unique normalized solution of the inverse monodromy problem under consideration. Then, by Theorem 2.3,

$$\text{trace}\{H_1(y)J\} \ge 0 \ \text{and} \ \det H_1(y) = 0 \ \text{for a.e.} \ y \in [0,\ell_1].$$

Therefore, by Lemma 5.2,

$$\det H(x) = \varphi_1'(x)^2 \det H_1(\varphi_1(x)) = 0$$

and

$$\text{trace}\{H(x)J\} = \varphi_1'(x)\text{trace}\{H_1(\varphi_1(x))J\} \ge 0$$

for a.e. $x \in [0,\ell]$.

Now suppose conversely that these last two conditions (i.e., the right hand side of (5, 6)) are met. Then, by Lemma 5.1, $H_1(y) = H(\psi_1(y))\psi_1'(y)$ will also meet these conditions for a.e. $y \in [0, \ell_1]$. Moreover, $H_1(y)$ is a normalized solution of the inverse monodromy problem for $U(\lambda)$. Therefore, by Theorem 2.4, $U \in \mathcal{E}(a, 0)$, where

$$\begin{aligned} a &= \int_0^{\ell_1} \text{trace}\{H_1(y)J\}dy \\ &= \int_0^{\ell_1} \text{trace}\{H(\psi_1(y))J\}\psi_1'(y)dy \\ &= \int_0^{\ell} \text{trace}\{H(x)J\}dx. \end{aligned}$$

□

There is of course an analogue of Theorem 5.2 for monodromy matrices $U \in \mathcal{E}(0, a) \cap \mathcal{U}(J)$ which can be established in much the same way. The statement is:

Theorem 5.3 *Let $U \in \mathcal{E} \cap \mathcal{U}(J)$ with $U(0) = I_2$ and let $H(x)$ be a solution of the inverse monodromy problem for the canonical differential system* (1.9) *with monodromy matrix $U(\lambda)$ such that* trace $H(x) > 0$ *for a.e. $x \in [0, \ell]$ Then:*

(5.8)
$$U \in \mathcal{E}(0, a) \Longleftrightarrow \text{trace } H(x)J \leq 0 \;\; and \;\; \det H(x) = 0 \quad for\ a.e.\ x \in [0, \ell].$$

Moreover, in this setting,

(5.9)
$$a = -\int_0^{\ell} \text{trace}\{H(s)J\}ds.$$

6 Three Classes of J-Inner mvf's and Some Examples

Let $\mathcal{S}^{p\times q}$ denote the Schur class of $p \times q$ mvf's which are both analytic and contractive in $\mathbb{C}_+$ and, for $p \leq q$, let $\mathcal{S}_{out}^{p\times q}$ denote the intersection of this class with the set of $p \times q$ mvf's which are also outer in $H_\infty^{p\times q}$ in the usual sense.

A mvf $U \in \mathcal{U}(J)$ is said to belong to the class $\mathcal{U}_S(J)$ of A-singular J-inner mvf's if $U = h^{-1}g$, where $h \in \mathcal{S}_{out}^{1\times 1}$ and $g \in \mathcal{S}_{out}^{2\times 2}$.

A mvf $U \in \mathcal{U}(J)$ is said to belong to the class $\mathcal{U}_{rR}(J)$ of right A-regular J-inner mvf's if $U\widetilde{U}^{-1} \in \mathcal{U}(J)$ for some $\widetilde{U} \in \mathcal{U}_S(J) \Longrightarrow$ that $\widetilde{U}$ is constant.

In [ArD1] these two classes and a third class: the class $\mathcal{U}_{sR}(J)$ of strongly regular J-inner mvf's, were characterized in terms of the associated de Branges space $\mathcal{H}(U)$, i.e., in terms of the RKHS (reproducing kernel Hilbert space) $\mathcal{H}(U)$ of[3] 2×1 vector valued functions with RK (reproducing kernel)

$$K_\omega(\lambda) = \frac{J - U(\lambda)JU(\omega)^*}{\rho_\omega(\lambda)}, \tag{6.1}$$

where

$$\rho_\omega(\lambda) = -2\pi i(\lambda - \overline{\omega}). \tag{6.2}$$

In particular it was shown there that if $U \in \mathcal{U}(J)$ and[4]

$$\mathcal{L}_U = \mathcal{H}(U) \cap L_2^2(\mathbb{R}), \tag{6.3}$$

then:

$$(1) \quad U \in \mathcal{U}_S(J) \Longleftrightarrow \mathcal{L}_U = \{0\}. \tag{6.4}$$

$$(2) \quad U \in \mathcal{U}_{rR}(J) \Longleftrightarrow \mathcal{L}_U \text{ is dense in } \mathcal{H}(U). \tag{6.5}$$

$$(3) \quad U \in \mathcal{U}_{sR}(J) \Longleftrightarrow \mathcal{H}(U) \subset L_2^2(\mathbb{R}). \tag{6.6}$$

Notice that if $J = \pm I_2$, then $\mathcal{U}(J) = \mathcal{U}_{sR}(J)$. For the original definition of $\mathcal{U}_{sR}(J)$ when J is diagonal, see p. 405 of [ArD1].

It follows from these characterizations that

$$\mathcal{U}(J) \cap L_\infty^{2\times 2}(\mathbb{R}) \subseteq \mathcal{U}_{sR}(J) \subseteq \mathcal{U}(J) \cap L_2^{2\times 2}\left(\frac{d\mu}{1+\mu^2}\right) \subseteq \mathcal{U}_{rR}(J), \tag{6.7}$$

where the last inclusion on the right depends upon the fact that $\mathcal{L}_U$ is dense in $\mathcal{H}(U)$ when

$$U \in \mathcal{U}(J) \cap L_2^{2\times 2}\left(\frac{d\mu}{1+\mu^2}\right). \tag{6.8}$$

Lemma 6.1 *Let $U \in \mathcal{E} \cap \mathcal{U}(J)$ with $U(0) = I_2$ and let $H(x)$ be a solution of the inverse monodromy problem for the canonical differential system* (1.9) *with monodromy matrix $U(\lambda)$ such that* trace $H(x) > 0$ *for a.e. $x \in [0, \ell]$. Then*

$$U \in \mathcal{U}_S(J) \Longleftrightarrow \text{trace } H(x)J = 0 \text{ and } \det H(x) = 0 \quad \text{for a.e. } x \in [0, \ell]. \tag{6.9}$$

---

[3]In [ArD1] these definitions are formulated for $m \times m$ mvf's and not just for 2×2, as here.

[4]This definition is meaningful since every vector valued function in $\mathcal{H}(U)$ belongs to the class of mermomorphic functions with bounded Nevanlinna characteristic in each half plane $\mathbb{C}_+$ and $\mathbb{C}_- = \overline{\mathbb{C}}_+$ and has nontangential boundary limits from $\mathbb{C}_+$ and $\mathbb{C}_-$ which agree a.e. on $\mathbb{R}$. The space $\mathcal{L}_U$ is the intersection of these boundary values with $L_2^2(\mathbb{R})$.

Proof: A mvf $U \in \mathcal{E} \cap \mathcal{U}(J)$ belongs to the class $U_S(J)$ of singular J-inner mvf's if and only if $U \in \mathcal{E}(0,0)$. The rest is immediate from Theorem 5.2. □

Theorem 6.1 *Let $U \in \mathcal{E}(a,0) \cap \mathcal{U}(J)$ with $U(0) = I_2$ and let $H(x)$ be a normalized solution of the inverse monodromy problem for the canonical differential system* (1.9) *with monodromy matrix $U(\lambda)$. Then $U \in \mathcal{U}_{rR}(J)$ if and only if*

$$\int_{\ell-\varepsilon}^{\ell} \mathrm{trace}\{H(x)J\}dx > 0 \quad \textit{for every} \quad \varepsilon \in (0,\ell].$$

Proof: Let $U(\lambda) = U_1(\lambda)U_2(\lambda)$, where

$$U_1 \in \mathcal{E}(a,0) \cap \mathcal{U}_{rR}(J) \ \text{ with } \ U_1(0) = I_2$$

and

$$U_2 \in \mathcal{E} \cap \mathcal{U}_S(J) \ \text{ with } \ U_2(0) = I_2 \ \text{ and } \ U_2(\lambda) \not\equiv \text{constant}.$$

Let $H^{(k)}(x)$, $0 \le x \le \ell_k$, denote the unique normalized solution of the inverse monodromy problem for the canonical differential system (1.9) with monodromy matrix $U_k(\lambda)$ for $k = 1, 2$. Then

$$\ell_k = -\mathrm{trace}\{iU_k'(0)J\}, \qquad k = 1, 2,$$

and

$$H(x) = \begin{cases} H^{(1)}(x) & \text{for a.e. } x \in [0,\ell_1] \\ H^{(2)}(x-\ell_1) & \text{for a.e. } x \in [\ell_1,\ell], \ \text{where } \ell = \ell_1 + \ell_2. \end{cases}$$

Thus, in view of Lemma 6.1,

$$\int_{\ell_1}^{\ell} \mathrm{trace}\{H(x)J\}dx = 0.$$

Conversely, by running the construction backwards, it is clear that if

$$\int_{\ell-\varepsilon}^{\ell} \mathrm{trace}\{H(s)J\}ds = 0$$

for some $\varepsilon \in (0,\ell]$, then $U(\lambda)$ admits a factorization of the form $U(\lambda) = U_1(\lambda)U_2(\lambda)$, where $U_k \in \mathcal{E} \cap \mathcal{U}(J)$ and $U_2 \in \mathcal{U}_S(J)$. □

Conclusions analogous to Lemma 6.1 and Theorem 6.1 for $U \in \mathcal{E}(a,a) \cap \mathcal{U}(-\mathcal{J})$ with $U(0) = I_2$ are presented in an unpublished manuscript of Krein and Langer [KL].

Examples

We turn next to some simple examples of monodromy matrices, one from each of the three classes introduced, and exhibit the solutions of the corresponding inverse monodromy problems. In all the examples, the given monodromy matrix $U \in \mathcal{E}(a, 0) \cap \mathcal{U}(J)$ for some $a \geq 0$ and the exhibited Hermitian $H(x)$ is normalized and therefore unique, by Theorem 2.3.

Example 1 Let $U_1(\lambda) = I_2 + i\lambda vv^*J$, where v is a unit vector in $\mathbb{C}^2$ that is J neutral, i.e., $v^*v = 1$ and $v^*Jv = 0$.

Discussion: Clearly $U_1 \in \mathcal{E}(0,0) \cap \mathcal{U}(J)$. Thus $U_1 \in \mathcal{U}_S(J)$. (Or on other grounds, $\mathcal{H}(U) = \{cv : c \in \mathbb{C}\}$ and hence the conclusion follows from (6.4).) Let

$$U_1(x, \lambda) = I_2 + i\lambda vv^*xJ, \qquad 0 \leq x \leq 1.$$

It is then readily checked that

$$U_1(x, \lambda) = \exp\{i\lambda vv^*xJ\}$$

and hence that

$$\frac{\partial U_1}{\partial x}(x, \lambda) = i\lambda U_1(x, \lambda) H^{(1)}(x) J$$

with

$$H^{(1)}(x) = vv^* \quad \text{for } x \in [0, 1].$$

Thus $U_1(\lambda) = U_1(1, \lambda)$ is the monodromy matrix for the canonical differential system (1.9) with constant Hermitian $H^{(1)}(x) = vv^*$ on $[0, 1]$. Moreover,

$$\det H^{(1)}(x) = 0, \qquad \text{trace}\{H^{(1)}(x)\} = v^*v = 1$$

and

$$\text{trace}\{H^{(1)}(x)J\} = \text{trace}\{v^*Jv\} = 0,$$

as it should be in view of Lemma 6.1.

Example 2 Let V be a 2×2 unitary matrix such that

$$V^*JV = j = \begin{bmatrix} 1 & 0 \\ 0 & -1 \end{bmatrix}$$

and let

$$U_2(\lambda) = V \begin{bmatrix} e^{i\lambda a} & 0 \\ 0 & 1 \end{bmatrix} V^*$$

for some $a > 0$.

Discussion: It is readily checked that

$$\frac{J - U_2(\lambda)JU_2(\omega)^*}{-2\pi i(\lambda - \overline{\omega})} = V \begin{bmatrix} k_\omega^a(\lambda) & 0 \\ 0 & 0 \end{bmatrix} V^*,$$

where

$$k_\omega^a(\lambda) = \frac{1}{2\pi}\int_0^a e^{i(\lambda-\overline{\omega})t}dt$$

is the reproducing kernel of the one-sided Paley-Wiener space of

$$\left\{\hat{f}(\lambda) = \int_0^a e^{i\lambda x} f(x)dx : \int_0^a |f(x)|^2 dx < \infty\right\}$$

with respect to the standard inner product in $L_2(\mathbb{R})$:

$$\langle \hat{f}, k_\omega^a\rangle = \int_{-\infty}^{\infty} \hat{f}(\mu)\overline{k_\omega^a(\mu)}d\mu = \int_0^a f(t)e^{i\omega t}dt = \hat{f}(\omega).$$

Thus $\mathcal{H}(U_2) \subset L_2^2(\mathbb{R})$ and hence, by (6.6), $U \in \mathcal{U}_{sR}(J)$. Let

$$U_2(x,\lambda) = V\begin{bmatrix} e^{i\lambda x} & 0 \\ 0 & 1\end{bmatrix}V^* \quad \text{for} \quad x \in [0,a].$$

Then it is readily checked that

$$\frac{\partial U_2}{\partial x}(x,\lambda) = i\lambda U_2(x,\lambda)H^{(2)}(x)J,$$

where

$$H^{(2)}(x) = V\begin{bmatrix} 1 & 0 \\ 0 & 0\end{bmatrix}V^*.$$

Thus,

$$\det\ H^{(2)}(x) = 0,$$

$$\text{trace}\{H^{(2)}(x)\} = \text{trace}\left\{\begin{bmatrix} 1 & 0 \\ 0 & 0\end{bmatrix}\right\} = 1$$

and

$$\text{trace}\{H^{(2)}(x)J\} = \text{trace}\left\{\begin{bmatrix} 1 & 0 \\ 0 & 0\end{bmatrix}V^*JV\right\} = \text{trace}\left\{\begin{bmatrix} 1 & 0 \\ 0 & 0\end{bmatrix}j\right\} = 1$$

for $x \in [0,a]$.

Example 3 Let $U_3(\lambda) = U_1(\lambda)U_2(\lambda)$.

Discussion: It is easily verified that

$$H^{(3)}(x) = \begin{cases} H^{(1)}(x) & \text{for } 0 < x < 1 \\ H^{(2)}(x-1) & \text{for } 1 < x < 1+a \end{cases}$$

is a normalized solution of the inverse monodromy problem for the canonical differential system (1.9) with monodromy matrix $U_3(\lambda)$. Moreover, since

$$\det\ H^{(3)}(x) = 0 \quad \text{and} \quad \text{trace}\{H^{(3)}(x)J\} \geq 0 \quad \text{for a.e.} \quad x \in [0, a+1]$$

and

$$\int_0^{a+1} \text{trace}\{H^{(3)}(x)J\}dx = a,$$

it follows from Theorem 5.2 that $U_3 \in \mathcal{E}(a, 0)$, as is selfevident from the given formula.

Let

$$K_\omega^{(j)}(\lambda) = \frac{J - U_j(\lambda)JU_j(\omega)^*}{\rho_\omega(\lambda)}, \tag{6.10}$$

denote the reproducing kernel for $\mathcal{H}(U_j)$. Then clearly

$$\frac{U_3(\lambda) - U_3(0)}{\lambda}\xi = 2\pi i K_0^{(3)}(\lambda)J\xi$$

belongs to $\mathcal{H}(U_3)$ for every choice of $\xi \in \mathbb{C}^2$. On the other hand, the decomposition

$$\begin{aligned}\frac{U_3(\lambda) - U_3(0)}{\lambda}\xi &= \frac{U_1(\lambda) - U_1(0)}{\lambda}U_2(\lambda)\xi + U_1(0)\frac{U_2(\lambda) - U_2(0)}{\lambda}\xi \\ &= ivv^*JU_2(\lambda)\xi + \frac{e^{i\lambda a} - 1}{\lambda}V\begin{bmatrix} 1 & 0 \\ 0 & 0 \end{bmatrix}V^*\xi,\end{aligned}$$

displays the fact that this function does not belong to $L_2(R)$ if $\xi \neq 0$. Therefore, $U_3(\lambda)$ is not strongly regular. It is right A-regular by definition and it is easily seen that

$$\int_\delta^{a+1} \text{trace}\{H^{(3)}(x)J\}dx > 0$$

for every $\delta \in [0, a + 1)$, as it should by Theorem 6.1.

An Instructive Calculation

It is an instructive exercise in the theory of reproducing kernel Hilbert spaces to verify that $\mathcal{L}_U$ is dense in $\mathcal{H}(U)$ in Example 3. To begin with, since

$$K_\omega^{(1)}(\lambda) = \frac{vv^*}{2\pi}$$

it follows readily from the standard formula

$$\|K_\omega^{(i)}u\|^2_{\mathcal{H}(U_i)} = u^*K_\omega^{(i)}u \tag{6.11}$$

which is valid for each of the reproducing kernel Hilbert spaces $\mathcal{H}(U_i)$, $i = 1, 2, 3$, that

$$\|v\|^2_{\mathcal{H}(U_1)} = 2\pi.$$

Next, let

$$v_1 = V\begin{bmatrix} 1 \\ 0 \end{bmatrix} \text{ and } v_2 = V\begin{bmatrix} 0 \\ 1 \end{bmatrix}. \tag{6.12}$$

Then,

$$\begin{aligned} 0 &= v^*Jv \\ &= v^*VjV^*v \\ &= v^*V\begin{bmatrix} 1 & 0 \\ 0 & 0 \end{bmatrix}V^*v - v^*V\begin{bmatrix} 0 & 0 \\ 0 & 1 \end{bmatrix}V^*v \\ &= |v^*v_1|^2 - |v^*v_2|^2. \end{aligned}$$

Thus, as $v^*v = 1$ and V is unitary, we see that

$$|v^*v_1| = |v^*v_2| \neq 0. \tag{6.13}$$

Then, since

$$\begin{aligned} K_\omega^{(3)}(\lambda) &= K_\omega^{(1)}(\lambda) + U_1(\lambda)K_\omega^{(2)}(\lambda)U_1(\omega)^* \\ &= K_\omega^{(1)}(\lambda) + U_1(\lambda)V\begin{bmatrix} k_\omega^a(\lambda) & 0 \\ 0 & 0 \end{bmatrix}V^*U_1(\omega)^* \end{aligned}$$

and

$$U_1(0) = I_2,$$

it follows that

$$K_0^{(3)}(\lambda)v_2 = K_0^{(1)}v_2 = \frac{vv^*v_2}{2\pi}.$$

Therefore, by an application of (6.11) with $i = 3$, we obtain

$$\begin{aligned} \|K_0^{(3)}v_2\|_{\mathcal{H}(U_3)} &= \frac{|v^*v_2|^2}{4\pi^2}\|v\|^2_{\mathcal{H}(U_3)} \\ &= v_2^*K_0^{(3)}(0)v_2 = \frac{|v^*v_2|^2}{2\pi}. \end{aligned}$$

This leads easily to the conclusion that

$$\|v\|^2_{\mathcal{H}(U_3)} = 2\pi = \|v\|^2_{\mathcal{H}(U_1)},$$

and hence that $\mathcal{H}(U_1)$ is included isometrically in $\mathcal{H}(U_3)$, since $v^*v_2 \neq 0$.

The next step is to use the already established isometric inclusion of $\mathcal{H}(U_1)$ in $\mathcal{H}(U_3)$ to verify that

$$\mathcal{H}(U_3) = \mathcal{H}(U_1) \oplus U_1\mathcal{H}(U_2) \tag{6.14}$$

and

$$\|U_1 f\|^2_{\mathcal{H}(U_3)} = \|f\|^2_{\mathcal{H}(U_2)}$$

for every $f \in \mathcal{H}(U_2)$. This is a standard argument which goes back to de Branges; see e.g., [dB5], [dB6] and Theorem 5.2 of [AlD] which is adapted from de Branges.

Lemma 6.2 *Let* $g \in \mathcal{H}(e_a)$ *with* $g(0) = 1$, *let* $g_1(\lambda) = \{g(\lambda) - g(0)\}/i\lambda$ *and let*

$$u_1 = (v^* v_1)^{-1} v_1.$$

Then the vector valued function

$$f_1(\lambda) = v + U_1(\lambda) V \begin{bmatrix} g_1(\lambda) & 0 \\ 0 & 0 \end{bmatrix} V^* u_1 \tag{6.15}$$

belongs to $\mathcal{H}(U_3) \cap L_2^2(\mathbb{R})$.

Proof: Clearly it is enough to show that the function

$$\begin{aligned} v + i\lambda v v^* J V \begin{bmatrix} g_1(\lambda) & 0 \\ 0 & 0 \end{bmatrix} V^* u_1 \\ = v \left\{ 1 + v^* J V \begin{bmatrix} g(\lambda) - g(0) & 0 \\ 0 & 0 \end{bmatrix} V^* u_1 \right\} \end{aligned}$$

belongs to $L_2^2(\mathbb{R})$. But this is indeed the case, since

$$\begin{aligned} 1 - v^* J V \begin{bmatrix} 1 & 0 \\ 0 & 0 \end{bmatrix} V^* u_1 &= 1 - v^* J V j V^* u_1 \\ &= 1 - v^* u_1 = 0, \end{aligned}$$

by construction. □

The next step is to show that the domain of the operator of multiplication by λ is dense in $\mathcal{H}(e_a)$.

Lemma 6.3 *Let* $f \in \mathcal{H}(e_a)$. *Then for every* $\varepsilon > 0$ *there exists an* $h \in \mathcal{H}(e_a)$ *such that* $\lambda h \in \mathcal{H}(e_a)$ *and*

$$\|f - h\|_2 \leq \varepsilon.$$

Proof: By assumption, f admits a representation of the form

$$f(\lambda) = \int_0^a e^{i\lambda t} \varphi(t) dt$$

for some $\varphi \in L_2(0, a)$. Let

$$h(\lambda) = \int_0^a e^{i\lambda t} \psi(t) dt$$

where $\psi \in C^1[0, a]$ has a continuous derivative ψ', $\psi(0) = 0$, $\psi(a) = 0$ and

$$\|\varphi - \psi\|_2 \leq \varepsilon/(2\pi)^{\frac{1}{2}}.$$

Then,

$$\|f - h\|_2 = \sqrt{2\pi}\,\|\varphi - \psi\|_2 \leq \varepsilon,$$

by the Plancherel formula, and

$$i\lambda h(\lambda) = \int_0^a \left(\frac{\partial}{\partial t} e^{i\lambda t}\right) \psi(t)dt = -\int_0^a e^{i\lambda t}\psi'(t)dt$$

belongs to $\mathcal{H}(e_a)$. □

It is now easy to complete the verification of condition (6.5): Let $f \in \mathcal{H}(U_3)$. Then, in view of the (orthogonal) decomposition (6.14), f must be of the form

$$f(\lambda) = dv + U_1(\lambda)V \begin{bmatrix} b(\lambda) & 0 \\ 0 & 0 \end{bmatrix} V^* u_1$$

for some choice of $d \in \mathbb{C}$ and $b \in \mathcal{H}(e_a)$. Thus,

$$f(\lambda) - df_1(\lambda) = U_1(\lambda)V \begin{bmatrix} b(\lambda) - dg_1(\lambda) & 0 \\ 0 & 0 \end{bmatrix} V^* u_1.$$

Let

$$f_2(\lambda) = U_1(\lambda)V \begin{bmatrix} h(\lambda) & 0 \\ 0 & 0 \end{bmatrix} V^* u_1,$$

where $h \in \mathcal{H}(e_a)$ is chosen so that $\lambda h \in \mathcal{H}(e_a)$ and

$$\|b - dg_1 - h\|_2 \leq \varepsilon |v^* v_1|.$$

Then,

$$df_1 + f_2 \in \mathcal{H}(U_3) \cap L_2^2(\mathbb{R})$$

and

$$\|f - df_1 - f_2\|_{\mathcal{H}(U_3)} = \frac{\|b - dg_1 - h\|_2}{|v^* v_1|} \leq \varepsilon,$$

as claimed.

Addendum

We remark that the monodromy matrix in Example 1 is a special case of the following general result which is also discussed in the manuscript [KL] of Krein and Langer for $U \in \mathcal{E}(a, a) \cap \mathcal{U}(-\mathcal{J})$.

Theorem 6.2 *Let $U \in \mathcal{E}(a, 0) \cap \mathcal{U}(J)$ with $U(0) = I_2$ and let $H(x)$ be a normalized solution of the inverse monodromy problem for the canonical differential system* (1.9) *with monodromy matrix $U(\lambda)$. Then $U(\lambda)$ admits a factorization $U(\lambda) = U_1(\lambda)U_2(\lambda)$ with a singular inner factor*

$$U_1(\lambda) = I_2 + i\lambda\delta v v^* J,$$

where

$$v^*v = 1, \quad v^*Jv = 0 \quad \text{and} \quad \delta > 0,$$

if and only if

$$H(x) = vv^* \quad \text{for} \quad x \in [0, \delta].$$

The proof is left to the reader. The strategy is much the same as the Proof of Theorem 6.2 except that now the singular factor appears on the left instead of the right.

7 Another Parametrization

In this section we shall consider another parametrization that is based on exponential type that is useful when the monodromy matrix and the matrizant are right A-regular. The latter is guaranteed when the monodromy matrix is strongly regular; see Theorem 3.7 of [ArD1] and Theorem 0.1 of [ArD3].

Lemma 7.1 *Let* $U \in \mathcal{E}(a, 0) \cap \mathcal{U}(J)$ *with* $U(0) = I_2$ *and let* $H(x)$ *be a solution of the inverse monodromy problem for the canonical differential system* (1.9) *with monodromy matrix* $U(\lambda)$. *Suppose further that*

(7.1) $$\text{trace}\{H(x)J\} > 0 \quad \text{for a.e.} \quad x \in [0, \ell]$$

and let

(7.2) $$\varphi_2(x) = \int_0^x \text{trace}\{H(s)J\}ds \quad \text{for} \quad x \in [0, \ell].$$

Then:

(1) $\varphi_2(x)$ *is a strictly increasing absolutely continuous function of* x *on the interval* $[0, \ell]$ *and*

(7.3) $$\varphi_2(\ell) = a.$$

(2) *The inverse function* $x = \psi_2(t)$ *to* $t = \varphi_2(x)$ *is a strictly increasing absolutely continuous function of* t *on the interval* $[0, a]$ *and*

(7.4) $$1 = \varphi_2'(\psi_2(t))\psi_2'(t) = \text{trace}\{H(\psi_2(t))J\}\psi_2'(t)$$

for a.e. point $t \in [0, a]$.

(3) *The mvf*

(7.5) $$H_2(t) = H(\psi_2(t))\psi_2'(t) \quad \text{for a.e.} \quad t \in [0, a]$$

meets the conditions

(7.6) $$\text{trace}\{H_2(t)J\} = 1 \quad \text{and} \quad \det H_2(t) = 0 \quad \text{for a.e.} \quad t \in [0, a].$$

Proof: The proof of (1), (2) and the first half of (3) is pretty much the same as the Proof of Lemma 5.1. The evaluation (7.3) is obtained from Theorem 5.2. The second half of (3) then follows from Theorem 5.2 and the formula for $H_2(t)$. Theorem 5.2 is applicable because the condition trace$\{H(x)J\} > 0$ for a.e. $x \in [0, \ell]$ guarantees that trace$\{H(x)\} > 0$ for a.e. $x \in [0, \ell]$. □

We remark that the previous lemma guarantees that if $U \in \mathcal{E}(a,0) \cap \mathcal{U}(J)$ with $U(0) = I_2$ and if there exists a solution $H(x)$ of the considered inverse monodromy problem which meets condition (7.1), then there exists a solution $H_2(t)$, $0 \le t \le a$, which meets the conditions in (7.6).

Lemma 7.2 *Let A be a constant* 2×2 *matrix. Then the following are equivalent:*

(1) det $A = 0$ *and* trace $A = 1$.

(2) $A^2 = A$, *but* $A \neq 0$ *and* $A \neq I_2$.

(3) *A has two eigenvalues:* $\lambda_1 = 1$ *and* $\lambda_2 = 0$.

Proof: Let λ be an eigenvalue of A. Then the assumption $A^2 = A$ implies that $\lambda^2 = \lambda$ and hence that there are five possible Jordan decompositions of A (up to order):

$$A = P\begin{bmatrix} 1 & a \\ 0 & 1 \end{bmatrix} P^{-1} \text{ with } a \neq 0, \quad A = I_2,$$

$$A = P\begin{bmatrix} 0 & a \\ 0 & 0 \end{bmatrix} P^{-1} \text{ with } a \neq 0, \quad A = 0,$$

and

$$A = P\begin{bmatrix} 1 & 0 \\ 0 & 0 \end{bmatrix} P^{-1}.$$

But only the last one is compatible with each one of the stated three conditions. □

Theorem 7.1 *Let* $U \in \mathcal{E} \cap \mathcal{U}(J)$ *with* $U(0) = I_2$ *and let* $H(t)$ *be a solution of the inverse monodromy problem for the canonical differential system* (1.9) *with monodromy matrix* $U(\lambda)$ *such that the mvf*

$$Q(t) = H(t)J$$

meets the condition

$$0 \neq Q(t)^2 = Q(t) \neq I_2 \quad \text{for a.e. } t \in [0, a]. \tag{7.7}$$

Then $U \in \mathcal{E}(a, 0)$ *and there exists a* 2×1 *mvf* $h(t)$ *such that*

$$H(t) = h(t)h(t)^* \text{ and } h(t)^* J h(t) = 1 \text{ for a.e. } t \in [0, a]. \tag{7.8}$$

Moreover, there is at most one solution $H(t)$ *of this inverse monodromy problem which meets condition* (7.8).

Proof: By Lemma 7.2,

$$\det H(t) = 0 \text{ and } \operatorname{trace}\{H(t)J\} = 1 \text{ for a.e. } t \in [0, a]. \tag{7.9}$$

Therefore,

$$H(t) = h(t)h(t)^*$$

for some 2×1 mvf $h(t)$ and

$$\begin{aligned} \operatorname{trace}\{H(t)J\} &= \operatorname{trace}\{h(t)h(t)^*J\} \\ &= h(t)^*Jh(t) = 1 \end{aligned}$$

for a.e. $t \in [0, a]$ as claimed.

Next, $U \in \mathcal{E}(a, 0)$, by Theorem 5.2, and hence, by Theorem 2.3, there exists a unique normalized solution $H_1(x)$, $0 \leq x \leq \ell$, for the inverse monodromy problem under consideration. Let

$$\varphi_1(t) = \int_0^t \operatorname{trace}\{H(s)\}ds = \int_0^t h(s)^*h(s)ds \text{ for } t \in [0, a].$$

Then

$$\varphi_1'(t) = h(t)^*h(t) \geq h(t)^*Jh(t) = 1 \text{ for a.e. } t \in [0, a], \tag{7.10}$$

and

$$H(t) = H_1(\varphi_1(t))\varphi_1'(t) \text{ for a.e. } t \in [0, a]. \tag{7.11}$$

Thus, $H(t)$ is uniquely determined by its trace, $\varphi_1(t)$. On the other hand, in view of (7.10), the function $x = \varphi_1(t)$, $0 \leq t \leq a$, admits a strictly increasing absolutely continuous inverse $t = \psi_1(x)$, for $0 \leq x \leq \ell$ and

$$H_1(x) = H(\psi_1(x))\psi_1'(x) \quad \text{for a.e. } \quad x \in [0, \ell],$$

where $\ell = \varphi_1(a)$. Thus,

$$\psi_1'(x) = \operatorname{trace}\{H(\psi_1(x)J\}\psi_1'(x) = \operatorname{trace}\{H_1(x)J\}$$

and

$$\ell = -i \operatorname{trace}\{U'(0)J\}$$

are uniquely determined by $U(\lambda)$. Therefore, so is $\varphi_1(t)$, $0 \leq t \leq a$, as needed. □

Remark 7.1 In her thesis [M1] (see also [M2]), I.V. Mikhailova showed that if $U \in \mathcal{E}(a, 0) \cap \mathcal{U}(J)$ with $U(0) = I_2$ and if there exists a constant $c > 0$ such that

$$J - U(i\nu)JU(i\nu)^* \leq cI_2 \quad \text{for} \quad \nu > 0, \tag{7.12}$$

then there exists a solution $H(t)$ of the inverse monodromy problem for the canonical differential system (1.9) with this monodromy matrix $U(\lambda)$ such that $Q(t) = H(t)J$ meets the condition (7.7). In particular, a mvf $U \in \mathcal{E}(a,0) \cap \mathcal{U}(J)$ will satisfy the bound (7.12) if $U(\lambda)$ is bounded on $\mathbb{R}$.

Theorem 7.2 *Let* $U \in \mathcal{E}(a,0) \cap \mathcal{U}(J)$ [*resp.* $U \in \mathcal{E}(0,a) \cap \mathcal{U}(J)$] *with* $U(0) = I_2$. *Then there is at most one solution* $H(t)$ *of the inverse monodromy problem for the canonical differential system* (1.9) *with monodromy matrix* $U(\lambda)$ *such that*

$$\text{trace}\{H(t)J\} = 1 \quad [\textit{resp. } \text{trace}\{H(t)J\} = -1] \tag{7.13}$$

for a.e. $t \in [0,a]$.

Proof: The inequality

$$\text{trace}\{H(t)\} \geq \text{trace}\{H(t)J\} \quad [\text{resp. trace}\{H(t)\} \geq -\text{trace}\{H(t)J\}]$$

together with assumption (7.12), guarantees that Theorem 5.2 [resp. Theorem 5.3] is applicable and hence that

$$\det\{H(t)\} = 0 \quad \text{for a.e. } t \in [0,a]$$

in both of the cases under consideration. The desired conclusion now follows immediately from Theorem 7.1 if $U \in \mathcal{E}(a,0)$ and by similar arguments if $U \in \mathcal{E}(0,a)$. □

References

[AlD] D. Alpay and H. Dym, Hilbert spaces of analytic functions, inverse scattering, and operator models I, *Integral Equations Operator Theory* **7** (1984), 589–641.

[ArD1] D.Z. Arov and H. Dym, J-inner matrix functions, interpolation and inverse problems for canonical systems, I: Foundations, *Integral Equations Operator Theory* **29** (1997), 373–454.

[ArD2] D.Z. Arov and H. Dym, On three Krein extension problems and some generalizations, *Integral Equations Operator Theory* **31** (1998), 1–91.

[ArD3] D.Z. Arov and H. Dym, J-inner matrix functions, interpolation and inverse problems for canonical systems, II: The inverse monodromy problem, *Integral Equations Operator Theory*, in press.

[ArD4] D.Z. Arov and H. Dym, J-inner matrix functions, interpolation and inverse problems for canonical systems, III: More on the inverse monodromy problem, *Integral Equations Operator Theory*, in press.

[dB1] L. de Branges, Some Hilbert spaces of entire functions, *Trans. Amer. Math. Soc* **96** (1960), 259–295.

[dB2] L. de Branges, Some Hilbert spaces of entire functions II, *Trans. Amer. Math. Soc* **99** (1961), 118–152.

[dB3] L. de Branges, Some Hilbert spaces of entire functions III, *Trans. Amer. Math. Soc* **100** (1961), 73–115.

[dB4] L. de Branges, Some Hilbert spaces of entire functions IV, *Trans. Amer. Math. Soc* **105** (1962), 43–83.

[dB5] L. de Branges, Some Hilbert spaces of analytic functions I, *Trans. Amer. Math. Soc* **106** (1963), 445–468.

[dB6] L. de Branges, Some Hilbert spaces of analytic functions II, *J. Math. Anal. Appl* **11** (1965), 44–72.

[dB7] L. de Branges, *Hilbert spaces of entire functions*, Prentice-Hall, Englewood Cliffs, N.J., 1968.

[Br] M.S. Brodskii, Triangular and Jordan Representations of Linear Operators, *Trans. Math. Monographs*, vol. 32, *Amer. Math. Soc., Providence, R.I.*, 1971.

[GK] I.C. Gohberg and M.G. Krein, Theory and Applications of Volterra Operators in Hilbert Space, *Trans. Monographs*, vol. 24, *Amer. Math. Society, Providence, R.I.*, 1970.

[GoGo] M.L. Gorbachuk and V.I. Gorbachuk, *M.G. Krein's Lectures on Entire Operators*, Birkhäuser, Basel, 1997.

[Kr] M.G. Krein, Theory of entire functions of exponential type, *Izv. Akad. Nauk SSSR, Ser. Mat* **11** (1947), no. 4, 309–326.

[KL] M.G. Krein and H. Langer, *Continuation of Hermitian positive definite functions and related questions*, unpublished manuscript.

[M1] I.V. Mikhailova, On the correspondence between two classes of entire J-inner matrix functions, *Dokl. Akad. Nauk*, Ukrain, SSR Ser. A, 1983, no. 4, 26–29.

[M2] I.V. Mikhailova, *The Theory of Entire J-Expansive Matrix Functions and its Application to Inverse Problems*, Ph.D. Thesis, Institute for Low Temperature Physics and Engineering, Kharkov, 1984.

[Po] V.P. Potapov, The multiplicative structure of J-contractive matrix functions, *Amer. Math. Soc. Transl.* (2) **15** (1960), 131–243.

Department of Mathematics
South-Ukranian Pedagogical University
270020 Odessa, Ukraine

Department of Mathematics
The Weizmann Institute of Science
Rehovot 76100, Israel

AMS Classification Numbers: 30E05, 30D99, 34A55, 34L40, 47A56, 47A57.

Operator Theory:
Advances and Applications, Vol. 122
© 2001 Birkhäuser Verlag Basel/Switzerland

Interpolation and Commutant Lifting for Multipliers on Reproducing Kernel Hilbert Spaces

Joseph A. Ball, Tavan T. Trent and Victor Vinnikov

Dedicated to M.A. Kaashoek on the occasion of his 60th birthday

We obtain an explicit representation formula and a Nevanlinna-Pick-type interpolation theorem for the multiplier space of the reproducing kernel space $\mathcal{H}(k_d)$ of analytic functions on the d-dimensional complex unit ball with reproducing kernel $k_d(z, w) = 1/(1 - \langle z, w \rangle)$. More generally, if k is a positive kernel on a set Ω such that $1/k$ has 1 positive square, then there is an embedding b of Ω into the ball $\mathbb{B}^d$ (where d is the number of negative squares of $1/k$) such that any multiplier W for $\mathcal{H}(k)$ lifts to a multiplier F for the space $\mathcal{H}(k_d)$ on the ball ($W = F \circ b$). As a corollary, multipliers for $\mathcal{H}(k)$ also have explicit realization formulas and satisfy a Nevanlinna-Pick-type interpolation theorem. All the results in fact extend to the case of matrix- and operator-valued multipliers and left tangential Nevanlinna-Pick interpolation. Contractive multiplier solutions of a given set of interpolation conditions correspond to unitary extensions of a partially defined isometric operator; hence a technique of Arov and Grossman can be used to give a linear-fractional parametrization for the set of all such interpolants. A more abstract formulation of the analysis leads to a commutant lifting theorem for such multipliers. In particular, we obtain a new proof of a result of Quiggin giving sufficient conditions on a kernel k for a Nevanlinna-Pick-type interpolation theorem to hold for the space of multipliers on a reproducing kernel Hilbert space $\mathcal{H}(k)$.

1 Introduction

The classical Nevanlinna-Pick interpolation theorem (see [30] and [32]) states that: *given points $z_1, \dots, z_n$ in the unit disk $\mathbb{D}$ in the complex plane and complex numbers $w_1, \dots, w_n$, there exists a function $f(z)$ analytic on the unit disk with $|f(z)| \leq 1$ on $\mathbb{D}$ ($f \in H^\infty$ with $\|f\|_\infty \leq 1$) if and only if the associated Pick matrix $[\frac{1-w_i\overline{w}_j}{1-z_i\overline{z}_j}]_{i,j=1,\dots,n}$ is positive-semidefinite.* A major new insight on the problem was introduced by D. Sarason [40] who identified the space of bounded analytic functions on the unit disk with multiplication operators acting on the Hilbert space H^2, or equivalently, as the commutant of the shift operator $f(z) \to zf(z)$ acting on H^2. The interpolation problem then can be identified

The first author was supported by the NSF under grant DMS-9500912 and by the Mathematical Sciences Research Institute.

The third author is an incumbent of Graham and Rhona Beck career development chair, Weizmann Institute.

with a commutant lifting problem, i.e., the problem of lifting an operator in the commutant of the restriction $P_{\mathcal{M}} M_{z|\mathcal{M}}$ of M_z to a $*$-invariant subspace $\mathcal{M}$ for M_z to an operator in the commutant of M_z on H^2. His solution of the problem was ultimately generalized by B. Sz.-Nagy and C. Foias to what is now known as the Commutant Lifting Theorem. More recently, several authors have been looking at a more general formulation of the problem in the context of a general reproducing kernel Hilbert space. Specifically, suppose that $k(z, w)$ is a complex-valued function on the Cartesian product $\Omega \times \Omega$ of a set Ω with itself which is a *positive kernel* on Ω in the sense that

$$\sum_{i,j=1}^{n} k(z_i, z_j)\overline{c_i} c_j \geq 0$$

for all finite collections of complex scalars $c_1, \ldots, c_n$ and points $z_1, \ldots, z_n$ in Ω for $n = 1, 2, \ldots$. Then one can define a positive-semidefinite inner product on finite linear combinations of the functions $k_w := k(\cdot, w)$ $(w \in \Omega)$ by

$$\left\langle \sum_{j=1}^{n} c_j k_{w_j}, \sum_{i=1}^{n} c_i k_{w_i} \right\rangle_{\mathcal{H}} = \sum_{i,j=1}^{n} k(w_i, w_j)\overline{c_i} c_j.$$

Modding out by any elements of norm zero and taking a completion gives a Hilbert space $\mathcal{H}(k)$ which can be identified with a space of complex-valued functions f on Ω for which

$$f(w) = \langle f, k_w \rangle_{\mathcal{H}(k)}$$

(i.e., k can be expressed in the form $k(z, w) = h(z)h(w)^*$ where $h(z) \in \mathcal{L}(\mathcal{H}(k), \mathbb{C})$ $(z \in \Omega)$ is the linear functional of evaluation at z). The space of multipliers $\mathcal{M}_k$ is the space of functions F for which $Ff \in \mathcal{H}(k)$ for each $f \in \mathcal{H}(k)$, and the multiplier norm is the norm of the corresponding multiplication operator on $\mathcal{H}(k)$. In the case where $k(z, w) = \frac{1}{1 - z\overline{w}}$ with Ω equal to the unit disk $\mathbb{D}$, the space $\mathcal{H}(k)$ is the Hardy space H^2 on the unit disk $\mathbb{D}$, the space of multipliers $\mathcal{M}_k$ is equal to H^∞ with multiplier norm equal to the infinity norm, and the assertion of the Nevanlinna-Pick theorem in this notation is: *there is a multiplier F with multiplier norm at most 1 which satisfies the interpolation conditions $F(z_i) = w_i$ for $i = 1, \ldots, n$ if and only if the matrix $[k(z_i, z_j)(1 - w_i\overline{w_j})]_{i,j=1,\ldots,n}$ is positive-semidefinite.* A natural question is to determine for which kernels k this "Nevanlinna-Pick theorem" holds.

A number of authors have worked on this question of determining which kernels satisfy the Nevanlinna-Pick theorem. In particular, J. Agler showed that the theorem holds true for the Dirichlet space (see [1]) and the Sobolev space (see [2]), and obtained a sufficient condition on the kernel k for the Nevanlinna-Pick theorem to hold, namely: that the inverse kernel should have a decomposition of the form $\frac{1}{k(z,w)} = a(z)\overline{a(w)} - k'(z, w)$ where $z \to a(z)$ is a scalar-valued function and $k'(z, w)$ is a positive kernel. P. Quiggin [38], inspired by the results of Agler, later simplified Agler's analysis, showed that Agler's condition is also necessary for the setting of scalar-valued interpolation. As applications of this general result he verified that the Dirichlet and the Sobolev space satisfy the Nevanlinna-Pick theorem. S. McCullough (see [27] and [28]) introduced the notion of complete Nevanlinna-Pick kernel (the Nevanlinna-Pick theorem holds for interpolation of matrix-valued interpolation values by matrix-valued multipliers) and showed that Agler's condition is necessary and sufficient for the kernel to be a complete Nevanlinna-Pick kernel; closely related results have now been given by Agler and J. McCarthy [3]. A number of recent papers (see [3], [36], [9], [10], [22], [37]) have proved that the Nevanlinna-Pick theorem holds true for the kernel $k_d(z,w) = \frac{1}{1-\langle z,w\rangle}$ defined on the d-dimensional ball $\Omega = \mathbb{B}^d$ in $\mathbb{C}^d$; some of these authors obtained their results as an averaging of statements actually holding in a noncommutative Fock space setting. (As a trivial reminder, the reproducing kernel for the Hardy space H^2 on $\mathbb{B}^d$ equals $[\frac{1}{1-\langle z,w\rangle}]^d$.) The Agler-McCarthy paper [3] also showed that the kernel $k_d(z,w) = \frac{1}{1-\langle z,w\rangle}$ has a certain universal property, namely, a multiplier for any complete Nevanlinna-Pick kernel lifts to a multiplier for the space $\mathcal{H}(k_d)$ (with d equal to the number of negative squares of $\frac{1}{k}$). There has also been independent work recovering much of the familiar function and interpolation theory for the Hardy space $H^2 = \mathcal{H}(k_1)$ for the particular case where the reciprocal kernel $\frac{1}{k(z,w)}$ has exactly 1 positive square and 1 negative square (see [5], [7], [8], [31]), thereby unifying the disk and the half-plane Hardy space theories. Another main point of these papers is the elaboration of the forms of unitary versus isometric versus coisometric colligations in this setting.

In the present paper we recover more general versions of a number of these results by a different method of proof. For the case of k_d on $\Omega = \mathbb{B}^d$, we show how contractive-multiplier solutions of an interpolation problem correspond to unitary extensions of an explicit, partially defined isometric operator determined by the interpolation data. These unitary operators are of the block form

$$(1.1) \qquad U = \begin{bmatrix} A & B \\ C & D \end{bmatrix} = \begin{bmatrix} A_1 & B_1 \\ \vdots & \vdots \\ A_d & B_d \\ C & D \end{bmatrix} : \begin{bmatrix} \mathcal{H} \\ \mathbb{C} \end{bmatrix} \to \begin{bmatrix} \oplus_1^d \mathcal{H} \\ \mathbb{C} \end{bmatrix}$$

for an auxiliary Hilbert space $\mathcal{H}$ (the state space of the colligation U). The contractive multiplier $F(z)$ corresponding to the unitary extension (1.1) is given by the explicit formula

$$(1.2) \qquad F(z) = D + C(I - Z(z)A)^{-1}Z(z)B$$

where we have set $Z(z) = [z_1 I_{\mathcal{H}} \dots z_d I_{\mathcal{H}}] : (\oplus_1^d \mathcal{H}) \to \mathcal{H}$ for $z = (z_1, \dots, z_d) \in \mathbb{B}^d$. The formula (1.2) suggests that $F(z)$ is the transfer function of a multivariable, linear system; there are plans to pursue these system theory connections in a future publication [16]. (A more abstract version of our method of proof for the one-dimensional case ($d = 1$) has been formulated as the Abstract Interpolation Problem by V.E. Katsnelson, A.Ya. Kheifets and P.M. Yuditskii (see [26]); a survey article delineating several different several-variable settings fitting with these ideas is [13].) By considering the case where one imposes interpolation conditions at all points of $\mathbb{B}^d$, we see that in fact a representation of the form (1.2) holds for any contractive multiplier. A variation on our technique of extending a partially defined isometry leads to an embedding $z \to b(z) = (b_1(z), \dots, b_d(z))$ of Ω into $\mathbb{B}^d$ whenever $k(z, w)$ is a positive definite kernel on Ω such that $\frac{1}{k}$ has 1 positive square and d negative squares, i.e., when $\frac{1}{k}$ can be written as

$$\frac{1}{k(z,w)} = a(z)\left(1 - \sum_{k=1}^{d} b_k(z)\overline{b_k(w)}\right)\overline{a(w)}$$

for some scalar functions $a(z), b_1(z), \dots, b_d(z)$ on Ω. In this situation, any multiplier $W(z)$ on $\mathcal{H}(k)$ has the form

$$W(z) = F \circ b(z)$$

where F defined on $\mathbb{B}^d$ is a multiplier for the space $\mathcal{H}(k_d)$; this recovers one of the results in [3]. We formulate our interpolation results for the left tangential interpolation problem, a more general version than the full matrix-valued interpolation conditions considered in [3]. We are also able to adapt a procedure of [11] (also discussed in [26]) to obtain a linear-fractional parametrization for the set of all contractive-multiplier solutions of a given left tangential interpolation problem. We also consider the right tangential interpolation problem where the solution criterion is formally different and much less explicit, a situation similar to that for interpolation by Schur-Agler-class functions on the polydisk $\mathbb{D}^d$ studied in [15]. Finally we give a purely function-theoretic proof of the commutant lifting theorem for this setting which contains the result on left tangential interpolation as a special case; this result has been obtained by Clancy and McCullough [17] for the case of weighted Hardy spaces on the unit disk (the first example in Section 6 below) and by McCullough [29] for the general case by using different approaches. The same result can also be derived as an averaging of the noncommutative commutant lifting theorem (see [33], [35]) in the context of the Fock-space model for a d-tuple of contraction operators $(T_1, \dots, T_d)$ for which the row $[T_1 \ \dots \ T_d]$ is contractive (see [36], [9], [10] and particularly [22] for a good discussion of this approach).

All our results hold with the same proofs for $d = \infty$ with $\mathbb{B}^\infty$ taken to be the unit ball of ℓ^2. In fact one can use a coordinate-free notation replacing $\mathbb{B}^d$ by the unit ball $\mathbb{B}$ in a Hilbert space $\mathcal{C}$ and $\oplus_1^d \mathcal{H}$ in (1.1) by $\mathcal{C} \otimes \mathcal{H}$; then $Z(z) : \mathcal{C} \otimes \mathcal{H} \to \mathcal{H}$ (see (1.2)) is defined by $Z(z)(v \otimes h) = \langle v, \mathcal{J} z\rangle h$ for $z \in \mathbb{B}$, $v \in \mathcal{C}$, $h \in \mathcal{H}$, where $\mathcal{J}$ is a conjugation on $\mathcal{C}$, i.e., a mapping of $\mathcal{C}$ into itself satisfying $\mathcal{J}(v+w) = \mathcal{J}v + \mathcal{J}w$, $\mathcal{J}(cv) = \overline{c}\mathcal{J}v$, $\langle \mathcal{J}v, \mathcal{J}w\rangle = \langle w, v\rangle$, and $\mathcal{J}^2 = I_{\mathcal{C}}$ ($v, w \in \mathcal{C}$, $c \in \mathbb{C}$). However we shall mostly stick to the coordinate-dependent notation and assume for the ease of writing that d is finite.

The approach and results of this paper parallel those for the Schur-Agler class on the polydisk developed in [15] and [14]. The results on interpolation and commutant lifting in the present paper are more in the spirit of the classical univariable case than those in [15] and [14]; in the latter case, the hypotheses for existence of solutions calls for some extra parameters which are not explicitly computable in terms of the data. For other results on interpolation for analytic functions of several variables, see [19], [18], [6] and [4].

The paper is organized as follows. Section 2 identifies the structure of multipliers for the reproducing kernel space $\mathcal{H}(k_d)$ on the ball $\mathbb{B}^d$. Section 3 obtains the parallel results for a general reproducing kernel Hilbert space $\mathcal{H}(k)$ for which the reciprocal kernel $\frac{1}{k}$ has 1 positive square. In Section 4 the technique of proof is adapted to obtain the Nevanlinna-Pick interpolation theorem for left tangential interpolation problems for multipliers on such spaces, including a parametrization for the set of all interpolants. Also included here is our less explicit result on right tangential interpolation. Section 5 derives the commutant lifting theorem as a further elaboration of the method from [26]. Section 6 discusses examples (considered by Quiggin [38]) of reproducing kernel spaces for which the reciprocal kernel has 1 positive square — the Dirichlet space on the unit disk (more generally, a weighted Hardy space with a certain condition on the weights) and a space with a "separated" kernel on an interval; we exhibit explicitly an embedding $z \to b(z)$ of the corresponding set Ω into the unit ball in some naturally appearing Hilbert space. A space with a separated kernel on an interval is essentially a weighted first-order Sobolev space; it is an interesting open question whether higher order Sobolev spaces on an interval, or Sobolev spaces in more than one dimension, satisfy the property that the reciprocal of the reproducing kernel has one positive square.

2 The Multiplier Space for $\mathcal{H}(k_d)$

In this section we study the particular kernel $k_d(z, w) = \frac{1}{1-\langle z,w\rangle_{\mathbb{C}^d}}$ on the ball $\mathbb{B}^d = \{z = (z_1, \dots, z_d) \in \mathbb{C}^d \colon |z_1|^2 + \cdots + |z_d|^2 < 1\}$ and the associated reproducing kernel Hilbert space $\mathcal{H}(k_d)$. The Hilbert space $\mathcal{H}(k_d)$ can be viewed as the completion of the span of the function $k_d(\cdot, w)$ $(w \in \mathbb{B}^d)$ in the inner product

$$\langle k_d(\cdot, w), k_d(\cdot, w')\rangle = k_d(w', w). \tag{2.1}$$

For $\mathcal{E}$ an auxiliary Hilbert space, we consider the tensor product Hilbert space $\mathcal{H}(k_d) \otimes \mathcal{E}$ which can be viewed as $\mathcal{E}$-valued functions in $\mathcal{H}(k_d)$. For $\mathcal{E}$ and $\mathcal{E}_*$ two Hilbert spaces, the space of multipliers $\mathcal{M}_d(\mathcal{E}, \mathcal{E}_*)$ is defined as the space of all $\mathcal{L}(\mathcal{E}, \mathcal{E}_*)$-valued holomorphic function $z \to F(z)$ on $\mathbb{B}^d$ such that the multiplication operator $M_F \colon f(z) \to F(z)f(z)$ maps $\mathcal{H}(k_d) \otimes \mathcal{E}$ into $\mathcal{H}(k_d) \otimes \mathcal{E}_*$. A consequence of the closed graph theorem is that the multiplication operator M_F associated with any such multiplier F then is a bounded operator from $\mathcal{H}(k_d) \otimes \mathcal{E}$

into $\mathcal{H}(k_d) \otimes \mathcal{E}_*$. One of the main results of this section is the following characterization of the space $\mathcal{M}_d(\mathcal{E}, \mathcal{E}_*)$.

Theorem 2.1 *Let F be a holomorphic $\mathcal{L}(\mathcal{E}, \mathcal{E}_*)$-valued function on $\mathbb{B}^d$. Then the following statements are equivalent.*

(1) *$F \in \mathcal{M}_d(\mathcal{E}, \mathcal{E}_*)$ with $\|M_F\|_{op} \leq 1$.*

(2) *The kernel*

$$K_F(z, w) = \frac{I - F(z)F(w)^*}{1 - \langle z, w \rangle_{\mathbb{C}^d}}$$

is a positive, sesquianalytic $\mathcal{E}_*$-valued kernel on $\mathbb{B}^d \times \mathbb{B}^d$, i.e., there is an auxiliary Hilbert space $\mathcal{H}$ and a holomorphic $\mathcal{L}(\mathcal{H}, \mathcal{E}_*)$-valued function $H(z)$ on $\mathbb{B}^d$ so that

$$\frac{I - F(z)F(w)^*}{1 - \langle z, w \rangle} = H(z)H(w)^* \tag{2.2}$$

for all $z, w \in \mathbb{B}^d$.

(3a) *There exists a unitary operator*

$$V = \begin{bmatrix} A & B \\ C & D \end{bmatrix} : \begin{bmatrix} \mathcal{H} \\ \mathcal{E} \end{bmatrix} \to \begin{bmatrix} \oplus_1^d \mathcal{H} \\ \mathcal{E}_* \end{bmatrix}$$

such that

$$F(z) = D + C(I - Z(z)A)^{-1}Z(z)B \tag{2.3}$$

where we have set

$$Z(z) = [\, z_1 I_{\mathcal{H}} \;\; \dots \;\; z_d I_{\mathcal{H}} \,] : \oplus_1^d \mathcal{H} \to \mathcal{H} \tag{2.4}$$

for $z \in \mathbb{B}^d$.

(3b) *There exists an isometric operator*

$$V = \begin{bmatrix} A & B \\ C & D \end{bmatrix} : \begin{bmatrix} \mathcal{H} \\ \mathcal{E} \end{bmatrix} \to \begin{bmatrix} \oplus_1^d \mathcal{H} \\ \mathcal{E}_* \end{bmatrix}$$

of the form

$$V = \begin{bmatrix} A & B \\ C & D \end{bmatrix} = \begin{bmatrix} \widetilde{A} & B \\ \widetilde{C} & D \end{bmatrix} \begin{bmatrix} \alpha & 0 \\ 0 & I \end{bmatrix}$$

where $\alpha : \mathcal{H} \to \oplus_1^d \mathcal{H}$ is isometric and

$$\widetilde{V} = \begin{bmatrix} \widetilde{A} & B \\ \widetilde{C} & D \end{bmatrix} : \begin{bmatrix} \oplus_1^d \mathcal{H} \\ \mathcal{E} \end{bmatrix} \to \begin{bmatrix} \oplus_1^d \mathcal{H} \\ \mathcal{E}_* \end{bmatrix}$$

is unitary such that $F(z)$ is given as in (2.3) and (2.4).

Proof: Note that $F \in \mathcal{M}_d(\mathcal{E}, \mathcal{E}_*)$ with $\|M_F^*\|_{op} \leq 1$ if and only if

$$\left\| M_F^* \sum_{j=1}^{n} k(\cdot, w_j) e_{*j} \right\|^2 \leq \left\| \sum_{j=1}^{n} k(\cdot, w_j) \right\|^2$$

(where the norms are in the space $\mathcal{H}(k_d) \otimes \mathcal{E}_*$ and $\mathcal{H}(k_d) \otimes \mathcal{E}$ respectively) for any collection of points $w_1, \ldots, w_n \in \mathbb{B}^d$ and vectors $e_{*1}, \ldots, e_{*n} \in \mathbb{B}^d$. By using the reproducing kernel property, as is well known in the general theory of reproducing kernel Hilbert spaces, it is easy to check that

$$M_F^* \colon k(\cdot, w) e_* \to k(\cdot, w) F(w)^* e_*$$

Hence the condition that $\|M_F\|_{op} \leq 1$ is equivalent to

$$\left\| \sum_{j=1}^{n} k(\cdot, w_j) F(w_j)^* e_{*j} \right\|^2 \leq \left\| \sum_{j=1}^{n} k(\cdot, w_j) e_{*j} \right\|^2 .$$

Using the definition (2.1) of the inner produce for $\mathcal{H}(k_d)$ we see that this condition in turn is equivalent to

$$\sum_{j,j'=1}^{N} \left\langle \frac{I - F(w_{j'}) F(w_j)}{1 - \langle w_{j'}, w_j \rangle} e_{*j}, e_{*j'} \right\rangle \geq 0. \tag{2.5}$$

By a standard fact concerning reproducing kernel Hilbert spaces (see e.g. the introduction of Section 3.3 in [15]), the positivity condition (2.5) is equivalent to the factorization (2.2). We thus see that (1) is equivalent to (2) in Theorem 2.1.

Next, suppose that (2) holds. Thus there is an auxiliary Hilbert space $\mathcal{H}_0$ and a holomorphic function $H(z)$ with values in $\mathcal{L}(\mathcal{H}_0, \mathcal{E}_*)$ such that

$$I - F(z)F(w)^* = H(z)H(w)^* - \langle z, w \rangle H(z)H(w)^*.$$

We may rewrite this identity in the form

$$\left(\begin{bmatrix} \overline{z}_1 \\ \vdots \\ \overline{z}_d \end{bmatrix} H(z)^* \right)^* \left(\begin{bmatrix} \overline{w}_1 \\ \vdots \\ \overline{w}_d \end{bmatrix} H(w)^* \right) + I = H(z)H(w)^* + F(z)F(w)^*.$$

From this identity we see that the formula

$$V_0^* : \begin{bmatrix} \begin{bmatrix} \overline{w}_1 \\ \vdots \\ \overline{w}_d \end{bmatrix} H(w)^* e_* \\ e_* \end{bmatrix} \to \begin{bmatrix} H(w)^* \\ F(w)^* \end{bmatrix} e_* \tag{2.6}$$

extends uniquely in a well-defined way by linearity to define an isometric transformation from

$$\mathcal{D}_{V_0^*} = \text{span}\left\{ \begin{bmatrix} \begin{bmatrix} \overline{w}_1 \\ \vdots \\ \overline{w}_d \end{bmatrix} H(w)^* e_* \\ e_* \end{bmatrix} : w \in \mathbb{B}^d \text{and } e_* \in \mathcal{E}_* \right\} \tag{2.7}$$

onto

$$\mathcal{R}_{V_0^*} = \text{span}\left\{ \begin{bmatrix} H(w)^* \\ F(w)^* \end{bmatrix} e_* : w \in \mathbb{B}^d \text{ and } e_* \in \mathcal{E}_* \right\}.$$

We therefore can extend V_0^* to a unitary operator of the form

$$V^* = \begin{bmatrix} A^* & C^* \\ B^* & D^* \end{bmatrix} : \begin{bmatrix} \oplus_1^d \mathcal{H} \\ \mathcal{E}_* \end{bmatrix} \to \begin{bmatrix} \mathcal{H} \\ \mathcal{E} \end{bmatrix}$$

where $\mathcal{H}$ is a Hilbert space containing $\mathcal{H}_0$ as a subspace. Since V^* is an extension of V_0^* we have the identities

$$\begin{aligned} A^* \begin{bmatrix} \overline{w}_1 \\ \vdots \\ \overline{w}_d \end{bmatrix} H(w)^* e_* + C^* e_* &= H(w)^* e_* \\ B^* \begin{bmatrix} \overline{w}_1 \\ \vdots \\ \overline{w}_d \end{bmatrix} H(w)^* e_* + D^* e_* &= F(w)^* e_*. \end{aligned} \tag{2.8}$$

From the first of the identities (2.8) we see that

$$(I - A^* Z(w)^*) H(w)^* e_* = C^* e_*. \tag{2.9}$$

Since V is unitary, certainly A^* is contractive and $A^* Z(w)^*$ is strictly contractive for each $w \in \mathbb{B}^d$, so $(I - A^* Z(w)^*)^{-1}$ exists for each such w. Hence (2.9) can be rewritten as

$$H(w)^* e_* = (I - A^* Z(w)^*)^{-1} C^* e_*.$$

Then plugging this into the second equation in (2.8) gives us

$$B^* Z(w)^* (I - A^* Z(w)^*)^{-1} C^* e_* + D^* e_* = F(w)^* e_*$$

from which we get

$$F(w)^* = D^* + B^* Z(w)^* (I - A^* Z(w)^*)^{-1} C^*.$$

Taking adjoints and replacing w by $z \in \mathbb{B}^d$ then gives

$$F(z) = D + C(I - Z(z)A)^{-1} Z(z) B$$

and statement (3a) of Theorem 2.1 follows.

To prove statement (3b) proceed as follows. Choose an isometric embedding $\alpha = \begin{bmatrix} \alpha_1 \\ \vdots \\ \alpha_d \end{bmatrix} : \mathbb{C} \to \mathbb{C}^d$ of $\mathbb{C}$ into $\mathbb{C}^d$. Then also $\alpha \colon \mathbb{D} \to \mathbb{B}^d$. Under the assumption that the kernel K_F is positive semidefinite (statement (2) in Theorem 2.1), it follows as above that the formula

$$\widetilde{V}_0^* \colon \begin{bmatrix} \begin{bmatrix} \overline{w}_1 \\ \vdots \\ \overline{w}_d \end{bmatrix} H(w)^* e_* \\ e_* \end{bmatrix} \to \begin{bmatrix} \begin{bmatrix} \alpha_1 \\ \vdots \\ \alpha_d \end{bmatrix} H(w)^* e_* \\ F(w)^* e_* \end{bmatrix} \tag{2.10}$$

is an isometry from $\mathcal{D}_{V_0}$ as in (2.7) onto

$$\mathcal{R}_{\widetilde{V}_0} = \operatorname{span} \left\{ \begin{bmatrix} \begin{bmatrix} \alpha_1 \\ \vdots \\ \alpha_d \end{bmatrix} H(w)^* e_* \\ F(w)^* e_* \end{bmatrix} : w \in \mathbb{B}^d \text{ and } e_* \in \mathcal{E}_* \right\}.$$

Choose a unitary extension

$$\widetilde{V}^* = \begin{bmatrix} \widetilde{A}^* & \widetilde{C}^* \\ B^* & D^* \end{bmatrix} : \begin{bmatrix} \oplus_1^d \mathcal{H} \\ \mathcal{E}_* \end{bmatrix} \to \begin{bmatrix} \oplus_1^d \mathcal{H} \\ \mathcal{E} \end{bmatrix}$$

where $\mathcal{H}$ is a Hilbert space containing $\mathcal{H}_0$ as a subspace. Since $\widetilde{V}^*$ extends $\widetilde{V}_0^*$ we get the two identities

$$\begin{aligned} \widetilde{A}^* \begin{bmatrix} \overline{w}_1 \\ \vdots \\ \overline{w}_d \end{bmatrix} H(w)^* e_* + C^* e_* &= \begin{bmatrix} \alpha_1 \\ \vdots \\ \alpha_d \end{bmatrix} H(w)^* e_* \\ \widetilde{B}^* \begin{bmatrix} \overline{w}_1 \\ \vdots \\ \overline{w}_d \end{bmatrix} H(w)^* e_* + D^* e_* &= F(w)^* e_*. \end{aligned} \tag{2.11}$$

Let us set

$$\begin{bmatrix} A^* & C^* \\ B^* & D^* \end{bmatrix} = \begin{bmatrix} [\overline{\alpha}_1 & \ldots & \overline{\alpha}_d] & 0 \\ & 0 & & I \end{bmatrix} \begin{bmatrix} \widetilde{A}^* & \widetilde{C}^* \\ B^* & D^* \end{bmatrix}.$$

Then, following the same steps as in the proof of (3a) we arrive at $F(z) = D + C(I - Z(z)A)^{-1}Z(z)B$ with

$$\begin{bmatrix} A & B \\ C & D \end{bmatrix} = \begin{bmatrix} \widetilde{A} & B \\ \widetilde{C} & D \end{bmatrix} \begin{bmatrix} \begin{bmatrix} \alpha_1 \\ \vdots \\ \alpha_d \end{bmatrix} & 0 \\ 0 & I \end{bmatrix} : \begin{bmatrix} \mathcal{H} \\ \mathcal{E} \end{bmatrix} \to \begin{bmatrix} \oplus_1^d \mathcal{H} \\ \mathcal{E}_* \end{bmatrix}$$

and statement (3b) in Theorem 2.1 follows.

To complete the Proof of Theorem 2.1, it suffices to show that either one of statements (3a) or (3b) implies statement (2). Thus assume that the $\mathcal{L}(\mathcal{E}, \mathcal{E}_*)$-valued function $F(z)$ on $\mathbb{B}^d$ has a realization of the form (2.3) with $V = \begin{bmatrix} A & B \\ C & D \end{bmatrix}$ as in part (3a) of the Theorem. Expanding out $I - F(z)F(w)^*$ gives

$$\begin{aligned} I - F(z)F(w)^* & \\ = I - DD^* - C(I - Z(z)A)^{-1}Z(z)BD^* & \\ -DB^*Z(w)^*(I - A^*Z(w)^*)^{-1}C^* & \\ -C(I - Z(z)A)^{-1}Z(z)BB^*Z(w)^*(I - A^*Z(w)^*)^{-1}C^*. & \end{aligned} \tag{2.12}$$

Since by assumption $\begin{bmatrix} A & B \\ C & D \end{bmatrix}$ is a unitary block operator matrix (from $\mathcal{H} \oplus \mathcal{E}$ onto $[\oplus_1^d \mathcal{H}] \oplus \mathcal{E}_*$) we have the following identities:

$$\begin{aligned} I - DD^* &= CC^*, \quad BD^* = -AC^* \\ I - AA^* &= BB^*. \end{aligned}$$

Plugging these identities into (2.12) and carrying out the simplifying algebra then yields

$$\begin{aligned} I - F(z)F(w)^* = C(I - Z(z)A)^{-1}[I - Z(z)Z(w)^*] \\ (I - A^*Z(w)^*)^{-1}C^* \end{aligned}$$

from which we get

$$\frac{I - F(z)F(w)^*}{1 - \langle z, w \rangle} = H(z)H(w)^* \tag{2.13}$$

with

$$H(z) = C(I - Z(z)A)^{-1}. \tag{2.14}$$

On the other hand, suppose that $F(z)$ has a realization of the form (2.3) with $V = \begin{bmatrix} A & B \\ C & D \end{bmatrix}$ as in part (3b) of the theorem. Since by assumption

$$\begin{bmatrix} A & B \\ C & D \end{bmatrix} = \begin{bmatrix} \widetilde{A} & B \\ \widetilde{C} & D \end{bmatrix} \begin{bmatrix} \alpha & 0 \\ 0 & I \end{bmatrix}$$

with α an isometry and $\begin{bmatrix} \widetilde{A} & B \\ \widetilde{C} & D \end{bmatrix}$ unitary, we arrive at the following identities:

$$\begin{aligned} I - DD^* &= \widetilde{C}\widetilde{C}^*, \ BD^* = -\widetilde{A}\widetilde{C}^* \\ I - \widetilde{A}\widetilde{A}^* &= BB^*. \end{aligned}$$

Plugging these identities into (2.12) and carrying out the simplifying algebra in this case yields

$$\begin{aligned} I - F(z)F(w)^* &= \widetilde{C}(I - \alpha Z(z)\widetilde{A})^{-1}[I - \alpha Z(z)Z(w)^*\alpha^*] \\ & \quad (I - \widetilde{A}^* Z(w)^*\alpha^*)^{-1}\widetilde{C}^* \\ &= (1 - \langle z, w\rangle)\widetilde{C}(I - \alpha Z(z)\widetilde{A})^{-1}\alpha\alpha^* \\ & \quad (I - \widetilde{A}^* Z(w)^*\alpha^*)^{-1}\widetilde{C}^* \\ & \quad \widetilde{C}(I - \alpha Z(z)\widetilde{A})^{-1}[I - \alpha\alpha^*](I - \widetilde{A}^* Z(w)^*\alpha^*)\widetilde{C}^* \end{aligned}$$

from which we get

$$\frac{I - F(z)F(w)^*}{1 - \langle z, w\rangle} = H(z)H(w)^*$$

with $H(z)$ is this case given by

$$H(z) = [\widetilde{C}(I - \alpha Z(z)\widetilde{A})^{-1}\alpha \ \widetilde{C}(I - \alpha Z(z)\widetilde{A})^{-1}(I - \alpha\alpha^*) \otimes h(z)]$$

where the holomorphic $\mathcal{L}(\widetilde{\mathcal{H}}, \mathbb{C})$-valued function $h(z)$ is chosen to satisfy

$$\frac{1}{1 - \langle z, w\rangle} = h(z)h(w)^*.$$

Statement (2) of Theorem 2.1 now follows from either (3a) or (3b), and the proof is complete. □

Sometimes it is convenient to weaken the assumption that the colligations $\begin{bmatrix} A & B \\ C & D \end{bmatrix}$ be unitary. Let us say that the colligation

$$V = \begin{bmatrix} A & B \\ C & D \end{bmatrix} : \begin{bmatrix} \mathcal{H} \\ \mathcal{E} \end{bmatrix} \to \begin{bmatrix} \oplus_1^d \mathcal{H} \\ \mathcal{E}_* \end{bmatrix}$$

is *weakly coisometric* if it is contractive and V^* is isometric on the span of elements of the form

$$\text{(2.15)} \qquad \left\{ \begin{bmatrix} Z(w)^*(I - A^*Z(w)^*)^{-1}C^*e_* \\ e_* \end{bmatrix} : w \in \mathbb{B}^d, \ e_* \in \mathcal{E}_* \right\}.$$

Note that in the one-variable case, it is automatic that elements of the form (2.15) span the whole space as long as (C, A) is an observable pair of operators (i.e., the intersection over all nonnegative integers n of the kernels of CA^n is trivial — in the context of a colligation V, many authors use this condition as the definition of the colligation V being *closely outer connected*). In this situation the assumption of weakly coisometric is then the same as coisometric. In the multivariable case, however, it cannot be expected that elements of the form (2.15) span the whole space, since the span of (2.15) as a subspace of $\oplus_1^d \mathcal{H}$ is not invariant under the canonical projection operators P_k from $\oplus_1^d \mathcal{H}$ to the kth copy of $\mathcal{H}$ $(k = 1, \ldots, d)$. Nevertheless, we have the following result for weakly coisometric d-variable colligations.

Theorem 2.2 *Suppose that* $V: \mathcal{H} \oplus \mathcal{E} \to (\oplus_1^d \mathcal{H}) \oplus \mathcal{E}_*$ *is a weakly coisometric colligation, and let* $F(z) = D + C(I - Z(z)A)^{-1} Z(z) B$. *Then*

$$\frac{I - F(z)F(w)^*}{1 - \langle z, w \rangle} = H(z)H(w)^* \text{ with } H(z) = C(I - Z(z)A)^{-1}. \tag{2.16}$$

Proof: Note the identity

$$A^* Z(w)^* (I - A^* Z(w)^*)^{-1} C^* + C^* = (I - A^* Z(w)^*)^{-1} C^*$$

and that

$$B^* Z(w)^* (I - A^* Z(w)^*)^{-1} C^* + D^* = F(w)^*$$

by the definition of $F(z)$. Hence necessarily

$$\begin{aligned} V^* &= \begin{bmatrix} A^* & C^* \\ B^* & D^* \end{bmatrix} : \begin{bmatrix} Z(w)^*(I - A^* Z(w)^* C^* \\ I \end{bmatrix} \\ &\to \begin{bmatrix} (I - A^* Z(w)^*)^{-1} C^* \\ F(w)^* \end{bmatrix} e_* \end{aligned}$$

for each $e_* \in \mathcal{E}_*$. The hypothesis that V^* is isometric on elements of the form (2.15) forces the inner product identity

$$\begin{aligned} &\left\langle \begin{bmatrix} Z(w)^*(I - A^* Z(w)^*)^{-1} C^* \\ I \end{bmatrix} e_*, \begin{bmatrix} Z(z)^*(I - A^* Z(z)^*)^{-1} C^* \\ F(z)^* \end{bmatrix} e'_* \right\rangle \\ &= \left\langle \begin{bmatrix} (I - A^* Z(w)^*)^{-1} C^* \\ F(w)^* \end{bmatrix} e_*, \begin{bmatrix} (I - A^* Z(z)^*)^{-1} C^* \\ F(z)^* \end{bmatrix} e'_* \right\rangle \end{aligned} \tag{2.17}$$

for all $z, w \in \mathbb{B}^d$ and $e_*, E'_* \in \mathcal{E}_*$. This in turn works out to be equivalent to the identity (2.16). □

Remark: Note that if $\mathcal{E}$ and $\mathcal{E}_*$ have the same finite dimension and if $\mathcal{H}_0$ is finite dimensional then any unitary extension $V = \begin{bmatrix} A & B \\ C & D \end{bmatrix}$ of V_0 given by (2.6) necessarily requires the use of an infinite-dimensional Hilbert space $\mathcal{H}$ containing $\mathcal{H}_0$ as a subspace. On the other hand, in this finite-dimensional situation, we can always take $\mathcal{H}$ to be finite-dimensional (and even equal to $\mathcal{H}_0$) in arranging for a unitary extension $V\colon [\oplus_1^d \mathcal{H}] \oplus \mathcal{E} \to [\oplus_1^d \mathcal{H}] \oplus \mathcal{E}_*$ of the V_0 given by (2.10). This is the one advantage of the representation in (3b) over the representation in (3a) in Theorem 2.1. More flexibility while maintaining the representation for the kernel $\frac{I-F(z)F(w)^*}{1-\langle z,w\rangle}$ is attained if one simply works with a weakly coisometric V as in Theorem 2.2.

Note that Theorems 2.1 and 2.2 give realization formulas for $I - F(z)F(w)^*$ in terms of (A, B, C, D) if $F(z)$ has a unitary (or more generally, weakly coisometric) realization $F(z) = D + C(I - Z(z)A)^{-1}Z(z)B$. It is of interest to obtain such representations for the conjugated expression $I - F(\overline{z})^*F(\overline{w})$. It turns out that these expressions are more complicated, and parallel more closely the corresponding expressions for the Schur-Agler class on the polydisk $\mathbb{D}^d$ as given in [15]. The result is as follows.

Theorem 2.3 *Let F be a holomorphic $\mathcal{L}(\mathcal{E}, \mathcal{E}_*)$-valued function on $\mathbb{B}^d$. Then the following statements are equivalent.*

(1) $F \in \mathcal{M}_d(\mathcal{E}, \mathcal{E}_*)$ *with* $\|M_F\|_{op} \le 1$.

(2) *There is an auxiliary Hilbert space $\mathcal{H}_0$ and an analytic $\mathcal{L}(\oplus_1^d \mathcal{H}_0, \mathcal{E})$-valued function $H(z)$ such that*

$$I - F(\overline{z})^*F(\overline{w}) = H(z)[I - Z(\overline{z})^*Z(\overline{w})]H(w)^*$$

where $Z(z) = [z_1 I_{\mathcal{H}} \ \ldots \ z_d I_{\mathcal{H}}]$.

(3) *Statement* (3) *of Theorem* 2.1 *holds.*

Proof: By Theorem 2.1, we need only show that (2) $\Longrightarrow$ (3a) in Theorem 2.1 and that (3a) $\Longrightarrow$ (2).

Assume therefore that there is an analytic function $H(z)$ so that (2) holds in the statement of the theorem. Rewrite this in the form

$$H(z)Z(\overline{z})^*Z(\overline{w})H(w)^* + I = H(z)H(w)^* + F(\overline{z})^*F(\overline{w}).$$

Therefore we can define an isometry V_0 from

$$\mathcal{D}_{V_0} := \text{closed span}\left\{\begin{bmatrix} Z(\overline{w})H(w)^* \\ I \end{bmatrix} e\colon z \in \mathbb{B}^d,\ e \in \mathcal{E}\right\}$$

onto

$$\mathcal{R}_{V_0} := \text{closed span}\left\{\begin{bmatrix} H(w)^* \\ F(\overline{w}) \end{bmatrix} e\colon z \in \mathbb{B}^d,\ e \in \mathcal{E}\right\}$$

by

$$V_0\colon \begin{bmatrix} Z(\overline{w})H(w)^* \\ I \end{bmatrix} e \to \begin{bmatrix} H(w)^* \\ F(\overline{w}) \end{bmatrix} e.$$

We can always extend V_0 to a unitary operator

$$V = \begin{bmatrix} A & B \\ C & D \end{bmatrix} : \begin{bmatrix} \mathcal{H} \\ \mathcal{E} \end{bmatrix} \to \begin{bmatrix} \oplus_1^d \mathcal{H} \\ \mathcal{E}_* \end{bmatrix}$$

where $\mathcal{H}$ is a Hilbert space containing $\mathcal{H}_0$ as a subspace if necessary. We then get the system of operator equations

$$\begin{aligned} AZ(\overline{w})H(w)^* + B &= H(w)^* \\ CZ(\overline{w})H(w)^* + D &= F(\overline{w}). \end{aligned}$$

Solving the first equation for $H(w)^*$ gives $H(w)^* = (I - AZ(\overline{w}))^{-1}B$. Plug this back into the second equation to get

$$F(\overline{w}) = CZ(\overline{w})(I - AZ(\overline{w}))^{-1}B + D = D + C(I - Z(\overline{w})A)^{-1}Z(\overline{w})B$$

and it follows that $F(z)$ has a representation as in (3a) as required.

Conversely, suppose that $F(z)$ has a realization as in (2.3). Expanding out $I - F(\overline{z})^* F(\overline{w})$ and using the relations

$$I - D^*D = B^*B,\ D^*C = -B^*A,\ C^*D = -A^*B,\ C * C = I - A^*A$$

resulting from the assumption that $U = \begin{bmatrix} A & B \\ C & D \end{bmatrix}$ is unitary, we get

$$\begin{aligned} & I - F(\overline{z})^* F(\overline{w}) \\ & = I - D^*D - D^*C(I - Z(\overline{w})A)^{-1}Z(\overline{w})B \\ & \quad -B^*(I - Z(\overline{z})^*A^*)^{-1}Z(\overline{z})^*C^*D \\ & \quad -B^*(I - Z(\overline{z})^*A^*)^{-1}Z(\overline{z})^*C^*CZ(\overline{w})(I - AZ(\overline{w}))^{-1}B \\ & = B^*(I - Z(\overline{z})^*A^*)^{-1}\{(I - Z(\overline{z})^*A^*)(I - AZ(\overline{w})) \\ & \quad +(I - Z(\overline{z})^*A^*)AZ(\overline{w}) + Z(\overline{z})^*A^*(I - AZ(\overline{w}))^{-1} \\ & \quad -Z(\overline{z})^*(I - A^*A)Z(\overline{w})\}(I - AZ(\overline{w}))^{-1}B \\ & = B^*(I - Z(\overline{z})^*A^*)^{-1}\{I - Z(\overline{z})^*Z(\overline{w})\} \\ & \quad (I - AZ(\overline{w}))^{-1}B \end{aligned}$$

and hence the representation (1) holds with $H(z) = B^*(I - Z(\overline{z})^*A^*)^{-1}$. □

As a useful dual to the notion of weakly coisometric defined immediately preceding Theorem 2.2, we now define a contractive colligation $U = \begin{bmatrix} A & B \\ C & D \end{bmatrix} : \begin{bmatrix} \mathcal{H} \\ \mathcal{E} \end{bmatrix} \to \begin{bmatrix} \oplus_1^d \mathcal{H} \\ \mathcal{E}_* \end{bmatrix}$ to be *weakly isometric* if U is isometric on the span of elements of the form

(2.18) $$\left\{ \begin{bmatrix} Z(\overline{w})(I - AZ(\overline{w}))^{-1}B \\ I \end{bmatrix} e \colon w \in \mathbb{B}^d,\ e \in \mathcal{E} \right\}.$$

In the one-dimensional case, it is automatic that the span of the set (2.18) is dense in the whole space, as long as (A, B) is an approximately controllable pair of operators (i.e., the span over all nonnegative integers n of the images of the operators $A^n B$ is dense in the whole space $\mathcal{H}$ — in the context of colligations, many authors use this notion as the definition of the colligation V being *closely inner connected*). In such a case it follows that the notions of weakly isometric and isometric coincide. From the proof of (3a) $\Longrightarrow$ (2) we observe the following dual version of Theorem 2.2.

Theorem 2.4 *Suppose that* $U: \mathcal{H} \oplus \mathcal{E} \to (\oplus_1^d \mathcal{H}) \oplus \mathcal{E}_*$ *is a weakly isometric* d*-variable colligation. Then* $F(z) = D + C(I - Z(z)A)^{-1} Z(z) B \in \mathcal{M}_d(\mathcal{E}, \mathcal{E}_*)$ *and*

$$I - F(\overline{z})^* F(\overline{w}) = H(z)(I - Z(\overline{z})^* Z(\overline{w})) H(w)^* \tag{2.19}$$

with $H(z) = B^*(I - Z(\overline{w})^* A^*)^{-1}$ *and* $Z(z) = [z_1 I_{\mathcal{H}} \ \ldots \ z_d I_{\mathcal{H}}]$.

Proof: From the identities

$$AZ(\overline{w})(I - AZ(\overline{w}))^{-1} B + B = (I - AZ(\overline{w}))^{-1} B \tag{2.20}$$

$$CZ(\overline{w})(I - AZ(\overline{w}))^{-1} B + D = F(\overline{w}) \tag{2.21}$$

we see that

$$U: \begin{bmatrix} Z(\overline{w})(I - AZ(\overline{w}))^{-1} B \\ I \end{bmatrix} e \to \begin{bmatrix} (I - AZ(\overline{w}))^{-1} B \\ F(\overline{w}) \end{bmatrix} e.$$

Since U is an isometry on the span of (2.18), we conclude that

$$\left\langle \begin{bmatrix} Z(\overline{w})(I - AZ(\overline{w}))^{-1} B \\ I \end{bmatrix} e, \begin{bmatrix} Z(\overline{z})(I - AZ(\overline{z}))^{-1} B \\ I \end{bmatrix} e' \right\rangle$$
$$= \left\langle \begin{bmatrix} (I - AZ(\overline{w}))^{-1} B \\ F(\overline{w}) \end{bmatrix} e, \begin{bmatrix} (I - AZ(\overline{z}))^{-1} B \\ F(\overline{z}) \end{bmatrix} e' \right\rangle.$$

If one now expands out these inner product identities and regroups, one obtains the representation (2.19) as required. □

3 Multipliers for Nevanlinna-Pick-type Kernels

Suppose that $k = k(z, w)$ is a positive kernel on the set Ω (where Ω is just an abstract set of points), i.e., suppose that k can be expressed in the form

$$k(z, w) = h(z) h(w)^*$$

for some function $h(z)$ on Ω with values in $\mathcal{L}(\mathcal{H}, \mathbb{C})$ for some auxiliary Hilbert space $\mathcal{H}$. As in the previous section, in addition to the reproducing kernel Hilbert space $\mathcal{H}(k)$ associated with $k(\cdot,\cdot)$ (the elements of which consist of scalar-valued functions on Ω), we may consider the vector-valued versions $\mathcal{H}(k)\otimes\mathcal{E}$ consisting of $\mathcal{E}$-valued functions on Ω, where $\mathcal{E}$ is any Hilbert space. Given two Hilbert spaces, $\mathcal{E}$ and $\mathcal{E}_*$, denote by $\mathcal{M}_k(\mathcal{E},\mathcal{E}_*)$ the associated space of multipliers

$$\begin{aligned}\mathcal{M}_k(\mathcal{E},\mathcal{E}_*) &= \{W\colon\Omega\to\mathcal{L}(\mathcal{E},\mathcal{E}_*)\colon W(z)f(z)\in\mathcal{H}(k)\otimes\mathcal{E}_*\\ &\qquad \text{for all } f(z)\in\mathcal{H}(k)\otimes\mathcal{E}\},\end{aligned}$$

and for $W\in\mathcal{M}_k(\mathcal{E},\mathcal{E}_*)$ let M_W be the corresponding multiplication operator.

We shall be primarily concerned with the case where the inverse kernel $\frac{1}{k(z,w)}$ has one positive square, i.e. the case where there is a scalar function $a(z)$ and an operator-valued function $b(z)$ (with values in $\mathcal{L}(\mathcal{C},\mathbb{C})$ for some auxiliary Hilbert space $\mathcal{C}$) such that

$$\frac{1}{k(z,w)} = a(z)\overline{a(w)} - b(z)b(w)^* \tag{3.1}$$

for all $z,w\in\Omega$. More concretely, we may choose a basis for $\mathcal{C}$ and view $\mathcal{C}$ as equal to $\mathbb{C}^d$, where $d=\dim\mathcal{C}$. If $d=\infty$, we identify $\mathbb{C}^d$ with the space ℓ^2 of square-summable complex sequences. (All Hilbert spaces are assumed to be separable.) Most of the notation in the following exposition will be set with $d<\infty$ in mind, but the reader can check that all the statements hold true with $d=\infty$ with a minor change of notation; alternatively, it is possible to use throughout a coordinate-free notation as mentioned in the Introduction.

It will often be convenient to assume that $a(z)=1$ in the representation (3.1). This can be achieved after a minor normalization as follows. We assume that each point evaluation $f(\cdot)\to f(w)$ for $w\in\Omega$ on $\mathcal{H}(k)$ is nontrivial, and hence $k(\cdot,w)$ is not the zero element of $\mathcal{H}(k)$ (when this is the case, we shall say that the kernel k is *nondegenerate* over Ω). It follows that $\|k(\cdot,w)\|^2=k(w,w)>0$ from which we see that $0<\frac{1}{k(w,w)}=|a(w)|^2-b(w)b(w)^*$. We conclude that $a(z)\neq 0$ for all $z\in\Omega$. Then the transformation $f(\cdot)\to a(\cdot)f(\cdot)$ transforms $\mathcal{H}(k)$ unitarily onto a new reproducing kernel Hilbert space $\mathcal{H}(k')$ where $k'(z,w)=a(z)k(z,w)\overline{a(w)}$. Then $\frac{1}{k'(z,w)}=1-b'(z)b'(w)^*$ where $b'(z)=b(z)/a(z)$. In the sequel we shall usually assume that this normalization has already been done.

The following analogue of Theorem 2.1 characterizes the space $\mathcal{M}_k(\mathcal{E},\mathcal{E}_*)$, and identifies a universal property of $\mathcal{M}_d(\mathcal{E},\mathcal{E}_*)$ $(=\mathcal{M}_{k_d}(\mathcal{E},\mathcal{E}_*))$, the multiplier space associated with the particular kernel $k_d(z,w)=\frac{1}{1-\langle z,w\rangle}$ on the ball $\mathbb{B}^d$ discussed in Section 2.

Theorem 3.1 *Let $k(\cdot,\cdot)$ be a positive definite kernel on $\Omega\times\Omega$ with associated reproducing kernel Hilbert space $\mathcal{H}(k)$ and let $W(\cdot)$ be a $\mathcal{L}(\mathcal{E},\mathcal{E}_*)$-valued function on Ω. Then the following conditions are equivalent.*

(1) *$W\in\mathcal{M}_k(\mathcal{E},\mathcal{E}_*)$ with $\|M_W\|_{op}\le 1$.*

(2) *The kernel $K_W(z,w):=k(z,w)(I_{\mathcal{E}_*}-W(z)W(w)^*)$ is a positive $\mathcal{L}(\mathcal{E}_*)$-valued kernel on Ω.*

In case k is a nondegenerate kernel and $\frac{1}{k(\cdot,\cdot)}$ has 1 positive-square, i.e., in case there is an $\mathcal{L}(\mathcal{C}, \mathbb{C})$-valued function $b(z)$ (where $\mathcal{C}$ is some other Hilbert space) such that (possibly after a normalization of k) $\frac{1}{k(\cdot,\cdot)}$ can be expressed in the form

$$\frac{1}{k(z, w)} = 1 - b(z)b(w)^*, \tag{3.2}$$

then either (1) *or* (2) *is also equivalent to:*

(3) *There is a multiplier $F \in \mathcal{M}_d(\mathcal{E}, \mathcal{E}_*)$ (where $\mathcal{M}_d(\mathcal{E}, \mathcal{E}_*)$ is as in Theorem* 2.1 *and $d = \dim \mathbb{C}$) such that $W(z) = F \circ b(z)$ for all $z \in \Omega$, where $b(\cdot): \mathbb{C} \to \mathbb{C}^d$ is the function appearing in* (3.2).

Proof: The proof of (1) $\iff$ (2) follows as in the proof of (1) $\iff$ (2) in Theorem 2.1; this is a basic general feature of multipliers on reproducing kernel Hilbert spaces.

Let us therefore assume statement (2) of the Theorem and the condition (3.2) on the kernel k. The statement (2) gives us an $\mathcal{L}(\mathcal{H}_0, \mathcal{E}_*)$-valued function $H(z)$ so that

$$k_W(z, w) := k(z, w)(I - W(z)W(w)^*) = H(z)H(w)^* \tag{3.3}$$

for some auxiliary Hilbert space $\mathcal{H}_0$. Divide through by $k(z, w)$ and use the assumption (3.2) to get

$$I_{\mathcal{E}_*} - W(z)W(w)^* = H(z)((1 - b(z)b(w)^*) \otimes I_{\mathcal{E}})H(w)^*.$$

Rewrite this identity as

$$\begin{aligned} &[b(z) \otimes H(z)][b(w)^* \otimes H(w)^*] + I_{\mathcal{E}_*} \\ &= H(z)H(w)^* + W(z)W(w)^*. \end{aligned} \tag{3.4}$$

At this stage let us identify the Hilbert space $\mathcal{C}$ with $\mathbb{C}^d$ (where $d = \dim \mathcal{C}$). Then $\mathcal{C} \otimes \mathcal{H}_0$ can be identified with $\oplus_1^d \mathcal{H}_0$ and

$$b(w)^* \otimes H(w)^* = \begin{bmatrix} \overline{b}_1(w) \\ \vdots \\ \overline{b}_d(w) \end{bmatrix} H(w)^*: \mathcal{E}_* \to \oplus_1^d \mathcal{H}_0$$

where $b(z) = [b_1(z) \ \dots \ b_d(z)]: \mathbb{C}^d \to \mathbb{C}$ (so the components $b_j(\cdot)$ are complex-valued functions on Ω). From the identity (3.4) we see that the formula

$$V_0^*: \begin{bmatrix} b(w)^*H(w)^*e_* \\ e_* \end{bmatrix} \to \begin{bmatrix} H(w)^* \\ W(w)^* \end{bmatrix} e_* \tag{3.5}$$

extends by linearity and continuity to define a unique isometry from the domain

$$\mathcal{D}_{V_0^*} = \operatorname{span}\left\{\begin{bmatrix} b(w)^* H(w)^* e_* \\ e_* \end{bmatrix} : w \in \Omega \text{ and } e_* \in \mathcal{E}_*\right\}$$

onto the range

$$\mathcal{R}_{V_0^*} = \operatorname{span}\left\{\begin{bmatrix} H(w)^* \\ W(w)^* \end{bmatrix} e_* : w \in \Omega \text{ and } e_* \in \mathcal{E}_*\right\}.$$

Extend V_0^* to a unitary operator

$$V^* = \begin{bmatrix} A^* & C^* \\ B^* & D^* \end{bmatrix} : \begin{bmatrix} \oplus_1^d \mathcal{H} \\ \mathcal{E}_* \end{bmatrix} \to \begin{bmatrix} \mathcal{H} \\ \mathcal{E} \end{bmatrix}$$

where $\mathcal{H}$ is a Hilbert space containing $\mathcal{H}_0$ as a subspace. Since V^* extends V_0^* we obtain the system of equations

$$\begin{aligned} A^* b(w)^* H(w)^* e_* + C^* e_* &= H(w)^* e_* \\ B^* b(w)^* H(w)^* e_* + D^* e_* &= W(w)^* e_* \end{aligned} \tag{3.6}$$

for all $w \in \Omega$ and $e_* \in \mathcal{E}_*$. From the fact that $k(w, w) > 0$ for $w \in \Omega$ we see that

$$0 < \frac{1}{k(w, w)} = 1 - b(w)b(w)^*$$

and hence $b(w)^* \in \mathbb{B}^d$ for all $w \in \Omega$. Also, since the operator A^* is a block in the unitary operator matrix V^*, certainly $\|A\| \leq 1$. We conclude that $I - A^* b(w)^* : \mathcal{H} \to \mathcal{H}$ is invertible for all $w \in \Omega$. Hence we can solve for $H(w)^* e_*$ in the first of equations (3.6) to get

$$H(w)^* e_* = (I - A^* b(w)^*)^{-1} C^* e_*.$$

Substituting this into the second of equations (3.6) yields the formula for $W(w)^*$:

$$W(w)^* = D^* + B^* b(w)^* (I - A^* b(w)^*)^{-1} C^*.$$

Taking adjoints and replacing w by $z \in \Omega$ then gives

$$W(z) = D + C(I - b(z)A)^{-1} b(z) B. \tag{3.7}$$

By statement (3a) in Theorem 2.1 we see that the $\mathcal{L}(\mathcal{E}, \mathcal{E}_*)$-valued function F defined by $F(\omega) = D + C(I - Z(\omega)A)^{-1} Z(\omega) B$ (for $\omega \in \mathbb{B}^d$) is in the multiplier class $\mathcal{M}_d(\mathcal{E}, \mathcal{E}_*)$ associated with the kernel $k_d(\omega, \omega') = \frac{1}{1 - \langle \omega, \omega' \rangle}$ on $\mathbb{B}^d$. From the formula (3.7), it is clear that $W(z)$ is the composition $W(z) = F \circ b(z)$ where $b(\cdot)$ is a mapping from Ω into $\mathbb{B}^d$. Statement (3) in Theorem 3.1 now follows.

Conversely, suppose that there is a multiplier $F \in \mathcal{M}_d(\mathcal{E}, \mathcal{E}_*)$ for the ball kernel function $k_d(\omega, \omega') = \frac{1}{1 - \langle w, w' \rangle}$ such that $W = F \circ b$ where $b : \Omega \to \mathbb{B}^d$ is as in

(3.2). As explained in the proof of (2) $\Longrightarrow$ (3), we know that $b(\cdot)$ maps Ω into $\mathbb{B}^d$. Since F is a multiplier for k_d, we know that

$$\frac{I - F(\omega)F(\omega')^*}{1 - \langle \omega, \omega' \rangle} = H(\omega)H(\omega')^* \tag{3.8}$$

for some $\mathcal{L}(\mathcal{H}_0, \mathcal{E}_*)$-valued function $H(\cdot)$ holomorphic on $\mathbb{B}^d$, where $\mathcal{H}_0$ is some auxiliary Hilbert space, by Theorem 2.1. Substitution of $b(z)$ for ω and of $b(w)$ for ω' in (3.8) then gives

$$k(z, w)(I - W(z)W(w)^*) = (H \circ b(z))(H \circ b(w))^*$$

and statement (2) in Theorem 3.1 follows. This completes the Proof of Theorem 3.1. □

4 Interpolation by Multipliers

In this section we explore interpolation problems for elements of multiplier spaces associated with the type of kernels discussed in the previous Section 3. We assume therefore that we are given a nondegenerate, positive kernel $k(\cdot, \cdot)$ on Ω along with two Hilbert spaces $\mathcal{E}$ and $\mathcal{E}_*$. We now formulate a fairly general class of interpolation problems where the unknown interpolant is required to be in the multiplier class $\mathcal{M}_k(\mathcal{E}, \mathcal{E}_*)$.

The data set for the interpolation problem is as follows. We are given n points $w_1, \dots, w_n \in \Omega$ along with auxiliary input Hilbert spaces $\mathcal{M}_1, \dots, \mathcal{M}_n$. For each $j = 1, \dots, n$ we are also given operators $x_j \in \mathcal{L}(\mathcal{M}_j, \mathcal{E}_*)$ and $y_j \in \mathcal{L}(\mathcal{M}_j, \mathcal{E})$. The problem then is to find a multiplier $W(\cdot) \in \mathcal{M}_k(\mathcal{E}, \mathcal{E}_*)$ with $\|M_W\|_{op} \leq 1$ which satisfies the interpolation conditions

$$W(w_j)^* x = y \text{ for } j = 1, \dots, n. \tag{4.1}$$

The following result (due to McCullough [27] for the case of scalar Carathéodory interpolation, to Quiggin [38] for the scalar case ($\mathcal{E} = \mathcal{E}_* = \mathcal{M}_j = \mathbb{C}$ and $x_j = 1$) and to Agler and McCarthy [3] for the full matrix-valued case ($\mathcal{M}_j = \mathcal{E}$ and $x_j = I$ for all j), shows that the answer is completely analogous to classical Nevanlinna-Pick interpolation for the case where the kernel $k(\cdot, \cdot)$ is such that $\frac{1}{k(\cdot,\cdot)}$ has 1 positive square. In fact, as pointed out in [3] (see also [28]) a converse assertion holds as well, but we do not discuss the converse here.

Theorem 4.1 *Assume that the kernel $k(\cdot, \cdot)$ on $\Omega \times \Omega$ is such that*

$$\frac{1}{k(z, w)} = 1 - b(z)b(w)^*$$

for a strictly contractive matrix function

$$b(z) = [b_1(z) \ \dots \ b_d(z)] : \mathbb{C}^d \to \mathbb{C}.$$

Then there exists a multiplier $W \in \mathcal{M}_k(\mathcal{E}, \mathcal{E}_*)$ *with* $\|M_W\|_{op} \leq 1$ *which satisfies the interpolation conditions* (4.1) *if and only if the block operator matrix*

$$[P_{ij}] = [k(w_i, w_j)(x_i^* x_j - y_i^* y_j)]_{i,j=1,\ldots,n} \tag{4.2}$$

is positive semidefinite.

Proof: The necessity of the positive semidefiniteness of the Pick matrix (4.2) for there to exist an interpolating multiplier is true for any positive kernel k. Indeed, if $W(z)$ is an interpolant, then one easily checks that $M_W^* \colon k(\cdot, w_i)e_* \to k(\cdot, w_i)W(w_i)^* x_i = k(\cdot, w_i)y_i$. The positive semidefiniteness of the Pick matrix (4.2) is just a concrete expression of the contractiveness of the operator M_W on the span of the elements $k(\cdot, w_j)x_j m_j$ ($j = 1, \ldots, n$ and $m_j \in \mathcal{M}_j$) in $\mathcal{H}(k) \otimes \mathcal{E}_*$.

Conversely, suppose that the Pick matrix (4.2) is positive semidefinite. Then there exists a factorization

$$k(w_i, w_j)(x_i^* x_j - y_i^* y_j) = H_i H_j^* \tag{4.3}$$

where $H_i \in \mathcal{L}(\mathcal{H}_0, \mathcal{M})$ for some auxiliary Hilbert space $\mathcal{H}_0$. Dividing through by $k(w_i, w_j)$ gives

$$x_i^* x_j - y_i^* y_j = H_i^*(1 - b(w_i)b(w_j)^*)H_j^*$$

which we prefer to rearrange as

$$H_i^* b(w_i)^* b(w_j)^* H_j^* + x_i^* x_j = H_i^* H_j^* + y_i^* y_j. \tag{4.4}$$

As in the Proof of Theorem 3.1 we identify $b(w)^*$ with $b(w)^* \otimes I_{\mathcal{H}_0}$, an embedding of $\mathcal{H}_0$ into $\oplus_1^d \mathcal{H}_0$. Then the formula

$$V_0^* : \begin{bmatrix} b(w_j)^* H_j^* m_j \\ x_j m_j \end{bmatrix} \to \begin{bmatrix} H_j^* m_j \\ y_j m_j \end{bmatrix} \tag{4.5}$$

extends by linearity and continuity to define an isometric mapping of the domain space

$$\mathcal{D}_{V_0^*} = \text{span} \left\{ \begin{bmatrix} b(w_j)^* H_j^* m_j \\ x_j m_j \end{bmatrix} : j = 1, \ldots, n \text{ and } m_j \in \mathcal{M}_j \right\}$$

onto the range space

$$\mathcal{R}_{V_0^*} = \text{span} \left\{ \begin{bmatrix} H_j^* m_j \\ y_j m_j \end{bmatrix} : j = 1, \ldots, n \text{ and } m_j \in \mathcal{M}_j \right\}.$$

Extend the partially defined isometry V_0^* to a unitary operator

$$V^* = \begin{bmatrix} A^* & C^* \\ B^* & D^* \end{bmatrix} : \begin{bmatrix} \oplus_1^d \mathcal{H} \\ \mathcal{E}_* \end{bmatrix} \to \begin{bmatrix} \mathcal{H} \\ \mathcal{E} \end{bmatrix}$$

where $\mathcal{H}$ is a Hilbert space containing $\mathcal{H}_0$ as a subspace (and is necessarily infinite-dimensional if $\mathcal{E}$ and $\mathcal{E}_*$ are of the same finite dimension). As in the Proof of Theorem 3.1 we arrive at the system of equations

$$\begin{aligned} A^*b(w_j)^*H_j^*m_j + C^*x_jm_j &= H_j^*m_j \\ B^*b(w_j)^*H_j^*m_j + D^*x_jm_j &= y_jm_j \end{aligned} \tag{4.6}$$

Just as in the Proof of Theorem 3.1, it is clear that $I - A^*b(w_j)^*$ is invertible on $\mathcal{H}$. Hence we can solve the first of equations (4.6) for $H_j^*m_j$ to get

$$H_j^*m_j = (I - A^*b(w_j)^*)^{-1}C^*x_jm_j$$

for $j = 1, \ldots, n$ and all $m_j \in \mathcal{M}_j$. Substitution of this expression into the second of equations (4.6) gives

$$[D^* + B^*b(w_j)^*(I - A^*b(w_j)^*)^{-1}C^*]x_jm_j = y_jm_j. \tag{4.7}$$

By the result of Theorem 3.1, it is clear that the $\mathcal{L}(\mathcal{E}, \mathcal{E}_*)$-valued function $W(z)$ defined by $W(z) = D + C(I - b(z)A)^{-1}b(z)C$ is in the multiplier class $\mathcal{M}_k(\mathcal{E}, \mathcal{E}_*)$, since the operator

$$V_0 = \begin{bmatrix} A & B \\ C & D \end{bmatrix}$$

is unitary, as required in part (3a) of Theorem 2.1. Moreover the identity (4.7) simply says that this $W(z)$ meets the interpolation conditions $W(w_j)^*x_j = y_j$. Theorem 4.1 now follows. □

Note that Theorem 4.1 actually says more: contractive solutions $W(\cdot) \in \mathcal{M}_k(\mathcal{E}, \mathcal{E}_*)$ of the interpolation conditions (4.1) arise from unitary extensions of the partially defined isometry V_0 (or V_0^*) given by (4.5). We would like to know the converse: *any contractive solution of the interpolation conditions* (4.1) *in* $\mathcal{M}_k(\mathcal{E}, \mathcal{E}_*)$ *arises from a unitary extension* V *of* V_0 *given by* (4.5). This is the content of the next result.

Theorem 4.2 *Suppose that* $W(\cdot) \in \mathcal{M}_k(\mathcal{E}, \mathcal{E}_*)$ *is a contractive solution of the interpolation conditions* (4.1). *Then there is a unitary extension* V *of* V_0 *given by* (4.5) *so that* F *is of the form as constructed in the Proof of Theorem* 4.1.

Proof: Suppose that $W(\cdot)$ is an interpolant in $\mathcal{M}_k(\mathcal{E}, \mathcal{E}_*)$ with $\|M_W\|_{op} \leq 1$. By Theorem 3.1 W has a realization

$$W(z) = D + C(I - b(z)A)^{-1}b(z)B \tag{4.8}$$

where

$$V = \begin{bmatrix} A & B \\ C & D \end{bmatrix} : \begin{bmatrix} \mathcal{H} \\ \mathcal{E} \end{bmatrix} \to \begin{bmatrix} \oplus_1^d \mathcal{H} \\ \mathcal{E}_* \end{bmatrix}$$

is unitary. If we set $\widetilde{H}_i = C(I - b(w_i)A)^{-1}$ we see from (2.13) and (2.14) that

$$k(w_i, w_j)(I - W(w_i)W(w_j)^*) = \widetilde{H}_i \widetilde{H}_j^*.$$

Since W also satisfies the interpolation conditions $W(w_j)^* x_j = y_j$ we get from this

$$k(w_i, w_j)(x_i^* x_j - y_i^* y_j) = H_i H_j^* \tag{4.9}$$

where we have set

$$H_i = x_i^* \widetilde{H}_i = x_i^* C(I - b(w_i)A)^{-1}. \tag{4.10}$$

Hence H_i defined by (4.10) provides a factorization of the Pick matrix as in the starting point of the Proof of Theorem 4.1. Next note that the interpolation conditions satisfied by $W(\cdot)$ can be written out explicitly in terms of A, B, C, D as

$$W(w_j)^* x_j = [D^* + B^* b(w_j)^* (I - A^* b(w_j)^*)^{-1} C^*] x_j = y_j$$

for each $j = 1, \ldots, n$. From (4.10) we can rewrite each of these equations as the system

$$\begin{bmatrix} A^* & C^* \\ B^* & D^* \end{bmatrix} \begin{bmatrix} b(w_j)^* H_j^* \\ x_j \end{bmatrix} = \begin{bmatrix} H_j^* \\ y_j \end{bmatrix}.$$

We conclude that the operator

$$V^* = \begin{bmatrix} A^* & B^* \\ C^* & D^* \end{bmatrix}$$

is a unitary extension of the partially defined isometry V_0^* given by (4.5). Moreover, from the formula (4.8) for W, it is clear that the construction in the Proof of Theorem 4.1 using this unitary extension V of V_0 leads to the original interpolant W, as claimed. The theorem follows. □

Consider now the right interpolation problem: *given points* $\{z_1, \ldots, z_n\} \in \Omega$, *operators* $u_j \in \mathcal{L}(\mathcal{M}_j, \mathcal{E})$ *and* $v_j \in \mathcal{L}(\mathcal{M}_j, \mathcal{E}_*)$ *for* $j = 1, \ldots, n$ *(where* $\mathcal{M}_1, \ldots, \mathcal{M}_n$ *are auxiliary Hilbert spaces), find* $W \in \mathcal{M}_k(\mathcal{E}, \mathcal{E}_*)$ *such that*

$$F(z_i)u_i = v_i \textit{ for } i = 1, \ldots, n. \tag{4.11}$$

Let us consider first the case where the underlying domain Ω is the ball $\mathbb{B}^d$ and the kernel k is the kernel k_d. In the case $d = 1$, one can easily reduce this right interpolation problem to a left interpolation problem as follows. Rewrite the right interpolation condition (4.11) as

$$G(\overline{z}_i)^* u_i = v_i$$

with $G(z) := F(\overline{z})^* \in H^\infty(\mathcal{L}(\mathcal{E}_*, \mathcal{E})$. This is a left interpolation problem for G with data $w_j = \overline{z}_j, x_j = u_j, y_j = v_j$. If G is a solution of this left interpolation problem with H^∞ norm at most 1 and we set $F(z) = G(\overline{z})^*$, then F is an operator-valued H^∞ function again of norm at most 1 and solves the right interpolation conditions (4.11). Thus a theory for left interpolation for the unit ball of the multiplier class $\mathcal{M}_1(\mathcal{E}_*, \mathcal{E})$ is equivalent to a theory for right interpolation for the unit ball of the multiplier class $\mathcal{M}_1(\mathcal{E}, \mathcal{E}_*)$. However, as is suggested by comparing Theorems 2.1 and 2.2 with Theorems 2.3 and 2.4, for $d > 1$ there appears to be no reason why the space of multipliers on $\mathcal{H}(k_d)$ should be invariant under the transformation $f(z) \to \overline{f(\overline{z})}$. Hence also there is no a priori reason why in general $F(\overline{z})^*$ should belong to $\mathcal{M}_d(\mathcal{E}_*, \mathcal{E})$ if $F(z)$ is in $\mathcal{M}_d(\mathcal{E}, \mathcal{E}_*)$. This suggests that a theorem characterizing when solutions exist for the right interpolation problem for $\mathcal{M}_k(\mathcal{E}, \mathcal{E}_*)$ will be formally different from the result Theorem 4.1 for left interpolation. We next give such a result for right interpolation. As is to be expected, the result is much less explicit than that for the left interpolation problem.

Theorem 4.3 *Assume that the kernel $k(\cdot, \cdot)$ on $\Omega \times \Omega$ is such that $\frac{1}{k(z,w)} = 1 - b(z)b(w)^*$ for a block row matrix function $b(z)$ with d components, as in the hypotheses of Theorem* 4.1. *Then there is a multiplier $W \in \mathcal{M}_k(\mathcal{E}, \mathcal{E}_*)$ with $\|M_W\|_{op} \leq 1$ which satisfies the interpolation conditions* (4.11) *if and only if there exist an auxiliary Hilbert space $\mathcal{H}_0$ and operators $H_{jk}\colon \mathcal{M}_j \to \mathcal{H}_0$ (for $j = 1, \dots, n$ and $k = 1, \dots, d$) so that*

$$u_i^* u_j - v_i^* v_j = \sum_{k,\ell=1}^{d} (\delta_{k\ell} - \overline{b_k(z_i)} b_\ell(z_j)) H_{ik}^* H_{j\ell} \tag{4.12}$$

for $i, j = 1, \dots, n$, where $\delta_{k,\ell}$ is the Kronecker delta equal to 1 *for $k = \ell$ and* 0 *for $k \neq \ell$.*

Proof: Suppose first that there is a multiplier $W \in \mathcal{M}_k(\mathcal{E}, \mathcal{E}_*)$ with $\|M_W\|_{op} \leq 1$ meeting the interpolation conditions (4.11). By Theorem 3.1 there is a $\mathbb{B}^d$-multiplier $F \in \mathcal{M}_d(\mathcal{E}, \mathcal{E}_*)$ such that $W(z) = F \circ b(z)$. Also, by Theorem 2.3 we know that there is an auxiliary Hilbert space $\mathcal{H}$ and an analytic function $H(z) = [H_1(z) \;\; \dots \;\; H_d(z)]$ on $\mathbb{B}^d$ with values in $\mathcal{L}(\oplus_1^d \mathcal{H}, \mathcal{E})$ such that

$$I - F(\overline{\omega})^* F(\overline{\eta}) = H(\omega)(I - Z(\overline{\omega})^* Z(\overline{\eta})) H(\eta)^*$$

for $\omega, \eta \in \mathbb{B}^d$. Let $\omega = (\overline{b_1(z_i)}, \dots, \overline{b_d(z_i)})$ and $\eta = (\overline{b_1(z_j)}, \dots, \overline{b_d(z_j)})$ to conclude that

$$\begin{aligned} u_i^*(I - W(z_i)^* W(z_j)) u_j &= u_i^* H(\overline{b(z_i)}) \\ &\quad (I - Z \circ b(z_i)^* Z \circ b(z_j)) H(\overline{b(z_j)})^* u_j. \end{aligned} \tag{4.13}$$

Note that now (4.12) follows from (4.13) upon setting $H_{jk} = H_k(\overline{b(z_j)})^* u_j$ and using the interpolation conditions (4.11).

Conversely, suppose that there exist operators $H_{jk}\colon \mathcal{M}_j \to \mathcal{H}_0$ satisfying conditions (4.13). Define a mapping U_0 on elements of the form

$$\begin{bmatrix} b_1(z_j)H_{j1} + \cdots + b_d(z_j)H_{jd} \\ u_j \end{bmatrix} m_j \tag{4.14}$$

(where $m_j \in \mathcal{M}_j$ and $j = 1, \dots, n$) by

$$U_0\colon \begin{bmatrix} b_1(z_j)H_{j1} + \cdots + b_d(z_j)H_{jd} \\ u_j \end{bmatrix} m_j \to \begin{bmatrix} H_{j1} \\ \vdots \\ H_{jd} \\ v_j \end{bmatrix} m_j. \tag{4.15}$$

Then (4.13) implies that U_0 extends by linearity to a well-defined unitary transformation from the closed span of elements of the form (4.14) (a subspace of $\mathcal{H}_0 \oplus \mathcal{E}$) to the closed span of elements of the form on the right hand side of (4.15) for $m_j \in \mathcal{M}_j$ and $j = 1, \dots, n$ (a closed subspace of $(\oplus_1^d \mathcal{H}_0) \oplus \mathcal{E}_*$). Choose any unitary extension of U_0 to a unitary operator

$$U = \begin{bmatrix} A & B \\ C & D \end{bmatrix} = \begin{bmatrix} A_1 & B_1 \\ \vdots & \vdots \\ A_d & B_d \\ C & D \end{bmatrix} : \mathcal{H} \oplus \mathcal{E} \to (\oplus_1^d \mathcal{H}) \oplus \mathcal{E}_*$$

where $\mathcal{H}$ is a Hilbert space containing $\mathcal{H}_0$ as a subspace. Explicitly, since U extends U_0 we have the operator equations

$$\begin{bmatrix} A & B \\ C & D \end{bmatrix} \begin{bmatrix} b_1(z_j)H_{j1} + \cdots + b_d(z_j)H_{jd} \\ u_j \end{bmatrix} = \begin{bmatrix} H_{j1} \\ \vdots \\ H_{jd} \\ v_j \end{bmatrix} \tag{4.16}$$

for $j = 1, \dots, n$. Define an analytic operator-valued function F on $\mathbb{B}^d$ by $F(\omega) = D + C(I - Z(\omega)A)^{-1}Z(\omega)B$ for $\omega \in \mathbb{B}^d$. Then by Theorem 2.1 we know that $F \in \mathcal{M}_d(\mathcal{E}, \mathcal{E}_*)$ with multiplier norm at most equal to 1, and hence, if we set $W = F \circ b$, then W is a multiplier in $\mathcal{M}_k(\mathcal{E}, \mathcal{E}_*)$ with multiplier norm at most equal to 1 by Theorem 3.1. Moreover, from the first block row of (4.16) we see that

$$\begin{bmatrix} H_{1j} \\ \vdots \\ H_{dj} \end{bmatrix} = (I - A(Z \circ b)(z_j))^{-1} B u_j$$

from which the second block row implies

$$C(Z \circ b)(z_j)(I - A(Z \circ b)(z_j))^{-1} B u_j + D u_j = v_j$$

for $j = 1, \dots, n$, and the interpolation conditions (4.11) follow. □

A general result of Arov and Grossman (see [11]) describes how to parametrize the set of all unitary extensions of a given partially defined isometry V_0. This leads to the next result on parametrization of solutions of a left interpolation problem. To describe the result we need a few preliminaries. We assume that we are given the data set for an interpolation problem as in (4.1). Assume that the Pick matrix $[P_{ij}] = k(w_i, w_j)(x_i^* x_j - y_i^* y_j)$ is positive semidefinite, and hence has a factorization

$$P_{ij} = H_i H_j^* \tag{4.17}$$

for some operators $H_i \colon \mathcal{H}_0 \to \mathcal{M}_i$ as in the Proof of Theorem 4.1. Then $V_0^* \colon \mathcal{D}_{V_0^*} \to \mathcal{R}_{V_0^*}$ defined by (4.5) is a partially defined isometric transformation with inverse and adjoint $V_0 \colon \mathcal{R}_{V_0^*} \to \mathcal{D}_{V_0^*}$. Introduce the defect spaces

$$\Delta = \begin{bmatrix} \mathcal{H}_0 \\ \mathcal{E} \end{bmatrix} \ominus \mathcal{R}_{V_0^*}, \quad \Delta_* = \begin{bmatrix} \oplus_1^d \mathcal{H}_0 \\ \mathcal{E}_* \end{bmatrix} \ominus \mathcal{D}_{V_0^*} \tag{4.18}$$

and let $\widetilde{\Delta}$ be another copy of Δ and $\widetilde{\Delta}_*$ be another copy of Δ_* with unitary identification maps

$$i \colon \Delta \to \widetilde{\Delta}, \quad i_* \colon \Delta_* \to \widetilde{\Delta}_*.$$

Define a unitary operator $\mathcal{U}$ from $\mathcal{R}_{V_0^*} \oplus \Delta \oplus \widetilde{\Delta}_*$ onto $\mathcal{D}_{V_0^*} \oplus \Delta_* \oplus \widetilde{\Delta}$ by

$$\mathcal{U} \colon \begin{bmatrix} r \\ \delta \\ \widetilde{\delta}_* \end{bmatrix} \to \begin{bmatrix} V_0 r \\ i_*^{-1}(\widetilde{\delta}_*) \\ i(\delta) \end{bmatrix}. \tag{4.19}$$

By definition,

$$\mathcal{R}_{V_0^*} \oplus \Delta = \begin{bmatrix} \mathcal{H}_0 \\ \mathcal{E} \end{bmatrix}, \quad \mathcal{D}_{V_0^*} \oplus \Delta_* = \begin{bmatrix} \oplus_1^d \mathcal{H}_0 \\ \mathcal{E}_* \end{bmatrix}$$

and hence we may (and now do) identify $\begin{bmatrix} \mathcal{R}_{V_0^*} \\ \Delta \end{bmatrix}$ with $\begin{bmatrix} \mathcal{H}_0 \\ \mathcal{E} \end{bmatrix}$ and $\begin{bmatrix} \mathcal{D}_{V_0^*} \\ \Delta_* \end{bmatrix}$ with $\begin{bmatrix} \oplus_1^d \mathcal{H}_0 \\ \mathcal{E}_* \end{bmatrix}$. Let us define auxiliary Hilbert spaces

$$\mathcal{E}' = \begin{bmatrix} \mathcal{E} \\ \widetilde{\Delta}_* \end{bmatrix}, \quad \mathcal{E}'_* = \begin{bmatrix} \mathcal{E}_* \\ \widetilde{\Delta} \end{bmatrix} \tag{4.20}$$

and block-decompose $\mathcal{U}$ defined by (4.19) according to

$$\mathcal{U} = \begin{bmatrix} \mathbf{A} & \mathbf{B} \\ \mathbf{C} & \mathbf{D} \end{bmatrix} \colon \begin{bmatrix} \mathcal{H}_0 \\ \mathcal{E}' \end{bmatrix} \to \begin{bmatrix} \oplus_1^d \mathcal{H}_0 \\ \mathcal{E}'_* \end{bmatrix}.$$

Then, by statement (3a) of Theorem 2.1, the $\mathcal{L}(\mathcal{E}', \mathcal{E}'_*)$-valued function

$$\mathbf{F}(\omega) = \mathbf{D} + \mathbf{C}(I - Z(\omega)\mathbf{A})^{-1} Z(\omega)\mathbf{B} \tag{4.21}$$

(for ω in the d-dimensional ball $\mathbb{B}^d$) defines a k_d-multiplier, i.e., an element of $\mathcal{M}_d(\mathcal{E}', \mathcal{E}'_*)$. Due to the structure of $\mathcal{E}'$ and $\mathcal{E}'_*$ as defined in (4.20), for each $\omega \in \mathbf{B}^d$ we may decompose $\mathbf{F}(\omega)$ as a 2×2 block matrix

$$\mathbf{F}(\omega) = \begin{bmatrix} \mathbf{F}_{11}(\omega) & \mathbf{F}_{12}(\omega) \\ \mathbf{F}_{21}(\omega) & \mathbf{F}_{22}(\omega) \end{bmatrix} \tag{4.22}$$

where $\mathbf{F}_{11}(\omega) \in \mathcal{L}(\mathcal{E}, \mathcal{E}_*)$, $\mathbf{F}_{22}(\omega) \in \mathcal{L}(\widetilde{\Delta}_*, \widetilde{\Delta})$, etc. We are now ready to state our result concerning the parametrization of contractive multiplier solutions of a set of interpolation conditions of the form (4.1).

Theorem 4.4 *Suppose that we are given the data set for a collection of interpolation conditions of the form* (4.1). *Assume that the Pick matrix* (4.2) *is positive semidefinite, that we have chosen a fixed factorization* (4.17), *and that we have constructed the element* $\mathbf{F}(\omega) = \begin{bmatrix} \mathbf{F}_{11}(\omega) & \mathbf{F}_{12}(\omega) \\ \mathbf{F}_{21}(\omega) & \mathbf{F}_{22}(\omega) \end{bmatrix}$ *of* $\mathcal{M}_d(\mathcal{E} \oplus \widetilde{\Delta}_*, \mathcal{E}_* \oplus \widetilde{\Delta})$ *as in* (4.21) *and* (4.22). *Then a given* $\mathcal{L}(\mathcal{E}, \mathcal{E}_*)$*-valued function* $W(\cdot)$ *on* Ω *is in the multiplier space* $\mathcal{M}_k(\mathcal{E}, \mathcal{E}_*)$ *with* $\|M_W\|_{op} \leq 1$ *and satisfies the interpolation conditions* (4.1) *if and only if* $W(\cdot)$ *can be represented in the form*

$$\begin{aligned} W(z) = {} & (\mathbf{F}_{11} \circ b)(z) + (\mathbf{F}_{12} \circ b)(z)(I - G(z) \\ & (\mathbf{F}_{22} \circ b)(z))^{-1} G(z)(\mathbf{F}_{21} \circ b)(z) \end{aligned} \tag{4.23}$$

for some $G(\cdot) \in \mathcal{M}_k(\widetilde{\Delta}, \widetilde{\Delta}_*)$ *with* $\|M_G\| \leq 1$.

Proof: By Theorem 4.2 we know that any interpolant $F \in \mathcal{M}_k(\mathcal{E}, \mathcal{E}_*)$ with $\|M_F\|_{op} \leq 1$ is the characteristic function $F_V(z) = D + C(I - b(z)A)^{-1}b(z)B$ of a unitary extension

$$V = \begin{bmatrix} A & B \\ C & D \end{bmatrix} : \begin{bmatrix} \mathcal{H}' \oplus \mathcal{H}_0 \\ \mathcal{E} \end{bmatrix} \to \begin{bmatrix} \oplus_1^d (\mathcal{H}' \oplus \mathcal{H}_0) \\ \mathcal{E}_* \end{bmatrix}$$

of the isometry $V_0 \colon \mathcal{R}_{V_0^*} \to \mathcal{D}_{V_0^*}$ defined by (4.5). The first step is to describe all such unitary extensions V_0. This is handled by the following lemma. □

Lemma 4.5 *Let* $\mathcal{U} \colon \mathcal{H}_0 \oplus \mathcal{E} \oplus \widetilde{\Delta}_* \to (\oplus_1^d \mathcal{H}_0) \oplus \mathcal{E}_* \oplus \widetilde{\Delta}$ *be as given by* (4.19), *and let*

$$U = \begin{bmatrix} a & b \\ c & d \end{bmatrix} : \begin{bmatrix} \mathcal{H}' \\ \widetilde{\Delta} \end{bmatrix} \to \begin{bmatrix} \oplus_1^d \mathcal{H}' \\ \widetilde{\Delta}_* \end{bmatrix}$$

be any d-variable unitary colligation of the indicated form (where $\mathcal{H}'$ *is an arbitrary auxiliary Hilbert space). Define a d-variable unitary colligation*

$$V = \mathcal{F}_{\mathcal{U}}[U] = \begin{bmatrix} A & B \\ C & D \end{bmatrix} : \begin{bmatrix} \mathcal{H}' \oplus \mathcal{H}_0 \\ \mathcal{E} \end{bmatrix} \to \begin{bmatrix} \oplus_1^d (\mathcal{H}' \oplus \mathcal{H}_0) \\ \mathcal{E}_* \end{bmatrix}$$

implicitly via the feedback connection law: $\begin{bmatrix} A & B \\ C & D \end{bmatrix} \begin{bmatrix} h_0 \oplus h' \\ e \end{bmatrix} = \begin{bmatrix} h'_0 \oplus h'' \\ e_* \end{bmatrix}$ *if there is* $\widetilde{\delta} \in \widetilde{\Delta}$ *and* $\widetilde{\delta}_* \in \widetilde{\Delta}_*$ *for which the system of equations*

$$\mathcal{U} \begin{bmatrix} h_0 \\ e \\ \widetilde{\delta}_* \end{bmatrix} = \begin{bmatrix} h'_0 \\ e_* \\ \widetilde{\delta} \end{bmatrix}$$

$$U \begin{bmatrix} h' \\ \widetilde{\delta} \end{bmatrix} = \begin{bmatrix} h'' \\ \widetilde{\delta}_* \end{bmatrix}. \tag{4.24}$$

is satisfied (here we identify $\oplus_1^d(\mathcal{H}' \oplus \mathcal{H}_0)$ *with* $(\oplus_1^d \mathcal{H}') \oplus (\oplus_1^d \mathcal{H}_0)$ *in the natural way). Then* V *is a well-defined* d*-variable unitary colligation which extends* V_0*, and every unitary extension* $V : \mathcal{H}' \oplus \mathcal{H}_0 \oplus \mathcal{E} \to \oplus_1^d(\mathcal{H}' \oplus \mathcal{H}_0) \oplus \mathcal{E}_*$ *which extends* V_0 *is of this form for some unitary colligation* $U : \mathcal{H}' \oplus \widetilde{\Delta} \to (\oplus_1^d \mathcal{H}') \oplus \widetilde{\Delta}_*$.

Proof: Write $\mathcal{U}$ as a block matrix with respect to the decompositions $\mathcal{R}_{V_0^*} \oplus \Delta \oplus \widetilde{\Delta}_*$ on the input side and $\mathcal{D}_{V_0^*} \oplus \Delta_* \oplus \widetilde{\Delta}$ on the output side. From the definition (4.19) of $\mathcal{U}$ we see that the matrix for $\mathcal{U}$ corresponding to these decompositions has the form

$$\mathcal{U} = \begin{bmatrix} \alpha & 0 & 0 \\ 0 & 0 & \gamma \\ 0 & \beta & 0 \end{bmatrix} : \begin{bmatrix} \mathcal{R}_{V_0^*} \\ \Delta \\ \widetilde{\Delta}_0 \end{bmatrix} \to \begin{bmatrix} \mathcal{D}_{V_0^*} \\ \Delta_* \\ \widetilde{\Delta} \end{bmatrix}. \tag{4.25}$$

Since $\mathcal{U}$ is unitary, we see that α, β and γ are unitary as operators from their respective domain spaces to their respective range spaces. Incorporating the change of coordinates

$$\mathcal{H}_0 \oplus \mathcal{E} \cong \mathcal{R}_{V_0^*} \oplus \Delta \text{ and } (\oplus_1^d \mathcal{H}_0) \oplus \mathcal{E}_* \cong \mathcal{D}_{V_0^*} \oplus \Delta_*$$

we identify $\mathcal{H}' \oplus \mathcal{H}_0 \oplus \mathcal{E}$ with $\mathcal{H}' \oplus \mathcal{R}_{V_0^*} \oplus \Delta$ and $(\oplus_1^d(\mathcal{H}' \oplus \mathcal{H}_0)) \oplus \mathcal{E}_*$ with $(\oplus_1^d \mathcal{H}') \oplus \mathcal{D}_{V_0^*} \oplus \Delta_*$. Then the feedback connection law (4.24) translates to: $V(h' \oplus r \oplus \delta) = h'' \oplus d \oplus \delta_*$ *means that there exists* $\widetilde{\delta} \in \widetilde{\Delta}$ *and* $\widetilde{\delta}_* \in \widetilde{\Delta}_*$ *for which the system of equations*

$$\mathcal{U} \begin{bmatrix} r \\ \delta \\ \widetilde{\delta}_* \end{bmatrix} = \begin{bmatrix} d \\ \delta_* \\ \widetilde{\delta} \end{bmatrix}$$

$$U \begin{bmatrix} h' \\ \widetilde{\delta} \end{bmatrix} = \begin{bmatrix} h'' \\ \widetilde{\delta}_* \end{bmatrix}. \tag{4.26}$$

is satisfied. By using the explicit form (4.25) of $\mathcal{U}$ and writing $U = \begin{bmatrix} a & b \\ c & d \end{bmatrix}$, (4.26) collapses to the five equation

$$\begin{aligned} \alpha r = d \quad \gamma\widetilde{\delta}_* = \delta_* \quad \beta\delta &= \widetilde{\delta} \\ ah' + b\widetilde{\delta} = h'' \quad ch' + d\widetilde{\delta} &= \widetilde{\delta}_*. \end{aligned} \tag{4.27}$$

It is then easy to see that one can solve (4.27) uniquely for h'', d, δ_*, $\widetilde{\delta}$ and $\widetilde{\delta}_*$ in terms of the remaining variables h', r and δ. Suppressing $\widetilde{\delta}$ and $\widetilde{\delta}_*$ leaves us with

$$V \begin{bmatrix} h' \\ r \\ \delta \end{bmatrix} = \begin{bmatrix} h'' \\ d \\ \delta_* \end{bmatrix} \text{ with } V = \begin{bmatrix} a & 0 & b\beta \\ 0 & \alpha & 0 \\ \gamma c & 0 & \gamma d\beta \end{bmatrix} \tag{4.28}$$

From the formula (4.28) for V we see that V is unitary and is an extension of V_0.

Conversely, suppose that $\mathbf{V}: \mathcal{H}' \oplus \mathcal{H}_0 \oplus \mathcal{E} \to (\oplus_1^d(\mathcal{H} \oplus \mathcal{H}_0)) \oplus \mathcal{E}_*$ is any extension of V_0. Change variables so as to conform with the domain and range decompositions as used above. Then the matrix for $\mathbf{V}$ must have the form

$$V = \begin{bmatrix} \mathbf{a} & 0 & \mathbf{b} \\ 0 & \alpha & 0 \\ \mathbf{c} & 0 & \mathbf{d} \end{bmatrix}.$$

Then we can represent $\mathbf{V}$ in the form of V in (4.28) by choosing $a = \mathbf{a}$, $b = \mathbf{b}\beta^*$, $c = \gamma^*\mathbf{c}$ and $d = \gamma^*\mathbf{d}\beta^*$. The lemma follows. □

Continuation of the Proof of Theorem 4.4: Note that, in general, if

$$V = \begin{bmatrix} A & B \\ C & D \end{bmatrix} : \mathcal{H} \oplus \mathcal{E} \to (\oplus_1^d \mathcal{H}) \oplus \mathcal{E}_*$$

is a d-variable unitary colligation and z is a fixed point in Ω, then the value of the transfer function associated with V at the point $z \in \Omega$ acting on the vector $e \in \mathcal{E}$, namely

$$W_V(z) = D + C(I - b(z)A)^{-1}b(z)B : e \to e_*,$$

where here $b(z)$ represents the block row operator

$$[b_1(z)I_{\mathcal{H}} \ \dots \ b_d(z)I_{\mathcal{H}}] : \oplus_1^d \mathcal{H} \to \mathcal{H},$$

is the result of the feedback connection

$$\begin{aligned} \begin{bmatrix} A & B \\ C & D \end{bmatrix} \begin{bmatrix} x \\ e \end{bmatrix} &= \begin{bmatrix} x' \\ e_* \end{bmatrix} \\ x &= b(z)x'. \end{aligned} \tag{4.29}$$

Apply this general principal (4.29) to the feedback connection (4.24) defining V. Then we see that $W_V(z)e = e_*$ means that the system of equations

$$\begin{aligned} \mathcal{U}\begin{bmatrix} h_0 \\ e \\ \widetilde{\delta}_* \end{bmatrix} &= \begin{bmatrix} h_0' \\ e_* \\ \widetilde{\delta} \end{bmatrix} \\ U\begin{bmatrix} h' \\ \widetilde{\delta} \end{bmatrix} &= \begin{bmatrix} h'' \\ \widetilde{\delta}_* \end{bmatrix} \\ h_0 &= b(z)h_0' \\ h' &= b(z)h'' \end{aligned}$$

is satisfied for some choice of $h_0 \in \mathcal{H}_0$, $h_0' \in \oplus_1^d \mathcal{H}_0$, $h' \in \mathcal{H}'$, $h'' \in \oplus_1^d \mathcal{H}'$, $\widetilde{\delta} \in \widetilde{\Delta}$ and $\widetilde{\delta}_* \in \widetilde{\Delta}_*$. Two more applications of the general principal (4.29) show that these equations reduce to the more condensed form

$$\begin{aligned} W_{\mathcal{U}}(z)\begin{bmatrix} e \\ \widetilde{\delta}_* \end{bmatrix} &= \begin{bmatrix} e_* \\ \widetilde{\delta} \end{bmatrix} \\ W_U(z)\widetilde{\delta} &= \widetilde{\delta}_*. \end{aligned}$$

Elimination of $\widetilde{\delta}$ and $\widetilde{\delta}_*$ from these equations finally leads to the form (4.23) for $W(z) = W_V(z)$ and Theorem 4.4 follows.

Remark: An analogous but less explicit parametrization result can be given for the right interpolation problem. For each choice of operators $H_{jk}\colon \mathcal{M}_j \to \mathcal{H}_0$ (for $j = 1, \dots, n$ and $k = 1, \dots, d$) compatible with the interpolation data as in the hypothesis of Theorem 4.3, there is an associated linear-fractional map

$$G(z) \to \mathbf{W}_{11}(z) + \mathbf{W}_{12}(z)(I - G(z)\mathbf{W}_{22}(z))^{-1}G(z)\mathbf{W}_{21}(z)$$

which parametrizes all those interpolants W in $\mathcal{M}_k(\mathcal{E}, \mathcal{E}_*)$ with multiplier norm at most 1 which also are consistent with the prespecified choice of operators H_{jk}. One then must take the union of all these sets of interpolants over the set of all compatible collections of operators $\{H_{jk}\}$ to pick up the set of all interpolants of norm at most 1. The situation is analogous to the parametrization result given in [15] for an interpolation problem for the Schur-Agler class of functions defined on the polydisk $\mathbb{D}^d$.

5 The Commutant Lifting Theorem for $\mathcal{M}_k(\mathcal{E}, \mathcal{E}_*)$

Let k be a nondegenerate positive kernel with $\mathcal{H}(k)$ the associated reproducing kernel Hilbert space of functions on Ω. We shall assume in addition that $\frac{1}{k}$ has 1 positive-square. Then we may write

$$k(z, w) = \frac{1}{a(z)} \cdot \frac{1}{\overline{a(w)}} \cdot \frac{1}{(1 - \sum_{k=1}^d b_k(z)\overline{b_k(w)})}$$

where $z \to (b_1(z), \ldots, b_d(z))$ is a mapping of Ω into $\mathbb{B}^d$. If f is a multiplier for $\mathcal{H}(k)$ we denote by M_f the associated multiplication operator $M_f: h(z) \to f(z)h(z)$ on $\mathcal{H}(k)$. As above, we are also interested in the vector-valued version $\mathcal{H}(k) \otimes \mathcal{E}$ of $\mathcal{H}(k)$ (where $\mathcal{E}$ is an auxiliary Hilbert space). Note that if f is a multiplier for the scalar space $\mathcal{H}(k)$, then $f \cdot I_{\mathcal{E}}$ is certainly a multiplier for the vector-valued space; we denote by M_{fI} the associated multiplication operator $M_{fI}: H(z) \to f(z)H(z)$ on $\mathcal{H}(k) \times \mathcal{E}$. Let us say that a subspace $\mathcal{M}$ of $\mathcal{H}(k) \times \mathcal{E}$ is $*$-invariant if $(M_{fI})^* \mathcal{M} \subset \mathcal{M}$ for all scalar multipliers $f \in \mathcal{M}_k$. The commutant lifting theorem can be formulated as follows.

Theorem 5.1 *Let $k(z, w)$ be a nondegenerate positive kernel on Ω such that $\frac{1}{k}$ has 1 positive square, and $\mathcal{E}$ and $\mathcal{E}_*$ be auxiliary Hilbert spaces. Suppose that $\mathcal{M} \subset \mathcal{H} \otimes \mathcal{E}$ and $\mathcal{M}_* \subset \mathcal{H}(k) \otimes \mathcal{E}_*$ are $*$-invariant subspaces (so $(M^*_{fI}\mathcal{M} \subset \mathcal{M}$ and $(M_{fI})^* \mathcal{M}_* \subset \mathcal{M}_*$ for all scalar multipliers $f \in \mathcal{M}_k$). Suppose that $X \in \mathcal{L}(\mathcal{M}, \mathcal{M}_*)$ is a bounded linear operator from $\mathcal{M}$ to $\mathcal{M}_*$ which intertwines a compressed scalar multiplier on $\mathcal{M}$ with the corresponding compressed scalar multiplier on $\mathcal{M}_*$:*

$$X^*(M_{fI})^*|_{\mathcal{M}_*} = (M_{fI})^* X^* \tag{5.1}$$

for all scalar multipliers $f \in \mathcal{M}_k$. Then there exists an operator-valued multiplier $W \in \mathcal{M}_k(\mathcal{E}, \mathcal{E}_*)$ such that (i) $\|M_W\| \leq 1$ *and* (ii) $(M_W)^*|_{\mathcal{M}_*} = X^*$.

Example 1 Let $w_1, \ldots, w_n \in \Omega$, $x_j \in \mathcal{L}(\mathcal{M}_j, \mathcal{E}_*)$, $u_j \in \mathcal{L}(\mathcal{M}_j, \mathcal{E})$ for $j = 1, \ldots, n$ be the data set for a left interpolation problem

$$W(w_j)^* x_j = y_j \text{ for } j = 1, \ldots, n.$$

Set

$$\begin{aligned} \mathcal{M}_* &= \text{span}\,\{k_{w_j} x_j m_j : m_j \in \mathcal{M}_j \text{ and } j = 1, \ldots, n\} \\ \mathcal{M} &= \text{span}\,\{k_{w_j} y_j m_j : m_j \in \mathcal{M}_j \text{ and } j = 1, \ldots, n\}. \end{aligned}$$

From

$$(M_{fI})^* k_{w_j} x_j m_j = k_{w_j} x_j m_j \overline{f(w_j)} \tag{5.2}$$

we see that $\mathcal{M}_*$ (and similarly $\mathcal{M}$) is $*$-invariant. Define $X^*: \mathcal{M}_* \to \mathcal{M}$ by

$$X^*: k_{w_j} x_j m_j \to k_{w_j} y_j m_j. \tag{5.3}$$

The Pick matrix condition (that $[k(w_i, w_j)(x_i^* x_j - y_i^* y_j)]_{1 \leq i,j \leq n}$ is a positive-semidefinite matrix) is exactly what is needed to see that $\|X^*\| \leq 1$. From properties (5.2) and (5.3) it is easy to check the intertwining condition (5.1). Hence the hypotheses of the commutant lifting theorem are satisfied. Therefore, there exists a multiplier W with multiplier norm $\|M_W\|$ at most 1 such that $(M_W)^*|_{\mathcal{M}_*} = X^*$. Hence, in particular

$$(M_W)^* k_{w_j} x_j m_j = k_{w_j} y_j m_j. \tag{5.4}$$

But an independent computation

$$\begin{aligned}
\langle F, (M_W)^* k_{w_j} x_j m_j \rangle_{\mathcal{H}(k)\otimes\mathcal{E}} &= \langle WF, k_{w_j} x_j m_j \rangle_{\mathcal{H}(k)\otimes\mathcal{E}_*} \\
&= \langle W(w_j)F(w_j), x_j m_j \rangle_{\mathcal{E}_*} \\
&= \langle F(w_j), W(w_j)^* x_j m_j \rangle_{\mathcal{E}} \\
&= \langle F, k_{w_j} W(w_j)^* x_j m_j \rangle_{\mathcal{H}(k)\otimes\mathcal{E}}
\end{aligned}$$

shows that

$$(M_W)^* k_{w_j} x_j m_j = k_{w_j} W(w_j)^* x_j m_j. \tag{5.5}$$

Combining (5.4) and (5.5) gives $W(w_j)^* x_j = y_j$ and thus we pick up left interpolation as a byproduct of the commutant lifting theorem. This connection between Nevanlinna-Pick interpolation and commutant lifting goes back to the seminal paper of Sarason [40].

Example 2 One can also use the lifting theorem to prove a Toeplitz-corona theorem in this context. The result is as follows: *given multipliers $a_1, \ldots, a_n$ in $\mathcal{M}_k$ then there exist multipliers $f_1, \ldots, f_n \in \mathcal{M}_k$ such that*

$$a_1 f_1 + \cdots + a_n f_n = 1 \text{ and } \|f_1\|^2 + \cdots + \|f_n\|^2 \le \frac{1}{\delta^2}$$

for a given $\delta > 0$ if and only if

$$M_{a_1}(M_{a_1})^* + \cdots + M_{a_n}(M_{a_n})^* \ge \delta^2 I. \tag{5.6}$$

or, equivalently, if and only if

(5.7) $k(z, w)[a_1(z)\overline{a_1(w)} + \cdots + a_n(z)\overline{a_n(w)} - \delta^2]$ *is a positive kernel.*

To see the equivalence of (5.6) and (5.7), note that the span of $\{k_w : w \in \Omega\}$ is dense in $\mathcal{H}(k)$ and apply the Hermitian form induced by the operator in (5.6) to elements of this span. To obtain this result as an application of the commutant lifting Theorem 5.1, let $\mathcal{M}_*$ equal the orthogonal complement of the kernel of the operator $Y := [M_{a_1} \; \ldots \; M_{a_n}]$ in $\oplus_1^n \mathcal{H}(k)$, and let $\mathcal{M}$ equal all of $\mathcal{H}(k)$ (thus $\mathcal{E}_* = \mathbb{C}^n$ and $\mathcal{E} = \mathbb{C}$). The hypothesis (5.6) implies that the restriction of Y to $\mathcal{M}_*$ is invertible with norm bounded by $\frac{1}{\delta}$. Define $X : \mathcal{M} \to \mathcal{M}_*$ to be δY^{-1}. Then the hypotheses of the lifting theorem are satisfied and we get a lift

$W = [W_1 \ \dots \ W_n]^T : \mathcal{H}(k) \to \oplus_1^d \mathcal{H}(k)$. If we set $f_j = \frac{1}{\delta} W_j$, then $f_1, \dots, f_n$ satisfies the conclusions of the corona theorem. A reference for this approach in the classical case is [39].

Proof of Theorem 5.1: Without loss of generality we may assume that we have normalized the kernel $k(z, w)$ to have the form

$$k(z, w) = \frac{1}{1 - \sum_{k=1}^d b_k(z)\overline{b_k(w)}}.$$

We claim that the space $\mathcal{H}(k)$ then is identical to the space $\widetilde{\mathcal{H}}$ of function f on Ω defined by

$$\widetilde{\mathcal{H}} = \left\{ f : f(z) = \sum_{n=(n_1,\dots,n_d)\in\mathbb{Z}_+^d} a_n b^n(z) \text{ such that } \|f\|_{\widetilde{\mathcal{H}}}^2 = \sum_{n\in\mathbb{Z}_+^d} \frac{n!}{|n|!} |a_n|^2 < \infty \right\}.$$

Here, for $n = (n_1, \dots, n_d) \in \mathbb{Z}_+^d$,

$$n! = n_1! n_2! \dots n_d!, \quad |n| = n_1 + \dots + n_d \text{ so } |n|! = (n_1 + \dots + n_d)!$$

and $b^n(z)$ is shorthand for

$$b^n(z) = b_1(z)^{n_1} \dots b_d(z)^{n_d}.$$

To see that $\widetilde{\mathcal{H}} = \mathcal{H}(k)$, observe that

$$\left\{ \sqrt{\frac{|n|!}{n!}} b^n(z) : n \in \mathbb{Z}_+^d \right\} \tag{5.8}$$

is an orthonormal basis for $\widetilde{\mathcal{H}}$. By a general principal for reproducing kernel Hilbert spaces, the reproducing kernel for $\widetilde{\mathcal{H}}$ is then given by

$$\begin{aligned}
\widetilde{k}(z, w) &= \sum_{n\in\mathbb{Z}_+^d} \frac{|n|!}{n!} b^n(z)\overline{b^n(w)} \\
&= \sum_{N=0}^{\infty} \left[\sum_{k=1}^d b_k(z)\overline{b_k(w)} \right]^N \\
&= \frac{1}{1 - \sum_{k=1}^d b_k(z)\overline{b_k(w)}} \\
&= k(z, w)
\end{aligned}$$

and the claim follows. As a corollary we see that each element $f \in \mathcal{H}(k)$ is in fact of the form $f = h \circ b$ where $h \in \mathcal{H}(k_d)$, the reproducing kernel Hilbert space of analytic functions on the ball $\mathbb{B}^d$ studied in Section 2; the correspondence is simply: if $f = \sum_{n\in\mathbb{Z}_+^d} a_n b^n(z)$, set $h(z) = \sum_{n\in\mathbb{Z}_+^d} a_n z^n$ for $z = (z_1, \dots, z_d) \in \mathbb{B}^d$ (with $z^n = z_1^{n_1} \dots z_d^{n_d}$ if $n = (n_1, \dots, n_d)$). This gives another explanation for the result on the lifting of multipliers on $\mathcal{H}(k)$ (part (3) of Theorem 3.1).

Thus $\mathcal{H}(k)$ has a natural orthogonal decomposition

$$\mathcal{H}(k) = \oplus_{N=0}^{\infty} \mathcal{H}_N(k)$$

where

$$\mathcal{H}_N(k) = \text{ span } \{b^n(z) : n \in \mathbb{Z}_+^d \text{ with } |n| = N\}. \tag{5.9}$$

For a coordinate-free version of most of these ideas (for the case $k = k_d$ on the ball $\mathbb{B}^d$), we refer the reader to the recent paper of Arveson [12].

It is convenient to have a formula for the orthogonal projection of $\mathcal{H}(k)$ onto $\mathcal{H}_N(k)$ for each $N = 1, 2, \dots$. The following lemma gives the answer.

Lemma 5.2 *Let $M_{b_1}, \dots, M_{b_d}$ be the multiplier operators on $\mathcal{H}(k)$ defined as above. Then the orthogonal projection of $\mathcal{H}(k)$ onto the subspace $\mathcal{H}_N(k)$ given by (5.9) is given by*

$$P_{\mathcal{H}_N(k)} = \sum_{n\in\mathbb{Z}_+^d : |n|=N} \frac{N!}{n!} (M_b)^n \left(I - \sum_{k=1}^{d} M_{b_k} (M_{b_k})^* \right) (M_b)^{*n} \tag{5.10}$$

where $(M_b)^n$ is shorthand for

$$(M_b)^n = M_{b_1}^{n_1} \dots M_{b_d}^{n_d}.$$

Before proving the lemma, we need to introduce some additional useful notation. Denote by $\mathcal{F}_+(d)$ the set of all words (or strings of symbols) $i_1 i_2 \dots i_n$ with symbols $i_1, \dots, i_n$ from the set $\{1, \dots, d\}$ of the first d natural numbers. For a given word $i = i_1 \dots i_n$ in $\mathcal{F}_+(d)$, we denote the number n of symbols in i by $|i|$. Note that $\mathcal{F}_+(d)$ has the structure of a free semigroup with generators equal to the words of length one $(1, \dots, d)$, with identity element equal to the empty word $\emptyset$, and with semigroup multiplication operator equal to juxtaposition. There is a mapping $\mathcal{J}$ of $\mathcal{F}_+(d)$ into $\mathbb{Z}_+^d$ given by: *if $i = i_1 \dots i_n \in \mathcal{F}_+(d)$, then $\mathcal{J}(i) = n = (n_1, \dots, n_d)$ if $\#\{j : i_j = k\} = n_k$ for $k = 1, \dots, d$.* For a given $n \in \mathbb{Z}_+^d$, denote by $\mathcal{I}_d(n)$ the preimage of $\mathcal{J}$ in $\mathcal{F}_+(d)$, i.e., by definition

$$\mathcal{I}_d(n) = \{i = i_1 \dots i_{|n|} : \#\{j : i_j = k\} = n_k \text{ for } k = 1, \dots d\}.$$

Another useful combinatorial fact is the size of the set $\mathcal{I}_d(n)$, namely

$$\#\mathcal{I}_d(n) = \frac{|n|!}{n!}.$$

Let us introduce the additional shorthand notation: for $i = i_1 \dots i_n \in \mathcal{F}_+(d)$, $b_i(z)$ is shorthand for

$$b_i(z) = b_{i_1}(z) \dots b_{i_n}(z).$$

We use the same notation for the corresponding operators:

$$(M_b)_i = M_{b_{i_1}} \dots M_{b_{i_n}} \text{ for } i = i_1 \dots i_n \in \mathcal{F}_+(d).$$

Note that since the operators $M_{b_1}, \dots, M_{b_d}$ pairwise commute, we have that

$$\sum_{i \in \mathcal{F}_+(d) : |i| = N} (M_b)_i \Phi (M_b)_i^* = \sum_{n \in \mathbb{Z}_+^d : |n| = N} \frac{N!}{n!} (M_b)^n \Phi (M_b)^{*n}$$

for any operator Φ on $\mathcal{H}(k)$. Therefore the formula for the projection operator in Lemma 5.2 can alternatively be written as

$$P_{\mathcal{H}_N(k)} = \sum_{i \in \mathcal{F}_+(d) : |i| = N} (M_b)_i \left(I - \sum_{k=1}^{d} M_{b_k} (M_{b_k})^* \right) (M_b)_i^*. \tag{5.11}$$

Proof of Lemma 5.2: From the explicit form of the orthonormal basis (5.8) (or by using the general formula for $(M_{b_k})^*$ derived in the proof of Lemma 3.8 in [12]), one can see that the action of $(M_{b_k})^*$ on a monomial $b^m(z)$ (with $m = (m_1, \dots, m_d)$) is given by

$$\begin{aligned} &((M_{b_k})^* b^m)(z) \\ &= \begin{cases} 0 & \text{if } m_k = 0, \\ \frac{m_k}{|m|} b_1(z)^{m_1} \cdots b_k(z)^{m_k - 1} \cdots b_d(z)^{m_d} & \text{otherwise.} \end{cases} \end{aligned} \tag{5.12}$$

From (5.12) we easily see that

$$\left(\left(I - \sum_{k=0}^{d} M_{b_k} (M_{b_k})^* \right) b^m \right)(z) = \begin{cases} 1 = z^0 & \text{if } m = (0, \dots, 0) \\ 0 & \text{otherwise.} \end{cases}$$

It follows that the projection $P_{\mathcal{H}_0(k)}$ onto the constant functions in $\mathcal{H}(k)$ is given by

$$P_{\mathcal{H}_0(k)} = I - \sum_{k=1}^{d} M_{b_k} (M_{b_k})^*. \tag{5.13}$$

This proves the case $N = 0$ of the lemma.

A similar application of (5.12) verifies that in general we have

$$\sum_{k=1}^{d} M_{b_k} P_{\mathcal{H}_N(k)} (M_{b_k})^* = P_{\mathcal{H}_{N+1}(k)}. \tag{5.14}$$

Inductively assume that (5.11) holds true. Note that we have already verified the validity of the case $N = 0$. We then use (5.14) to compute

$$
\begin{aligned}
P_{\mathcal{H}_{N+1}(k)} &= \sum_{k=1}^{d} M_{b_k}\Bigg[\sum_{i\in\mathcal{F}_+(d):|i|=N} (M_b)_i \\
&\qquad \left(I - \sum_{k=1}^{d} M_{b_k}(M_{b_k})^*\right)(M_b)_i^*\Bigg](M_{b_k})^* \\
&= \sum_{i\in\mathcal{F}_+(d):|i|=N+1} (M_b)_i\left(I - \sum_{k=1}^{d} M_{b_k}(M_{b_k})^*\right)(M_b)_i^*
\end{aligned} \tag{5.15}
$$

and the lemma follows.

As a corollary we have the following stability result.

Corollary 5.3 *For $M_{b_1}, \ldots, M_{b_d}$ equal to the multiplier operators on $\mathcal{H}(k)$ as above and for h any fixed element of $\mathcal{H}(k)$, we have*

$$
\lim_{M\to\infty}\left\|\sum_{n\in\mathbb{Z}_+^d:|n|=M}\frac{M!}{n!}(M_b)^{*n}h\right\|^2 = \lim_{M\to\infty}\sum_{i\in\mathcal{F}_+(d):|i|=M}\|(M_b)_i^*h\|^2 = 0.
$$

Proof: Note that

$$
\begin{aligned}
P_{\oplus_{N=M}^{\infty}\mathcal{H}_N(k)} &= I - \sum_{N=0}^{M-1} P_{\mathcal{H}_N(k)} \\
&= I - \sum_{N=0}^{M-1}\Bigg[\sum_{i\in\mathcal{F}_+(d):|i|=N}(M_b)_i(M_b)_i^* \\
&\qquad - \sum_{i\in\mathcal{F}_+(d):|i|=N+1}(M_b)_i(M_b)_i^*\Bigg] \\
&= \sum_{i\in\mathcal{F}_+(d):|i|=M}(M_b)_i(M_b)_i^*.
\end{aligned}
$$

Thus

$$
\begin{aligned}
\sum_{i\in\mathcal{F}_+(d):|i|=M}\|(M_b)_i^*h\|^2 &= \left\langle\sum_{i\in\mathcal{F}_+(d):|i|=M}(M_b)_i(M_b)_i^*h, h\right\rangle \\
&= \|P_{\oplus_{N=M}^{\infty}\mathcal{H}_N(k)}h\|^2.
\end{aligned}
$$

Since

$$\bigcap_{M\geq 0}[\oplus_{N=M}^{\infty}\mathcal{H}_N(k)] = \{0\}$$

it follows that $\lim_{M\to\infty}\|P_{\oplus_{N=M}^{\infty}\mathcal{H}_N(k)}h\|^2 = 0$ and the corollary follows. □

Remark: Lemma 5.2 and Corollary 5.3 become easier if we use that $\mathcal{H}(k_d)$ (and hence also $\mathcal{H}(k)$) arises as the symmetric projection of the full Fock space; in the noncommutative Fock space setting, the analogues of Lemma 5.2 and Corollary 5.3 are relatively straightforward. For details on this Fock space setting we refer to [34], [20], [21] and [12].

Continuation of the Proof of Theorem 5.1: Now let us consider the general situation in the hypotheses of commutant lifting theorem (Theorem 5.1). For $k = 1, \ldots, d$ define operators

$$T_k = P_{\mathcal{M}}M_{b_kI}|_{\mathcal{M}}, \quad T_{*k} = P_{\mathcal{M}_*}M_{b_kI}|_{\mathcal{M}_*}$$

and set

$$P = I_{\mathcal{M}_*} - XX^* \geq 0.$$

Then for each $k = 1, \ldots, d$ we have

$$\begin{aligned} T_{*k}T_{*k}^* - T_{*k}XX^*T_{*k}^* &= T_{*k}(I - XX^*)T_{*k}^* \\ &= T_{*k}PT_{*k}^*. \end{aligned}$$

The intertwining property of X then gives

$$T_{*k}T_{*k}^* - XT_kT_k^*X^* = T_{*k}PT_{*k}^*.$$

Summing over $k = 1, \ldots, d$ gives then

$$\sum_{k=1}^{d} T_{*k}T_{*k}^* - X\left[\sum_{k=1}^{d} T_kT_k^*\right]X^* = \sum_{k=1}^{d} T_{*k}PT_{*k}^*. \tag{5.16}$$

The $N = 0$ case of Lemma 5.2 gives us that the projection $P_{\mathcal{E}_*}$ onto the constant functions in $\mathcal{H}(k)\otimes\mathcal{E}_*$ is given by

$$P_{\mathcal{E}_*} = I - \sum_{k=1}^{d}(M_{b_kI})(M_{b_kI})^*.$$

Recalling the definition of T_{*k} and using the $*$-invariance of $\mathcal{M}_*$, we see that (5.16) can be rewritten as

$$P_{\mathcal{M}_*}(I - P_{\mathcal{E}_*})|_{\mathcal{M}_*} - XP_{\mathcal{M}}(I - P_{\mathcal{E}})X^* = \sum_{k=1}^{d} T_{*k}PT_{*k}^*.$$

A slight rearrangement gives us

$$I_{\mathcal{M}_*} - P_{\mathcal{M}_*} P_{\mathcal{E}_*}|_{\mathcal{M}_*} - XX^* + XP_{\mathcal{M}} P_{\mathcal{E}} X^* = \sum_{k=1}^{d} T_{*k} P T_{*k}^*.$$

Recalling that $P = I - XX^*$, upon rearrangement this in turn condenses to

$$\sum_{k=1}^{d} T_{*k} P T_{*k}^* + P_{\mathcal{M}_*} P_{\mathcal{E}_*}|_{\mathcal{M}_*} = P + XP_{\mathcal{M}} P_{\mathcal{E}} X^*. \tag{5.17}$$

Since $P \geq 0$, we may factor P as $P = QQ^*$ for an operator $Q\colon \mathcal{H}_0 \to \mathcal{M}_*$ where $\mathcal{H}_0$ is an auxiliary Hilbert space. This operator identity (5.17) is equivalent to the equation

$$V_0^*\colon \begin{bmatrix} Q^* T_{*1}^* \\ \vdots \\ Q^* T_{*d}^* \\ P_{\mathcal{E}_*} \end{bmatrix} m_* \to \begin{bmatrix} Q^* \\ P_{\mathcal{E}} X^* \end{bmatrix} m_*$$

(where m_* is a generic element of $\mathcal{M}_*$) defining an isometry V_0^* from

$$\mathcal{D}_{V_0^*} = \begin{bmatrix} Q^* T_{*1}^* \\ \vdots \\ Q^* T_{*d}^* \\ P_{\mathcal{E}_*} \end{bmatrix} \mathcal{M}_* \subset \left(\oplus_1^d \mathcal{H}_0\right) \oplus \mathcal{E}_*$$

onto $\mathcal{R}_{V_0^*} = \begin{bmatrix} Q^* \\ P_{\mathcal{E}} X^* \end{bmatrix} \mathcal{M}_* \subset \mathcal{H}_0 \oplus \mathcal{E}$. Let us extend this isometry to a unitary operator

$$V^* = \begin{bmatrix} A^* & C^* \\ B^* & D^* \end{bmatrix} = \begin{bmatrix} A_1^* & \dots & A_d^* & C^* \\ B_1^* & \dots & B_d^* & D^* \end{bmatrix} \colon \begin{bmatrix} \oplus_1^d \mathcal{H} \\ \mathcal{E}_* \end{bmatrix} \to \begin{bmatrix} \mathcal{H} \\ \mathcal{E} \end{bmatrix}$$

where $\mathcal{H}$ is a Hilbert space containing $\mathcal{H}_0$ as a subspace. In particular, since V^* extends V_0^* we have

$$\begin{aligned} [A_1^* \ \dots \ A_d^*] \begin{bmatrix} Q^* T_{*1}^* \\ \vdots \\ Q^* T_{*d}^* \end{bmatrix} + C^* P_{\mathcal{E}_*}|_{\mathcal{M}_*} &= Q^* \\ [B_1^* \ \dots \ B_d^*] \begin{bmatrix} Q^* T_{*1}^* \\ \vdots \\ Q^* T_{*d}^* \end{bmatrix} + D^* P_{\mathcal{E}_*}|_{\mathcal{M}_*} &= P_{\mathcal{E}} X^*. \end{aligned} \tag{5.18}$$

Our candidate for the multiplier W satisfying the conclusion of the lifting theorem now is

$$W(z) = D + C(I - Z \circ b(z)A)^{-1} Z \circ b(z)B. \tag{5.19}$$

Note that since $V = \begin{bmatrix} A & B \\ C & D \end{bmatrix} : \mathcal{H} \oplus \mathcal{E} \to (\oplus_1^d \mathcal{H}) \oplus \mathcal{E}_*$ is unitary, it is clear from the results of Section 3 that W is a contractive multiplier in $\mathcal{M}_k(\mathcal{E}, \mathcal{E}_*)$. It remains only to verify the extension property $(M_W)^*|_{\mathcal{M}_*} = X^*$.

Toward this end we need some more preliminaries which we organize into a series of lemmas. Define a map Γ from $\mathcal{H}$ into functions on Ω by

$$\Gamma : x \to C(I - Z \circ b(z)A)^{-1} x. \tag{5.20}$$

Lemma 5.4 *The map Γ defined by* (5.20) *maps $\mathcal{H}$ contractively into $\mathcal{H}(k) \otimes \mathcal{E}_*$.*

Proof: Compute

$$\begin{aligned} (\Gamma x)(z) &= C(I - [b_1(z)A_1 + \cdots + b_d(z)A_d])^{-1} x \\ &= \sum_{N=0}^{\infty} C \left(\sum_{k=1}^{d} b_k(z) A_k \right)^N x \qquad (5.21) \\ &= \sum_{n \in \mathbb{Z}_+^d} \left\{ \sum_{i \in \mathcal{I}(n)} C A_i x \right\} b^n(z) \end{aligned}$$

where we have used the shorthand multivariable notation

$$A_i = A_{i_1} \dots A_{i_{|n|}} \text{ if } i = i_1 \dots i_{|n|} \in \mathcal{I}(n) \subset \mathcal{F}_+(d).$$

Hence, if the d-tuple $(A_1, \dots, A_d)$ of operators on $\mathcal{H}$ is pairwise commuting, (5.21) becomes

$$(\Gamma x)(z) = \sum_{n \in \mathbb{Z}_+^d} \left(\frac{|n|!}{n!} C A^n x \right) b^n(z)$$

where now we use the shorthand notation

$$A^n = A_1^{n_1} \dots A_d^{n_d} \text{ if } n = (n_1, \dots, n_d) \in \mathbb{Z}_+^d.$$

Hence, in case $(A_1, \dots, A_d)$ is pairwise commutative, by using that (5.8) is an orthonormal basis and the unitary property of $V = \begin{bmatrix} A & B \\ C & D \end{bmatrix}$, we can compute

$$\|\Gamma x\|_{\mathcal{H}(k)}^2 = \lim_{M \to \infty} \sum_{N=0}^{M} \sum_{n \in \mathbb{Z}_+^d : |n| = N} \|C A^n x\|^2$$

$$= \lim_{M\to\infty} \sum_{N=0}^{M} \sum_{n\in\mathbb{Z}_+^d:|n|=N} \langle (A^*)^n C^* C A^n x, x\rangle$$

$$= \lim_{M\to\infty} \sum_{N=0}^{M} \sum_{n\in\mathbb{Z}_+^d:|n|=N} \tag{5.22}$$

$$\langle (A^*)^n (I - A_1^* A_1 - \cdots - A_d^* A_d) A^n x, x\rangle$$

$$= \lim_{M\to\infty} \left\{ \|x\|^2 - \sum_{n\in\mathbb{Z}_+^d:|n|=M+1} \|A^n x\|^2 \right\} \leq \|x\|^2$$

and the lemma holds in this case.

In case the d-tuple $\{A_1, \ldots, A_d\}$ is not pairwise commutative, the analysis is similar. In this general case from the form of the orthonormal basis (5.8), by (5.21) and again by using the unitary property of V, we have

$$\|\Gamma x\|^2_{\mathcal{H}(k)} = \sum_{n\in\mathbb{Z}_+^d} \frac{n!}{|n|!} \left\| \sum_{i\in\mathcal{I}(n)} C A_i x \right\|^2$$

$$\leq \sum_{n\in\mathbb{Z}_+^d} \left\| \sum_{i\in\mathcal{I}(n)} C A_i x \right\|^2$$

$$= \lim_{M\to\infty} \sum_{N=0}^{M} \sum_{i\in\mathcal{F}_+(d):|i|=N} \langle (A_i)^* C^* C A_i x, x\rangle$$

$$= \lim_{M\to\infty} \sum_{N=0}^{M} \sum_{i\in\mathcal{F}_+(d):|i|=N}$$

$$\langle (A_i)^* (I - A_1^* A_1 - \cdots - A_d^* A_d) A_i x, x\rangle$$

$$= \lim_{M\to\infty} \left[\|x\|^2 - \sum_{i\in\mathcal{F}_+(d):|i|=M+1} \|A_i x\|^2 \right] \leq \|x\|^2$$

and the lemma follows for this case as well. □

Remark: Lemma 5.4 can also be found as Theorem 4.5 in [12] for the case where $(A_1, \ldots, A_d)$ is pairwise commutative. Our proof is somewhat different and perhaps more direct.

Lemma 5.5 *The operator* $\Psi := \Gamma^*|_{\mathcal{M}_*}$ *satisfies the equation*

$$A^* \begin{bmatrix} \Psi T_{*1}^* \\ \vdots \\ \Psi T_{*d}^* \end{bmatrix} + C^* P_{\mathcal{E}_*}|_{\mathcal{M}_*} = \Psi. \tag{5.23}$$

Proof: Compute, for $\Psi = \Gamma^*|_{\mathcal{M}_*}$,

$$\left\langle A^* \begin{bmatrix} \Psi T_{*1}^* \\ \vdots \\ \Psi T_{*d}^* \end{bmatrix} m_* + C^* P_{\mathcal{E}_*} m_*, h \right\rangle_{\mathcal{H}}$$

$$= \left\langle \left[\sum_{k=1}^{d} A_k^* \Gamma^* (M_{b_k I})^* + C^* P_{\mathcal{E}_*} \right] m_*, h \right\rangle_{\mathcal{H}} \tag{5.24}$$

$$= \left\langle m_*, \left[\sum_{k=1}^{d} M_{b_k I} \Gamma A_k + P_{\mathcal{E}_*} C \right] h \right\rangle_{\mathcal{H}(k) \otimes \mathcal{E}_*}$$

But we have

$$\begin{aligned} \left(\sum_{k=1}^{d} M_{b_k I} \Gamma A_k h \right)(z) &= \sum_{k=1}^{d} b_k(z) C (I - Z \circ b(z) A)^{-1} A_k h \\ &= C (I - Z \circ b(z) A)^{-1} Z \circ b(z) A h \\ &= -Ch + C (I - Z \circ b(z) A)^{-1} h \\ &= -Ch + (\Gamma h)(z). \end{aligned}$$

Thus

$$\begin{aligned} &\left\langle m_*, \left[\sum_{k=1}^{d} M_{b_k I} \Gamma A_k + P_{\mathcal{E}_*} C \right] h \right\rangle_{\mathcal{H}(k) \otimes \mathcal{E}_*} \\ &= \langle m_*, \Gamma h \rangle = \langle \Gamma^* m_*, h \rangle_{\mathcal{H}} = \langle \Psi m_*, h \rangle. \end{aligned} \tag{5.25}$$

By combining (5.25) with (5.24), the lemma follows. □

Lemma 5.6 *The solution* Ψ *to* (5.23) *is unique.*

Proof: Suppose that Ψ and Ψ' both solve (5.23). Then $\Psi_0 := \Psi - \Psi'$ satisfies the homogeneous equation

$$A^* \begin{bmatrix} \Psi_0 T_{*1}^* \\ \vdots \\ \Psi_0 T_{*d}^* \end{bmatrix} = \Psi_0. \tag{5.26}$$

Define $\mathcal{A}$ on $\mathcal{L}(\mathcal{H}(k) \otimes \mathcal{E}_*, \mathcal{H})$ by

$$\mathcal{A}(\Phi) = A_1^* \Phi M_{b_1 *}^* + \cdots + A_d^* \Phi M_{b_d}^*.$$

Note that the content of (5.26) is that

$$\mathcal{A}(\Psi_0 P_{\mathcal{M}_*})|\mathcal{M}_* = \Psi_0$$

Iteration of this gives

$$\mathcal{A}^N(\Psi_0 P_{\mathcal{M}_*})|\mathcal{M}_* = \Psi_0 \text{ for } N = 1, 2, 3, \ldots.$$

and hence trivially

$$\lim_{N\to\infty} \mathcal{A}^N(\Psi_0)m_* = \Psi_0 m_* \text{ weakly for all } m_* \in \mathcal{M}_*. \tag{5.27}$$

On the other hand, we claim that $\mathcal{A}^N(\Phi)$ tends to 0 as $N \to \infty$ in the weak operator topology for each $\Phi \in \mathcal{L}(\mathcal{H}(k) \otimes \mathcal{E}_*, \mathcal{H})$, i.e., for each such Φ, $x \in \mathcal{H}$ and $h \in \mathcal{H}(k) \otimes \mathcal{E}_*$, we have

$$\lim_{N\to\infty} \langle \mathcal{A}^N(\Phi)h, x\rangle_{\mathcal{H}} = 0. \tag{5.28}$$

The proof of this takes a number of steps as follows.

We note first that $\mathcal{A}^N(\Phi)$ is uniformly bounded in norm. Indeed, by the unitary property of $V = \begin{bmatrix} A & B \\ C & D \end{bmatrix}$ we are guaranteed that $A = \begin{bmatrix} A_1 \\ \vdots \\ A_d \end{bmatrix}$ is contractive. By the $N = 0$ case of Lemma 5.2, we have that $0 \leq P_{\mathcal{E}_*} = I - \sum_{k=1}^d M_{b_k} M_{b_k}^*$ and hence also $M_b := [M_{b_1} \ \ldots \ M_{b_d}]$ is contractive. It follows that

$$\|\mathcal{A}(\Phi)\| = \left\| [A_1 \ \ldots \ A_k] \begin{bmatrix} \Phi & & \\ & \ddots & \\ & & \Phi \end{bmatrix} \begin{bmatrix} (M_{b_1})^* \\ \vdots \\ (M_{b_d})^* \end{bmatrix} \right\| \leq \|\Phi\|$$

and the uniform boundedness claim follows by induction.

Secondly, we show that (5.28) holds for the case that $\Omega = \mathbb{B}^d$ with $k(z, w) = k_d(z, w) = \frac{1}{1-\langle z, w\rangle}$ and with h equal to an admissible kernel function. If $d < \infty$, we say that h is an *admissible kernel function* if h is of the form

$$h = (k_d)_w e_* \left(= \frac{1}{1 - \langle \cdot, w\rangle} e_* \right) \tag{5.29}$$

with $e_* \in \mathcal{E}_*$ and with $w = (w_1, \ldots, w_d) \in \mathbb{B}^d$ satisfying $|w_k| < \frac{1}{d}$ for $k = 1, \ldots, d$. If $d = \infty$, we say that h is an *admissible kernel function* if h is of the form (5.29) with with w in the unit ball of $\mathbb{B}^\infty = \ell^2$ restricted to be of the form $w = (w_1, w_2, \ldots)$ with $w_k = 0$ once k is larger than some finite n and with $w_{max} := \max_{k=1,\ldots,n} |w_k|$ satisfying $w_{max} < 1/n$. To see this for the case $d < \infty$, note that an inductive argument gives

$$\mathcal{A}^N(\Phi) = \sum_{i \in \mathcal{F}_+(d):|i|=N} (A^*)_i \Phi(M_z)_i^*.$$

Note also that $(M_{z_k})^*(k_d)_w = \overline{w}_k(k_d)_w$. Hence

$$\begin{aligned} \|\mathcal{A}^N(\Phi)(k_d)_w e_*\| &= \left\| \sum_{i \in \mathcal{F}_+(d):|i|=N} w_i A_i^* \Phi(k_d)_w e_* \right\| \\ &\leq \|\Phi(k_d)_w e_*\| \cdot \sum_{i \in \mathcal{F}_+(d):|i|=N} |w_i| \\ &\leq \|\Phi(k_d)_w e_*\| d^N |w_{max}|^N. \end{aligned}$$

By assumption $d|w_{max}| < 1$ and hence (5.28) holds for this case. The assertion for the case $d = \infty$ follows by a similar argument.

Note next that the span of the set of admissible kernel functions is dense in $\mathcal{H}(k_d) \otimes \mathcal{E}_*$. This fact combined with the uniform boundedness of $\mathcal{A}^N(\Phi)$ with respect to N verified above proves (5.28) for the case where $\Omega = \mathbb{B}^d$ with $k = k_d$. But in general we know that there is an isometric injection $h \circ b \to h$ of $\mathcal{H}(k)$ into $\mathcal{H}(k_d)$; one sees easily from this fact that (5.28) holds for $\mathcal{H}(k)$ if it holds for $\mathcal{H}(k_d)$.

Finally, apply (5.28) to the case $\Phi = \Psi_0 P_{\mathcal{M}_*}$ and combine with (5.27) to see that Ψ_0 must be zero and Lemma 5.6 follows. □

Conclusion of the Proof of Theorem 5.1: From the first of equations (5.18) and by Lemma 5.5, we see that both Q^* and $\Gamma^*|_{\mathcal{M}_*}$ satisfy (5.23). Lemma 5.6 then implies that $Q^* = \Gamma^*|\mathcal{M}_*$. It follows that the second of equations (5.18) may be rewritten as

$$B^* \begin{bmatrix} \Gamma^* T_{*1}^* \\ \vdots \\ \Gamma^* T_{*d}^* \end{bmatrix} + D^* P_{\mathcal{E}_*}|_{\mathcal{M}_*} = P_{\mathcal{E}} X^*|_{\mathcal{M}_*}.$$

We claim next that

$$P_{\mathcal{E}}(M_W)^*|_{\mathcal{M}_*} = P_{\mathcal{E}} X^* \tag{5.30}$$

where W is our candidate contractive multiplier lifting as in (5.19)

$$W(z) = D + C(I - Z \circ b(z)A)^{-1} Z \circ b(z) B.$$

Indeed, for $m_* \in \mathcal{M}_*$ and $e \in \mathcal{E}$ we have

$$\begin{aligned} \langle P_{\mathcal{E}} X^* m_*, e\rangle_{\mathcal{E}} &= \left\langle \left(D^* P_{\mathcal{E}_*} + B^* \begin{bmatrix} \Gamma^* T_{*1}^* \\ \vdots \\ \Gamma^* T_{*d}^* \end{bmatrix} \right) m_*, e \right\rangle_{\mathcal{E}} \\ &= \langle m_*, (D + [M_{b_1 I}\Gamma \ \ldots \ M_{b_d I}\Gamma] B) e\rangle_{\mathcal{H}(k)\otimes\mathcal{E}_*}. \end{aligned} \tag{5.31}$$

In more detail,

$$\begin{aligned} &\big([M_{b_1 I}\Gamma \ \ldots \ M_{b_d I}\Gamma] Be\big)(z) \\ &\quad = b_1(z) C(I - Z \circ b(z) A)^{-1} B_1 e + \cdots + \\ &\qquad b_d(z) C(I - Z \circ b(z) A)^{-1} B_d e \\ &\quad = C(I - Z \circ b(z))^{-1} Z \circ b(z) Be. \end{aligned} \tag{5.32}$$

Hence the right hand side of (5.31) becomes

$$\langle m_*, M_W e\rangle_{\mathcal{H}(k)\otimes\mathcal{E}_*} = \langle P_{\mathcal{E}} (M_W)^* m_*, e\rangle_{\mathcal{E}}$$

and (5.30) follows.

Apply (5.30) to $\Pi_{k=1}^d (M_{b_k I})^{*n_k} m_* = \Pi_{k=1}^d T_{*k}^{*n_k} m_* \in \mathcal{M}_*$ to get, for $e \in \mathcal{E}$,

$$\langle P_{\mathcal{E}} (M_W)^* (\Pi_{k=1}^d (M_{b_k I})^{*n_k} m_*), e\rangle = \langle P_{\mathcal{E}} X^* (\Pi_{k=1}^d (M_{b_k I})^{*n_k} m_*), e\rangle$$

from which we get

$$\langle (M_W)^* m_*, \Pi_{k=1}^d (M_{b_k I})^{n_k} e\rangle = \langle X^* m_*, \Pi_{k=1}^d (M_{b_k I})^{n_k} e\rangle$$

for all $n = (n_1, \ldots, n_d) \in \mathbb{Z}_+^d$. Since elements of the form $\Pi_{k=1}^d (M_{b_k I})^{n_k} e$ span $\mathcal{H}(k) \otimes \mathcal{E}$ (note the form of the orthonormal basis (5.8) for $\mathcal{H}(k)$), we conclude that $(M_W)^*|_{\mathcal{M}_*} = X^*$ and the theorem follows.

6 Examples and Applications

We shall consider now two classes of examples considered by Quiggin [38] of reproducing kernel spaces for which the reciprocal kernel has 1 positive square: weighted Hardy spaces on the unit disk and function spaces on an interval with a "separated" kernel. Our objective is to find explicitly a scalar function $a(z)$ and an operator-valued function $b(z)$ with values in $\mathcal{L}(\mathcal{C}, \mathbb{C})$ (for some "natural" Hilbert space $\mathcal{C}$) so that

$$\frac{1}{k(z,w)} = a(z)(1 - b(z)b(w)^*)\overline{a(w)}. \tag{6.1}$$

As our first example, let $\{w_n\}_{n=0}^{\infty}$ be a sequence of positive numbers, and let $H^2(\{w_n\})$ be the corresponding weighted Hardy space of analytic functions $f(z) = \sum_{n=0}^{\infty} f_n z^n$ on $\mathbb{D}$ with $\|f\|^2_{\{w_n\}} = \sum_{n=0}^{\infty} w_n |f_n|^2 < \infty$. We assume that $\phi(\zeta) = \sum_{n=0}^{\infty} \frac{1}{w_n}\zeta^n$ is analytic on $\mathbb{D}$; this ensures that $H^2(\{w_n\})$ is complete and is a reproducing kernel Hilbert space with reproducing kernel $k(z,w) = \phi(z\overline{w})$. We shall assume also that $\phi(\zeta)$ has no zeroes in $\mathbb{D}$.

Then we can write

$$\frac{1}{\phi(\zeta)} = w_0(1 - \alpha(\zeta))$$

where $\alpha(\zeta) = \sum_{n=1}^{\infty} \alpha_n \zeta^n$ is a holomorphic function on $\mathbb{D}$ with real Taylor coefficients satisfying $\alpha(\zeta) < 1$ for ζ real, $0 \leq \zeta < 1$. It follows that $\frac{1}{k(z,w)}$ has 1 positive square if and only if all the Taylor coefficients of $\alpha(\zeta)$ are nonnegative. If this is the case we can take any complex numbers β_n with $\beta_n \overline{\beta_n} = \alpha_n$ and define $\beta(\zeta) = \sum_{n=1}^{\infty} \beta_n \zeta^n$; we see that $\beta(\zeta)$ belongs to the closed unit ball in the usual Hardy space H^2 on $\mathbb{D}$. We then obtain (6.1) with $a(z) = \sqrt{w_0}$, $\mathcal{C} = H^2$, and

$$b(z)(f) = \langle f(\cdot), \mathcal{J}\beta(z\cdot)\rangle_{H^2}, \tag{6.2}$$

where $\mathcal{J}$ is the usual complex conjugation on H^2: $(\mathcal{J}f)(\cdot) = \overline{f(\overline{\cdot})}$ $(f \in H^2$, $z \in \mathbb{D})$.

In particular we can consider the Dirichlet space D of analytic functions $f(z)$ on $\mathbb{D}$ with $\|f\|_D^2 = \frac{1}{\pi}\int_{\mathbb{D}} |f'(z)|^2 dxdy + \sup_{r<1} \frac{1}{2\pi}\int_0^{2\pi} |f(re^{i\theta})|^2 d\theta < \infty$. This is a weighted Hardy space with the weight sequence $w_n = n+1$ so that $\phi(\zeta) = \frac{1}{\zeta}\log\frac{1}{1-\zeta}$ and a direct calculation shows that the Taylor coefficients of $\alpha(\zeta) = 1 - \frac{1}{w_0\phi(\zeta)}$ are all nonnegative. It is interesting to determine explicit necessary and sufficient conditions on the weight sequence w_n for the nonnegativity of α_n; as shown in [38], a sufficient condition is that the sequence w_n/w_{n+1} is non-decreasing.

Our second example is a reproducing kernel Hilbert space on a real interval $[x_0, x_1]$ with a "separated" kernel:

$$k(x,y) = \begin{cases} \phi(x)\psi(y) & \text{if } x \leq y, \\ \phi(y)\psi(x) & \text{if } y \leq x, \end{cases} \tag{6.3}$$

where ϕ, ψ are real valued functions on the interval that are nowhere vanishing and have everywhere the same sign and such that ϕ/ψ is non-decreasing. This kernel is positive since $k(x,y) = h(x)h(y)^*$ where $h(x)$ is a linear functional on $L^2([0, \psi(x_0)/\phi(x_0)])$ given by

$$h(x)(f) = \langle f, \phi(x)\chi_{[0,\psi(x)/\phi(x)]}\rangle,$$

where χ denotes the characteristic function of an interval. We see immediately that $\frac{1}{k(x,y)}$ has 1 positive square since we have a representation (6.1) with $a(x) = \frac{1}{\phi(x)}\sqrt{\phi(x_1)/\psi(x_1)}$, $\mathcal{C} = L^2([\phi(x_0)\psi(x_1)/\psi(x_0)\phi(x_1), 1])$, and

$$b(x)(f) = \langle f, \chi_{[\phi(x)\psi(x_1)/\psi(x)\phi(x_1),1]}\rangle. \tag{6.4}$$

Reproducing kernel spaces with a separated kernel on an interval are essentially first order Sobolev spaces. These first order Sobolev spaces as linear spaces are all equal to the linear space

$$W^{2,1} = \{f\colon [x_0, x_1] \to \mathbb{C}\colon f,\ f' \text{ absolutely continuous with } |f'|^2 \text{ integrable}\}.$$

The following theorem sums up the connection between reproducing kernel spaces with a separated kernel and first order weighted Sobolev spaces.

Theorem 6.1 *Consider two continuous real-valued weight functions w_0 and w_1 on the interval $[x_0, x_1]$ with w_1 strictly positive and continuously differentiable. The following statements are equivalent.*

1. *The (self-adjoint) differential operator $\mathbf{L}$ given by the differential expression*

$$L = -\frac{d}{dx} w_1 \frac{d}{dx} + w_0,$$

with domain

$$\mathrm{Dom}(\mathbf{L}) = \{f : [x_0, x_1] \to \mathbb{C} : f,\ f' \text{ absolutely continuous, } |f''|^2 \text{ integrable, } f'(x_0) = f'(x_1) = 0\}$$

in $L^2([x_0, x_1])$, is positive-definite:

$$\int_{x_0}^{x_1} (Lf)(x)\ \overline{f(x)}\, dx > 0 \text{ for all nontrivial } f \in \mathrm{Dom}(\mathbf{L}).$$

2. *The Sobolev form Q_{w_0,w_1} given by*

$$Q_{w_0,w_1}[f] = \int_{x_0}^{x_1} w_0(x)|f(x)|^2\, dx + \int_{x_0}^{x_1} w_1(x)|f'(x)|^2\, dx$$

is positive definite on $W^{2,1}$, i.e., $Q_{w_0,w_1}[f] > 0$ for any nontrivial $f \in W^{2,1}$.

3. *There exists a nowhere vanishing solution u of $Lu = 0$ with $\frac{u'(x_1)}{u(x_1)} \geq 0$ and $\frac{u'(x_0)}{u(x_0)} \leq 0$ with at least one of these inequalities being a strict inequality.*

If any (and hence all) of these conditions are satisfied and if we set $W^{2,1}_{w_0,w_1}[x_0, x_1]$ to be the linear space $W^{2,1}$ with norm $\|f\|^2_{W^{2,1}_{w_0,w_1}} := Q_{w_0,w_1}[f]$, then $W^{2,1}_{w_0,w_1}$ is a reproducing kernel Hilbert space with a separated kernel (6.3) *where ϕ and ψ are solutions of the differential equation $Lf = 0$ satisfying the boundary conditions $f'(x_0) = 0$ and $f'(x_1) = 0$, respectively, and normalized by $w_1(\psi\phi' - \phi\psi') = 1$.*

Conversely, suppose that ϕ, ψ are real-valued, continuously differentiable, nowhere vanishing functions on $[x_0, x_1]$ with $\phi'(x_0) = 0$, $\psi'(x_1) = 0$ and ϕ/ψ

strictly increasing. Then $W^{2,1}_{w_0,w_1}$ *is a reproducing kernel Hilbert space with reproducing kernel given by* (6.3), *where*

$$w_0 = \frac{\phi'\psi'' - \phi''\psi'}{(\psi\phi' - \phi\psi')^2}, \quad w_1 = \frac{1}{\psi\phi' - \phi\psi'}. \tag{6.5}$$

Proof: For $f \in \mathrm{Dom}(\mathbf{L})$, one can easily check that

$$\langle Lf, f\rangle_{L^2} = Q_{w_0,w_1}[f].$$

The implication (2) $\Longrightarrow$ (1) follows immediately from this identity.

We now show (1) $\Longrightarrow$ (2). Let $\mathcal{H}$ be the Hilbert space completion of $\mathrm{Dom}(\mathbf{L})$ with respect to the norm induced by the sesquilinear form Q_{w_0,w_1}. Suppose that $\{f_n\}$ is a Cauchy sequence (with respect to Q_{w_0,w_1}) in $\mathrm{Dom}(\mathbf{L})$. Since $\mathbf{L}$ has compact resolvent (see [23, p. 1330]) and 0 is not an eigenvalue of $\mathbf{L}$, it follows that

$$\langle Lf_n - Lf_m, f_n - f_m\rangle_{L^2} \geq \epsilon^2 \|f_n - f_m\|^2_{L^2}$$

for some $\epsilon > 0$, hence $\{f_n\}$ is a Cauchy sequence with respect to the L_2-norm. Since w_1 is strictly positive, from the fact that $\{f_n\}$ is Cauchy both with respect to Q_{w_0,w_1} and in L^2-norm, we see that $\{f_n'\}$ is Cauchy in L^2-norm, and hence converges to a function $g \in L^2$. Since $\{f_n\}$ also converges to a function f in L^2-norm, it follows that the limit function f is absolutely continuous with derivative $f' = g \in L^2$ and that $Q_{w_0,w_1}[f_n - f] \to 0$. Thus $\mathcal{H}$ can be identified with linear subspace of $W^{2,1}$ with the square of the norm defined by Q_{w_0,w_1}. Conversely, if $f \in W^{2,1}$, we can approximate f' in L^2-norm by a sequence g_n of absolutely continuous functions with derivative in L_2 and with $g(x_0) = g(x_1) = 0$. We then set $f_n(x) = f(x_0) + \int_{x_0}^x g_n(t)\,dt$ and observe that $Q_{w_0,w_1}[f_n - f] \to 0$ as $n \to \infty$. Thus $\mathcal{H}$ equals $W^{2,1}$ with the square of the norm defined by Q_{w_0,w_1}, and (2) follows; also we see that $W^{2,1}_{w_0,w_1}$ is complete and hence a Hilbert space.

If either (1) or (2) is satisfied, we can see either by a direct calculation or by using the theory of Sturm-Liouville boundary value problems that $W^{2,1}_{w_0,w_1}$ is a reproducing kernel Hilbert space with reproducing kernel given by (6.3) where ϕ and ψ are as in the statement of the theorem (the reproducing kernel is the Green's function for the differential operator $\mathbf{L}$, see, e.g., [23, p. 1329]). The fact that the kernel is positive and non-degenerate (since the space contains, e.g., the constants) implies that ϕ and ψ are nowhere vanishing and have everywhere the same sign; the normalization condition implies that $(\phi/\psi)' > 0$, hence ϕ/ψ is increasing.

To see now that (1) or (2) implies (3), take the required nowhere vanishing solution u to be either ϕ or ψ. The conditions $u'(x_1)/u(x_1) \geq 0$ and $u'(x_0)/u(x_0) \leq 0$ (with the condition that at least one of these inequalitities be strict) follow from the positivity of $\phi'\psi - \psi'\phi$, the fact that ϕ and ψ have the same sign, and the conditions $\phi'(x_0) = 0$, $\psi'(x_1) = 0$.

We next show that (3) implies (1). We set $\beta = -u'/u$. A straightforward verification shows that

$$L = \left(-\frac{d}{dx} + \beta\right) w_1 \left(\frac{d}{dx} + \beta\right).$$

Furthermore, for any $f \in \mathrm{Dom}(\mathbf{L})$, we have

$$\langle Lf, f\rangle_{L^2} = \langle Df, Df\rangle_{L^2} - w_1(x_1)\beta(x_1)|f(x_1)|^2 + w_1(x_0)\beta(x_0)|f(x_0)|^2$$

where D is the first order differential expression

$$D = w_1^{1/2}\left(\frac{d}{dx} + \beta\right).$$

The sum of the boundary terms is necessarily nonnegative. To show that $\mathbf{L}$ is positive-definite, it remains to show that $Df = 0$ for $f \in \mathrm{Dom}(\mathbf{L})$ implies that $f = 0$. From $Df = 0$, one gets that $f = cu$ with c a constant. If f (and hence c) were nonzero, it would follow that $u \in \mathrm{Dom}(\mathbf{L})$, contradicting the fact that either $u'(x_0)$ or $u'(x_1)$ is not zero.

For the converse part of the theorem, note that with w_0 and w_1 defined as in (6.5), ϕ and ψ satisfy

$$L\phi = 0, \quad L\psi = 0.$$

Applying (3) with $u = \phi$ or $u = \psi$ shows that Q_{w_0,w_1} is positive definite on $W^{2,1}$ and the direct part of the theorem tells us that the reproducing kernel of $W^{2,1}_{w_0,w_1}$ is given by (6.3) as required. □

Remark: The correspondence between factorization of differential operators, positivity and the existence of nowhere vanishing solutions is well known in the literature; see [25, Section 26].

It is an interesting question whether, for more general separated kernels

$$k(x, y) = \begin{cases} \phi(x)^*\Gamma\psi(y) & \text{if } x \le y, \\ \psi(x)^*\Gamma\phi(y) & \text{if } y \le x, \end{cases}$$

where ϕ, ψ are $\mathbb{C}^n$-valued functions on the interval $[x_0, x_1]$ and Γ is a constant self-adjoint $n \times n$ matrix, the reciprocal kernel has one positive square. It is not immediately obvious when such a kernel is positive. Such kernels are connected with higher order weighted Sobolev spaces on an interval analogously to Theorem 6.1. Another interesting question is whether, for Sobolev spaces on a domain in a higher dimensional space, the reciprocal of the reproducing kernel has one positive square. Notice in this connection that the Dirichlet space considered earlier in this section is the "analytic subspace" of a first order Sobolev space on the unit disk.

References

[1] J. Agler, Interpolation, unpublished preprint.

[2] J. Agler, Nevanlinna-Pick interpolation on Sobolev space, *Proc. Amer. Math. Soc.* **108** (1990), 341–351.

[3] J. Agler and J.E. McCarthy, Complete Nevanlinna-Pick kernels, preprint, 1997.

[4] D. Alpay, J.A. Ball and V. Belotnikov, On the bitangential interpolation problem for contractive valued functions in the polydisk, *J. Operator Theory*, to appear (available at http://www.math.vt.edu/people/ball/research/alpayballbol.pdf).

[5] D. Alpay, V. Bolotnikov, A. Dijksma and H. de Snoo, On some operator colligations and associated reproducing kernel Hilbert spaces, in *Operator Extensions, Interpolation of Functions and Related Topics: The International Conference on Operator Theory, Timisoara (Romania), June 1–5, 1992* (Ed. A. Gheondea, D. Timotin and F.-H. Vasilescu), Operator Theory: Advances and Applications vol. 61, Birkhäuser-Verlag, Basel, 1993, 1–27.

[6] D. Alpay, V. Boltnikov, A. Dijksma, J.R. Rovnyak and C. Sadosky, Espaces de Hilbert inclus contractivement dans l'espace de Hardy du bi-disque, *Comptes Rendus de l'Académie des Sciences: Série I, Mathématique* **326** (1998), 1365–1370.

[7] D. Alpay and H. Dym, On a new class of reproducing kernel spaces and a new generalization of the Iohvidov's laws, *Linear Algebra and Applications* **241/243** (1996), 3–84.

[8] D. Alpay and H. Dym, On a new class of structural reproducing kernel spaces, *J. Functional Analysis* **111** (1993), 1–28.

[9] A. Arias and G. Popescu, Noncommutative interpolation and Poisson transforms, preprint, 1997.

[10] A. Arias and G. Popescu, Noncommutative interpolation and Poisson transforms II, preprint, 1997.

[11] D.Z. Arov and L.Z. Grossman, Scattering matrices in the theory of unitary extensions of isometric operators, *Math. Nachrichten* **157** (1992), 105–123.

[12] W. Arveson, Subalgebras of C^*-algebras III: multivariable operator theory, *Acta Math.* **181** (1998), 159–228.

[13] J.A. Ball, Linear systems, operator model theory and scattering: multivariable generalizations, *Proceedings of International Conference on Operator Theory and its Applications to Scientific Industrial Problems (Winnipeg, 1998)*, to appear.

[14] J.A. Ball, W.S. Li, D. Timotin and T.T. Trent, A commutant lifting theorem on the polydisc: interpolation problems for the bidisc, *Indiana University Mathematics J.*, to appear.

[15] J.A. Ball and T.T. Trent, Unitary colligations, reproducing kernel Hilbert spaces, and Nevanlinna-Pick interpolation in several variables, *J. Functional Analysis* **157** (1998), 1–61.

[16] J.A. Ball and V. Vinnikov, Multivariable linear systems, scattering, unitary dilations and operator model theory for row contractions, in preparation.

[17] R.S. Clancy and S. McCullough, Projective modules and Hilbert spaces with a Nevanlinna-Pick kernel, *Proc. Amer. Math. Soc.* **126** (1998), 3299–3305.

[18] B. Cole, K. Lewis and J. Wermer, Pick conditions on a uniform algebra and von Neumann inequalities, *J. Functional Analysis* **107** (1992), 235–254.

[19] M. Cotlar and C. Sadosky, Nehari and Nevanlinna-Pick problems and holomorphic extensions in the polydisk in terms of restricted BMO, *J. Functional Analysis* **124** (1994), 205–210.

[20] K.R. Davidson and D.R. Pitts, Invariant subspaces and hyper-reflexivity for free semigroup algebras, *Proc. London Math. Soc.* **78** (1999). 401–430.

[21] K.R. Davidson and D.R. Pitts, The algebraic structure of non-commutative analytic Toeplitz algebras, *Math. Ann.* **311** (1998), 275–303.

[22] K.R. Davidson and D.R. Pitts, Nevanlinna-Pick interpolation for non-commutative analytic Toeplitz algebra, *Integral Equations and Operator Theory* **31** (1998), 321–337.

[23] N. Dunford and J.T. Schwartz, *Linear Operators, Part II: Spectral Theory*, Wiley Interscience, 1963.

[24] A.E. Frazho, Complements to models for noncommuting operators, *J. Functional Analysis* **59** (1984), 445–461.

[25] I.M. Gelfand and S.V. Fomin, *Calculus of Variations* (English translation of Russian original), Prentice-Hall, 1963.

[26] A.Ya. Kheifets and P.M. Yuditskii, An analysis and extension of V.P. Potapov's approach to interpolation problems with applications to the generalized bitangential Schur-Nevanlinna-Pick problem and j-inner-outer factorization, in *Matrix and Operator Valued Functions: The Vladimir Petrovich Potapov Memorial Volume* (Ed. I. Gohberg and L.A. Sakhnovich), Operator Theory: Advances and Applications vol. 72, Birkhäuser-Verlag, 1994, 133–161.

[27] S. McCullough, Carathéodory interpolation kernels, *Integral Equations and Operator Theory* **15** (1992), 43–71.

[28] S. McCullough, The local de Branges-Rovnyak construction and complete Nevanlinna-Pick kernels, in *Algebraic Methods in Operator Theory* (Ed. R. Curto and P.E.T. Jørgensen), Birkhäuser-Verlag, Boston, 1994, 15–24.

[29] S. McCullough, Commutant lifting on a two holed domain, *Integral Equations and Operator Theory*, to appear.

[30] R. Nevanlinna, Über beschränkte Funktionen die in gegebene Punkten vorgeschriebene Werte annehmen, *Ann. Acad. Sci. Fenn. Ser.* A **13** (1919), no. 1.

[31] A.A. Nudelman, Some generalizations of classical interpolation problems, in *Operator Extensions, Interpolation of Functions and Related Topics* (Ed. A. Gheondea, D. Timotin and F.-H. Vasilescu), Operator Theory: Advances and Applications vol. 61, Birkhäuser-Verlag, Basel, 1993, 171–188.

[32] G. Pick, Über die Beschränkungen analytischer Funktionen, welche durch vorgegebene Funktionswerte bewirkt werden, *Math. Ann.* **77** (1916), 7–23.

[33] G. Popescu, Multi-analytic operators and some factorization theorems, *Indiana Univ. Math J.* **38** (1989), 693–710.

[34] G. Popescu, Isometric dilations for infinite sequences of noncommuting operators, *Trans. Amer. Math. Soc.* **316** (1989), 523–536.

[35] G. Popescu, On intertwining dilations for sequences of noncommuting operators, *J. Math. Anal. Appl.* **167** (1992), 382–402.

[36] G. Popescu, Multi-analytic operators on Fock spaces, *Math. Ann.* **303** (1995), 31–46.

[37] G. Popescu, Interpolation problems in several variables, *J. Mathematical Analysis Applications*, to appear.

[38] P. Quiggin, For which reproducing kernel Hilbert spaces is Pick's theorem true? *Integral Equations and Operator Theory* **16** (1993), 244–266.

[39] M. Rosenblum, A corona theorem for countably many functions, *Integral Equations and Operator Theory* **3** (1980), 125–137.

[40] D. Sarason, Generalized interpolation in H^∞, *Trans. Amer. Math. Soc.* **127** (1967), 179–203.

Department of Mathematics
Virginia Tech
Blacksburg
Virginia 24061
email: ball@math.vt.edu

Department of Mathematics
University of Alabama
Tuscaloosa
Alabama 35406
email: ttrent@gp.as.ua.edu

Department of Theoretical Mathematics
Weizmann Institute of Science
Rehovot 76100
Israel
after February 1, 2000: Department of Mathematics
Ben-Gurion University of the Negev
Beer-Sheva 84105
Israel
email: vinnikov@wisdom.weizmann.ac.il

1991 *Mathematics Subject Classification.* Primary: 47A57; Secondary: 32E30, 47A20, 47A48.

Operator Theory:
Advances and Applications, Vol. 122
© 2001 Birkhäuser Verlag Basel/Switzerland

Sums of Idempotents and Logarithmic Residues in Matrix Algebras

H. Bart, T. Ehrhardt and B. Silbermann

Dedicated to M.A. Kaashoek on the occasion of his sixtieth birthday

Logarithmic residues are contour integrals of logarithmic derivatives of vector-valued analytic functions. In several matrix algebras, the set of logarithmic residues coincides with the set of sums of idempotents. The connected components of this set are identified.

1 Introduction

Let B be a (complex) Banach algebra. A logarithmic residue in B is a contour integral of the logarithmic derivative $f'(\lambda)f(\lambda)^{-1}$ of an analytic B-valued function $f(\lambda)$. The first to consider this type of expression was L. Mittenthal [M]. He gave (very restrictive) sufficient conditions for a logarithmic residue to be an idempotent (see also [B]).

As was observed in [BES2], each sum of idempotents in B is a logarithmic residue in B. The converse of this statement is not always true; a counterexample can be found in [BES3]. So, in general, the set of logarithmic residues is not identical with the set of sums of idempotents. However, in [BES3] it was also shown that in commutative Banach algebras, the set of logarithmic residues does coincide with the set of sums of idempotents. In the present paper, we shall see that the same is true for certain matrix algebras. Here follows a brief description of our results.

Section 2 is of a preliminary nature. In Section 3 we discuss the case when B is a Banach subalgebra of $\mathbb{C}^{m\times m}$ generated by a single matrix T. It is shown that the logarithmic residues in such an algebra are just the sums of spectral idempotents (=Riesz projections) associated with T. The algebras in question are of course commutative.

Section 4 contains a detailed discussion of a result obtained by R.E. Hartwig and M.S. Putcha [HP] and also by P.Y. Wu [Wu1]. The result in question is a characterization of sums of (rank one) idempotent matrices in terms of rank and trace. We give a new proof which, besides being quite transparent, puts things in a somewhat wider framework.

Section 5 is concerned with logarithmic residues of Fredholm operator valued functions. Such logarithmic residues are again sums of idempotents, and in fact of idempotents of finite rank. The proof is based on the characterization of sums of idempotent matrices in terms of rank and trace given by [HP] and [Wu1] and uses elements from Systems Theory (realizations). Specializing to matrix valued analytic functions we obtain a characterization of logarithmic residues as sums of idempotents in the full (non-commutative) matrix algebra $\mathbb{C}^{m\times m}$.

In Section 6 we study the algebra of block upper triangular matrices (arbitrary but fixed block structure). Extreme examples of such algebras are the full matrix algebras (one block case), already encountered in Section 5, and the algebras of upper triangular matrices (scalar blocks). Once more the logarithmic residues are just the sums of idempotents (that can even be chosen to have rank one). The proof of this result rests on a rank decomposition theorem for block upper triangular matrices [BW] which, in turn, is based on results from the theory of Integer Programming [PS]. The second main result in this section is concerned with the topological properties of the set of logarithmic residues in the Banach algebras of block upper triangular matrices. In particular we identify the connected components of this set.

A few remarks about notation. The set of integers is denoted by $\mathbb{Z}$, the complex plane by $\mathbb{C}$, the Banach algebra of all $m \times m$ complex matrices by $\mathbb{C}^{m\times m}$, and the linear space of complex column m-vectors by $\mathbb{C}^m$ (i.e., $\mathbb{C}^m = \mathbb{C}^{m\times 1}$). We write I_n for the $n \times n$ identity matrix. If A is a square matrix, then $|A|$ stands for the determinant of A, rank A for the rank of A, trace A for the trace of A, Ran A for the range of A and Ker A for the null space of A. Similar notations are used for linear operators (when applicable). Whenever useful, matrices are considered as linear operators in the customary manner. Finally, the symbol $\bigoplus$ and the superscript ^{T} are used to signal direct sums (of matrices or subspaces) and transposes of matrices, respectively.

Acknowledgement

The first author wishes to thank David C. Lay for stimulating discussions about the material contained in Section 6 of the present paper.

2 Preliminaries

In what follows, B will be a (complex) Banach algebra with unit element. In working with contour integrals, we shall employ bounded Cauchy domains (in $\mathbb{C}$) and their (positively oriented) boundaries. For a discussion of these notions, see, for instance, [TL].

Let D be a bounded Cauchy domain. The (positively oriented) boundary of D will be denoted by ∂D. We write $\mathcal{A}_\partial(D; B)$ for the set of all B-valued functions f with the following properties: f is defined and analytic on an open neighborhood of the closure $\overline{D}(= D \cup \partial D)$ of D and f takes invertible values on all of ∂D. For $f \in \mathcal{A}_\partial(D; B)$ one can define

$$\text{LogRes}(f; D) = \frac{1}{2\pi i} \int_{\partial D} f'(\lambda) f(\lambda)^{-1} d\lambda \tag{2.1}$$

where f' stands for the derivative of f.

The elements of the form (2.1) are the left logarithmic residues in B. More specifically, we call LogRes$(f; D)$ *the left logarithmic residue of f with respect to D.*

The expression (2.1) involves the *left* logarithmic derivative $f'(\lambda)f(\lambda)^{-1}$ of the function f. It is just as well possible to work with the *right* logarithmic derivative $f(\lambda)^{-1}f'(\lambda)$ of f. One then obtains the *right* logarithmic residue of f with respect to D.

There is a strong connection between logarithmic residues and sums of idempotents. The first hint in this direction comes from the following result taken from [BES2]; cf. also [BES1]. For the convenience of the reader and for later reference, we present it with a full proof.

Proposition 2.1 *A sum of idempotents in a Banach algebra B is a logarithmic residue in B.*

In fact, if s a sum of idempotents in B, then there exist a bounded Cauchy domain D and $f \in \mathcal{A}_\partial(D; B)$ such that

$$s = \frac{1}{2\pi i} \int_{\partial D} f'(\lambda) f(\lambda)^{-1} d\lambda = \frac{1}{2\pi i} \int_{\partial D} f(\lambda)^{-1} f'(\lambda) d\lambda. \tag{2.2}$$

So s is both the left and right logarithmic residue of f with respect to D.

Proof: Let $s = p_1 + \cdots + p_k$ be a sum of idempotents $p_1, \ldots, p_k$ in B. Choose different complex numbers $\alpha_1, \ldots, \alpha_k$, and put

$$D = \bigcup_{j=1}^{k} \left\{ z \in \mathbb{C} \,\middle|\, |z - \alpha_j| < \frac{1}{4} r \right\}, \quad U = \bigcup_{j=1}^{k} \left\{ z \in \mathbb{C} \,\middle|\, |z - \alpha_j| < \frac{1}{3} r \right\},$$

where $r = \min\{ |\alpha_i - \alpha_j| \,|\, i, j = 1, \ldots, k;\ i \neq j\}$. Then D is a bounded Cauchy domain and U is an open neighborhood of $\overline{D}$. Note that ∂D consists of a disjoint union of k positively oriented circles. Define $f : U \to B$ by

$$f(\lambda) = e - p_j + (\lambda - \alpha_j) p_j, \qquad |\lambda - \alpha_j| < \frac{1}{3} r; \ \ j = 1, \ldots, k.$$

Then $f \in \mathcal{A}_\partial(D; B)$ and a straightforward computation shows that, for $|\lambda - \alpha_j| < \frac{1}{3} r, \ \ j = 1, \ldots, k,$

$$f'(\lambda) f(\lambda)^{-1} = f(\lambda)^{-1} f'(\lambda) = \frac{1}{\lambda - \alpha_j} p_j.$$

It follows that (2.2) is satisfied, and the proof is complete. □

Elaborating on the above argument, we note that $p_1 + \cdots + p_k$ was obtained as a logarithmic residue of a function f defined on a domain consisting of k connected components. In [E], it is shown that a sum of idempotents in a Banach algebra B can be obtained as the logarithmic residue of an *entire* B-valued function, i.e., a B-valued function defined and analytic on all of $\mathbb{C}$.

As is shown in [BES3], in commutative Banach algebras the set of logarithmic residues coincides with the set of sums of idempotents. In the next sections, we shall see that the same is true for certain matrix algebras. For generalizations of the results to logarithmic residues of *meromorphic* functions, see [BES3], Section 7, and Remark 7.3 there in particular.

3 Matrix Algebras Generated by a Single Matrix

In this section, we consider Banach algebras generated by a single matrix. Such a Banach algebra is commutative. So, by [BES3], Theorem 6.1, the set of logarithmic residues and the set of sums of idempotents coincide. We shall give a direct proof of this fact involving matrix techniques. Along the way, some additional details will be presented. We begin with a technical lemma.

Lemma 3.1 *Let N be a nilpotent $m \times m$ matrix, and let $f_0, \dots, f_{m-1}$ be complex-valued functions, defined and analytic on an open subset U of $\mathbb{C}$. Assume f_0 is not identically zero on U and write*

$$F(\lambda) = \sum_{k=0}^{m-1} f_k(\lambda) N^k. \tag{3.1}$$

Then F is an analytic $m \times m$ matrix valued function on U and, for $\lambda \in U$, the matrix $F(\lambda)$ is invertible if and only if $f_0(\lambda) \neq 0$. Moreover, the logarithmic derivative of F is given by

$$F'(\lambda)F(\lambda)^{-1} = \frac{f_0'(\lambda)}{f_0(\lambda)} I_m - \frac{d}{d\lambda} \sum_{k=1}^{m-1} \frac{1}{k} A(\lambda)^k, \tag{3.2}$$

where A is some $m \times m$ matrix valued function defined and analytic on the open set $U_0 = \{\lambda \in U \mid f_0(\lambda) \neq 0\}$.

The identity (3.2) holds on U_0 which is exactly the domain of the logarithmic derivative of F. Note that (because of commutativity) there is no need to distinguish here between left and right logarithmic derivatives.

Proof: Clearly $f_0(\lambda) = 0$ implies $F(\lambda)^m = 0$, hence $F(\lambda)$ is not invertible. Thus we focus on those $\lambda \in U$ for which $f_0(\lambda) \neq 0$.

For $\lambda \in U_0$ (i.e. $\lambda \in U$ and $f_0(\lambda) \neq 0$), write

$$A(\lambda) = -\sum_{k=1}^{m-1} \frac{f_k(\lambda)}{f_0(\lambda)} N^k.$$

Then $A(\lambda)^m = 0$, and so $I_m - A(\lambda)$ is invertible with inverse

$$(I_m - A(\lambda))^{-1} = \sum_{k=0}^{m-1} A(\lambda)^k.$$

Now $F(\lambda) = f_0(\lambda)(I_m - A(\lambda))$. It follows that, for $\lambda \in U_0$, $F(\lambda)$ is invertible and

$$F(\lambda)^{-1} = \frac{1}{f_0(\lambda)}(I_m - A(\lambda))^{-1} = \frac{1}{f_0(\lambda)} \sum_{k=0}^{m-1} A(\lambda)^k.$$

For the derivative F' of F, we have

$$F'(\lambda) = f_0'(\lambda)(I_m - A(\lambda)) - f_0(\lambda)A'(\lambda),$$

and so the logarithmic derivative of F is given by

$$\begin{aligned} F'(\lambda)F(\lambda)^{-1} &= \frac{f_0'(\lambda)}{f_0(\lambda)} I_m - A'(\lambda) \sum_{k=0}^{m-1} A(\lambda)^k \\ &= \frac{f_0'(\lambda)}{f_0(\lambda)} I_m - \sum_{k=0}^{m-1} \frac{1}{k+1} \frac{d}{d\lambda} A(\lambda)^{k+1}. \end{aligned}$$

This proves the lemma. □

Theorem 3.2 *Let T be an $m \times m$ matrix, and let B_T be the Banach subalgebra of $\mathbb{C}^{m \times m}$ generated by T. For a matrix L in B_T, the following statements are equivalent:*

(i) *L is a logarithmic residue in B_T,*

(ii) *L is a sum of idempotents in B_T,*

(iii) *L is a sum of spectral idempotents associated with T.*

Apart from the direct proof that we shall give (not referring to [BES3]), the new aspect is the connection with spectral idempotents. For information on spectral idempotents (also called Riesz projections), see [GLR], Section 2.4 and [GGK], Section 1.2.

Proof: The spectral idempotents of T belong to B_T. Hence (iii) implies (ii). The implication (ii) $\Rightarrow$ (i) is immediate from Proposition 2.1. It remains to prove that

(i) implies (iii). Here we may assume T is a Jordan matrix. The justification for this is, of course, that each matrix can be brought into Jordan form by an appropriate similarity transformation.

Since T is (assumed to be) a Jordan matrix, we can write T as a direct sum

$$T = \bigoplus_{j=1}^{s} (\alpha_j I_{m_j} + N_j)$$

where $\alpha_1, \ldots, \alpha_s$ are the different eigenvalues of T and $N_1, \ldots, N_s$ are nilpotent matrices of order $m_1, \ldots, m_s$, respectively (so $m = m_1 + \cdots + m_s$). The spectral idempotents of T are the matrices

$$\bigoplus_{j=1}^{s} d_j I_{m_j}$$

with $d_1, \ldots, d_s$ either zero or one.

Now, let L be a logarithmic residue in B_T, say $L = \text{LogRes}(F; D)$ where D is a Cauchy domain and $F \in \mathcal{A}_\partial(D; B_T)$. The above decomposition of T induces one for the function F:

$$F(\lambda) = \bigoplus_{j=1}^{s} F_j(\lambda)$$

with $F_j \in \mathcal{A}_\partial(D; B_{N_j})$, $j = 1, \ldots, s$. The functions $F_1, \ldots, F_s$ are of the sort considered in Lemma 3.1. So for each of them we have an identity of the type (3.2). Now the second term in the right hand side of (3.2) obviously has a primitive, so it vanishes upon integration. It follows that

$$\text{LogRes}(F; D) = \bigoplus_{j=1}^{s} n_j I_{m_j},$$

with $n_1, \ldots, n_s$ non-negative integers. Hence $\text{LogRes}(F; D)$ is indeed a sum of spectral idempotents of T. □

Corollary 3.3 *Let T be an $m \times m$ matrix, and let B_T be the Banach algebra of $\mathbb{C}^{m\times m}$ generated by T. Then the set of logarithmic residues in B_T is a discrete (hence closed) subset of B_T.*

Proof: Let $P_1, \ldots, P_s$ be the different spectral idempotents of T. What we need to show is that the set of all sums $k_1 P_1 + \ldots k_s P_s$ with $k_1, \ldots, k_s$ non-negative integers is a discrete subset of B_T. But this is obvious from the fact that $P_1, \ldots, P_s$ are linearly independent elements in $\mathbb{C}^{m\times m}$. Indeed, identify $\mathbb{C}^{m\times m}$ with $\mathbb{C}^{(m^2)}$ in such a way that $P_1, \ldots, P_s$ correspond to the first s unit vectors in $\mathbb{C}^{(m^2)}$, respectively. □

Remark 3.4 Elaborating on Lemma 3.1 and its proof, we let $A(\lambda)$ be as in the latter and take a closer look at the matrix function

$$G(\lambda) = \sum_{k=1}^{m-1} \frac{1}{k} A(\lambda)^k$$

featuring in the identity (3.2). The point that we want to make is that

$$G(\lambda) = \sum_{k=1}^{m-1} \frac{(-1)^k}{k} \left(\sum_{j=1}^{m-1} \frac{f_j(\lambda)}{f_0(\lambda)} N^j \right)^k = \sum_{j=1}^{m-1} g_j(\lambda) N^j$$

with

$$g_j(\lambda) = \sum_{\substack{k_1, k_2, \ldots, k_j = 0, 1, \ldots \\ k_1 + 2k_2 + \cdots + jk_j = j}} \frac{(-1)^{k_1 + \cdots + k_j}}{k_1 + \cdots + k_j} \frac{(k_1 + \cdots + k_j)!}{k_1! \ldots k_j!} \frac{f_1(\lambda)^{k_1} \ldots f_j(\lambda)^{k_j}}{f_0(\lambda)^{k_1 + \cdots + k_j}}.$$

We leave the proof of this as a (not completely trivial) exercise.

It is perhaps interesting to note that for fixed j, the number of terms in the expression for $g_j(\lambda)$ is the number of partitions of j into integer parts. This number is equal to the coefficient of x^j in

$$\prod_{p=1}^{\infty} \frac{1}{1 - x^p},$$

see [S], Section 3.12 and the references given there.

It is illuminating to carry the argument a little further. From (3.1), we have

$$F'(\lambda) = \sum_{k=0}^{m-1} f_k'(\lambda) N^k.$$

In general we do not have an adequate expression for $F(\lambda)^{-1}$, but in the case when N is the nilpotent $m \times m$ lower triangular Jordan block such an expression does exist:

$$F(\lambda)^{-1} = \sum_{k=0}^{m-1} \frac{(-1)^k}{f_0(\lambda)^{k+1}} \begin{vmatrix} f_1(\lambda) & f_0(\lambda) & 0 & \cdots & 0 \\ f_2(\lambda) & f_1(\lambda) & f_0(\lambda) & \ddots & \vdots \\ \vdots & \ddots & \ddots & \ddots & 0 \\ f_{k-1}(\lambda) & & \ddots & \ddots & f_0(\lambda) \\ f_k(\lambda) & f_{k-1}(\lambda) & \cdots & f_2(\lambda) & f_1(\lambda) \end{vmatrix} N^k,$$

where the determinant of the (vacuous) 0×0 matrix (occurring when $k = 0$) is understood as being 1. For the logarithmic derivative of F, i.e. the left hand side of (3.2), one now gets

$$\sum_{j=0}^{m-1}\sum_{k=0}^{j}\frac{(-1)^k}{f_0(\lambda)^{k+1}}f'_{j-k}(\lambda)\begin{vmatrix} f_1(\lambda) & f_0(\lambda) & 0 & \dots & 0 \\ f_2(\lambda) & f_1(\lambda) & f_0(\lambda) & \ddots & \vdots \\ \vdots & \ddots & \ddots & \ddots & 0 \\ f_{j-1}(\lambda) & & \ddots & \ddots & f_0(\lambda) \\ f_j(\lambda) & f_{j-1}(\lambda) & \dots & f_2(\lambda) & f_1(\lambda) \end{vmatrix} N^j.$$

Combining this with what we saw above, we obtain

$$\sum_{k=0}^{j}\frac{(-1)^k}{f_0(\lambda)^{k+1}}f'_{j-k}(\lambda)\begin{vmatrix} f_1(\lambda) & f_0(\lambda) & 0 & \dots & 0 \\ f_2(\lambda) & f_1(\lambda) & f_0(\lambda) & \ddots & \vdots \\ \vdots & \ddots & \ddots & \ddots & 0 \\ f_{j-1}(\lambda) & & \ddots & \ddots & f_0(\lambda) \\ f_j(\lambda) & f_{j-1}(\lambda) & \dots & f_2(\lambda) & f_1(\lambda) \end{vmatrix}$$

$$= -\frac{d}{d\lambda}\sum_{\substack{k_1,k_2,\dots,k_j=0,1,\dots \\ k_1+2k_2+\dots+jk_j=j}}\frac{(-1)^{k_1+\dots+k_j}}{k_1+\dots+k_j}\frac{(k_1+\dots+k_j)!}{k_1!\dots k_j!}\frac{f_1(\lambda)^{k_1}\dots f_j(\lambda)^{k_j}}{f_0(\lambda)^{k_1+\dots+k_j}}.$$

This identity can also be proved directly and once it has been obtained, it can be used to give an alternative Proof of Lemma 3.1. The basis for the latter is that by a similarity transformation any nilpotent matrix can be changed into a direct sum of nilpotent lower triangular Jordan blocks.

At the background of all this is the following expression involving a Toeplitz-Hessenberg matrix determinant:

$$\begin{vmatrix} a_1 & a_0 & 0 & \dots & 0 \\ a_2 & a_1 & a_0 & \dots & \vdots \\ \vdots & \ddots & \ddots & \ddots & 0 \\ a_{n-1} & & \ddots & \ddots & a_0 \\ a_n & a_{n-1} & \dots & a_2 & a_1 \end{vmatrix} = \sum_{k=1}^{n}(-1)^{n-k}a_0^{n-k}\sum_{\substack{\mu_1,\dots,\mu_k=1,2,\dots \\ \mu_1+\dots+\mu_k=n}} a_{\mu_1}\dots a_{\mu_k}$$

$$= (-1)^n\sum_{\substack{k_1,k_2,\dots,k_n=0,1,\dots \\ k_1+2k_2+\dots+nk_n=n}}(-1)^{k_1+\dots+k_n}a_0^{n-(k_1+\dots+k_n)}$$

$$\frac{(k_1+\dots+k_n)!}{k_1!\dots k_n!}a_1^{k_1}\dots a_n^{k_n}.$$

The first identity can be established by an induction argument; the second is a matter of bookkeeping (cf. [R], page 189, Problem 3). □

We conclude this section by observing that Theorem 3.2 and Corollary 3.3 admit a generalization to the case when T is a finite rank operator on a possibly *infinite* dimensional Banach space X. We leave the details to the reader (cf. the first part of the Proof of Theorem 5.1 below). For arbitrary operators in an infinite dimensional context, the general results on commutative Banach algebras hold (cf. [BES3]), but the connection with spectral idempotents is not clear.

4 Rank, Trace and Decomposition of Matrices

In this section we deal with the following result.

Theorem 4.1 *Let A be an $m \times m$ matrix. Then A is a sum of idempotent $m \times m$ matrices if and only if the trace of A is an integer larger than or equal to the rank of A. In that case, the idempotent matrices may all be chosen to have rank one, have a range contained in that of A and have a null space containing that of A.*

Apart from the remark about the null spaces, this theorem is identical to [Wu1], Theorem 1. The first part of the theorem (so the mere characterization of sums of idempotent matrices) is a particular case of [HP], Theorem 1. Thus, basically, Theorem 4.1 is known. Here we present a new and quite transparent proof which actually yields a somewhat more general result.

Proof: The proof of the only if part of the characterization of sums of idempotent matrices given in Theorem 4.1 is easy (see [Wu1] and [HP]): Note that rank and trace of idempotent matrices coincide; hence, if $A = P_1 + \cdots + P_k$ is a sum of idempotents, then $\text{rank } A \leq \text{rank } P_1 + \cdots + \text{rank } P_k = \text{trace } P_1 + \cdots + \text{trace } P_k = \text{trace } A \in \mathbb{Z}$.

For the if part we reason as follows. Put $r = \text{rank } A$, and assume $t = \text{trace } A$ is an integer larger than or equal to r. Then A can be written in the form

$$A = \sum_{j=1}^{t} u_j v_j^T$$

with $u_1, \ldots, u_t \in \text{Ran } A$ and $\text{Ker } A \subset \text{Ker } v_j^T$, $j = 1, \ldots, t$. (In fact, the vectors $u_1, \ldots, u_t, v_1, \ldots, v_t$ can be chosen in such a way that $u_1, \ldots, u_r$ and $v_1, \ldots, v_r$ are linearly independent, and $u_{r+1}, \ldots, u_t, v_{r+1}, \ldots, v_t$ are all zero.) Replacing $u_1, \ldots, u_t$ by appropriate linear combinations of these same vectors $u_1, \ldots, u_t$, and replacing $v_1, \ldots, v_t$ by appropriate linear combinations of these same vectors $v_1, \ldots, v_t$, one can ensure that

$$u_j^T v_j = u_1^T v_1, \qquad j = 1, \ldots, t.$$

For a detailed argument (and additional material), see below, Lemma 4.3 in particular. Observe now that

$$t = \text{trace } A = \text{trace}\left(\sum_{j=1}^{t} u_j v_j^T\right) = \sum_{j=1}^{t} u_j^T v_j = t u_1^T v_1.$$

Hence the numbers $u_1^T v_1, \ldots, u_t^T v_t$ are all equal to one. But then the matrices $u_1 v_1^T, \ldots, u_t v_t^T$ are rank one idempotents. Also their ranges are contained in Ran A, and Ker A is contained in their null spaces. This completes the proof. □

In the above Proof of Theorem 4.1, certain details were left out for crispness of presentation. We shall now focus on these details and, in passing, put Theorem 4.1 in a somewhat wider framework. The next observation is crucial for what follows.

Lemma 4.2 *Let $a_1, a_2, b_1, b_2 \in \mathbb{C}^m$, and suppose that (at least) one of the numbers $a_1^T b_2$, $a_2^T b_1$ and $a_1^T b_1 - a_2^T b_2$ is non-zero. Then, given a complex number μ, there exist vectors $\tilde{a}_1, \tilde{a}_2, \tilde{b}_1, \tilde{b}_2 \in \mathbb{C}^m$ such that*

(i) *$\tilde{a}_1$ and $\tilde{a}_2$ are linear combinations of a_1 and a_2,*

(ii) *$\tilde{b}_1$ and $\tilde{b}_2$ are linear combinations of b_1 and b_2,*

(iii) $\tilde{a}_1 \tilde{b}_1^T + \tilde{a}_2 \tilde{b}_2^T = a_1 b_1^T + a_2 b_2^T$,

(iv) $\tilde{a}_1^T \tilde{b}_1 = \mu$.

Without the condition on $a_1^T b_2$, $a_2^T b_1$ and $a_1^T b_1 - a_2^T b_2$, the lemma is not true. For more about this, see Proposition 4.7 at the end of this section.

Proof: First, assume $a_1^T b_2 \neq 0$. Put

$$\alpha = \frac{a_1^T b_1 - \mu}{a_1^T b_2},$$

and write $\tilde{a}_1 = a_1$, $\tilde{a}_2 = \alpha a_1 + a_2$, $\tilde{b}_1 = b_1 - \alpha b_2$ and $\tilde{b}_2 = b_2$. Then clearly (i) and (ii) are satisfied. Also, a straightforward computation shows that (iii) holds. Finally, $\tilde{a}_1^T \tilde{b}_1 = a_1^T (b_1 - \alpha b_2) = a_1^T b_1 - \alpha a_1^T b_2 = a_1^T b_1 - (a_1^T b_1 - \mu) = \mu$, so (iv) is satisfied too.

The argument in case $a_2^T b_1 \neq 0$ is similar. Thus we may assume that $a_1^T b_2 = a_2^T b_1 = 0$ and $a_1^T b_1 - a_2^T b_2 \neq 0$. In that situation, we proceed as follows. Put

$$\beta = \frac{a_1^T b_1 - \mu}{a_1^T b_1 - a_2^T b_2},$$

$\tilde{a}_1 = (1-\beta)a_1 + \beta a_2$, $\tilde{a}_2 = a_1 - a_2$, $\tilde{b}_1 = b_1 + b_2$ and $\tilde{b}_2 = \beta b_1 - (1-\beta) b_2$. Then clearly (i) and (ii) are satisfied. Also, a straightforward computation shows that (iii) holds too. Finally, $\tilde{a}_1^T \tilde{b}_1 = ((1-\beta)a_1 + \beta a_2)^T (b_1 + b_2) = (1-\beta) a_1^T b_1 + \beta a_2^T b_2 = a_1^T b_1 - \beta(a_1^T b_1 - a_2^T b_2) = \mu$. Thus (ii) is satisfied, and the proof is complete. □

Lemma 4.3 *Let* $a_1, \dots, a_n,\ b_1, \dots, b_n \in \mathbb{C}^m$. *Then there exist* $u_1, \dots, u_n$, $v_1, \dots, v_n \in \mathbb{C}^m$ *such that*

(1) $u_1, \dots, u_n$ *are linear combinations of* $a_1, \dots, a_n$,

(2) $v_1, \dots, v_n$ *are linear combinations of* $b_1, \dots, b_n$,

(3) $\sum_{j=1}^n u_j v_j^T = \sum_{j=1}^n a_j b_j^T$,

(4) $u_j^T v_j = u_1^T v_1, \qquad j = 1, \dots, n.$

Proof: The proof is by induction. For $n = 1$, things are obvious. So, let us assume $n \geq 2$. If $a_j^T b_j = a_1^T b_1,\ j = 1, \dots, n$, then one can take $u_j = a_j,\ v_j = b_j,\ j = 1, \dots, n$. Therefore consider the situation where $a_j^T b_j \neq a_1^T b_1$ for some j among $1, \dots, n$. Without loss of generality we may suppose that $j = 2$, i.e., $a_2^T b_2 \neq a_1^T b_1$. But then there exist $\tilde{a}_1, \tilde{a}_2, \tilde{b}_1, \tilde{b}_2$ such that (i)–(iv) of Lemma 4.2 are satisfied, where for μ we take the number

$$\mu = \frac{1}{n} \sum_{j=1}^n a_j^T b_j.$$

Now put $u_1 = \tilde{a}_1$ and $v_1 = \tilde{b}_1$. Then $u_1^T v_1 = \mu$. For $j = 3, \dots, n$, we write $\tilde{a}_j = a_j, \tilde{b}_j = b_j$. Then

$$\sum_{j=2}^n \tilde{a}_j \tilde{b}_j^T = -u_1 v_1^T + \sum_{j=1}^n a_j b_j^T. \tag{4.1}$$

By induction hypothesis, we may assume that there exist $u_2, \dots, u_n$ and $v_2, \dots, v_n$ in $\mathbb{C}^m$ such that

(1′) $u_2, \dots, u_n$ are linear combinations of $\tilde{a}_2, \dots, \tilde{a}_n$,

(2′) $v_2, \dots, v_n$ are linear combinations of $\tilde{b}_2, \dots, \tilde{b}_n$,

(3′) $\sum_{j=2}^n u_j v_j^T = \sum_{j=2}^n \tilde{a}_j \tilde{b}_j^T$,

(4′) $u_j^T v_j = u_2^T v_2, \qquad j = 2, \dots, n.$

The vectors $u_1, \dots, u_n, v_1, \dots, v_n$ now have the desired properties (1)–(4). To see this, we argue as follows.

The vector $u_1 = \tilde{a}_1$, is a linear combination of a_1 and a_2, and the vector v_1 is a linear combination of b_1 and b_2. Since $\tilde{a}_2$ is a linear combination of a_1 and a_2, and $\tilde{a}_j = a_j,\ j = 3, \dots, n$, it follows from (1′) that $u_2, \dots, u_n$ are linear combinations of $a_1, \dots, a_n$. Similarly, $v_2, \dots, v_n$ are linear combinations of $b_1, \dots, b_n$. So (1) and (2) are satisfied. Combining (4.1) and (3′), one immediately arrives at

$$\sum_{j=2}^n u_j v_j^T = -u_1 v_1^T + \sum_{j=1}^n a_j b_j^T$$

which is just a reformulation of (3). Taking traces, one also obtains

$$\sum_{j=2}^{n} u_j^T v_j = -u_1^T v_1 + \sum_{j=1}^{n} a_j^T b_j = -\mu + n\mu = (n-1)\mu.$$

From (4′) it is obvious that the first term in these identities is equal to $(n-1)\mu_2^T v_2$. So $\mu_2^T v_2 = \mu$. Since $\mu_1^T v_1 = \mu$ too, it is now clear from (4′) that (4) holds, and the proof is complete. □

In the next corollary, which contains the basic observation underlying Theorem 4.1, we consider an additive decomposition of a given matrix A into matrices having rank at most one and satisfying a certain quadratic relation.

Corollary 4.4 *Let A be an $m \times m$ matrix, let n be a positive integer larger than or equal to the rank of A, and put*

$$\rho = \frac{1}{n}\text{trace } A. \tag{4.2}$$

Then there exist $m \times m$ matrices $Q_1, \ldots, Q_n$ having rank at most one, such that

(i) $A = \sum_{j=1}^{n} Q_j$,
(ii) $\text{Ran } Q_j \subset \text{Ran } A, \qquad j = 1, \ldots, n.$
(iii) $\text{Ker } A \subset \text{Ker } Q_j, \qquad j = 1, \ldots, n.$
(iv) $\text{trace } Q_j = \rho$ *and* $Q_j^2 = \rho Q_j, \qquad j = 1, \ldots, n.$

Observe that if the conclusion of the corollary holds, then necessarily rank $A \leq \text{rank} Q_1 + \ldots \text{rank } Q_n \leq n$.

Proof: Write A in the form

$$A = \sum_{j=1}^{n} a_j b_j^T$$

with $a_1, \ldots, a_n \in \text{Ran } A$ and $\text{Ker } A \subset \text{Ker } b_j^T j = 1, \ldots, n$ (cf. the Proof of Theorem 4.1). Choose $u_1, \ldots, u_n, v_1, \ldots, v_n \in \mathbb{C}^m$ such that (1)–(4) of Lemma 4.3 are satisfied, and put $Q_j = u_j v_j^T$, $j = 1, \ldots, n$. Then Q_j has rank at most one and (i)–(iii) are satisfied. Observe that

$$\rho = \frac{1}{n}\text{trace } A = \frac{1}{n}\text{trace}\left(\sum_{j=1}^{n} u_j v_j^T\right) = \frac{1}{n}\sum_{j=1}^{n} v_j^T u_j.$$

Together with (4) of Lemma 4.3, this gives trace $Q_j = v_j^T u_j = \rho$, $j = 1, \ldots, n$. Also, $Q_j^2 = (u_j v_j^T)(u_j v_j^T) = u_j(v_j^T u_j)v_j^T = \rho Q_j$, and the proof is complete. □

In the next two remarks, we elaborate on Corollary 4.4, thereby making the connection with idempotents. We distinguish between the case when trace $A \neq 0$ and the case when trace $A = 0$.

Remark 4.5 Let A be an $m \times m$ matrix, and suppose trace $A \neq 0$ (hence $A \neq 0$). Let n be an integer larger than or equal to rank A – automatically positive in this case – and define ρ by (4.2). Then $\rho \neq 0$. Choose $Q_1, \dots, Q_n$ with properties as in Corollary 4.4. Since trace $Q_j = \rho \neq 0$, we have rank $Q_j = 1$, $j = 1, \dots, n$. Put $P_j = \frac{1}{\rho} Q_j$. Then $P_1, \dots, P_n$ are rank one idempotents and

$$A = \rho \sum_{j=1}^{n} P_j. \tag{4.3}$$

Also (ii) and (iii) of Corollary 4.4 are satisfied with Q_j replaced by P_j. The case when $\rho = 1$ is of special interest because we see from (4.3) that then A is a sum of rank one idempotents. Note that n can be chosen in such a way that $\rho = 1$ if and only if trace A is an integer and rank $A \leq$ trace A (cf. Theorem 4.1).

Remark 4.6 Let A be an $m \times m$ matrix and let n be a positive integer. Suppose trace $A = 0$. Then the number ρ given by (4.2) is equal to zero, and hence the matrices $Q_1, \dots, Q_n$ featuring in Corollary 4.4 are nilpotent with order of nilpotency at most two. Since a sum of nilpotent matrices always has zero trace, we may conclude the following: *An $m \times m$ matrix A is a sum of nilpotent matrices if and only if* trace $A = 0$. *In that case the nilpotent matrices may all be chosen to have rank one, have a range contained in that of A, have a null space containing that of A and have order of nilpotency at most two.* Here also a connection with idempotents is apparent. Indeed, as is easily seen, a rank one nilpotent matrix can always be written as a difference of two rank one idempotent matrices.

Further material on sums of idempotent matrices or sums of nilpotent matrices can be found in the survey article [Wu2]. There also the question of finding the minimal number of terms in the additive decomposition is discussed. For the case of idempotent matrices, see in particular [Wu1], [LMR] and [Wa].

Finally we come back to Lemma 4.2. The assumptions in the following proposition describe essentially the situation where the assumptions of Lemma 4.2 are *not* fulfilled. This result makes clear that the conditions in that lemma are really needed.

Proposition 4.7 *Let $a_1, \dots, a_n, b_1, \dots, b_n \in \mathbb{C}^m$, let λ be a non-zero complex number, and assume*

$$a_i^T b_j = \lambda \delta_{ij}, \qquad i, j = 1, \dots, n,$$

where δ_{ij} is the Kronecker delta. Further, let $u_1, \dots, u_n, v_1, \dots, v_n \in \mathbb{C}^m$, and suppose

$$\sum_{j=1}^{n} u_j v_j^T = \sum_{j=1}^{n} a_j b_j^T.$$

Then $u_i^T v_j = \lambda\delta_{ij}$ $(= a_i^T b_j)$, $i, j = 1, \dots, n$.

As will be clear from the proof, the conditions of this proposition can only be met when $m \geq n$. The requirement $\lambda \neq 0$ is essential: Indeed, consider the following identities:

$$\begin{pmatrix} 1 \\ 0 \end{pmatrix} (0\ 1) + \begin{pmatrix} 1 \\ 0 \end{pmatrix} (0\ 1) = \begin{pmatrix} 0 & 2 \\ 0 & 0 \end{pmatrix} = \begin{pmatrix} 1 \\ 0 \end{pmatrix} (1\ 1) + \begin{pmatrix} 1 \\ 0 \end{pmatrix} (-1\ 1).$$

Proof: Introduce

$$A = \begin{pmatrix} a_1^T \\ \vdots \\ a_n^T \end{pmatrix}, \qquad U = \begin{pmatrix} u_1^T \\ \vdots \\ u_n^T \end{pmatrix},$$

$$B = (b_1 \dots b_n), \qquad V = (v_1 \dots v_n).$$

Then A and U are $n \times m$ matrices, B and V are $m \times n$ matrices, and

$$\begin{aligned} BA &= \sum_{j=1}^{n} b_j a_j^T = \left(\sum_{j=1}^{n} a_j b_j^T \right)^T \\ &= \left(\sum_{j=1}^{n} u_j v_j^T \right)^T = \sum_{j=1}^{n} v_j u_j^T = VU. \end{aligned}$$

Since $\lambda \neq 0$, we have

$$m + \operatorname{rank}(\lambda I_n - AB) = n + \operatorname{rank}(\lambda I_m - BA).$$

This is a well-known matrix identity (see, e.g., [GGK], Section III.4). It follows that

$$\begin{aligned} \operatorname{rank}(\lambda I_n - AB) &= n - m + \operatorname{rank}(\lambda I_m - BA) \\ &= n - m + \operatorname{rank}(\lambda I_m - VU) \\ &= \operatorname{rank}(\lambda I_n - UV). \end{aligned}$$

Now the (i, j)-th entry in the product AB is $a_i^T b_j$. So our assumptions on the vectors $a_1, \dots, a_n, b_1, \dots, b_n$ can be rephrased as $AB = \lambda I_n$ (hence $n \leq m$). In other words, $\operatorname{rank}(\lambda I_n - AB) = 0$. But then rank $(\lambda I_n - UV) = 0$ too, that is $UV = \lambda I_n$. Since the (i, j)-th entry in the product UV is $u_i^T v_j$, the proof is complete. □

The basic observation behind the above proof is the following: *If A and U are $n \times m$ matrices, B and V are $m \times n$ matrices, $VU = BA$ and $AB = I_n$, then $UV = I_n$ too.* This does not generalize to operators on infinite dimensional spaces. To see this, let A and B be equal to the identity operator I on the sequence space l_2, and define U and V on l_2 by

$$U(x_1, x_2, \ldots) = (0, x_1, x_2, \ldots), \quad V(x_1, x_2, \ldots) = (x_2, x_3, \ldots).$$

Then $VU = I = BA = AB$, but $UV \neq I$.

5 Logarithmic Residues of Matrix and Fredholm Operator Valued Functions

Sums of idempotents in the full matrix algebra $\mathbb{C}^{m \times m}$ were studied in [HP], [Wu1], [LMR] and [Wa]; cf. also Section 4. Here we make the connection with logarithmic residues. We begin by discussing things in the somewhat wider framework of logarithmic residues of Fredholm operator valued functions. A detailed study of this type of logarithmic residues will be given in [BES4].

Let X be a (possibly infinite dimensional) Banach space. A bounded linear operator $T : X \to X$ is called a *Fredholm operator* if the null space Ker T of T is finite dimensional and the range space Ran T of T has finite codimension in X. These conditions imply that Ran T is closed. The set of all Fredholm operators on X will be denoted by $\Phi(X)$. It is well-known that $\Phi(X)$ is an open subset of the Banach algebra $\mathcal{B}(X)$ of all bounded linear operators on X.

A *projection operator* on X is just an idempotent in $\mathcal{B}(X)$. Such an operator has finite rank if and only if it is the sum of rank one projection operators on X. For finite rank operators, the concept of the trace is well-defined.

Theorem 5.1 *Let X be a Banach space, and let L be a bounded linear operator on X. The following statements are equivalent:*

(i) *L is a left logarithmic residue of an analytic $\mathcal{B}(X)$-valued function taking its values in $\Phi(X)$;*

(ii) *L is a right logarithmic residue of an analytic $\mathcal{B}(X)$-valued function taking its values in $\Phi(X)$;*

(iii) *L is a sum of (rank one) finite rank projection operators on X;*

(iv) *L has finite rank and the trace of L is an integer larger than or equal to the rank of L.*

With respect to (iii) we note that if L is a sum of rank one projection operators on X, the number of terms in such a sum is equal to trace L. This follows from the additivity of the trace in combination with the fact that for a finite rank projection operator, trace and rank coincide.

Proof: We begin by proving that (iv) implies (iii). From the assumption that L has finite rank, one can easily deduce the existence of two closed subspaces M and N of X with the following properties: M has finite dimension, N has finite codimension in X, $X = M \bigoplus N$, $\operatorname{Ran} L \subset M$ and $N \subset \operatorname{Ker} L$. Thus with respect to the decomposition $X = M \bigoplus N$, the operator L has the representation

$$L = \begin{pmatrix} L_M & 0 \\ 0 & 0 \end{pmatrix},$$

where L_M is the restriction of L to M.

Since M is finite dimensional, L_M can be identified with a square matrix. Also rank L_M = rank $L \leq$ trace L = trace L_M. In particular, trace L_M is an integer. On the other hand we get (iii) as soon as we have written L_M as a sum of rank one idempotent matrices. Thus, as far as the implication (iv) $\Rightarrow$ (iii) is concerned, we have reduced everything to the matrix case. That case, however, has already been dealt with in [HP] and [Wu1]; cf. also Theorem 4.1. So (iv) indeed implies (iii).

Next assume that (iii) is satisfied, say $L = P_1 + \cdots + P_s$ where $P_1, \ldots, P_s$ are finite rank (or even rank one) projection operators on X. In the Proof of Proposition 2.1, we saw that L is the left/right logarithmic residue of a function which can be constructed in a simple way out of $P_1, \ldots, P_s$. In the present situation where $P_1, \ldots, P_s$ have finite rank, the function in question is easily seen to be Fredholm operator valued. This covers the implications (iii) $\Rightarrow$ (i) and (iii) $\Rightarrow$ (ii). We add that L can even be obtained as the left/right logarithmic residue of an *entire* Fredholm operator valued function (cf. [E]).

Now let us turn to the implications (i) $\Rightarrow$ (iv) and (ii) $\Rightarrow$ (iv). We focus on (i) $\Rightarrow$ (iv); for (ii) $\Rightarrow$ (iv) the argument is analogous.

Let $L = \operatorname{LogRes}(F; D)$ where D is a bounded Cauchy domain and where $F \in \mathcal{A}_\partial(D; \mathcal{B}(X))$ takes its values in $\Phi(X)$. Thus F is an analytic Fredholm operator valued function. For the theory on the functions of that type, see [GGK] and the references given there. This theory forms the underpinning of the argument given below.

Since F takes invertible values on the boundary ∂D of D, we have that $F(\lambda)$ is invertible on all of D with the possible exception of a finite number of points where $F(\lambda)^{-1}$ has a pole. This enables us to make a reduction to the case where D is an open disc centered at the origin, $F(\lambda)$ is invertible for $0 \neq \lambda \in D$ and $F(\lambda)^{-1}$ has a pole at the origin. Let $P(\lambda)$ be the principal part of the Laurent expansion of $F(\lambda)^{-1}$ at the origin. Then the difference of $F(\lambda)^{-1}$ and $P(\lambda)$ is analytic, and we get

$$L = \frac{1}{2\pi i} \int_{\partial D} F'(\lambda) F(\lambda)^{-1} d\lambda = \frac{1}{2\pi i} \int_{\partial D} F'(\lambda) P(\lambda) d\lambda. \tag{5.1}$$

It is important to note that $F(\lambda)^{-1}$ has not only a pole at the origin but is even finite meromorphic there. This means that the coefficients in the principal part $P(\lambda)$ are of finite rank. As only a finite number of non-zero finite rank operators are involved, there exist two closed subspaces M and N of X with the following properties: M has finite dimension, N has finite codimension in X, $X = M \bigoplus N$, Ran $P(\lambda) \subset M$ and $N \subset$ Ker $P(\lambda)$, the latter two inclusions holding for $\lambda \neq 0$. With respect to the decomposition $X = M \bigoplus N$, the operator $P(\lambda)$ is of the form

$$P(\lambda) = \begin{pmatrix} W(\lambda) & 0 \\ 0 & 0 \end{pmatrix}, \tag{5.2}$$

where $W(\lambda)$ can be identified with a rational $m \times m$ matrix function ($m = \dim M$) having a unique pole at the origin and vanishing at infinity. This enables us to employ some elements from Systems Theory (cf. [BGK] or [GGK] and the references given there).

The $m \times m$ rational matrix function $W(\lambda)$ admits what is called a realization:

$$W(\lambda) = C(\lambda I_n - A)^{-1} B, \tag{5.3}$$

where A, B and C are appropriately chosen matrices. If we take the realization (5.3) to be minimal (i.e., the matrix A has smallest possible size), then the eigenvalues of A coincide with the poles of $W(\lambda)$ and (5.3) holds outside the spectrum of A. Here this means that A is nilpotent and (5.3) is valid for all non-zero λ. Combining (5.1)–(5.3), we see that Ker $B \bigoplus N$ is contained in Ker L, and so L has finite rank not exceeding the rank of B. Since B is an $n \times m$ matrix, we may conclude that rank $L \leq n$.

We now focus on n, the order of the matrix A in the minimal realization (5.3). It is known that n is equal to the McMillan degree of $W(\lambda)$. This fact can also be expressed by saying that n is equal to the number of poles of $W(\lambda)$ counted according to local degree (cf. [BGK]), also called pole multiplicity. In the present situation, this simply means that n is the pole multiplicity of the origin as a pole of $W(\lambda)$. Combining this with (5.2) and using the fact that $P(\lambda)$ is the principal part of the Laurent expansion of $F(\lambda)^{-1}$ at the origin, we obtain that n is equal to the pole multiplicity of the origin as a pole of $F(\lambda)^{-1}$. The latter, however, is equal to the zero multiplicity of the origin as a zero of $F(\lambda)$. By a result in [GS] it follows that

$$n = \text{trace}\left(\frac{1}{2\pi i}\int_{\partial D} F'(\lambda)F(\lambda)^{-1}d\lambda\right)$$

(cf. also [BKL] and [GGK]). In other words $n =$ trace L and we get rank $L \leq$ trace L, as desired. □

Theorem 5.1 provides a situation where the conditions on the function under consideration imply that the corresponding logarithmic residue is a sum of idempotents. Another such situation occurs when the function is commutative in the sense that its values commute with one another. To see this note that such a function can be viewed as one which takes its values in a commutative Banach algebra and apply [BES3], Theorem 6.1.

The following result is a special case of Theorem 5.1 ($X = \mathbb{C}^m$). We emphasize that the equivalence of (c) and (d) can be found already in [HP] and [Wu1]. The new element is the connection with logarithmic residues.

Corollary 5.2 *Let L be an $m \times m$ matrix, i.e. let $L \in \mathbb{C}^{m\times m}$. The following statements are equivalent:*

(a) *L is a left logarithmic residue in $\mathbb{C}^{m\times m}$;*

(b) *L is a right logarithmic residue in $\mathbb{C}^{m\times m}$;*

(c) *L is a sum of (rank one) idempotents in $\mathbb{C}^{m\times m}$;*

(d) *The trace of L is an integer larger than or equal to the rank of L.*

It is clear from this result that in $\mathbb{C}^{m\times m}$ the set of *left* logarithmic residues coincides with the set of *right* logarithmic residues. In Section 6 (where we shall generalize Corollary 5.2), we shall investigate the topological properties of these (coinciding) sets in a somewhat broader context.

Corollary 5.2 deals with the Banach algebra of all bounded linear operators on a *finite* dimensional (in fact m-dimensional) Hilbert space. How is the situation for an *infinite* dimensional Hilbert space H? The answer is implied by a result in [PT]: each bounded linear operator on such a Hilbert space can be written as a sum of (five) idempotent bounded linear operators. Hence, on account of Proposition 2.1 (cf. also [E]), in the Banach algebra $\mathcal{B}(H)$ of all bounded linear operators on an infinite dimensional Hilbert space H, the set of (left/right) logarithmic residues and the set of sums of idempotents coincide; in fact they are both equal to $\mathcal{B}(H)$. So there is a striking contrast between the finite and the infinite dimensional case.

We conclude this section with a simple example.

Example 5.3 Let

$$A = \begin{pmatrix} 2 & 0 & 0 \\ 0 & 0 & 1 \\ 0 & 0 & 0 \end{pmatrix}.$$

Then rank A = trace $A = 2$. So, by Corollary 5.2, the matrix A is a left/right logarithmic residue in $\mathbb{C}^{3\times 3}$ and A is a sum of idempotent matrices. In fact,

$$A = \begin{pmatrix} 1 & 0 & \frac{1}{2} \\ 1 & 0 & \frac{1}{2} \\ 0 & 0 & 0 \end{pmatrix} + \begin{pmatrix} 1 & 0 & -\frac{1}{2} \\ -1 & 0 & \frac{1}{2} \\ 0 & 0 & 0 \end{pmatrix} \tag{5.4}$$

is a decomposition of A into rank one idempotents. Note that A is upper triangular, but that the matrices in the right hand side of (5.4) are not. Theorem 6.1 below makes clear that A is not a logarithmic residue in the Banach algebra of upper triangular 3×3 matrices and that it is impossible to write A as a sum of upper triangular idempotent 3×3 matrices.

6 The Algebra of Block Upper Triangular Matrices

From now on $m, m_1, \ldots, m_k$ will be arbitrary, but given, positive integers,

$$m = m_1 + \cdots + m_k. \tag{6.1}$$

If A is an $m \times m$ matrix, we can partition A, writing it as a block matrix

$$A = (A_{ij})_{i,j=1}^{k}, \qquad A_{ij} \in \mathbb{C}^{m_i \times m_j}. \tag{6.2}$$

We say that A is *block upper triangular* with respect to the additive decomposition (6.1) if $A_{ij} = 0$ whenever $i > j$. The set of all $m \times m$ matrices that are block upper triangular with respect to (6.1) will be denoted by $\mathbb{C}_{upper}^{m_1,\ldots,m_k}$. Note that $\mathbb{C}_{upper}^{m_1,\ldots,m_k}$ is a Banach subalgebra of $\mathbb{C}^{m\times m}$. For $k = 1$ and (hence) $m_1 = m$, this Banach subalgebra of $\mathbb{C}^{m\times m}$ coincides with $\mathbb{C}^{m\times m}$; for $k = m$ and (hence) $m_1 = \cdots = m_k = 1$, it is equal to the Banach subalgebra of $\mathbb{C}^{m\times m}$ consisting of all upper triangular matrices.

Returning to the general situation, let $A \in \mathbb{C}_{upper}^{m_1,\ldots,m_k}$ and write A in the block form (6.2). Following the terminology of [BW], we call A *block upper rectangular of type l* if $A_{ij} = 0$ whenever $i > l$ or $j < l$. Here l is one of the integers $1, \ldots, k$. Also, by a *simple block submatrix* of A, we mean a submatrix of the form

$$A[s,t] = (A_{ij})_{i,j=s}^{t} = \begin{bmatrix} A_{ss} & \ldots & A_{s,t-1} & A_{st} \\ 0 & & & A_{s+1,t} \\ \vdots & \ddots & & \vdots \\ 0 & \ldots & 0 & A_{tt} \end{bmatrix}.$$

Here s and t are integers, $1 \le s \le t \le k$. Note that $A[j,j] = A_{jj},\ j = 1, \ldots, k$. Also observe that if $A \in \mathbb{C}_{upper}^{m_1,\ldots,m_k}$ is invertible as an $m \times m$ matrix, (i.e. $|A| \neq 0$), then $A^{-1} \in \mathbb{C}_{upper}^{m_1,\ldots,m_k}$ too.

We are now ready to formulate the main result of this section (which, by the way, contains Corollary 5.2 as a special case).

Theorem 6.1 *Let $L \in \mathbb{C}_{upper}^{m_1,\ldots,m_k}$. The following statements are equivalent:*

(i) *L is a left logarithmic residue in $\mathbb{C}_{upper}^{m_1,\ldots,m_k}$;*

(ii) *L is a right logarithmic residue in $\mathbb{C}_{upper}^{m_1,\ldots,m_k}$;*

(iii) *L is a sum of idempotents in $\mathbb{C}_{upper}^{m_1,\ldots,m_k}$;*

(iv) *L is a sum of rank one idempotents in $\mathbb{C}_{upper}^{m_1,\ldots,m_k}$;*

(v) *For each pair of integers $s, t,\ 1 \le s \le t \le k$, the trace of the simple block submatrix $L[s,t]$ of L is an integer larger than or equal to the rank of $L[s,t]$, in other words:*

$$\operatorname{rank} L[s,t] \le \operatorname{trace} L[s,t] \in \mathbb{Z}, \qquad 1 \le s \le t \le k.$$

Proof: Clearly (iv) implies (iii). By Proposition 2.1, we have that (iii) implies both (i) and (ii). Suppose (i) is satisfied, and write $L = \text{LogRes}(f; D)$ where D is a bounded Cauchy domain and $f \in \mathcal{A}_\partial(D; \mathbb{C}_{upper}^{m_1,\dots,m_k})$. Let s and t be integers, $1 \le s \le t \le k$. Then

$$L[s,t] = \frac{1}{2\pi i}\int_{\partial D} f'_{[s,t]}(\lambda) f_{[s,t]}(\lambda)^{-1} d\lambda$$

where $f_{[s,t]}(\lambda) = f(\lambda)[s,t]$ is the simple block submatrix of $f(\lambda) \in \mathbb{C}_{upper}^{m_1,\dots,m_k}$ determined by s and t. So, in particular, $L[s,t]$ is the left logarithmic residue of a matrix valued analytic function. But then, by Corollary 5.2, the trace of $L[s,t]$ is an integer larger than or equal to the rank of $L[s,t]$. Hence (i) implies (v). Similarly, (ii) implies (v) too.

It remains to prove that (v) implies (iv). This is more complicated and the argument will rely crucially on a rank decomposition theorem obtained in [BW].

Suppose (v) is satisfied and write L in the block form

$$L = (L_{ij})_{i,j=1}^k, \qquad L_{i,j} \in \mathbb{C}^{m_i \times m_j}.$$

For $j = 1, \dots, k$, put $\tau_j = \text{trace } L_{jj}$. Then $\tau_1, \dots \tau_k$ are non-negative integers and, for $1 \le s \le t \le k$,

$$\text{rank } L[s,t] \le \text{trace } L[s,t] = \sum_{j=s}^{t} \text{trace } L_{jj} = \sum_{j=s}^{t} \tau_j.$$

Thus the Rank Decomposition Theorem from [BW] applies and L can be written as

$$L = \sum_{l=1}^{k} R(l) \tag{6.3}$$

where, for $l = 1, \dots, k$, the matrix $R(l)$ is block upper rectangular of type l and rank $R(l) \le \tau_l$.

Write

$$R(l) = (R(l)_{ij})_{i,j=1}^k, \qquad R(l)_{ij} \in \mathbb{C}^{m_i \times m_j}.$$

Since $R(l)$ is block upper rectangular of type l, we have $R(l)_{jj} = 0$ for $j \ne l$. This gives

$$L_{jj} = \sum_{l=1}^{k} R(l)_{jj} = R(j)_{jj},$$

and it follows that trace $R(j)$ = trace $R(j)_{jj}$ = trace $L_{jj} = \tau_j \ge$ rank $R(j)$. Hence Theorem 4.1 applies. The conclusion is that $R(j)$ can be written as a sum of rank one idempotents that may all be chosen to have their ranges contained in the range of $R(j)$ and to have their null spaces containing the null space of $R(j)$. In the present situation where $R(j)$ is block upper rectangular of type j, this means that the idempotents can be chosen to have their last $m_{j+1}+\cdots+m_k$ rows zero and to have their first $m_1+\cdots+m_{j-1}$ columns zero. In other words, $R(j)$ can be written as a sum of rank one idempotents that are block upper rectangular of type j and hence belong to $\mathbb{C}_{upper}^{m_1,\ldots,m_k}$. In view of (6.3) the proof is now complete. For later use, however, we add that not only can $R(j)$ be written as a sum of rank one idempotents that are block upper rectangular of type j, but, in addition, the number of terms in such a sum is necessarily equal to $\tau_j = \text{trace } R(j)$. So the matrix L can be written as a sum of rank one idempotents belonging to $\mathbb{C}_{upper}^{m_1,\ldots,m_k}$, among them $\tau_1 = \text{trace } L_{11}$ block upper rectangular of type 1, $\tau_2 = \text{trace } L_{22}$ block upper rectangular of type 2, ..., and $\tau_k = \text{trace } L_{kk}$ block upper rectangular of type k. □

From Theorem 6.1 it is clear that in the Banach algebra $\mathbb{C}_{upper}^{m_1,\ldots,m_k}$ the set of left logarithmic residues coincides with the set of right logarithmic residues. We shall now investigate the topological properties of these coinciding sets. Note that in the scalar case where we work with the Banach algebra $\mathbb{C}$ ($m = k = m_1 = 1$), the set of logarithmic residues is just the set of non-negative integers. In the non-scalar case, the situation is more complex.

It is convenient to introduce some notation. The symbol $\mathcal{L}_{upper}^{m_1,\ldots,m_k}$ will stand for the set of all left/right logarithmic residues in $\mathbb{C}_{upper}^{m_1,\ldots,m_k}$. For $\tau_1, \ldots, \tau_k$ non-negative integers, we define the set $\mathcal{L}_{\tau_1,\ldots,\tau_k}^{m_1,\ldots,m_k}$ as follows: $A = (A_{ij})_{i,j=1}^{k}$, written in the partitioned form (6.2), belongs to $\mathcal{L}_{\tau_1,\ldots,\tau_k}^{m_1,\ldots,m_k}$ if and only if $A \in \mathcal{L}_{upper}^{m_1,\ldots,m_k}$ and trace $A_{jj} = \tau_j$, $j = 1, \ldots, k$. Finally, for $l = 1, \ldots, k$, let

$$\mathcal{L}_{(l)}^{m_1,\ldots,m_k} = \mathcal{L}_{\delta_{1l},\ldots,\delta_{kl}}^{m_1,\ldots,m_k},$$

where δ_{jl} denotes the Kronecker delta.

A few observations (evident from the material presented so far) are in order:

(1) $\mathcal{L}_{0,\ldots,0}^{m_1,\ldots,m_k} = \{0\}$,

(2) The set $\mathcal{L}_{(l)}^{m_1,\ldots,m_k}$ consists of all rank one idempotents that are block upper rectangular of type l,

(3) $$\mathcal{L}_{\tau_1,\ldots,\tau_k}^{m_1,\ldots,m_k} = \underbrace{\mathcal{L}_{(1)}^{m_1,\ldots,m_k} + \cdots + \mathcal{L}_{(1)}^{m_1,\ldots,m_k}}_{\tau_1 \text{ terms}} + \underbrace{\mathcal{L}_{(2)}^{m_1,\ldots,m_k} + \cdots + \mathcal{L}_{(2)}^{m_1,\ldots,m_k}}_{\tau_2 \text{ terms}} + \cdots + \underbrace{\mathcal{L}_{(k)}^{m_1,\ldots,m_k} + \cdots + \mathcal{L}_{(k)}^{m_1,\ldots,m_k}}_{\tau_k \text{ terms}},$$

(4) $\mathcal{L}_{upper}^{m_1,\ldots,m_k} = \bigcup_{\tau_1,\ldots,\tau_k=0,1\ldots} \mathcal{L}_{\tau_1,\ldots,\tau_k}^{m_1,\ldots,m_k}$.

The union in the right hand side of (4) is a disjoint union. In fact, the distance between the sets $\mathcal{L}^{m_1,\dots,m_k}_{\tau_1,\dots,\tau_k}$ and $\mathcal{L}^{m_1,\dots,m_k}_{\tau'_1,\dots,\tau'_k}$ (taken with respect to any of the usual matrix norms) is larger than or equal to the number

$$\max_{j=1,\dots,k} \frac{|\tau_j - \tau'_j|}{m_j}. \tag{6.4}$$

To see this, we argue as follows.

Recall that for an $n\times n$ matrix T, the trace is equal to the sum of the eigenvalues of T (counted according to multiplicities). From this it is clear that $|\text{trace } T| \le n\|T\|$ (see also the second paragraph of the proof of Theorem 6.2 below). Now let $A \in \mathcal{L}^{m_1,\dots,m_k}_{\tau_1,\dots,\tau_k}$, $A' \in \mathcal{L}^{m_1,\dots,m_k}_{\tau'_1,\dots,\tau'_k}$, and write

$$A = \left(A_{ij}\right)_{i,j=1}^{k}, \qquad A_{ij} \in \mathbb{C}^{m_i\times m_j},$$

$$A' = \left(A'_{ij}\right)_{i,j=1}^{k}, \qquad A'_{ij} \in \mathbb{C}^{m_i\times m_j}.$$

Then $\|A - A'\| \ge \|A_{jj} - A'_{jj}\| \ge \frac{1}{m_j}|\text{trace}(A_{jj} - A'_{jj})| = \frac{1}{m_j}|\tau_j - \tau'_j|$.

We now come to our next theorem.

Theorem 6.2 *Let $\mathcal{L}^{m_1,\dots,m_k}_{upper}$ and $\mathcal{L}^{m_1,\dots,m_k}_{\tau_1,\dots,\tau_k}$ be the subsets of the Banach algebra $\mathbb{C}^{m_1,\dots,m_k}_{upper}$ introduced above. Then the following statements hold true:*

(i) *The sets $\mathcal{L}^{m_1,\dots,m_k}_{upper}$ and $\mathcal{L}^{m_1,\dots,m_k}_{\tau_1,\dots,\tau_k}$ are closed and have empty interior;*

(ii) *The $m\times m$ zero matrix is an isolated point of $\mathcal{L}^{m_1,\dots,m_k}_{upper}$ and, in the case $m = m_1 + \dots + m_k \ge 2$, it is the only such point of $\mathcal{L}^{m_1,\dots,m_k}_{upper}$. In fact, when $m \ge 2$ and the numbers $\tau_1, \dots, \tau_k$ are not all zero, the set $\mathcal{L}^{m_1,\dots,m_k}_{\tau_1,\dots,\tau_k}$ does not have isolated points;*

(iii) *The sets $\mathcal{L}^{m_1,\dots,m_k}_{\tau_1,\dots,\tau_k}$ are the connected components of $\mathcal{L}^{m_1,\dots,m_k}_{upper}$.*

In the scalar case ($m = k = m_1 = 1$), the set of logarithmic residues (being the set of non-negative integers) consists of isolated points only. More generally, the set of logarithmic residues in any commutative Banach algebra consists of isolated points only. For details and additional results, see [BES3].

It is convenient to let the Proof of Theorem 6.2 be preceded by a remark.

Remark 6.3 The Banach algebra $\mathbb{C}^{m_1,\dots,m_k}_{upper}$ is a Banach subalgebra of $\mathbb{C}^{m\times m}$ ($m = m_1 + \dots + m_k$). In the latter, two rank one idempotents are always similar. How is the situation in $\mathbb{C}^{m_1,\dots,m_k}_{upper}$? To answer this question, we need to make some preparations.

Let $P \in \mathbb{C}^{m_1,\dots,m_k}_{upper}$ be a rank one idempotent, and write P in the partitioned form

$$P = \left(P_{ij}\right)_{i,j=1}^{k}, \qquad P_{ij} \in \mathbb{C}^{m_i\times m_j}. \tag{6.5}$$

Then $P_{ij} = 0$ whenever $i > j$. Clearly $P_{11}, \dots, P_{kk}$ are idempotent matrices, and for such matrices rank and trace coincide. It follows that

$$1 = \text{rank } P \leq \sum_{j=1}^{k} \text{rank } P_{jj} = \sum_{j=1}^{k} \text{trace } P_{jj} = \text{trace } P = 1.$$

Hence there is precisely one l among the integers $1, \dots, k$ such that $P_{ll} \neq 0$. We denote this integer by $l(P)$.

Now let $Q \in \mathbb{C}_{upper}^{m_1,\dots,m_k}$ be another rank one idempotent. We claim that the following statement holds true: *P and Q are similar in $\mathbb{C}_{upper}^{m_1,\dots,m_k}$ if and only if $l(P) = l(Q)$.* The only if part of this statement is trivial. So we concentrate on the if part.

For the time being, let us ignore the fact that P and Q belong to $\mathbb{C}_{upper}^{m_1,\dots,m_k}$, but let us consider P and Q just as $m \times m$ matrices. Write $P = uv^T$ with $u, v \in \mathbb{C}^m$. Since P is a rank one idempotent, we have

$$u_1v_1 + \dots + u_mv_m = 1,$$

where $u^T = (u_1, \dots, u_m)$, $v^T = (v_1, \dots, v_m)$. Choose p such that $u_pv_p \neq 0$, and define the $m \times m$ matrix S_P by

$$S_P e_j = \begin{cases} u, & j = p, \\ e_j - \frac{v_j}{v_p} e_p, & j \neq p. \end{cases}$$

Here $e_1, \dots, e_m$ are the unit vectors in $\mathbb{C}^m$. So e_j has all entries zero except for the j-th which is equal to one. One verifies easily that S_P is invertible and that S_P^{-1} is given by

$$e_j^T S_P^{-1} = \begin{cases} v^T, & j = p, \\ e_j^T - u_j v^T, & j \neq p. \end{cases}$$

It is now clear that $S_P^{-1} P S_P = S_P^{-1} uv^T S_P = e_p e_p^T$. So S_P is a specific invertible matrix establishing the similarity between the rank one idempotents P and $e_p e_p^T$. An analogous construction can be carried out for Q. This gives the matrix S_Q.

Now recall that $P \in \mathbb{C}_{upper}^{m_1,\dots,m_k}$. Put $l = l(P)$ and, with P_{ij} as in (6.5), consider the matrices

$$(P_{ij})_{i,j=1}^{l-1}, \qquad (P_{ij})_{i,j=l+1}^{k}$$

These matrices are idempotent and have zero trace. So they are zero matrices. We conclude that $P_{ij} = 0$ whenever $i > l$ or $j < l$. In other words, P is block upper rectangular of type l. With u and v as above (so $P = uv^T$), we have

$$u_j = 0\,, \qquad j > m_1 + \dots + m_l,$$

$$v_i = 0\,, \qquad i \leq m_1 + \dots + m_{l-1}.$$

But then, since $u_p v_p \neq 0$,

$$m_1 + \cdots + m_{l(P)-1} + 1 \leq p \leq m_1 + \cdots + m_{l(P)}. \tag{6.6}$$

It is now obvious that the matrix S_P constructed above (and its inverse) belong to $\mathbb{C}_{upper}^{m_1,\ldots,m_k}$. In fact $I_m - S_P$ and $I_m - S_P^{-1}$ are block upper rectangular of type l. Analogous statements hold true for S_Q.

By our construction $S_P^{-1} P S_P = e_p e_p^T$ with p satisfying (6.6). Analogously, $S_Q^{-1} Q S_Q = e_q e_q^T$ where

$$m_1 + \cdots + m_{l(Q)-1} + 1 \leq q \leq m_1 + \cdots + m_{l(Q)}. \tag{6.7}$$

Combining (6.6) and (6.7), and assuming that $l(Q) = l(P) = l$, we obtain

$$m_1 + \cdots + m_{l-1} + 1 \leq p, q \leq m_1 + \cdots + m_l. \tag{6.8}$$

Let Π be the (permutation) matrix which results from I_m by interchanging the p-th and q-th column (or row). Then $\Pi = \Pi^T = \Pi^{-1}$ and $e_p e_p^T = \Pi^{-1} e_q e_q^T \Pi$. It follows from (6.8) that Π belongs to $\mathbb{C}_{upper}^{m_1,\ldots,m_k}$. Actually $I_m - \Pi$ is again block upper rectangular of type l.

Finally, put $S = S_P \Pi^{-1} S_Q^{-1}$. Then S is invertible and $S^{-1} P S = Q$. Also S and S^{-1} belong to $\mathbb{C}_{upper}^{m_1,\ldots,m_k}$ and, in fact, both $I_m - S$ and $I_m - S^{-1}$ are block upper rectangular of type l.

Proof of Theorem 6.2: To see that $\mathcal{L}_{upper}^{m_1,\ldots,m_k}$ and $\mathcal{L}_{\tau_1,\ldots,\tau_k}^{m_1,\ldots,m_k}$ are closed, use the characterization of logarithmic residues given in Theorem 6.1(v), the continuity of the trace and the lower semicontinuity of the rank (cf. the proof of [Wu1], Corollary 2). From the fact that the matrices in $\mathcal{L}_{upper}^{m_1,\ldots,m_k}$ have integer trace, it is evident that the sets $\mathcal{L}_{upper}^{m_1,\ldots,m_k}$ and $\mathcal{L}_{\tau_1,\ldots,\tau_k}^{m_1,\ldots,m_k}$ have empty interior. This proves (i).

Next we shall demonstrate that the zero matrix is an isolated point of $\mathcal{L}_{upper}^{m_1,\ldots,m_k}$. For this we need a general observation about matrices. Let T be an $n \times n$ matrix, and let $\lambda_1, \ldots, \lambda_r$ be the non-zero eigenvalues of T counted according to multiplicity. Then $\lambda_1 + \cdots + \lambda_r = \operatorname{trace} T$ and $r \leq \operatorname{rank} T$. Hence $|\operatorname{trace} T| \leq |\lambda_1| + \cdots + |\lambda_r| \leq r\|T\| \leq \|T\| \cdot \operatorname{rank} T$. Now let $0 \neq A \in \mathcal{L}_{upper}^{m_1,\ldots,m_k}$. Then $1 \leq \operatorname{rank} A \leq \operatorname{trace} A$. Hence

$$\operatorname{trace} A \leq \|A\| \cdot \operatorname{rank} A \leq \|A\| \cdot \operatorname{trace} A.$$

Dividing by trace A, we see that $\|A\| \geq 1$.

We proceed by showing that, in case $m \geq 2$ and $(\tau_1, \ldots, \tau_k) \neq (0, \ldots, 0)$, the set $\mathcal{L}_{\tau_1,\ldots,\tau_k}^{m_1,\ldots,m_k}$ does not contain isolated points. By the description of $\mathcal{L}_{\tau_1,\ldots,\tau_k}^{m_1,\ldots,m_k}$ given in (3) earlier in this section, it suffices to prove that the sets $\mathcal{L}_{(l)}^{m_1,\ldots,m_k}$ do not have isolated points.

Let l be one of the integers $1, \ldots, k$, and let $P \in \mathcal{L}_{(l)}^{m_1,\ldots,m_k}$. Then P is a rank one idempotent which is block upper rectangular of type l. Also $l(P) = l$. Here we use the notation introduced in Remark 6.3. From this remark we know that there exists an integer p among $1, \dots, m$ such that P and $e_p e_p^T$ are similar in $\mathbb{C}_{upper}^{m_1,\dots,m_k}$. Write

$$P = S^{-1} e_p e_p^T S$$

with $S \in \mathbb{C}_{upper}^{m_1,\dots,m_k}$ an invertible matrix. Now $l(e_p e_p^T) = l(P) = l$, so $e_p e_p^T \in \mathcal{L}_{(l)}^{m_1,\dots,m_k}$. By assumption $m \geq 2$. So $p < m$ or $p = m > 1$. In case $p < m$, we put

$$P_\varepsilon = S^{-1}(e_p e_p^T + \varepsilon e_p e_m^T) S;$$

in case $p = m > 1$

$$P_\varepsilon = S^{-1}(e_m e_m^T + \varepsilon e_1 e_m^T) S.$$

In either case $P_\varepsilon \in \mathcal{L}_{(l)}^{m_1,\dots,m_k}$ and $P_\varepsilon \to P$ as $\varepsilon \to 0$. Since $P_\varepsilon \neq P$ whenever $\varepsilon \neq 0$, the desired result follows.

Finally we turn to statement (iii). Consider $\mathcal{L}_{\tau_1,\dots,\tau_k}^{m_1,\dots,m_k}$ for some fixed choice of $\tau_1, \dots, \tau_k$, and put

$$\mathcal{L}' = \bigcup_{\substack{\sigma_1,\dots,\sigma_k = 0,1,\dots \\ (\sigma_1,\dots,\sigma_k) \neq (\tau_1,\dots,\tau_k)}} \mathcal{L}_{\sigma_1,\dots,\sigma_k}^{m_1,\dots,m_k}.$$

Then, by the observations involving (6.4), the distance between $\mathcal{L}_{\tau_1,\dots,\tau_k}^{m_1,\dots,m_k}$ and $\mathcal{L}'$ is larger than or equal to $1/m$. Also

$$\mathcal{L}_{upper}^{m_1,\dots,m_k} = \mathcal{L}_{\tau_1,\dots,\tau_k}^{m_1,\dots,m_k} \cup \mathcal{L}',$$

and so $\mathcal{L}_{\tau_1,\dots,\tau_k}^{m_1,\dots,m_k}$ is both open and closed in $\mathcal{L}_{upper}^{m_1,\dots,m_k}$. Thus it suffices to prove that $\mathcal{L}_{\tau_1,\dots,\tau_k}^{m_1,\dots,m_k}$ is connected. But then, in view of (1) and (3) above, it is enough to show that the sets $\mathcal{L}_{(l)}^{m_1,\dots,m_k}$ are arcwise connected.

Let l be one of the integers $1, \dots, k$, and take P_0 and P_1 in $\mathcal{L}_{(l)}^{m_1,\dots,m_k}$. Then $l(P_0) = l(P_1) = l$. Hence, by Remark 6.3, there exists an invertible $S \in \mathbb{C}_{upper}^{m_1,\dots,m_k}$ such that $P_1 = S^{-1} P_0 S$. Now let L be a (suitable) logarithm of S. From (for instance) the description of L as a polynomial in S it is clear that $L \in \mathbb{C}_{upper}^{m_1,\dots,m_k}$. But then $e^{tL} \in \mathbb{C}_{upper}^{m_1,\dots,m_k}$ too. Write

$$P(t) = e^{-tL} P_0 e^{tL}.$$

Then $P(t) \in \mathcal{L}_{(l)}^{m_1,\dots,m_k}$, $P(0) = P_0$, $P(1) = P_1$ and $P(t)$ depends continuously on t. The proof is now complete.

Our last theorem in this paper addresses the issue of *left* versus *right* logarithmic residues.

Theorem 6.4 *Let L and R be logarithmic residues in $\mathbb{C}_{upper}^{m_1,\dots,m_k}$, i.e., let $L, R \in \mathcal{L}_{upper}^{m_1,\dots,m_k}$. The following statements are equivalent:*

(i) *There exist a bounded Cauchy domain D and a function f belonging to $\mathcal{A}_\partial(D; \mathbb{C}_{upper}^{m_1,\dots,m_k})$ such that L is the left and R is the right logarithmic residue of f with respect to D, i.e.,*

$$L = \frac{1}{2\pi i}\int_{\partial D} f'(\lambda)f(\lambda)^{-1}d\lambda, \tag{6.9}$$

$$R = \frac{1}{2\pi i}\int_{\partial D} f(\lambda)^{-1}f'(\lambda)d\lambda; \tag{6.10}$$

(ii) *There exist non-negative integers $\tau_1, \dots, \tau_k$ such that $L, R \in \mathcal{L}_{\tau_1,\dots,\tau_k}^{m_1,\dots,m_k}$, i.e., L and R belong to the same connected component of $\mathcal{L}_{upper}^{m_1,\dots,m_k}$.*

The proof, especially the part dealing with the implication (ii) $\Rightarrow$ (i), will provide additional information about the freedom in choosing D and the possibility of imposing extra conditions on f. Note that in the one block case ($k = 1, m_1 = m$) where the Banach algebra under consideration is $\mathbb{C}^{m\times m}$, statement (ii) amounts to the requirement that L and R have the same trace. For a generalization of this result to logarithmic residues of Fredholm operator valued functions, see [BES4]. In the case of scalar blocks ($k = m,\ m_j = 1$) where the Banach algebra under consideration is that of the upper triangular $m \times m$ matrices, statement (ii) amounts to the requirement that L and R have the same diagonal.

Proof: Let D and f be as in (i). For s and t integers, $1 \le s \le t \le k$, we have

$$L[s,t] = \frac{1}{2\pi i}\int_{\partial D} f'_{[s,t]}(\lambda)f_{[s,t]}(\lambda)^{-1}d\lambda,$$

$$R[s,t] = \frac{1}{2\pi i}\int f_{[s,t]}(\lambda)^{-1}f'_{[s,t]}(\lambda)d\lambda,$$

where (as in the Proof of Theorem 6.1), $f_{[s,t]}(\lambda) = f(\lambda)[s,t]$ is the simple block submatrix of $f(\lambda) \in \mathbb{C}_{upper}^{m_1,\dots,m_k}$ determined by s and t. Taking traces and using the well-known commutativity property of the trace, one gets

$$\begin{aligned} \text{trace } L[s,t] &= \frac{1}{2\pi i}\int_{\partial D} \text{trace}(f'_{[s,t]}(\lambda)f_{[s,t]}(\lambda)^{-1})d\lambda \\ &= \frac{1}{2\pi i}\int_{\partial D} \text{trace}(f_{[s,t]}(\lambda)^{-1}f'_{[s,t]}(\lambda))d\lambda \\ &= \text{trace } R[s,t]. \end{aligned}$$

In particular, trace $L[j,j] =$ trace $R[j,j],\ j = 1,\dots,k$. Now, for $j = 1,\dots,k$, put $\tau_j =$ trace $L[j,j] =$ trace $R[j,j]$. Then $L, R \in \mathcal{L}_{\tau_1,\dots,\tau_k}^{m_1,\dots,m_k}$. This proves that (i) implies (ii).

Next assume (ii) holds, and let $\tau_1,\dots,\tau_k$ be non-negative integers such that $L, R \in \mathcal{L}_{\tau_1,\dots,\tau_k}^{m_1,\dots,m_k}$. Put $\tau = \tau_1 + \dots + \tau_k (=$ trace $L =$ trace $R)$. We shall prove a modified (and a priori stronger) version of (i): *Given a bounded Cauchy domain*

D and different complex numbers $z_1, \ldots, z_\tau$ in D, there exists a function $f : \mathbb{C} \to \mathbb{C}_{upper}^{m_1,\ldots,m_k}$ with the following properties:

(a) *f is entire, i.e., analytic on all of $\mathbb{C}$,*

(b) *$f(\lambda)$ is invertible whenever $\lambda \neq z_1, \ldots, z_\tau$; in particular f belongs to $\mathcal{A}_\partial(D; \mathbb{C}_{upper}^{m_1,\ldots,m_k})$,*

(c) *$f(\lambda)^{-1}$ has simple poles at $z_1, \ldots, z_\tau$,*

(d) *L is the left and R is the right logarithmic residue of f with respect to D, i.e.,* (6.9) *and* (6.10) *are satisfied.*

The argument is as follows.

From Theorem 6.1 and its proof, we know that we can write L and R in the form

$$L = \sum_{j=1}^{k}\sum_{i=1}^{\tau_j} P_{ji}, \qquad R = \sum_{j=1}^{k}\sum_{i=1}^{\tau_j} Q_{ji} \tag{6.11}$$

with $P_{ji}^2 = P_{ji}$, $Q_{ji}^2 = Q_{ji}$, rank P_{ji} = rank $Q_{ji} = 1$ and P_{ji}, Q_{ji} block upper rectangular of type j. The latter implies that $l(P_{ji}) = l(Q_{ji}) = j$ where we use the notation of Remark 6.3. By that same remark P_{ji} and Q_{ji} are similar in $\mathbb{C}_{upper}^{m_1,\ldots,m_k}$. So there exists an invertible $S_{ji} \in \mathbb{C}_{upper}^{m_1,\ldots,m_k}$ such that $Q_{ji} = S_{ji}^{-1} P_{ji} S_{ji}$.

Write $\{z_1, \ldots, z_\tau\}$ as $\{z_{ji} \mid i = 1, \ldots, \tau_j, j = 1, \ldots, k\}$. As an intermediate step in the proof we shall now construct an entire $\mathbb{C}_{upper}^{m_1,\ldots,m_k}$-valued function a such that $a(\lambda)$ is invertible for all $\lambda \in \mathbb{C}$ and, in addition,

$$a(z_{ji}) = S_{ji}, \qquad i = 1, \ldots, \tau_j;\ j = 1, \ldots, k,$$

$$a'(z_{ji}) = 0, \qquad i = 1, \ldots, \tau_j;\ j = 1, \ldots, k.$$

Let T_{ji} be a (suitable) logarithm of S_{ji}. From (for instance) the description of T_{ji} as a polynomial in S_{ji} it is clear that $T_{ji} \in \mathbb{C}_{upper}^{m_1,\ldots,m_k}$. Choose scalar polynomials g_{ji} such that

$$\begin{aligned} g_{ji}(z_{st}) &= \begin{cases} 1, & (j,i) = (s,t), \\ 0, & (j,i) \neq (s,t), \end{cases} \\ g_{ji}'(z_{st}) &= 0, \end{aligned}$$

and introduce

$$a_{ji}(\lambda) = e^{g_{ji}(\lambda)T_{ji}}.$$

Then a_{ji} is an entire $\mathbb{C}_{upper}^{m_1,\ldots,m_k}$-valued function, $a_{ji}(\lambda)$ is invertible for all $\lambda \in \mathbb{C}$ and

$$\begin{aligned} a_{ji}(z_{st}) &= \begin{cases} S_{ji}, & (j,i) = (s,t), \\ I_m, & (j,i) \neq (s,t), \end{cases} \\ a_{ji}'(z_{st}) &= 0. \end{aligned}$$

For a one can now choose a product of the functions a_{ji}, i.e.,

$$a(\lambda) = \prod_{j=1}^{k} \prod_{i=1}^{\tau_j} a_{ji}(\lambda),$$

where the order in which the product is taken is irrelevant.

So far for the intermediate step. Let h be an entire $\mathbb{C}^{m_1,\dots,m_k}_{upper}$-valued function with the following properties: $h(\lambda)$ is invertible on all of $\mathbb{C}$ except in the points z_{ji} where $h(\lambda)^{-1}$ has simple poles and, for $\varepsilon > 0$ sufficiently small,

$$\frac{1}{2\pi i} \int_{|\lambda - z_{ji}|=\varepsilon} h'(\lambda)h(\lambda)^{-1} d\lambda = \frac{1}{2\pi i} \int_{|\lambda - z_{ji}|=\varepsilon} h(\lambda)^{-1} h'(\lambda) d\lambda = P_{ji}.$$

The existence of such a function has been established in [E] in the more general context of the study of logarithmic residues in arbitrary Banach algebras. Write $f(\lambda) = h(\lambda)a(\lambda)$. Then $f : \mathbb{C} \to \mathbb{C}^{m_1,\dots,m_k}_{upper}$ is a function which obviously satisfies (a), (b) and (c). It remains to show that (d) is satisfied too.

Observe that

$$\begin{aligned}
&\frac{1}{2\pi i} \int_{|\lambda - z_{ji}|=\varepsilon} f'(\lambda) f(\lambda)^{-1} d\lambda \\
&= \frac{1}{2\pi i} \int_{|\lambda - z_{ji}|=\varepsilon} h'(\lambda)h(\lambda)^{-1} d\lambda \\
&\quad + \frac{1}{2\pi i} \int_{|\lambda - z_{ji}|=\varepsilon} h(\lambda)a'(\lambda)a(\lambda)^{-1}h(\lambda)^{-1} d\lambda \\
&= P_{ji} + \frac{1}{2\pi i} \int_{|\lambda - z_{ji}|=\varepsilon} h(\lambda)a'(\lambda)a(\lambda)^{-1}h(\lambda)^{-1} d\lambda.
\end{aligned}$$

Since $h(\lambda)^{-1}$ has a simple pole at z_{ji} and $a'(z_{ji}) = 0$, the last integral vanishes. So

$$\frac{1}{2\pi i} \int_{|\lambda - z_{ji}|=\varepsilon} f'(\lambda) f(\lambda)^{-1} d\lambda = P_{ji}, \quad i = 1, \dots, \tau_j; \; j = 1, \dots, k. \tag{6.12}$$

Analogously we have

$$\begin{aligned}
&\frac{1}{2\pi i} \int_{|\lambda - z_{ji}|=\varepsilon} f(\lambda)^{-1} f'(\lambda) d\lambda \\
&= \frac{1}{2\pi i} \int_{|\lambda - z_{ji}|=\varepsilon} a(\lambda)^{-1} a'(\lambda) d\lambda \\
&\quad + \frac{1}{2\pi i} \int_{|\lambda - z_{ji}|=\varepsilon} a(\lambda)^{-1}h(\lambda)^{-1}h'(\lambda)a(\lambda) d\lambda \\
&= \frac{1}{2\pi i} \int_{|\lambda - z_{ji}|=\varepsilon} a(\lambda)^{-1}h(\lambda)^{-1}h'(\lambda)a(\lambda) d\lambda = S_{ji}^{-1} P_{ji} S_{ji}.
\end{aligned}$$

Note here that $h(\lambda)^{-1}$ has a simple pole at z_{ji} and that the coefficient of the Laurent expansion of $h(\lambda)^{-1}h'(\lambda)$ at z_{ji} is

$$\frac{1}{2\pi i}\int_{|\lambda - z_{ji}|=\varepsilon} h(\lambda)^{-1}h'(\lambda)d\lambda = P_{ji}.$$

Since $S_{ji}^{-1}P_{ji}S_{ji} = Q_{ji}$, it follows that

$$\frac{1}{2\pi i}\int_{|\lambda - z_{ji}|=\varepsilon} f(\lambda)^{-1}f'(\lambda)d\lambda = Q_{ji}, \quad i = 1,\dots,\tau_j; \quad j = 1,\dots,k. \tag{6.13}$$

Now

$$\frac{1}{2\pi i}\int_{\partial D} f'(\lambda)f(\lambda)^{-1}d\lambda = \sum_{j=1}^{k}\sum_{i=1}^{\tau_j}\frac{1}{2\pi i}\int_{|\lambda - z_{ji}|=\varepsilon} f'(\lambda)f(\lambda)^{-1}d\lambda,$$

$$\frac{1}{2\pi i}\int_{\partial D} f(\lambda)^{-1}f'(\lambda)d\lambda = \sum_{j=1}^{k}\sum_{i=1}^{\tau_j}\frac{1}{2\pi i}\int_{|\lambda - z_{ji}|=\varepsilon} f(\lambda)^{-1}f'(\lambda)d\lambda,$$

and the desired result (d) follows by combining (6.11)–(6.13) □

The results in this section have a counterpart for the Banach subalgebras of $\mathbb{C}^{m\times m}$ consisting of block diagonal matrices (with given block size). We leave the details to the reader.

References

[B] H. Bart, Spectral properties of locally holomorphic vector-valued functions, *Pacific J.Math.* **52** (1974), 321–329.

[BES1] H. Bart, T. Ehrhardt and B. Silbermann, Zero sums of idempotents in Banach algebras, *Integral Equations and Operator Theory* **19** (1994), 125–134.

[BES2] H. Bart, T. Ehrhardt and B. Silbermann, Logarithmic residues in Banach algebras, *Integral Equations and Operator Theory* **19** (1994), 135–152.

[BES3] H. Bart, T. Ehrhardt and B. Silbermann, Logarithmic residues, generalized idempotents and sums of idempotents in Banach algebras, *Integral Equations and Operator Theory* **29** (1997), 155–186.

[BES4] H. Bart, T. Ehrhardt and B. Silbermann, Logarithmic residues of some special classes of analytic vector valued functions, forthcoming.

[BGK] H. Bart, I. Gohberg and M.A. Kaashoek, *Minimal Factorization of Matrix and Operator Functions*, Operator Theory: Advances and Applications, vol. 1, Birkhäuser, Basel, 1979.

[BKL] H. Bart, M.A. Kaashoek and D.C. Lay, The integral formula for the reduced algebraic multiplicity of meromorphic operator functions, *Proceedings Edinburgh Mathematical Society* **21** (1978), 65–72.

[BW] H. Bart and A.P.M. Wagelmans, An Integer Programming Problem and Rank Decomposition of Block Upper Triangular Matrices, forthcoming.

[E] T. Ehrhardt, Finite sums of idempotents and logarithmic residues on connected domains, *Integral Equations and Operator Theory* **21** (1995), 238–242.

[GGK] I. Gohberg, S. Goldberg and M.A. Kaashoek, *Classes of Linear Operators*, vol. 1, Operator Theory: Advances and Applications, vol. 49, Birkhäuser, Basel, 1990.

[GLR] I. Gohberg, P. Lancaster and L. Rodman, *Invariant Subspaces of Matrices with Applications*, Canadian Mathematical Society Series of Monographs and Advanced Texts, John Wiley and Sons, New York etc., 1986.

[GS] I.C. Gohberg and E.I. Sigal, An operator generalization of the logarithmic residue theorem and the theorem of Rouché, *Mat. Sbornik 84* (126) (1971), 607–629 (Russian); English Transl.: *Math. USSR Sbornik* **13** (1971), 603–625.

[HP] R.E. Hartwig and M.S. Putcha, When is a matrix a sum of idempotents? *Linear and Multilinear Algebra* **26** (1990), 279–286.

[LMR] C. Laurie, B. Mathes and H. Radjavi, Sums of idempotents, *Linear Algebra Appl.* **208/209** (1994), 175–197.

[M] L. Mittenthal, Operator valued analytic functions and generalizations of spectral theory, *Pacific J.Math.* **24** (1968), 119–132.

[PS] C.H. Papadimitriou and K. Steiglitz, *Combinatorial Optimization, Algorithms and Complexity*, Prentice-Hall, inc., Englewood Cliffs, New Jersey, 1982.

[PT] C. Pearcy and D. Topping, Sums of small numbers of idempotents, *Michigan Math. J.* **14** (1967), 453–465.

[R] J. Riordan, *Combinatorial Identities*, John Wiley and Sons, New York etc., 1968.

[S] N.J.A. Sloane, *A Handbook of Integer Sequences*, Academic Press, New York and London, 1973.

[TL] A.E. Taylor and D.C. Lay, *Introduction to Functional Analysis*, Second Edition, John Wiley and Sons, New York, 1980.

[Wa] J.-H. Wang, The length problem for a sum of idempotents, *Linear Algebra Appl.* **215** (1995), 135–159.

[Wu1] P.Y. Wu, Sums of idempotent matrices, *Linear Algebra Appl.* **142** (1990), 43–54.

[Wu2] P.Y. Wu, Additive combinations of special operators, In: *Functional Analysis and Operator Theory*, Banach Center Publications, vol. 30, Warsaw, 1994, 337–361.

H. Bart
Econometrisch Instituut
Erasmus Universiteit Rotterdam
Postbus 1738
3000 DR Rotterdam
The Netherlands

T. Ehrhardt and B. Silbermann
Fakultät für Mathematik
Technische Universität Chemnitz
09107 Chemnitz
Germany

Operator Theory:
Advances and Applications, Vol. 122
© 2001 Birkhäuser Verlag Basel/Switzerland

Generalized Nevanlinna Functions with Polynomial Asymptotic Behaviour at Infinity and Regular Perturbations

Vladimir Derkach, Seppo Hassi and Henk De Snoo

Dedicated to Rien Kaashoek on the occasion of his sixtieth birthday

Let $\mathbf{N}_\kappa$ be the class of meromorphic functions $Q(z)$ defined on $\mathbb{C} \setminus \mathbb{R}$ with $Q(\bar{z}) = \overline{Q(z)}$, and such that on its domain of holomorphy the kernel

$$\mathsf{N}_Q(z, w) = \frac{Q(z) - \overline{Q(w)}}{z - \bar{w}}, \; z \neq \bar{w}, \quad \mathsf{N}_Q(z, \bar{z}) = Q'(z),$$

has a finite number κ of negative squares. The class $\mathbf{N}_\kappa$ plays an important role in the extension theory of symmetric operators S acting on Pontryagin spaces. The additional assumption that one of the so-called Q-functions $Q(z)$ of S has the asymptotic behaviour $Q(z) = P(z) + o(z)$ as $z \widehat{\to} \infty$, with $P(z)$ a real polynomial, gives rise to a model, where the selfadjoint extensions of S can be interpreted as regular perturbations of a selfadjoint operator A in a Pontryagin space. The associated operator theoretical facts are explained and spectral theoretical characterizations for such regular perturbations will be described by means of the underlying spectral measures.

1 Introduction

Let $\mathbf{N}_\kappa$ be the class of meromorphic functions as defined above, i.e., the class of generalized Nevanlinna functions with κ negative squares. Each function $Q(z) \in \mathbf{N}_\kappa$ has a representation by means of a symmetric operator S with defect numbers $(1, 1)$ and a selfadjoint extension A of S with a nonempty resolvent set $\rho(A)$ in a Pontryagin space $(\mathfrak{H}, [\cdot, \cdot])$ with κ negative squares. Conversely, to every such S and a selfadjoint extension A, $\rho(A) \neq \emptyset$, of S in $\mathfrak{H}$ one can associate a generalized Nevanlinna function $Q(z)$, the Q-function of S and A. As in the Hilbert space case the selfadjoint extensions of S in $\mathfrak{H}$ can be parametrized by $A(\tau)$, $\tau \in \mathbb{R} \cup \{\infty\}$. The spectral theoretical analysis of the extensions $A(\tau)$ can be carried out by means of their Q-functions $Q_\tau(z)$, which are obtained via the linear fractional transforms

$$(1.1) \qquad Q_\tau(z) = \frac{Q(z) - \tau}{\tau Q(z) + 1}, \quad \tau \in \mathbb{R} \cup \{\infty\},$$

cf. [4], [16]. If for some (and then precisely for one) $Q_\tau(z)$ the following condition

$$\lim_{z \widehat{\to} \infty} \frac{Q_\tau(z)}{z} = 0$$

fails to hold, then $A(\tau)$ is a true linear relation of the form

$$S_F = S \dot{+} (\{0\} \oplus \operatorname{span}\{\omega\}), \quad \omega \in \mathfrak{H}, \tag{1.2}$$

which is the so-called generalized Friedrichs extension, while the other selfadjoint extensions of S are operators, which can be written as rank one perturbations of each other:

$$A(\alpha) = A + \alpha[\cdot, \omega]\omega, \quad \omega \in \mathfrak{H}, \quad \alpha \in \mathbb{R}, \tag{1.3}$$

cf. [4]. Let us mention that one arrives at situations like in (1.3), for instance by considering a (generalized) perturbation of a selfadjoint operator A in a Hilbert space with an element ω lying outside $\mathfrak{H}$, in some scale space $\mathfrak{H}_{-k} \supset \mathfrak{H}$, $k \in \mathbb{Z}_+$, which is dual to the graph of $|A|^{k/2}$, cf. e.g. [21]. In several recent papers perturbations like in (1.3) for a selfadjoint operator A in a Hilbert space have been considered, both for semibounded and non-semibounded A, with $\omega \in \mathfrak{H}, \mathfrak{H}_{-1}, \mathfrak{H}_{-2}$; see [1], [9], [13], [14], [15], [22], and the references therein. These three cases all admit interpretations by means of selfadjoint extensions in the original Hilbert space. Similar results in the case of Pontryagin space operators have been worked out recently in [4].

More specific results for the perturbations (1.3) can be given by introducing some further conditions on $\omega \in \mathfrak{H}$. In particular, the condition $\omega \in \operatorname{dom}|A|^{k/2}$, $k \in \mathbb{Z}_+$, generally speaking provides smoothness of ω and was studied in the Hilbert space case in [12]. If A is a selfadjoint operator in a Pontryagin space $\mathfrak{H}$, then ω may be positive $[\omega, \omega] > 0$, negative $[\omega, \omega] < 0$, or neutral $[\omega, \omega] = 0$. The results in the cases $[\omega, \omega] > 0$ and $[\omega, \omega] < 0$ are analogous to the Hilbert space case (cf. [4]), and the main theoretical interest is in the case $[\omega, \omega] = 0$. Concrete examples of such situations appear, for example, in quantum mechanical models of mathematical physics.

Let $\mathfrak{H}$ be a Pontryagin space, let A be a selfadjoint operator in $\mathfrak{H}$, and assume that $\omega \in \mathfrak{H}$. Then the spectral theoretical results for the perturbations $A(\alpha)$ in (1.3) and the "limiting perturbation" S_F given by (1.2) can be derived by means of the function

$$Q(z) = \gamma + [(A - z)^{-1}\omega, \omega], \quad z \in \rho(A), \quad \gamma \in \mathbb{R}, \quad \omega \in \mathfrak{H},$$

and its linear fractional transforms $Q_\tau(z)$ in (1.1). If ω is neutral, then the selfadjoint relation S_F has a nontrivial root subspace

$$\mathsf{S}_\infty(S_F) = \{ g \in \mathfrak{H} : \{0, g\} \in S_F^n \text{ for some } n \in \mathbb{N} \}, \tag{1.4}$$

which in general contains the multivalued part mul S_F as a proper subspace. We call the perturbations of A regular if the root subspace $\mathsf{S}_\infty(S_F)$ is a nondegenerate subspace of $\mathfrak{H}$. Observe that the perturbations $A(\alpha)$, $\alpha \in \mathbb{R} \cup \{\infty\}$, are completely determined by ω. Let us mention that, if, for instance, $[\omega, \omega] = 0$, then ∞ is necessarily a so-called critical point of S_F. If $[\omega, \omega] = 0$, but $\omega \notin \mathrm{dom} A$, then ∞ is a singular critical point of S_F (i.e., the spectral family of S_F is unbounded in the strong topology of $\mathfrak{H}$). In fact, it is a consequence of some results in H. Langer [19] that for a selfadjoint relation H in $\mathfrak{H}$ with a nonempty resolvent set $\rho(H)$, the root subspace $\mathsf{S}_\infty(H)$ is nondegenerate if and only if ∞ is not a singular critical point of H. Hence, regular perturbations correspond precisely to the case where ∞ is not a singular critical point of S_F.

In this paper regular perturbations will be studied, various characterizations for such perturbations will be given, and the associated spectral theoretical properties will be derived. In Section 2 some basic facts concerning the class $\mathbf{N}_\kappa$ of generalized Nevanlinna functions are given. The multiplicities of the generalized poles of nonpositive type for the sum of generalized Nevanlinna functions are studied in Section 3; in fact, necessary and sufficient conditions are obtained for the total multiplicity to be additive, see Theorem 3.4. In Section 4 it is proved that the Q-function $Q_F(z)$ of S_F and S related to regular perturbations is characterized by the following asymptotic property

$$Q_F(z) = P(z) + o(z), \quad z \widehat{\to} \infty, \tag{1.5}$$

where $P(z)$ is a real polynomial, see Theorem 4.1. Regular perturbations with $[\omega, \omega] = 0$ are also smooth in the sense of [12], i.e., in this case regularity automatically guarantees that $\omega \in \mathrm{dom} A^k$ for some $k > 0$. If, for instance, $\mathfrak{H}$ is finite-dimensional or if $[\omega, \omega] \neq 0$, and in particular if $\mathfrak{H}$ is a Hilbert space, then the perturbations are regular. In Section 5 the regularity of the perturbations is characterized in terms of asymptotic expansions of the associated Q-functions $Q_\tau(z)$, $\tau \in \mathbb{R} \cup \{\infty\}$, see Theorem 5.2. The results in this section are related to certain subclasses of $\mathbf{N}_\kappa$; see [6]. In particular, generalizations are obtained for the corresponding Hilbert space results in [12]. The spectral theoretical characterizations of regular perturbations in Section 6 are based on a recent factorication result concerning generalized Nevanlinna functions $Q(z) \in \mathbf{N}_\kappa$ and the associated integral representation, cf. [4], [7].

The regularity of the rank one perturbations (1.3) has been defined in terms of the root subspace $\mathsf{S}_\infty(S_F)$ of the generalized Friedrichs extension S_F in (1.2). In this paper the regularity has been characterized in terms of various properties of the Q-functions $Q_\tau(z)$, $\tau \in \mathbb{R} \cup \{\infty\}$, and also in terms of the element ω. The case of nonregular rank one perturbations is complicated; when $\kappa = 1$ a full description is contained in [5]. Some of the main results in the present paper have also independent interest, since they can be interpreted as criteria for the regularity of the critical point ∞ of a selfadjoint relation in a Pontryagin space.

2 Preliminaries

The class $\mathbf{N}_\kappa$ of generalized Nevanlinna functions with κ negative squares, consists of functions $Q(z)$, meromorphic on $\mathbb{C}\backslash\mathbb{R}$, which are real, i.e. $Q(\bar{z}) = \overline{Q(z)}$, $z \in \rho(Q)$, and for which the kernel

$$
(2.1) \qquad \begin{cases} \mathsf{N}_Q(z,w) = \frac{Q(z)-\overline{Q(w)}}{z-\bar{w}}, & z, w \in \rho(Q), \quad z \neq \bar{w}, \\ \mathsf{N}_Q(z,\bar{z}) = Q'(z), & z \in \rho(Q), \end{cases}
$$

has κ negative squares. Here $\rho(Q)$ denotes the domain of holomorphy of $Q(z)$ in $\mathbb{C}$. The kernel condition means that for each choice of $n \in \mathbb{N}$, and $z_1, \dots, z_n \in \rho(Q)$, the $n \times n$ hermitian matrix

$$
(\mathsf{N}_Q(z_i, z_j))_{i,j=1}^n
$$

has at most, and for at least one such choice precisely, κ negative eigenvalues, counting multiplicities.

Let $\mathfrak{H}$ be a Pontryagin space with κ negative squares and let S be a symmetric operator in $\mathfrak{H}$ with defect numbers $(1, 1)$. Moreover, let A be a selfadjoint extension of S in $\mathfrak{H}$ with $z_0 \in \rho(A)$. Choose a nontrivial vector $\chi(z_0) \in \ker(S^* - z_0)$ and extend it analytically as follows:

$$
(2.2) \qquad \chi(z) = (I + (z - z_0)(A - z)^{-1})\chi(z_0), \quad z \in \rho(A).
$$

Then $\chi(z) \in \ker(S^* - z)$ and S can be recovered from A and $\chi(z)$ via

$$
(2.3) \qquad S = \{\{f, g\} \in A : [g - \bar{z}f, \chi(z)] = 0\}.
$$

The operator S is said to be simple if every $z \in \mathbb{C}\backslash\mathbb{R}$ is a point of regular type of S, and

$$
\Pi(S) := \overline{\text{span}}\{\chi(z) : z \in \mathbb{C}\backslash\mathbb{R}\} = \mathfrak{H}.
$$

The Q-function $Q(z)$ associated to the pair (S, A) is defined as a solution of

$$
(2.4) \qquad \frac{Q(z) - \overline{Q(z_0)}}{z - \bar{z}_0} = [\chi(z), \chi(z_0)],
$$

cf. [16]. It is uniquely defined up to a real constant and belongs to $\mathbf{N}_{\kappa'}$ for some $\kappa' \leq \kappa$. It follows from (2.2) and (2.4) that

$$
(2.5) \qquad Q(z) = s + \bar{z}_0[v, v] + (z - \bar{z}_0)[(I + (z - z_0)(A - z)^{-1})v, v],
$$

with $s \in \mathbb{R}$ and $v = \chi(z_0)$. The representation (2.5) is called *minimal* if $\Pi(S) =$ $(\overline{\text{span}}\{\chi(z) : z \in \rho(A)\} =)\mathfrak{H}$. It is known that the operator S is simple if and only if the representation (2.5) is minimal. If the representation (2.5) is minimal, then the equality $\kappa' = \kappa$ holds. In the general case κ' coincides with the dimension $\kappa_-(\Pi(S))$ of a maximal negative subspace of $\Pi(S)$ (see [16]).

Each function $Q(z) \in \mathbf{N}_\kappa$ has an operator representation of the form (2.5). In fact, the reproducing kernel Pontryagin space $\mathfrak{P}(Q)$ associated with (2.1) has κ negative squares, and in $\mathfrak{P}(Q)$ there exist a simple symmetric operator $S(Q)$ with defect numbers $(1, 1)$ and a selfadjoint extension $A(Q)$ of $S(Q)$ with a nonempty resolvent set $\rho(A)$, such that $Q(z)$ is the Q-function of $S(Q)$ and $A(Q)$, see for instance [8]. For every minimal representation (2.5) the linear relations S and A are uniquely determined up to isometric isomorphisms, $\rho(A)$ coincides with the domain of holomorphy $\rho(Q)$ of $Q(z)$ in $\mathbb{C}$. The selfadjoint extension A is an operator if and only if

$$\lim_{y\to\infty} \frac{Q(iy)}{y} = 0. \tag{2.6}$$

The statement (2.6) was originally proved in [16, Satz 4.1]. In a (minimal) representation (2.5) A need not be an operator, i.e. its multivalued part mul A can be nontrivial. In fact, there may be a nontrivial root subspace at ∞:

$$\mathsf{S}_\infty(A) = \{g \in \mathfrak{H} : \{0, g\} \in A^n \text{ for some } n \in \mathbb{N}\},$$

which is necessarily finite dimensional with $\dim \mathsf{S}_\infty(A) \le 2\kappa + 1$, cf. [2]. The next proposition describes the situation, where the equality $\mathsf{S}_\infty(A) = \mathfrak{P}(Q)$ holds. First, the following notations will be introduced.

Let $P(z) = \sum_{i=0}^n c_i z^i$, $c_n \neq 0$, be a real polynomial and associate with $P(z)$ the $n \times n$ matrix B (in fact the Bezoutian Bez $(P, 1)$) by

$$B = (c_{i+j-1})_{i,j=1}^n,$$

where $c_k = 0$ if $k > n$. Define the linear space $\mathfrak{P}_{n-1}$ of polynomials of formal degree $n-1$ with the inner product

$$[f(\cdot), g(\cdot)] = g^* B^{-1} f, \tag{2.7}$$

where $f, g \in \mathbb{C}^n$ denote vectors, whose components are the coefficients of the polynomials $f(z)$ and $g(z)$, respectively.

Proposition 2.1 *Let $P(z) = \sum_{i=0}^n c_i z^i$, $c_n \neq 0$, be a real polynomial. If c_n is positive (negative), then the kernel $\mathsf{N}_P(\cdot, \cdot)$ of $P(z)$ has $[(n+1)/2]$ positive (negative) squares, and $[n/2]$ negative (positive) squares, respectively. The corresponding reproducing kernel Pontryagin space is equal to $\mathfrak{P}_{n-1}$. In this model space, the closed symmetric operator $S(P)$ associated to $P(z)$ is nondensely defined and its selfadjoint extension $A(P)$ has a nontrivial multivalued part. Furthermore, the root subspace $\mathsf{S}_\infty(A(P))$ coincides with $\mathfrak{P}_{n-1}$.*

Proof: Denote $Z_{n-1} = (1, z, \dots, z^{n-1})$ and $W_{n-1} = (1, w, \dots, w^{n-1})$. Then for $f(\cdot), g(\cdot) \in \mathfrak{P}_{n-1}$, we have $f(z) = Z_{n-1} f$, $g(w) = W_{n-1} g$, and a straightforward calculation yields

$$\mathsf{N}_P(w,z) = \frac{P(w) - P(\bar{z})}{w - \bar{z}} = W_{n-1} B Z_{n-1}^*, \quad w, z \in \mathbb{C}. \tag{2.8}$$

It follows from (2.7) and (2.8) that

$$[f(\cdot), \mathsf{N}_P(\cdot, z)] = Z_{n-1} f = f(z). \tag{2.9}$$

This shows that $\mathfrak{P}_{n-1}$ is a reproducing kernel space with the kernel $\mathsf{N}_P(w,z)$, which according to (2.8) has $\kappa_+ = [(n+1)/2]$ positive and $\kappa_- = [n/2]$ negative squares, when for instance $c_n > 0$; by (2.7) κ_+ and κ_- coincide with the numbers of positive and negative squares of the space $\mathfrak{P}_{n-1}$, respectively. Define the linear relations $S(P)$ and $A(P)$ in $\mathfrak{P}_{n-1}$ via

$$S(P) = \{\{f(w), wf(w)\} : f(w), wf(w) \in \mathfrak{P}_{n-1}\}, \tag{2.10}$$

$$\begin{aligned} A(P) \ &= \ \{\{f(w), wf(w) + c\} : f(w), \\ & \qquad wf(w) + c \in \mathfrak{P}_{n-1}, \ c \in \mathbb{C}\}. \end{aligned} \tag{2.11}$$

Then $S(P)$ is the multiplication operator in $\mathfrak{P}_{n-1}$ and hence it is symmetric. Moreover, $(\operatorname{dom} S(P))^{\perp}$ is spanned by a constant function, while the defect subspaces of $S(P)$ are spanned by

$$\chi_P(z) = \mathsf{N}_P(\cdot, \bar{z}), \quad z \in \mathbb{C}. \tag{2.12}$$

Clearly, $A(P)$ is a proper closed symmetric extension of $S(P)$ and hence it is selfadjoint. It is easy to check that

$$R_z f(w) = \frac{f(w) - f(z)}{w - z}, \quad f(\cdot) \in \mathfrak{P}_{n-1}, \quad z, w \in \mathbb{C}. \tag{2.13}$$

This implies that $\sigma(A(P)) = \{\infty\}$ and, further, that $\mathsf{S}_\infty(A(P)) = \mathfrak{P}_{n-1}$.

Identity (2.13) applied to $\chi_P(z_0)$ gives $(z - z_0) R_z \chi_P(z_0) = \chi_P(z) - \chi_P(z_0)$, so that (2.2) holds. According to (2.9), $[\chi_P(z), \chi_P(z_0)] = \mathsf{N}_P(\bar{z}_0, \bar{z})$, thus also (2.4) is satisfied. Therefore, $P(z)$ is a Q-function for $S(P)$, $A(P)$. □

The notion of generalized poles or zeros plays an important role in the theory of generalized Nevanlinna functions. Let $Q(z) \in \mathbf{N}_\kappa$. Assume that $\alpha \in \mathbb{R}$ and let $\kappa_\alpha = \kappa_\alpha(Q) \geq 0$ be the smallest integer such that

$$-\infty < \lim_{z \widehat{\to} \alpha} (z - \alpha)^{2\kappa_\alpha + 1} Q(z) \leq 0. \tag{2.14}$$

Let $\kappa_\infty = \kappa_\infty(Q) \geq 0$ be the smallest integer such that

$$0 \leq \lim_{z \widehat{\to} \infty} \frac{Q(z)}{z^{2\kappa_\infty + 1}} < \infty. \tag{2.15}$$

If $\kappa_\alpha(Q) > 0$ or $\kappa_\infty(Q) > 0$, then $\alpha \in \mathbb{R}$ or ∞ is said to be a generalized pole of nonpositive type of $Q(z)$ with multiplicity $\kappa_\alpha(Q)$ or $\kappa_\infty(Q)$, respectively. Analogously one defines the generalized zeros $\beta \in \mathbb{R} \cup \{\infty\}$ of nonpositive type of $Q(z)$ and their multiplicities $\pi_\infty(Q)$, cf. e.g. [20]. In fact, if $Q(z) \not\equiv 0$ then the generalized zeros are precisely the generalized poles of the transform $-1/Q(z)$ of $Q(z) \in \mathbf{N}_\kappa$, and $\pi_\beta(Q) = \kappa_\alpha(-1/Q)$. The next result is due to M.G. Kreĭn and H. Langer, cf. [18].

Theorem 2.2 *Let $Q(z) \in \mathbf{N}_\kappa$. Then $Q(z)$ has poles in $\mathbb{C}^+$ and generalized poles of nonpositive type in $\mathbb{R} \cup \{\infty\}$ of total multiplicity κ. Moreover, if $Q(z)$ is not trivial, $Q(z)$ has zeros in $\mathbb{C}^+$ and generalized zeros of nonpositive type in $\mathbb{R} \cup \{\infty\}$ of total multiplicity κ.*

In the sequel the set of all the (generalized) poles of $Q(z)$, including ∞, will be denoted by $\tilde{\sigma}_0(Q)$. Observe, that for a polynomial $P(z)$ (cf. Proposition 2.1) the only possible (generalized) pole is ∞, i.e. $\tilde{\sigma}_0(P) = \{\infty\}$.

3 Multiplicity of the Generalized Poles for the Sum of $\mathbf{N}_\kappa$-Functions

Let $Q_1(z) \in \mathbf{N}_{\kappa_1}$ and $Q_2(z) \in \mathbf{N}_{\kappa_2}$ be generalized Nevanlinna functions. Then the sum $Q_1(z) + Q_2(z)$ is again a generalized Nevanlinna function. A model space for the sum can be built from the model spaces for each of the functions separately. For $Q_i(z) \in \mathbf{N}_{\kappa_i}$, $i = 1, 2$, there are minimal representations with model spaces $\mathfrak{H}_i$ with selfadjoint relations A_i in $\mathfrak{H}_i$ and symmetric restrictions S_i:

$$S_i = \{\{f, g\} \in A_i : [g - \bar{z}f, \chi_i(z)] = 0\},$$

where

$$\chi_i(z) = (I + (z - z_0)(A_i - z)^{-1})\chi(z_0),$$

and $z_0 \in \rho(A_1) \cap \rho(A_2)$. In the orthogonal sum space $\mathfrak{H}_1 \oplus \mathfrak{H}_2$, the orthogonal sum relation $A_1 \oplus A_2$ is selfadjoint. The orthogonal sum $S_1 \oplus S_2$ is symmetric and has defect numbers $(2, 2)$. In order to realize the sum $Q_1(z) + Q_2(z)$ in the space $\mathfrak{H}_1 \oplus \mathfrak{H}_2$ a symmetric extension of $S_1 \oplus S_2$ has to be introduced (see [10] for the Hilbert space case).

Proposition 3.1 *Let $Q_1(z) \in \mathbf{N}_{\kappa_1}$ and $Q_2(z) \in \mathbf{N}_{\kappa_2}$. The sum $Q_1(z) + Q_2(z)$ is the Q-function associated with $A_1 \oplus A_2$ and the symmetric extension $\tilde{S}$ of $S_1 \oplus S_2$ given by*

$$\tilde{S} = \left\{\{f, g\} \in A_1 \oplus A_2 : \left[g - zf, \begin{pmatrix} \chi_1(\bar{z}) \\ \chi_2(\bar{z}) \end{pmatrix}\right] = 0\right\}, \tag{3.1}$$

so that $Q_1(z) + Q_2(z) \in \mathbf{N}_\kappa$ with $\kappa \leq \kappa_1 + \kappa_2$. Moreover, $\kappa = \kappa_1 + \kappa_2$ if and only if the orthogonal complement of $\Pi(\tilde{S})$ is a positive subspace in $\mathfrak{H}_1 \oplus \mathfrak{H}_2$.

Proof: The proof of the first statement can be given by a straightforward adaptation of the arguments in [10]. To prove the second statement it is enough to observe that the space $(\mathfrak{H}_1 \oplus \mathfrak{H}_2) \ominus \Pi(\tilde{S})$ is positive if and only if $\kappa_-(\Pi(\tilde{S})) = \kappa_1 + \kappa_2$. □

A different characterization for the equality $\kappa = \kappa_1 + \kappa_2$ can be given in terms of the asymptotic properties of the generalized Nevanlinna functions $Q_1(z)$ and $Q_2(z)$ at the generalized poles of nonpositive type.

Lemma 3.2 *Let $Q_1(z) \in \mathbf{N}_{\kappa_1}$, $Q_2(z) \in \mathbf{N}_{\kappa_2}$, and let $\alpha \in \mathbb{R} \cup \{\infty\}$. Then*

$$\kappa_\alpha(Q_1 + Q_2) = \kappa_\alpha(Q_1) + \kappa_\alpha(Q_2) \tag{3.2}$$

if and only if

$$\min\{\kappa_\alpha(Q_1), \kappa_\alpha(Q_2)\} = 0, \tag{3.3}$$

and

$$\kappa_\alpha(Q_1 + Q_2) = 1, \quad \textit{if } \max\{\kappa_\alpha(Q_1), \kappa_\alpha(Q_2)\} = 1. \tag{3.4}$$

Proof: For any $\alpha \in \mathbb{R} \cup \{\infty\}$ it follows from (2.14) and (2.15), that

$$\kappa_\alpha(Q_1 + Q_2) \leq \max\{\kappa_\alpha(Q_1), \kappa_\alpha(Q_2)\}, \tag{3.5}$$

with equality when

$$|\kappa_\alpha(Q_1) - \kappa_\alpha(Q_2)| > 1. \tag{3.6}$$

Clearly, it follows from (3.5) that

$$\kappa_\alpha(Q_1 + Q_2) \leq \kappa_\alpha(Q_1) + \kappa_\alpha(Q_2). \tag{3.7}$$

Now assume that (3.2) holds. Then (3.3) follows from (3.5), while (3.4) is obvious. Conversely, assume that (3.3) holds. In case (3.6), equality prevails in (3.5), which means that (3.2) holds. If (3.6) is not satisfied, then

$$|\kappa_\alpha(Q_1) - \kappa_\alpha(Q_2)| \leq 1.$$

If $\kappa_\alpha(Q_1) = \kappa_\alpha(Q_2) = 0$, then (3.2) holds due to (3.5). Otherwise,

$$\max\{\kappa_\alpha(Q_1), \kappa_\alpha(Q_2)\} = 1$$

and (3.2) follows from (3.4). □

Next we treat the conditions (3.3) and (3.4) in Lemma 3.2 in more detail.

Lemma 3.3 *Let $Q_1(z) \in \mathbf{N}_{\kappa_1}$, $Q_2(z) \in \mathbf{N}_{\kappa_2}$, and let $\alpha \in \mathbb{R} \cup \{\infty\}$. Assume that $\kappa_\alpha(Q_1) = 1$ and $\kappa_\alpha(Q_2) = 0$. Then*

$$\kappa_\alpha(Q_1 + Q_2) = 0, \tag{3.8}$$

if and only if

$$0 < \lim_{y \to 0} iyQ_1(\alpha + iy) \leq - \lim_{y \to 0} iyQ_2(\alpha + iy) \quad (< \infty), \tag{3.9}$$

when $\alpha \in \mathbb{R}$, and

$$0 > \lim_{y \to \infty} \frac{Q_1(iy)}{iy} \geq - \lim_{y \to \infty} \frac{Q_2(iy)}{iy} \quad (> -\infty), \tag{3.10}$$

when $\alpha = \infty$.

Proof: The identity $\kappa_\alpha(Q_1 + Q_2) = 0$ means that

$$-\infty < \lim_{y \to 0} iy \left(Q_1(\alpha + iy) + Q_2(\alpha + iy)\right) \leq 0, \tag{3.11}$$

if $\alpha \in \mathbb{R}$, and that

$$\infty > \lim_{y \to \infty} \frac{Q_1(iy) + Q_2(iy)}{iy} \geq 0, \tag{3.12}$$

if $\alpha = \infty$. Since $\kappa_\alpha(Q_2) = 0$ the corresponding limits for $Q_2(z)$, and thus also for $Q_1(z)$, are finite. Now, the first estimates in (3.9) and (3.10) follow from $\kappa_\alpha(Q_1) = 1$, while the second estimates are obtained from (3.11) and (3.12), respectively. The estimates inside the brackets in (3.9) and (3.10), are of course due to $\kappa_\alpha(Q_2) = 0$. □

The next theorem gives a criterion for the equality $\kappa(Q_1 + Q_2) = \kappa(Q_1) + \kappa(Q_2)$, by means of the location and the asymptotics of the generalized poles of $Q_1(z)$ and $Q_2(z)$.

Theorem 3.4 *Let $Q_1(z) \in \mathbf{N}_{\kappa_1}$ and $Q_2(z) \in \mathbf{N}_{\kappa_2}$. Then $Q_1(z) + Q_2(z) \in \mathbf{N}_\kappa$ with $\kappa = \kappa_1 + \kappa_2$ if and only if*

$$\tilde{\sigma}_0(Q_1) \cap \tilde{\sigma}_0(Q_2) = \emptyset \tag{3.13}$$

and

$$\begin{aligned} &\kappa_\alpha(Q_1 + Q_2) = 1, \quad \text{if } \max\{\kappa_\alpha(Q_1), \kappa_\alpha(Q_2)\} = 1, \\ &\alpha \in \mathbb{R} \cup \{\infty\}. \end{aligned} \tag{3.14}$$

Moreover, if (3.13) holds, then the second condition (3.14) does not hold if and only if

$$0 < \lim_{y \to 0} iyQ_j(\alpha + iy) \leq - \lim_{y \to 0} iyQ_k(\alpha + iy) \quad (< \infty),$$

for some $\alpha \in \mathbb{R}$, *or*

$$0 > \lim_{y\to\infty} \frac{Q_j(iy)}{iy} \geq -\lim_{y\to\infty} \frac{Q_k(iy)}{iy} \quad (> -\infty),$$

for $\alpha = \infty$ *with* $j \neq k,\ j,k = 1,2$.

Proof: According to Theorem 2.2 the total multiplicity of the (generalized) poles of $Q(z) \in \mathbf{N}_\kappa$ is equal to κ. Clearly, for nonreal poles the conditions reduce to (3.13). Thus, we may suppose that $Q_1(z)$ and $Q_2(z)$ have no poles in $\mathbb{C}\setminus\mathbb{R}$. The estimate (3.5) shows that

$$\tilde{\sigma}_0(Q_1+Q_2) \subset \tilde{\sigma}_0(Q_1) \cup \tilde{\sigma}_0(Q_2),$$

and, moreover, that $\kappa = \kappa_1 + \kappa_2$ if and only if $\kappa_\alpha(Q_1+Q_2) = \kappa_\alpha(Q_1) + \kappa_\alpha(Q_2)$ for all $\alpha \in \mathbb{R}\cup\{\infty\}$. Now the first statement of the proposition follows from Lemma 3.2.

To prove the second part, assume (3.13), so that (3.3) holds. The condition (3.14) is not satisfied precisely when for some $\alpha \in \mathbb{R}\cup\{\infty\}$,

$$\kappa_\alpha(Q_1+Q_2) = 0, \text{ and } \kappa_\alpha(Q_j) = 1 \text{ for either } j = 1 \text{ or } j = 2.$$

Therefore, the second statement of the proposition follows from Lemma 3.3. □

Earlier, some sufficient results for this equality were obtained in Kreĭn-Langer [17] and Daho-Langer [3]. As a direct consequence of Theorem 3.4 one obtains the following result which includes the scalar version of [3, Proposition 3.2] and, in particular, [17, Satz 1.13].

Proposition 3.5 *Let* $Q_j(z) \in \mathbf{N}_{\kappa_j}$, $j = 1,2$, *and assume that*

$$Q_k(\alpha+iy) = o(1/y), \quad y \to 0, \quad \text{if } \alpha \in \tilde{\sigma}_0(Q_j)\cap\mathbb{R},$$

(3.15)
$$k \neq j; \quad j,k = 1,2,$$

(3.16)
$$Q_k(iy) = o(y), \quad y\to\infty, \quad \text{if } \infty \in \tilde{\sigma}_0(Q_j).$$

Then $\tilde{\sigma}_0(Q_1)\cap\tilde{\sigma}_0(Q_2)\cap\mathbb{R} = \emptyset$. *Moreover, if* $Q_1(z)$ *and* $Q_2(z)$ *do not have any common nonreal poles, then* $\tilde{\sigma}_0(Q_1)\cap\tilde{\sigma}_0(Q_2) = \emptyset$ *and* $Q_1(z)+Q_2(z) \in \mathbf{N}_{\kappa_1+\kappa_2}$.

Proof: The assumption in (3.15) means that $\lim_{y\to\infty} yQ_k(\alpha+iy) = 0$. Therefore, $\kappa_\alpha(Q_k) = 0$ if $\alpha \in \tilde{\sigma}_0(Q_j)\cap\mathbb{R}$ and the estimates for $Q_j(z)$ in Theorem 3.4 fail to hold. Similarly, (3.16) implies that $\kappa_\infty(Q_k) = 0$ if $\infty \in \tilde{\sigma}_0(Q_j)$ and that the estimates for $Q_j(z)$ in Theorem 3.4 do not hold. Hence, $\tilde{\sigma}_0(Q_1)\cap\tilde{\sigma}_0(Q_2)\cap\mathbb{R} = \emptyset$, and the second assertion follows now from Theorem 3.4. □

4 Polynomial Behaviour at Infinity

Let $Q(z)$ be a generalized Nevanlinna function in $\mathbf{N}_\kappa$. If the condition (2.6) does not hold the selfadjoint extension A of S in the corresponding model space $\mathfrak{H}$ is multivalued, i.e. $\operatorname{mul} A \neq \{0\}$. Then also the root subspace $\mathsf{S}_\infty(A)$ is nontrivial; it may be degenerate or nondegenerate. The nondegeneracy of $\mathsf{S}_\infty(A)$ can be described by means of the asymptotic behaviour of $Q(z)$ as $z \widehat{\rightarrow} \infty$.

Theorem 4.1 *Let S be a simple symmetric operator in a Pontryagin space $\mathfrak{H}$ with defect numbers* $(1, 1)$, *and let A be a selfadjoint extension of S. Let $Q(z)$ be the Q-function of S and A. Then the root subspace $\mathsf{S}_\infty(A)$ is a nontrivial nondegenerate subspace of $\mathfrak{H}$ if and only if there exists a real polynomial $P(z)$ with* $\deg P(z) \geq 1$, *such that*

$$Q(z) = P(z) + o(z), \quad z \widehat{\rightarrow} \infty. \tag{4.1}$$

Proof: Assume that $\mathsf{S}_\infty(A)$ is a nontrivial nondegenerate subspace of $\mathfrak{H}$. Then $\mathfrak{H}$ has an orthogonal sum decomposition $\mathfrak{H} = \mathfrak{H}_1 \oplus \mathfrak{H}_2$ in terms of two Pontryagin spaces $\mathfrak{H}_1 = \mathsf{S}_\infty(A)$ and $\mathfrak{H}_2$. This decomposition reduces the selfadjoint relation A into an orthogonal sum $A = A_1 \oplus A_2$, where A_1 corresponds to $\mathsf{S}_\infty(A)$ as in Proposition 2.1 and A_2 is an operator. Due to (2.5) the orthogonal decomposition of A leads to a sum decomposition of $Q(z)$:

$$Q(z) = Q_1(z) + Q_2(z), \tag{4.2}$$

where $Q_1(z)$ is a polynomial with $\deg Q_1(z) \geq 1$ and $Q_2(z) = o(z)$, $z \widehat{\rightarrow} \infty$. Hence (4.1) is satisfied with $P(z) = Q_1(z)$.

Conversely, assume (4.1) for some real polynomial $P(z)$ with $\deg P(z) \geq 1$. Then $Q(z)$ admits the decomposition (4.2) with $Q_1(z) = P(z)$ and $Q_2(z) = o(z)$. According to Proposition 2.1, $Q_1(z)$ belongs to $\mathbf{N}_{\kappa_1}$ for some κ_1. Moreover, $Q_2(z) = Q(z) + (-P(z))$ belongs to $\mathbf{N}_{\kappa_2}$ for some κ_2 according to Proposition 2.1 applied to $-P(z)$ and the first part of Proposition 3.1. Associate to $Q_1(z)$ and $Q_2(z)$ minimal representations with simple symmetric operators S_1, S_2 and selfadjoint extensions A_1, A_2 in reproducing kernel Pontryagin spaces $\mathfrak{H}_1$, $\mathfrak{H}_2$, respectively, such that $Q_1(z)$ and $Q_2(z)$ are their Q-functions. Then $Q(z) = Q_1(z) + Q_2(z)$ is the Q-function associated with $A_1 \oplus A_2$ and the symmetric extension $\tilde{S}$ given by (3.1). In order to show that $\tilde{S}$ is simple, assume that there are vectors $h_1 \in \mathfrak{H}_1$ and $h_2 \in \mathfrak{H}_2$, such that

$$[h_1, \chi_1(\bar{z})] + [h_2, \chi_2(\bar{z})] = 0, \quad z \in \rho(A_1 \oplus A_2).$$

Then the reproducing kernel property (cf. (2.9)) shows that $[h_1, \chi_1(\bar{z})]$ is a polynomial and that $[h_2, \chi_2(\bar{z})] = o(1)$, since A_2 is an operator, cf. (2.2), (2.6). This means that

$$[h_1, \chi_1(\bar{z})] = [h_2, \chi_2(\bar{z})] = 0, \quad z \in \rho(A_1 \oplus A_2). \tag{4.3}$$

Since S_1 and S_2 are simple, (4.3) implies that $h_1 = 0$, $h_2 = 0$. Hence, $\tilde{S}$ is also simple, and A and $A_1 \oplus A_2$ are unitary equivalent. This implies that $\mathsf{S}_\infty(A)$ is a nontrivial nondegenerate subspace of $\mathfrak{H}$. □

In Proposition 2.1 it was shown that for a polynomial the identity $\mathsf{S}_\infty(A(P)) = \mathfrak{P}(P)$ holds. Also the converse is true. In fact, if $\mathsf{S}_\infty(A(Q)) = \mathfrak{P}(Q)$ holds, then $\rho(Q) = \mathbb{C}$, i.e. ∞ is the only generalized pole of $Q(z)$. Now clearly $\mathsf{S}_\infty(A(P))$ is nondegenerate, and by Theorem 4.1 $Q(z) = o(z) + P(z)$, where $o(z) \in \mathbf{N}$ is entire. Thus, $o(z) \equiv c$ and $Q(z)$ is a polynomial.

The next result is a simple consequence of Proposition 2.1 and Proposition 3.5; cf. also [17, Lemma 3.3].

Corollary 4.2 *Let the assumptions be as in Theorem* 4.1. *Assume that* $Q(z) = P(z) + o(z)$ $(z\widehat{\rightarrow}\infty)$ *belongs to* $\mathbf{N}_\kappa$, *where* $P(z) = \sum_{i=0}^n c_i z^i$, $c_n \neq 0$, *is a real polynomial. Then* $\kappa = \kappa(P) + \kappa(o)$. *Moreover,* $\dim \mathsf{S}_\infty(A) = \deg P(z)$ (≥ 0) *and*

$$\kappa_\infty(Q) = \kappa(P) = \kappa_-(\mathsf{S}_\infty(A)) = \begin{cases} [\frac{n}{2}], & \text{if } c_n > 0, \\ [\frac{n+1}{2}], & \text{if } c_n < 0. \end{cases} \tag{4.4}$$

Proof: The equality $\kappa = (\kappa(Q) =) \kappa(P) + \kappa(o)$ follows from Proposition 3.5. The Proof of Theorem 4.1 shows that $\mathsf{S}_\infty(A) = \mathsf{S}_\infty(A_1)$. Thus,

$$\dim \mathsf{S}_\infty(A) = \dim \mathsf{S}_\infty(A_1) = \deg P(z)$$

by Proposition 2.1. Clearly, $\kappa_\infty(Q) = \kappa_\infty(P) = \kappa(P)$ and hence the remaining identities in (4.4) are obtained from Proposition 2.1. □

5 The Subclasses Induced via Polynomial Asymptotics to $\mathbf{N}_{\kappa,0}$

In this section the subclasses induced in $\mathbf{N}_{\kappa,0}$ by the generalized Nevanlinna functions of the form

$$Q_F(z) = P(z) + o(z) \in \mathbf{N}_\kappa, \tag{5.1}$$

via the linear fractional transform

$$Q(z) = -\frac{1}{Q_F(z)}, \tag{5.2}$$

will be studied, and their role in the extension theory of symmetric operators will be explained. Functions of the form (5.1) are examples of Q-functions related to the generalized Friedrichs extensions of nondensely defined symmetric operators in a Pontryagin space, cf. [4], [5]. In this section several characterizations for this class of symmetric operators will be given.

Recall (see [4]) that $Q(z)$ is said to belong to the Kac class $\mathbf{N}_{\kappa,1}$ of generalized Nevanlinna functions, if $Q(z) \in \mathbf{N}_\kappa$ and

$$\lim_{y\to\infty} \frac{Q(iy)}{y} = 0, \quad \int_\eta^\infty \frac{|\operatorname{Im} Q(iy)|}{y}\, dy < \infty, \tag{5.3}$$

and that $Q(z) \in \mathbf{N}_{\kappa,0}$, if

$$Q(z) = \gamma + O(1/z), \quad z\widehat{\to}\infty, \quad Q(z) \in \mathbf{N}_\kappa.$$

According to [4] each function $Q(z) \in \mathbf{N}_{\kappa,1}$ has the following operator representation:

$$\begin{aligned} &Q(z) = \gamma + ((A-z)^{-1}\omega, \omega), \quad z \in \rho(A)\ (\neq \emptyset), \\ &\gamma \in \mathbb{R}, \quad \omega \in \mathfrak{H}_{-1}, \end{aligned} \tag{5.4}$$

where $\mathfrak{H}_{-1} \supset \mathfrak{H}$ denotes a Pontryagin space of generalized elements. Here $Q(z) \in \mathbf{N}_{\kappa,0}$ if and only if $\omega \in \mathfrak{H}$, in which case the duality $(\cdot,\cdot)$ in (5.4) reduces to the inner product of $\mathfrak{H}$. Functions of the form (5.1) appear as Q-functions of symmetric operators S that arise from generalized rank one perturbations of selfadjoint operators in $\mathfrak{H}$:

$$A(\alpha) = A + \alpha(\cdot,\omega)\omega, \quad \alpha \in \mathbb{R}, \quad \omega \in \mathfrak{H}_{-1}. \tag{5.5}$$

As in the Hilbert space case the symmetric operator S associated to (5.5) is given by

$$S = \{f \in \operatorname{dom} A : (f,\omega) = 0\}. \tag{5.6}$$

Again, if $Q(z) \in \mathbf{N}_{\kappa,0}$, or equivalently if $\omega \in \mathfrak{H}$, then the generalized perturbations in (5.5) reduce to rank one perturbations in (1.3), and S is nondensely defined. Associated to (5.5) there is a generalized Friedrichs extension S_F of S, which can be seen as the limiting perturbation of $A(\alpha)$'s as $\alpha \to \infty$, cf. [11]. The extension S_F of S need not be an operator; in fact, S_F is an operator if and only if $\omega \notin \mathfrak{H}$. The Q-functions $Q_\tau(z)$ associated to S and $A(\alpha)$, $\alpha \in \mathbb{R} \cup \{\infty\}$, are connected via a linear fractional transform, cf. (1.1). In fact, the connection between the parameters τ and α is given by $\alpha = \tau/(1+\tau\gamma)$, cf. [4]. The functions $Q_F(z)$ of the form (5.1) always correspond to S_F, while their transforms $-1/Q_F(z)$ belong to the class $\mathbf{N}_{\kappa,0}$. However, as will become clear below, in this way only certain further subclasses of $\mathbf{N}_{\kappa,0}$ will be obtained.

The assumption $Q(z) \in \mathbf{N}_{\kappa,1}$ implies that the limit $\lim_{z\widehat{\to}\infty} Q(z) = \gamma \in \mathbb{R}$ exists. Moreover, if $Q(z) \in \mathbf{N}_{\kappa,0}$ ($\subset \mathbf{N}_{\kappa,1}$), then also the limit

$$\lim_{z\widehat{\to}\infty} -z(Q(z)-\gamma) = [\omega,\omega]$$

is finite and belongs to $\mathbb{R}$. The function $Q(z) \in \mathbf{N}_{\kappa,0}$ is said to be from the class $\mathbf{N}_{\kappa,-2n}$, $n \in \mathbb{N}$, if there are real numbers γ and $s_0, \dots, s_{2n-1}$ such that the function

$$Q^{\{2n\}}(z) := z^{2n} \left(Q(z) - \gamma + \sum_{j=0}^{2n-1} \frac{s_j}{z^{j+1}} \right)$$

is $O(1/z)$; or, equivalently, $Q^{\{2n\}}(z)$ belongs to the class $\mathbf{N}_{\kappa',0}$ for some $\kappa' \leq \kappa$. The function $Q(z) \in \mathbf{N}_{\kappa,0}$ is said to be from the class $\mathbf{N}_{\kappa,-2n+1}$, $n \in \mathbb{N}$, if $Q^{\{2n\}}(z)$ belongs to the class $\mathbf{N}_{\kappa',1}$ for a choice of $s_0, \dots, s_{2n-1}$ and some $\kappa' \leq \kappa$. With these definitions, a result of Kreĭn and Langer, see [17, Satz 1.10], can be stated as follows.

Proposition 5.1 *Let $Q(z)$ be the Q-function of S and A, and assume that $Q(z) \in \mathbf{N}_{\kappa,0}$. Then $Q(z) \in \mathbf{N}_{\kappa,-2n}$, $n \in \mathbb{N}$, if and only if in (5.4) $\omega \in \operatorname{dom} A^n$. In this case $Q(z)$ has the asymptotic expansion*

$$Q(z) = \gamma - \sum_{i=0}^{2n} \frac{s_j}{z^{j+1}} + o\left(\frac{1}{z^{2n+1}}\right), \quad z\widehat{\rightarrow}\infty,$$

where the numbers $s_j \in \mathbb{R}$, $j = 0, \dots, 2n$, are given by

$$s_j = [A^j\omega, \omega], \quad s_{n+j} = [A^j\omega, A^n\omega], \quad j = 0, \dots, n.$$

Let $Q(z)$ belong to $\mathbf{N}_{\kappa,-2n+1}$. Then the function $Q^{\{2n\}}(z)$ belongs to the class $\mathbf{N}_{\kappa',1}$ for some $\kappa' \leq \kappa$ and, therefore, one can choose $s_{2n-1} \in \mathbb{R}$ such that $Q^{\{2n\}}(z) = o(1)$ as $z\widehat{\rightarrow}\infty$. Let us emphasize that every function $Q(z)$ from the class $\mathbf{N}_{\kappa,-n}$ (both for odd and even n) has the following asymptotic expansion

$$Q(z) = \gamma - \frac{s_0}{z^1} - \dots - \frac{s_n}{z^{n+1}} + o\left(\frac{1}{z^{n+1}}\right), \quad z\widehat{\rightarrow}\infty. \tag{5.7}$$

Theorem 5.2 *Let $Q(z)$ be the Q-function of S and A with $Q(z) \in \mathbf{N}_{\kappa,0}$ and let S_F be the generalized Friedrichs extension of S, whose Q-function is $Q_F(z)$. Then the following assertions are equivalent:*

(i) *There is a polynomial $P_{n+1}(z)$ of degree $n+1$ such that*

$$Q_F(z) = \frac{1}{\gamma - Q(z)} = P_{n+1}(z) + o(z); \tag{5.8}$$

(ii) *$Q(z)$ has the following asymptotic expansion with $s_n \neq 0$:*

$$Q(z) = \gamma - \frac{s_n}{z^{n+1}} - \dots - \frac{s_{2n}}{z^{2n+1}} + o\left(\frac{1}{z^{2n+1}}\right), \quad z\widehat{\rightarrow}\infty; \tag{5.9}$$

(iii) *$Q(z)$ has the operator representation $Q(z) = \gamma + [(A - z)^{-1}\omega, \omega]$ with $\omega \in \operatorname{dom} A^n$, $[A^i\omega, \omega] = 0$ for $i = 0, \dots, n-1$, and $[A^n\omega, \omega] \neq 0$.*

Proof: Assume that $Q(z)$ has the asymptotic expansion (5.9). Then

$$z^{n+1}(\gamma - Q(z)) = s_n + \cdots + \frac{s_{2n}}{z^n} + o\left(\frac{1}{z^n}\right), \quad z\widehat{\to}\infty. \tag{5.10}$$

This implies

$$\begin{aligned} \frac{Q_F(z)}{z^{n+1}} &= \left(s_n + \cdots + \frac{s_{2n}}{z^n} + o\left(\frac{1}{z^n}\right)\right)^{-1} \\ &= d_0 + \frac{d_1}{z} + \cdots + \frac{d_n}{z_n} + o\left(\frac{1}{z^n}\right), \quad z\widehat{\to}\infty, \end{aligned} \tag{5.11}$$

for some real $d_0 = 1/s_n, d_1, \dots, d_n$. Hence one obtains the equality (5.8) with $P_{n+1}(z) = d_0 z^{n+1} + \cdots + d_n$.

Conversely, the equality (5.8) implies (5.11), which, in turn, yields (5.10). This proves the equivalence (i) $\Longleftrightarrow$ (ii). The equivalence (ii) $\Longleftrightarrow$ (iii) is now implied by Proposition 5.1. □

Each $Q(z) \in \mathbf{N}_{\kappa,0}$ has the representation (5.4) with $\omega \in \mathfrak{H}$ and an asymptotic expansion as in Proposition 5.1. Observe, that if $[\omega, \omega] = 0$ and $\omega \notin \operatorname{dom} A$, then the conditions (i)–(iii) in Theorem 5.2 do not hold. Therefore, in this case the root subspace $\mathsf{S}_\infty(S_F)$ is degenerate and the perturbations $A(\alpha)$ are not regular. The following result gives a simple criterion for the nondegeneracy of the root subspace $\mathsf{S}_\infty(S_F)$ by means of the subclasses $\mathbf{N}_{\kappa,-n}$.

Corollary 5.3 *Let the assumptions be as in Theorem 5.2 and let $Q(z) \in \mathbf{N}_{\kappa,-n}$ and $s_j = 0$ if $j < n$, $s_n \neq 0$. Then the root subspace $\mathsf{S}_\infty(S_F)$ is nondegenerate if and only if $Q(z) \in \mathbf{N}_{\kappa,-2n}$. In this case* $\dim(\mathsf{S}_\infty(S_F)) = n+1$, $\kappa_\infty(Q_F) = [(n+1)/2]$ *if $s_n > 0$ and $\kappa_\infty(Q_F) = [n/2] + 1$ if $s_n < 0$.*

Proof: It follows from (5.7) and the assumptions of the corollary that $Q(z)$ admits the asymptotic expansion

$$Q(z) = \gamma - \frac{s_n}{z^{n+1}} + o\left(\frac{1}{z^{n+1}}\right), \quad z\widehat{\to}\infty. \tag{5.12}$$

This implies that

$$Q_F(z) = \frac{1}{\gamma - Q(z)} = \frac{z^{n+1}}{s_n} + o(z^{n+1}), \quad z\widehat{\to}\infty. \tag{5.13}$$

Now assume that $\mathsf{S}_\infty(S_F)$ is nondegenerate. Then by Theorem 4.1 $Q_F(z) = R(z) + o(z)$, where $R(z)$ is a polynomial. This representation and (5.13) imply that

$\deg R(z) = n+1$. In view of Theorem 5.2, $Q(z)$ admits the asymptotic expansion (5.9) and, therefore, $Q(z) \in \mathbf{N}_{\kappa,-2n}$.

Conversely, assume that $Q(z) \in \mathbf{N}_{\kappa,-2n}$. Then $Q(z)$ admits the asymptotic expansion (5.9). Hence, by Theorem 4.1 and Theorem 5.2, the space $\mathsf{S}_\infty(S_F)$ is nondegenerate. It follows from (5.8) and Proposition 2.1 that $\dim(\mathsf{S}_\infty(S_F)) = \deg P(z) = n+1$. The last statement is now implied by Corollary 4.2. □

The subclasses $\mathbf{N}_{\kappa,-2n}$ appearing in Theorem 5.2 and Corollary 5.3 reflect the spectral behaviour of the perturbations in (5.5). The next result shows that for each such perturbation $A(\alpha)$ with α finite, the corresponding Q-functions all belong to the one and the same subclass $\mathbf{N}_{\kappa,-2n}$.

Proposition 5.4 *Let $Q_\alpha(z)$ be the Q-function of $A(\alpha)$, $\alpha \in \mathbb{R}$, and S in (5.6). If $Q_\alpha(z) \in \mathbf{N}_{\kappa,-2n}$ for some $\alpha \in \mathbb{R}$, then $Q_\alpha(z) \in \mathbf{N}_{\kappa,-2n}$ and the asymptotic expansions in (5.7) have the same order for every $\alpha \in \mathbb{R}$. Moreover, if for some $\alpha \in \mathbb{R}$ the moments $s_0(Q_\alpha), \dots, s_k(Q_\alpha)$, $k \le 2n$, are equal to zero, then the same holds for the moments of every $Q_\alpha(z)$ with $\alpha \in \mathbb{R}$.*

The proof of this result can be carried out along the lines of [4, Proposition 6.1]; for the Hilbert space case see also [12, Corollary 4.3]. The next result generalizes [12, Theorem 5.2] and [12, Theorem 5.3] for the present situation.

Theorem 5.5 *Assume that the Q-function $Q(z)$ of S and A belongs to $\mathbf{N}_{\kappa,-2n}$. Then*

$$\text{dom } A(\alpha)^{n+1} = \text{dom } A^{n+1}, \quad \alpha \in \mathbb{R}. \tag{5.14}$$

Conversely, if for two different selfadjoint operator extensions A_1 and A_2 of S the inclusion

$$\text{dom } A_1^{n+1} \supset \text{dom } A_2^{n+1}$$

is satisfied, then the identity (5.14) holds and the corresponding Q-functions all belong to $\mathbf{N}_{\kappa,-2n}$.

Moreover, the perturbations $A(\alpha)$ in (5.5) are regular if and only if $[A^k\omega, \omega] \neq 0$ for some $k \le n$.

Proof: The proof of the first part is carried out along the same lines as in [12], using Proposition 5.4 and [4, Theorem 6.3]. The last statement follows from Corollary 5.3. □

6 Spectral Characterizations of Regular Rank One Perturbations

In this section the regularity of rank one perturbations will be characterized in purely analytic terms, i.e. in terms of the spectral measures of the Q-functions of the perturbations. For this purpose recall the factorized integral representation for generalized Nevanlinna functions, cf. [4, Corollary 3.4], [7]. Let $Q(z) \in \mathbf{N}_\kappa$, $Q(z) \not\equiv 0$, and let $p(z)$ and $q(z)$ be real normed polynomials, nonnegative on $\mathbb{R}$, formed by the finite (generalized) poles and zeros (of nonpositive type) of $Q(z)$. Then there exists a (unique) Nevanlinna function $Q^0(z) \in \mathbf{N}$, such that

$$Q(z) = \frac{p(z)}{q(z)}\, Q^0(z). \tag{6.1}$$

Therefore, each function $Q(z) \in \mathbf{N}_\kappa$ admits the unique integral representation

$$Q(z) = \frac{p(z)}{q(z)} \left(a + bz + \int_{\mathbb{R}} \left(\frac{1}{t-z} - \frac{t}{t^2+1} \right) d\sigma(t) \right), \tag{6.2}$$

where the polynomials $p(z)$ and $q(z)$ are as in (6.1), $a \in \mathbb{R}$, $b \geq 0$, and $\sigma(t)$ is a nondecreasing left-continuous function on $\mathbb{R}$ satisfying

$$\int_{\mathbb{R}} \frac{d\sigma(t)}{t^2+1} < \infty. \tag{6.3}$$

Conversely, every function of the form (6.2) belongs to $\mathbf{N}_\kappa$, where

$$\kappa = \frac{1}{2} \max\{\deg p,\ \deg q\},$$

cf. [4, Proposition 3.2].

Let $Q(z)$ belong to $\mathbf{N}_\kappa$ and consider the asymptotic behaviour of $Q(z)$ as $z \widehat{\rightarrow} \infty$. It is an immediate consequence of the integral representation (6.2) that the multiplicity $\kappa_\infty = \kappa_\infty(Q)$ of ∞ as a generalized pole of $Q(z)$ determines the following growth conditions for $Q(z)$ as $z \widehat{\rightarrow} \infty$:

$$Q(z) = O(z^{2\kappa_\infty + 1}), \quad z \widehat{\rightarrow} \infty, \tag{6.4}$$

and

$$Q(z) \neq o(z^{2\kappa_\infty - 1}), \quad z \widehat{\rightarrow} \infty, \tag{6.5}$$

if $Q(z) \not\equiv 0$. The asymptotic behaviour specified in (5.1) for functions in $\mathbf{N}_\kappa$ is associated to the nondegeneracy of the root subspace $\mathsf{S}_\infty(S_F)$ of the generalized Friedrichs extension S_F at ∞ by Theorem 4.1. The next result shows that the asymptotics in (5.1) and the nondegeneracy of $\mathsf{S}_\infty(S_F)$ are connected to certain integrability properties of the measure $d\sigma(t)$ in the integral representation (6.2) of the associated Q-function $Q_F(z)$. They are determined by the growth of $Q_F(z)$ at ∞, cf. (6.4), (6.5).

Theorem 6.1 *Let the assumptions be as in Theorem* 5.2, *and let* $\kappa_\infty = \kappa_\infty(Q_F) > 0$. *Then the root subspace* $\mathsf{S}_\infty(S_F)$ *is nondegenerate if and only if in the integral representation* (6.2) *of* $Q_F(z)$ *the measure* $d\sigma(t)$ *satisfies*

$$\int_{\mathbb{R}} t^{2(\kappa_\infty - 1)}\, d\sigma(t) < \infty. \tag{6.6}$$

Proof: Assume that $\mathsf{S}_\infty(S_F)$ is nondegenerate. Then Theorem 5.2 and the integral representation (6.2) give

$$(6.7)\qquad \begin{aligned} Q_F(z) &= P(z)+Q_2(z) \\ &= \frac{p(z)}{q(z)}\left(a+bz+\int_{\mathbb{R}}\left(\frac{1}{t-z}-\frac{t}{t^2+1}\right)d\sigma(t)\right). \end{aligned}$$

Here $Q_2(z)=o(z)$ so that $\kappa_\infty(Q_2)=0$ and thus it has a factorization of the form

$$(6.8)\qquad Q_2(z)=\frac{c(z)}{d(z)}\,Q_2^0(z),\quad \deg c(z)\le \deg d(z).$$

Denote by $d\sigma_2(t)$ the measure associated to $Q_2(z)$ as in (6.2). Then the generalized Stieltjes inversion formula applied to the representations of $Q_F(z)$ in (6.7) yields

$$(6.9)\qquad \frac{c(t)}{d(t)}\,d\sigma_2(t)=\frac{p(t)}{q(t)}\,d\sigma(t),$$

for all $t\in\mathbb{R}$ with $|t|$ large enough. Since deg $c(z)\le$ deg $d(z)$ and deg $p(z)-$ deg $q(z)=2\kappa_\infty$, (6.3) and (6.9) imply

$$\int_{\mathbb{R}}\frac{t^{2\kappa_\infty}}{t^2+1}\,d\sigma(t)<\infty.$$

This gives (6.6).

Conversely, assume that (6.6) holds. The integral representation (6.2) for $Q_F(z)$ can be rewritten as

$$Q_F(z)=\frac{p(z)}{q(z)}\left(\int_{\mathbb{R}}\frac{1+tz}{t-z}\frac{d\sigma(t)}{1+t^2}+a+bz\right),\quad a\in\mathbb{R},\quad b\ge 0.$$

Here deg $p(z)-$ deg $q(z)=2\kappa_\infty$. Let $p(z)=p_1(z)p_2(z)$ be factorized such that deg $p_1(z)=$ deg $q(z)$ and deg $p_2(z)=2\kappa_\infty$. Now a simple calculation, using the identity

$$\frac{p_2(z)-p_2(t)}{t-z}=h(t,z),\quad \deg_{t,z}h(t,z)=2\kappa_\infty-1,$$

leads to

$$(6.10)\qquad Q_F(z)=\frac{p_1(z)}{q(z)}\int_{\mathbb{R}}\frac{1+tz}{t-z}\,d\mu(t)+\frac{k(z)}{q(z)},$$

where deg $p(z)\le$ deg $k(z)\le 1+$ deg $p(z)$ and

$$\int_{\mathbb{R}}d\mu(t)=\int_{\mathbb{R}}\frac{p_2(t)}{t^2+1}\,d\sigma(t)<\infty.$$

Clearly, one can write $k(z) = k_1(z) + q(z)P(z)$, where all the polynomials are real and satisfy deg $k_1(z) <$ deg $q(z)$, deg $P(z) =$ deg $k(z) -$ deg $q(z) \leq 2\kappa_\infty + 1$. Substituting this form of $k(z)$ in (6.10) shows that $Q_F(z)$ admits the representation

$$Q_F(z) = P(z) + Q_2(z); \quad Q_2(z) = o(z), \quad z\widehat{\to}\infty.$$

Therefore, by Theorem 4.1, the root subspace $\mathsf{S}_\infty(S_F)$ is nondegenerate. This completes the proof. □

Next regular perturbations of A are characterized by means of the corresponding Q-function in (5.4). For this purpose, recall the following result concerning the classes $\mathbf{N}_{\kappa,-l}$, $l \in \mathbb{Z}_+$, of generalized Nevanlinna functions, based on the integral representation (6.2) (see [6]):

Proposition 6.2 *Let a generalized Nevanlinna function $Q(z)$ admit the integral representation* (6.2) *and let ∞ be the generalized zero of nonpositive type of $Q(z)$ of multiplicity $\pi_\infty = \pi_\infty(Q)$. Then $Q(z) \in \mathbf{N}_{\kappa,-l}$ if and only if*

$$\int_{\mathbb{R}} (1 + |t|)^{l-2\pi_\infty} d\sigma(t) < \infty. \tag{6.11}$$

For each $\gamma \in \mathbb{R}$, the function $Q(z) \in \mathbf{N}_\kappa$ also admits the unique integral representation

$$\begin{aligned} Q(z) &= \gamma + \frac{q(z)}{p(z)} \\ &\left(\alpha + \beta z + \int_{\mathbb{R}} \left(\frac{1}{t-z} - \frac{t}{t^2+1}\right) d\rho(t)\right), \end{aligned} \tag{6.12}$$

where $\alpha \in \mathbb{R}$, $\beta \geq 0$, and $\rho(t)$ is a nondecreasing left-continuous function on $\mathbb{R}$ satisfying

$$\int_{\mathbb{R}} \frac{d\rho(t)}{t^2+1} < \infty. \tag{6.13}$$

Theorem 6.3 *Let the assumptions be as in Theorem 5.2. Assume that $\pi_\infty = \pi_\infty(Q - \gamma) > 0$ and let $n \in \mathbb{Z}_+$ be the largest integer such that*

$$d = \lim_{z\widehat{\to}\infty} z^{n+1}(Q(z) - \gamma) \quad (\in \mathbb{R})$$

exists. Then the root subspace $\mathsf{S}_\infty(S_F)$ is nondegenerate if and only if $d \neq 0$ and in the integral representation (6.12) *of $Q(z)$ the measure $d\rho(t)$ satisfies*

$$\int_{\mathbb{R}} (1 + |t|)^{2(n-\pi_\infty)} d\rho(t) < \infty. \tag{6.14}$$

Proof: Since $\pi_\infty > 0$, the integer n is well defined and the definition of π_∞ shows that, in fact, $2\pi_\infty - 2 \leq n \leq 2\pi_\infty$. It follows from the asymptotic expansion (5.7) and Theorem 5.2 that the root subspace $\mathsf{S}_\infty(S_F)$ is nondegenerate if and only if $d \neq 0$ and $Q(z) \in \mathbf{N}_{\kappa,-2n}$. By Proposition 6.2, $Q(z) \in \mathbf{N}_{\kappa,-2n}$ or equivalently, $Q(z) - \gamma \in \mathbf{N}_{\kappa,-2n}$, if and only if (6.14) is satisfied. This proves the theorem. □

The connection of Theorem 6.3 to the rank one perturbations of A induced by $\omega \in \mathfrak{H}$ is via $Q(z) = \gamma + [(A - z)^{-1}\omega, \omega]$, see Theorem 5.2. In view of Proposition 5.4, a characterization for the regularity of the rank one perturbations, similar to the one in Theorem 6.3, can be based on the Q-function of each perturbation $A(\alpha)$, $\alpha \neq 0$.

The above results can also be formulated as criteria for the regularity or nonsingularity of ∞ as a spectral point of a selfadjoint relation A in $\mathfrak{H}$; see [5], [6] for some related results and applications in this direction.

Acknowledgements

Part of the present research was supported by the Academy of Finland and the Dutch Organization for Scientific Research NWO (SWON).

References

[1] S. Albeverio and P. Kurasov, Singular perturbations of differential operators (Solvable Schrödinger type operators), (to appear).

[2] J. Bognar, Indefinite inner product spaces, *Ergeb. Math. Grenzgeb., Bd.* 78, Springer-Verlag, New York-Heidelberg, 1974.

[3] K. Daho and H. Langer, "Matrix functions of the class $\mathbf{N}_\kappa$", *Math. Nachr.* **120** (1985), 275–294.

[4] V. Derkach, S. Hassi and H.S.V. de Snoo,"Operator models associated with Kac subclasses of generalized Nevanlinna functions", Methods of Functional Analysis and Topology, **5** (1999), 65–87.

[5] V. Derkach, S. Hassi and H.S.V. de Snoo, "Rank one perturbations in a Pontryagin space with one negative square", (in preparation).

[6] V. Derkach, S. Hassi and H.S.V. de Snoo, "Asymptotic expansions of generalized Nevanlinna functions", (in preparation).

[7] A. Dijksma, H. Langer, A. Luger and Yu. Shondin, "A factorization result for generalized Nevanlinna functions of the class $\mathbf{N}_\kappa$", preprint 1998.

[8] A. Dijksma, H. Langer and H.S.V. de Snoo, "Eigenvalues and pole functions of Hamiltonian systems with eigenvalue depending boundary conditions", *Math. Nachr.* **161** (1993), 107–154.

[9] F. Gesztesy and B. Simon, "Rank one perturbations at infinite coupling", *J. Funct. Anal.* **128** (1995), 245–252.

[10] S. Hassi, M. Kaltenbäck and H.S.V. de Snoo, "The sum of matrix Nevanlinna functions and selfadjoint extensions in exit spaces", *Oper. Theory Adv. Appl.* **103** (1998), 137–154.

[11] S. Hassi and H.S.V. de Snoo, "On rank one perturbations of selfadjoint operators", Integral Equations Operator Theory, **29** (1997), 288–300.

[12] S. Hassi, H.S.V. de Snoo and A.D.I. Willemsma, "Smooth rank one perturbations of selfadjoint operators", *Proc. Amer. Math. Soc.* **126** (1998), 2663–2675.

[13] P. Jonas and H. Langer, "Some questions in the perturbation theory of J-nonnegative operators in Kreĭn spaces", *Math. Nachr.* **114** (1983), 205–226.

[14] P. Jonas and H. Langer, "Selfadjoint extensions of a closed linear relation of defect one in a Kreĭn space", *Oper. Theory Adv. Appl.* **80** (1995), 176–205.

[15] A. Kiselev and B. Simon, "Rank one perturbations with infinitesimal coupling", *J. Funct. Anal.* **130** (1995), 345–356.

[16] M.G. Kreĭn and H. Langer, "Über die Q-function eines π-hermiteschen Operators in Raume Π_κ", *Acta. Sci. Math. (Szeged)* **34** (1973), 191–230.

[17] M.G. Kreĭn and H. Langer, "Über einige Fortsetzungsprobleme, die eng mit der Theorie hermitescher operatoren im Raume Π_κ zusammenhangen. I. Einige Funktionenklassen und ihre Dahrstellungen", *Math. Nachr.* **77** (1977), 187–236.

[18] M.G. Kreĭn and H. Langer, "Some propositions on analytic matrix functions related to the theory of operators in the space Π_κ", *Acta Sci. Math.* **43** (1981), 181–205.

[19] H. Langer, "Spectral functions of definitizable operators in Kreĭn spaces", Functional analysis, Proceedings of a conference held at Dubrovnik, Yugoslavia 1981; Lecture Notes in Mathematics 948, Springer-Verlag, Berlin, 1982, 1–46.

[20] H. Langer, "A characterization of generalized zeros of negative type of functions of the class $\mathbf{N}_\kappa$", *Oper. Theory Adv. Appl.* **17** (1986), 201–212.

[21] Yu.G. Shondin, "Quantum-mechanical models in R_n associated with extensions of the energy operator in Pontryagin space", *Theor. Math. Phys.* **74** (1988), 220–230.

[22] B. Simon, "Spectral analysis of rank one perturbations and applications", in J. Feldman, R. Froese and L.M. Rosen (editors), Proceedings on Mathematical Quantum Theory II: Schrödinger operators, CRM Proceedings and Lecture Notes, vol. 8, *Amer. Math. Soc.*, Providence, R.I., 1995.

Vladimir Derkach
Department of Mathematics
Donetsk State University
Universitetskaya str. 24
340055 Donetsk
Ukraine
derkach@univ.donetsk.ua

Seppo Hassi
Department of Statistics
University of Helsinki
PL 54, 00014 Helsinki
Finland
hassi@cc.helsinki.fi

Henk de Snoo
Department of Mathematics
University of Groningen
Postbus 800, 9700 AV Groningen
Nederland
desnoo@math.rug.nl

AMS Classification numbers: Primary 46C20, 47A55; Secondary 47A57, 47B15, 47B25.

Operator Theory:
Advances and Applications, Vol. 122
© 2001 Birkhäuser Verlag Basel/Switzerland

Extensions of Matrix-valued Inner Products on Modules and the Inversion Formula for Block Toeplitz Matrices

Robert L. Ellis and Israel Gohberg

Dedicated to Marinus A. Kaashoek with admiration and friendship on the occasion of his sixtieth birthday.

In this paper we continue the development started in [EGL] of Hilbert modules over the algebra of $r \times r$ complex matrices, with the ultimate goal of obtaining a new proof of the Gohberg-Heinig inversion formula for positive definite block Toeplitz matrices. The approach was motivated by ideas from prediction theory as suggested in [GL].

Introduction

In this paper we develop an abstract version of a structure appearing in prediction theory for stochastic stationary processes. (See [GL].) This structure associates with every positive definite block Toeplitz matrix two orthonormal bases that are used to construct the inverse of the matrix. We first consider the simpler, scalar case and then proceed to the block case.

We begin with a sequence $x_0, x_1, \ldots$ of vectors in a complex vector space, along with a positive definite scalar product $\langle\, ,\, \rangle$ on the span V of $x_0, \ldots, x_n$ for some positive integer n. It is assumed that $\langle x_k, x_j \rangle$ depends only on the difference $j-k$, which means that the Gram matrix $(\langle x_k, x_j \rangle)_{0 \le j,k \le n}$ is a positive definite Toeplitz matrix. Then we extend the scalar product to span $\{x_0, x_1, \ldots\}$ in such a way that for all $j, k \ge 0$, $\langle x_k, x_j \rangle$ still depends only on $j-k$, and so that the best approximation of x_0 in span$\{x_1, \ldots, x_m\}$ for the norm determined by the scalar product does not improve when m is increased beyond n. In the process, two orthonormal bases are constructed, one for the completion of span $\{x_0, x_1, \ldots\}$ and the other for the orthogonal complement of V in the completion.

In the proof of the formula for the inverse of a given positive definite Toeplitz matrix $(\gamma_{j-k})_{0 \le j,k \le n}$, we use the preceding structure for the sequence $\{e_0, e_1, \ldots\}$ in $\ell_2(0, \infty)$ with the scalar product determined by specifying

$$\langle e_k, e_j \rangle = \gamma_{j-k} \qquad (0 \le j, k \le n)$$

For the $r \times r$ block case, we replace the sequence in a complex vector space by a sequence $x_0, x_1, \ldots$ in a module over the algebra $\mathcal{M}$ of $r \times r$ complex matrices. Correspondingly, we replace the scalar product by a positive definite $r \times r$

matrix-valued inner product $\langle\,,\,\rangle$ on the module spanned by $x_0, \ldots, x_n$. We assume that $\langle x_k, x_j\rangle$ depends only on $j-k$. Then the matrix-valued inner product is extended to the module spanned by $x_0, x_1, \ldots$ so that not only does $\langle x_k, x_j\rangle$ still depend only on $j-k$, but also so that the appropriate type of best approximation of x_0 in the submodule spanned by $x_1, \ldots, x_m$ does not improve when m is increased beyond n. As in the scalar case, two orthonormal bases are constructed that are used to produce the formula for the inverse of a positive definite block Toeplitz matrix.

1 The Scalar Case

Throughout this section, $\mathcal{X}$ will be a complex vector space and $x_0, x_1, \ldots$ will be a linearly independent sequence of vectors in $\mathcal{X}$, by which we mean that $x_0, x_1, \ldots, x_m$ are linearly independent for every nonnegative integer m. The subspace of $\mathcal{X}$ spanned by $x_0, x_1, \ldots$ will be denoted by $\operatorname{span}\{x_0, x_1, \ldots\}$.

For a given positive integer n, a positive definite scalar product $\langle\,,\,\rangle$ on $\operatorname{span}\{x_0, x_1, \ldots\}$ is said to be *n-informative* for the sequence $x_0, x_1, \ldots$ if for every $k \geq 0$, the projection of x_0 onto $\operatorname{span}\{x_1, \ldots, x_{n+k}\}$ is in $\operatorname{span}\{x_1, \ldots, x_n\}$. This may also be expressed by saying that for every $k \geq 0$, the best approximation of x_0 in $\operatorname{span}\{x_1, \ldots, x_{n+k}\}$ is the same as the best approximation of x_0 in $\operatorname{span}\{x_1, \ldots, x_n\}$.

A positive definite scalar product $\langle\,,\,\rangle$ is *stationary* for $x_0, x_1, \ldots$ if $\langle x_k, x_j\rangle$ depends only on the difference $j-k$. Clearly if $\langle\,,\,\rangle$ is both n-informative and stationary for $x_0, x_1, \ldots$, then for all nonnegative integers m and k, the projection of x_m onto $\operatorname{span}\{x_{m+1}, \ldots, x_{m+n+k}\}$ is in $\operatorname{span}\{x_{m+1}, \ldots, x_{m+n}\}$.

The motivation for the preceding definitions comes from prediction theory for stationary stochastic processes. (See [GL] and [L].) There the vectors $x_0, x_1, \ldots$ represent random vairables and $\langle x_k, x_j\rangle$ is the expected value of $\bar{x}_j x_k$. The projection of x_0 onto $\operatorname{span}\{x_1, \ldots, x_n\}$ is called the *forward prediction* of x_0 by $x_1, \ldots, x_n$. The *backward prediction* of x_n by $x_0, x_1, \ldots, x_{n-1}$ is defined analogously.

1.1 Extension of a Scalar Product

In this subsection, our main concern is the following: Suppose n is a given positive integer and $\langle\,,\,\rangle$ is a positive definite scalar product on $\operatorname{span}\{x_0, \ldots, x_n\}$ with the property that $\langle x_k, x_j\rangle$ depends only on $j-k$ for $0 \leq j, k \leq n$. We would like to extend $\langle\,,\,\rangle$ to a positive definite scalar product on $\operatorname{span}\{x_0, x_1, \ldots\}$ that is both n-informative and stationary for $x_0, x_1, \ldots$. In Theorem 1.2 we will prove that such an extension always exists and is unique. The following lemma will be used. We denote by $G(x_0, \ldots, x_n)$ the Gram matrix $(\langle x_k, x_j\rangle)_{0\leq j,k\leq n}$. The conjugate of a complex number v is denoted by $\bar{v}$.

Lemma 1.1 *Let $\langle\, , \,\rangle$ be a positive definite scalar product on span$\{x_0, \ldots, x_n\}$ with*

$$\langle x_k, x_j \rangle = \gamma_{j-k} \text{ for } 0 \le j, k \le n. \tag{1.1}$$

a. *The best approximation of x_0 in span$\{x_1, \ldots, x_n\}$ for $\langle\, , \,\rangle$ is the vector $-\sum_{j=1}^{n} v_j x_j$, where $v_1, \ldots, v_n$ are the unique complex numbers that satisfy*

$$G(x_0, \ldots, x_n) \begin{pmatrix} 1 \\ v_1 \\ \vdots \\ v_n \end{pmatrix} = \begin{pmatrix} \nu \\ 0 \\ \vdots \\ 0 \end{pmatrix} \tag{1.2}$$

for some positive number ν.

b. *The best approximation of x_n in span$\{x_0, \ldots, x_{n-1}\}$ for $\langle\, , \,\rangle$ is the vector $-\sum_{j=0}^{n-1} \bar{v}_{n-j} x_j$, where $v_1, \ldots, v_n$ are as in part* (a).

Proof:

a. Let $-\sum_{j=1}^{n} v_j x_j$ be the best approximation of x_0 in span$\{x_1, \ldots, x_n\}$ for $\langle\, , \,\rangle$. This means that

$$\begin{aligned} &\left\langle x_0 + \sum_{j=1}^{n} v_j x_j,\ x_0 + \sum_{j=1}^{n} v_j x_j \right\rangle \\ &\le \left\langle x_0 + \sum_{j=1}^{n} t_j x_j,\ x_0 + \sum_{j=1}^{n} t_j x_j \right\rangle \end{aligned} \tag{1.3}$$

for all complex numbers $t_1, \ldots, t_n$. Then $x_0 + \sum_{j=1}^{n} v_j x_j$ is orthogonal to $x_1, \ldots, x_n$, so

$$\langle x_0, x_k \rangle + \sum_{j=1}^{n} v_j \langle x_j, x_k \rangle = 0 \qquad (1 \le k \le n)$$

This implies that

$$\gamma_k + \sum_{j=1}^{n} v_j \gamma_{k-j} = 0 \qquad (1 \le k \le n) \tag{1.4}$$

where $\gamma_{-n}, \ldots, \gamma_n$ are as in (1.1). The equations in (1.4) can be written as

$$\begin{pmatrix} \gamma_1 & \gamma_0 & \cdots & \gamma_{-n+1} \\ \gamma_2 & \gamma_1 & \cdots & \gamma_{-n+2} \\ \vdots & & & \\ \gamma_n & \gamma_{n-1} & \cdots & \gamma_0 \end{pmatrix} \begin{pmatrix} 1 \\ v_1 \\ \vdots \\ v_n \end{pmatrix} = \begin{pmatrix} 0 \\ 0 \\ \vdots \\ 0 \end{pmatrix} \tag{1.5}$$

Letting $\nu = \gamma_0 + \gamma_{-1} v_1 + \cdots + \gamma_{-n} v_n$, we have

$$\begin{pmatrix} \gamma_0 & \gamma_{-1} & \cdots & \gamma_{-n} \\ \gamma_1 & \gamma_0 & \cdots & \gamma_{-n+1} \\ \vdots & & & \\ \gamma_n & \gamma_{n-1} & \cdots & \gamma_0 \end{pmatrix} \begin{pmatrix} 1 \\ v_1 \\ \vdots \\ v_n \end{pmatrix} = \begin{pmatrix} \nu \\ 0 \\ \vdots \\ 0 \end{pmatrix} \tag{1.6}$$

which is the same as (1.2). Since $G(x_0, \ldots, x_n)$ is positive definite,

$$\nu = (1\ \bar{v}_1 \ldots \bar{v}_n) G(x_0, \ldots, x_n) \begin{pmatrix} 1 \\ v_1 \\ \vdots \\ v_n \end{pmatrix} > 0$$

Since (1.2) implies (1.5), which determines the unique best approximation of x_0 in $\operatorname{span}\{x_1, \ldots, x_n\}$, there can be only one positive number ν for which there are solutions $v_1, \ldots, v_n$ of (1.2).

b. The best approximation of x_n in $\operatorname{span}\{x_0, \ldots, x_{n-1}\}$ is the vector $-\sum_{j=0}^{n-1} w_j x_j$ such that

$$\begin{aligned} &\left\langle x_n + \sum_{j=0}^{n-1} w_j x_j,\ x_n + \sum_{j=0}^{n-1} w_j x_j \right\rangle \\ &\leq \left\langle x_n \sum_{j=0}^{n-1} t_j x_j,\ x_n + \sum_{j=0}^{n-1} t_j x_j \right\rangle \end{aligned} \tag{1.7}$$

for all complex numbers $t_0, \ldots, t_{n-1}$. Then $x_n + \sum_{j=0}^{n-1} w_j x_j$ is orthogonal to $x_0, \ldots, x_{n-1}$, i.e.

$$\langle x_n, x_k \rangle + \sum_{j=0}^{n-1} w_j \langle x_j, x_k \rangle = 0 \qquad (0 \leq k \leq n-1)$$

This implies that

$$\begin{pmatrix} \gamma_0 & \gamma_{-1} & \cdots & \gamma_{-n} \\ \gamma_1 & \gamma_0 & \cdots & \gamma_{-n+1} \\ \vdots & & & \\ \gamma_{n-1} & \gamma_{n-2} & \cdots & \gamma_{-1} \end{pmatrix} \begin{pmatrix} w_0 \\ \vdots \\ w_{n-1} \\ 1 \end{pmatrix} = \begin{pmatrix} 0 \\ \vdots \\ 0 \\ 0 \end{pmatrix}$$

Letting $\omega = \gamma_n w_0 + \cdots + \gamma_1 w_{n-1} + \gamma_0$, we find that

$$\begin{pmatrix} \gamma_0 & \gamma_{-1} & \cdots & \gamma_{-n} \\ \gamma_1 & \gamma_0 & \cdots & \gamma_{-n+1} \\ \vdots & & & \\ \gamma_n & \gamma_{n-1} & \cdots & \gamma_0 \end{pmatrix} \begin{pmatrix} w_0 \\ \vdots \\ w_{n-1} \\ 1 \end{pmatrix} = \begin{pmatrix} 0 \\ \vdots \\ 0 \\ \omega \end{pmatrix} \tag{1.8}$$

Then

$$\omega = (\overline{w}_0 \dots \overline{w}_{n-1}\ 1) G(x_0, \dots, x_n) \begin{pmatrix} w_0 \\ \vdots \\ w_{n-1} \\ 1 \end{pmatrix} > 0$$

Let R be the $(n+1) \times (n+1)$ matrix defined by

$$R = \begin{pmatrix} 0 & & & 1 \\ & & \cdot & \\ & \cdot & & \\ 1 & & & 0 \end{pmatrix}$$

Then $R^2 = I$ and (1.6) implies that

$$RG(x_0, \dots, x_n) RR \begin{pmatrix} 1 \\ v \\ \vdots \\ v_n \end{pmatrix} = R \begin{pmatrix} \nu \\ 0 \\ \vdots \\ 0 \end{pmatrix}$$

so that

$$RG(x_0, \dots, x_n) R \begin{pmatrix} v_n \\ \vdots \\ v_1 \\ 1 \end{pmatrix} = \begin{pmatrix} 0 \\ \vdots \\ 0 \\ \nu \end{pmatrix}$$

Since

$$RG(x_0, \dots, x_n) R = \overline{G(x_0, \dots, x_n)}$$

it follows that

(1.9) $$G(x_0, \dots, x_n) \begin{pmatrix} \bar{v}_n \\ \vdots \\ \bar{v}_1 \\ 1 \end{pmatrix} = \begin{pmatrix} 0 \\ \vdots \\ 0 \\ \nu \end{pmatrix}$$

Comparing (1.9) and (1.8), we see that $\omega = \nu$ and $w_k = \overline{v_{n-k}}$ for $0 \leq k \leq n-1$, which proves part b.

□

1.2 Connections with Prediction Theory and the Extension Theorem

Lemma 1.1 is related to prediction theory for stationary stochastic processes, as developed in [GL]. There one has a sequence $\{X_k \mid k \geq 0\}$ of random variables with a family of probability distributions for the joint occurrances of values for finite subsets of $\{X_k \mid k \geq 0\}$. It is assumed that $\{X_k \mid k \geq 0\}$ is stationary, which means that the expected value of $\overline{X}_k X_j$ depends only on the difference $j-k$. When values of $X_1, \ldots, X_n$ are given, the *backward prediction* of X_0 is $-\sum_{k=1}^{n} v_k X_k$, where $v_1, \ldots, v_n$ are the complex numbers such that

$$\mathcal{E}\left(\left|X_0 + \sum_{k=1}^{n} v_k X_k\right|^2\right) \leq \mathcal{E}\left(\left|X_0 + \sum_{k=1}^{n} t_k X_k\right|^2\right) \tag{1.10}$$

for all complex numbers $t_1, \ldots, t_n$. Here $\mathcal{E}(\ )$ denotes the expected value of a random variable. If we define a scalar product by specifying

$$\langle X_j, X_k \rangle = \mathcal{E}(\overline{X}_k X_j) \qquad (j, k \geq 0) \tag{1.11}$$

then $X_0, X_1, \ldots$ are stationary for $\langle\ ,\ \rangle$ and (1.10) becomes the same as (1.3). The number $\mathcal{E}(\overline{X}_k, X_j)$ is called the *covariance* of X_j and X_k. Similarly, when values of $X_0, \ldots, X_{n-1}$ are given, the *forward prediction* of X_n is $-\sum_{k=0}^{n-1} w_k X_k$, where $w_0, \ldots, w_{n-1}$ are complex numbers such that

$$\mathcal{E}\left(\left|X_n + \sum_{k=0}^{n-1} w_k X_k\right|^2\right) \leq \mathcal{E}\left(\left|X_n + \sum_{k=0}^{n-1} t_k X_k\right|^2\right)$$

for all complex numbers $t_0, \ldots, t_{n-1}$. This corresponds to (1.7). From the results of Lemma 1.1 we can therefore state that the forward prediction of X_n from $X_0, \ldots, X_{n-1}$ can be obtained from the backward prediction of X_0 from $X_1, \ldots, X_n$ and *vice versa.* One only has to reverse the order of the coefficients $v_1, \ldots, v_n$ for the backward prediction and take their complex conjugates to obtain the coefficients for the forward prediction.

The following problem was considered in [GL]. Suppose that $X_0, X_1, \ldots$ is a sequence of random variables and the covariances $\mathcal{E}(\overline{X}_k X_j)$ are specified only for $0 \leq j, k \leq n$ with the value depending only on $j-k$. How should the covariance $\mathcal{E}(\overline{X}_k X_j)$ be specified for other values of j, k? The point of view adopted in [GL] was to specify the new covariances so that for any $k > n$, backward prediction of X_0 from $X_1, \ldots, X_k$ should be the same as the original backward prediction of X_0 from $X_1, \ldots, X_n$. In that way no new "information" is used. This was the motivation for the definition of n-informative.

Theorem 1.2 *Let $\mathcal{X}$ be a vector space, and let $x_0, x_1, \ldots$ be linearly independent vectors in $\mathcal{X}$. Let $\langle\ ,\ \rangle$ be a positive definite scalar product with $\langle x_k, x_j \rangle = \gamma_{j-k}$ for $0 \leq j, k \leq n$. Then there is a unique extension of $\langle\ ,\ \rangle$ to a positive definite* *scalar product on* span$\{x_0, x_1, \ldots\}$ *that is both* n*-informative and stationary for* $x_0, x_1, \ldots$. *This extension is obtained by specifying that*

$$\langle x_k, x_j\rangle = \gamma_{j-k} \qquad (j, k = 0, 1, \ldots)$$

where $\gamma_{-n}, \ldots, \gamma_n$ *are the given entries in* $G(x_0, \ldots, x_n)$ *and*

$$\overline{\gamma_{-(k+1)}} = \gamma_{k+1} = -\gamma_k v_1 - \cdots - \gamma_{k-n+1} v_n \qquad (k = n, n+1, \ldots)$$

Here $v_1, \ldots, v_n$ *are the solutions of* (1.2).

Proof: For any integers k and ℓ with $0 \leq k \leq \ell$ let

$$\mathcal{X}_k^\ell = \text{span}\{x_k, \ldots, x_\ell\}$$

We will prove by induction that for any $k \geq n$ the scalar product can be extended to $\mathcal{X}_1^k$ so that $\langle x_i, x_j\rangle$ depends only on $j - i$ for $0 \leq i, j \leq k$ and the best approximation of x_0 in $\mathcal{X}_1^k$ is the same as in $\mathcal{X}_1^n$.

For $k = n$ there is nothing to prove. For the inductive step suppose that $k \geq n$ and $\langle\ ,\ \rangle$ has been extended to $\mathcal{X}_0^k$ with the described properties. Lemma 1.1(a) implies that in order to extend $\langle\ ,\ \rangle$ from $\mathcal{X}_0^k$ to $\mathcal{X}_0^{k+1}$ with the desired properties, we must define $\gamma_{k+1} = \langle x_0, x_{k+1}\rangle$ and $\gamma_{-k-1} = \overline{\gamma_{k+1}}$ so that

$$\begin{pmatrix} \gamma_0 & \gamma_{-1} & \cdots & \gamma_{-k-1} \\ \gamma_1 & \gamma_0 & \cdots & \gamma_{-k} \\ \vdots & & & \\ \gamma_{k+1} & \gamma_k & \cdots & \gamma_0 \end{pmatrix} \begin{pmatrix} 1 \\ v_1 \\ \vdots \\ v_n \\ 0 \\ \vdots \\ 0 \end{pmatrix} = \begin{pmatrix} \nu \\ 0 \\ \vdots \\ 0 \end{pmatrix} \tag{1.12}$$

Thus we define

$$\gamma_{k+1} = -\gamma_k v_1 - \cdots - \gamma_{k-n+1} v_n$$

To prove that the extension is positive definite, we consider any vector

$$x = \sum_{j=0}^{k+1} a_j x_j$$

in $\mathcal{X}_0^{k+1}$ and write

$$x = y + z$$

where

$$y = a_0 \left(x_0 + \sum_{j=1}^{n} v_j x_j \right)$$

and

$$z = \sum_{j=1}^{n} (a_j - a_0 v_j) x_j + \sum_{j=n+1}^{k+1} a_j x_j$$

Then y is orthogonal to z since $x_0 + \sum_{j=1}^{n} v_j x_j$ is orthogonal to $x_1, \ldots, x_{k+1}$ by (1.12). The Pythagorean Theorem therefore implies that

$$\langle x, x \rangle = \langle y, y \rangle + \langle z, z \rangle$$

Since $x_0, \ldots, x_{k+1}$ are stationary for $\langle\ ,\ \rangle$, it follows that

$$\langle z, z \rangle = \langle w, w \rangle$$

where

$$w = \sum_{j=0}^{n-1} (a_{j+1} - a_0 v_{j+1}) x_j + \sum_{j=n}^{k} a_{j+1} x_j$$

Since $\langle y, y \rangle \geq 0$ and $\langle w, w \rangle \geq 0$ by the inductive hypothesis, we conclude that $\langle x, x \rangle \geq 0$. Also $\langle x, x \rangle = 0$ only if $\langle y, y \rangle = 0$ and $\langle w, w \rangle = 0$, which imply that $y = w = 0$. The linear independence of $x_0, \ldots, x_k$ then implies that $a_{j+1} - a_0 v_{j+1} = 0$ for $0 \leq j \leq n-1$ and $a_{j+1} = 0$ for $n \leq j \leq k$. It follows that $z = 0$, so $x = y + z = 0$. This completes the induction. □

Theorem 1.2 can be interpreted in terms of band sequences and the maximum distance problem considered in [BEGL]. Since the extension is an n-informatory extension, for any $k > n$ the best approximation of x_0 in span$\{x_1, \ldots, x_k\}$ is the same as the best approximation $-\sum_{j=1}^{n} v_j x_j$ of x_0 in span$\{x_1, \ldots, x_n\}$. Thus, by Lemma 1.1(b), the best approximation of x_k in span$\{x_0, \ldots, x_{k-1}\}$ is $-\sum_{j=0}^{n-1} \bar{v}_{n-j} x_{k-n+j}$. With different notation, this is the same as statement (2) in Proposition 2.4 in [BEGL]. It follows that $\{x_k \mid k \geq 0\}$ is an n-band sequence and hence for any $k > n$, $x_0, \ldots, x_k$ is a solution of the maximum distance problem in $\mathbf{C}^{k+1}$ when $\langle x_j, x_\ell \rangle$ is specified for $|\ell - j| \leq n$.

1.3 Properties of the Extended Scalar Product

Next we will obtain several properties of the extended scalar product that will be useful in proving the Gohberg-Semencul formula. For completeness we first prove a well-known result (See [L], for example.) that will be employed in proving the first property.

Lemma 1.3 *Let T_n be a positive definite Toeplitz matrix and let $v_1, \ldots, v_n$ be numbers satisfying*

$$T_n \begin{pmatrix} 1 \\ v_1 \\ \vdots \\ v_n \end{pmatrix} = \begin{pmatrix} \nu \\ 0 \\ \vdots \\ 0 \end{pmatrix}$$

for some positive number ν. Then the polynomial

$$p(z) = 1 + v_1 z + \cdots + v_n z^n$$

has no zeros in the closed unit disk.

Proof: Let α be any zero of p and write

$$1 + v_1 z + \cdots + v_n z^n = (z - \alpha)$$

$$(-\alpha^{-1} + w_1 z + \cdots + w_{n-2} z^{n-2} + v_n z^{n-1})$$

Then $\alpha \neq 0$ and by equating coefficients of like powers of z, we have

$$b + \alpha c = d \tag{1.13}$$

where

$$b = \begin{pmatrix} 1 \\ v_1 \\ \vdots \\ v_n \end{pmatrix}, \quad c = \begin{pmatrix} -\alpha^{-1} \\ w_1 \\ \vdots \\ w_{n-2} \\ v_n \\ 0 \end{pmatrix}, \quad d = \begin{pmatrix} 0 \\ -\alpha^{-1} \\ w_1 \\ \vdots \\ w_{n-2} \\ v_n \end{pmatrix}$$

The fact that T_n is Toeplitz and positive definite implies that

$$c^* T_n c = d^* T_n d > 0$$

By (1.6)

$$b^* T_n b = (1 \bar{v}_1 \ldots \bar{v}_n) \begin{pmatrix} \nu \\ 0 \\ \vdots \\ 0 \end{pmatrix} = \nu \tag{1.14}$$

and

$$\bar{\alpha} c^* T_n b = \bar{\alpha}(-\bar{\alpha}^{-1}\ \overline{w}_1 \ldots \overline{w}_{n-2}\ \bar{v}_n\ 0) \begin{pmatrix} \nu \\ 0 \\ \vdots \\ 0 \end{pmatrix} = -\nu \tag{1.15}$$

and by (1.13)

$$(b^* + \overline{\alpha} c^*) T_n (b + \alpha c) = d^* T_n d = c^* T_n c$$

Expanding the left side and using (1.14) and (1.15), we find that

$$\nu - \nu - \nu + |\alpha|^2 c^* T_n c = c^* T_n c$$

Therefore

$$|\alpha|^2 - 1 = \frac{\nu}{c^* T_n c} > 0$$

which proves that p has no zeros in the closed unit disk. □

Theorem 1.4 *Let $\mathcal{X}$ be a vector space and let $x_0, x_1, \ldots$ be linearly independent vectors in $\mathcal{X}$. Let $\langle\, ,\, \rangle$ be a positive definite scalar product on span$\{x_0, x_1, \ldots\}$ that is both n-informative and stationary for $x_0, x_1, \ldots$. Then the Gram matrix $T = (\langle x_k, x_j \rangle)_{j,k\geq 0}$ defines an invertible bounded linear operator on $\ell_2(0,\infty)$, and its inverse admits the factorization*

$$T^{-1} = \nu^{-1} L L^* \tag{1.16}$$

where L is the invertible lower triangular Toeplitz matrix with first column $(1\ v_1 \ldots v_n 0 \ldots)^T$. Here $v_1, \ldots, v_n$ and ν are the solutions of (1.2).

Proof: Let T and L be as defined in the theorem. Then L is a bounded operator on $\ell_2(0,\infty)$. We will prove that L is invertible by considering the polynomial

$$p(z) = 1 + v_1 z + \cdots + v_n z^n$$

By Lemma 1.3, p has no zeros in the closed unit disk. Therefore $1/p$ is analytic in an open disk containing the closed unit disk and hence has the form

$$\frac{1}{p(z)} = \sum_{k=0}^{\infty} c_k z^k$$

with

$$\sum_{k=0}^{\infty} |c_k| < \infty$$

This implies that the lower-triangular Toeplitz matrix C with first column $(c_0\ c_1\ c_2 \ldots)^T$ defines a bounded linear operator on $\ell_2(0,\infty)$, and the fact that

$$p(z) \sum_{k=0}^{\infty} c_k z^k = 1$$

implies that L is invertible with C being the inverse of L.

Since each column of L has only finitely many nonzero entries, the product TL is defined, and (1.12) implies that

$$TL = U \tag{1.17}$$

where U is an upper triangular matrix with all diagonal entries being equal to the positive number ν. Since $\gamma_{-k} = \overline{\gamma_k}$, (1.12) also implies that the entries in the jth row of L^*T are the complex conjugates of the entries in the jth column of TL. Therefore

$$L^*T = U^* \tag{1.18}$$

From (1.17) and (1.18) it follows that

$$L^*TL = L^*U \quad \text{and} \quad L^*TL = U^*L$$

Since L^*U is upper-triangular with diagonal entries equal to ν, and U^*L is lower-triangular with diagonal entries equal to ν, we conclude that

$$L^*TL = \nu I \tag{1.19}$$

It follows that T is an invertible bounded operator on $\ell_2(0, \infty)$ and

$$T^{-1} = \nu^{-1}LL^*$$

□

In Theorem 1.4, L is a lower triangular banded matrix with band width n, that is, its (j, k) entry is 0 unless $0 \le j-k \le n$. Then L^* is an upper triangular banded matrix with bandwidth n. It follows from (1.16) that T^{-1} is a banded matrix with band width n; its (j, k) entry is 0 unless $|j - k| \le n$.

The statement in Theorem 1.4 that T is an invertible bounded linear operator, along with a theorem of Bari [B], implies that $x_0, x_1, \ldots$ is a Riesz basis of the completion $\mathcal{H}$ of span$\{x_0, x_1, \ldots\}$ for the extended scalar product. This means, by definition, that there is an orthonormal basis $\{\varphi_0, \varphi_1, \ldots\}$ of $\mathcal{H}$ and an invertible bounded operator $A : \mathcal{H} \to \mathcal{H}$ such that $x_k = A\varphi_k$ $(k \ge 0)$. It follows that every vector x in $\mathcal{H}$ has a unique representation in the form

$$x = \sum_{k=0}^{\infty} a_k x_k \tag{1.20}$$

where $(a_0\ a_1 \ldots)^T \in \ell_2(0, \infty)$. (See Section VI.2 of [GK].) It follows from the continuity of T that the scalar product on $\mathcal{H}$ is given by

$$\langle x, y\rangle = b^*Ta \tag{1.21}$$

where $x = \sum_{k=0}^{\infty} a_k x_k$ and $y = \sum_{k=0}^{\infty} b_k x_k$.

In the next theorem, we obtain orthonormal bases for $\mathcal{H}$ and the orthogonal complement of span$\{x_0, \ldots, x_n\}$ in $\mathcal{H}$.

Theorem 1.5 *Let $\mathcal{X}$ be a vector space and let $x_0, x_1, \ldots$ be linearly independent vectors in $\mathcal{X}$. Let $\langle\,,\,\rangle$ be an n-informatory stationary positive definite scalar product on V = span$\{x_0, x_1, \ldots\}$ and let $\mathcal{H}$ be the completion of span$\{x_0, x_1, \ldots\}$ for $\langle\,,\,\rangle$. Define*

$$y_k = x_k + \sum_{j=1}^{n} v_j x_{k+j} \qquad (k = 0, 1, \ldots) \tag{1.22}$$

and

$$z_k = x_k + \sum_{j=1}^{n} \bar{v}_j x_{k-j} \qquad (k = n+1, n+2, \ldots) \tag{1.23}$$

Then $\{\nu^{-1/2}y_0, \nu^{-1/2}y_1, \ldots,\}$ *is an orthonormal basis for* $\mathcal{H}$, *and* $\{\nu^{-1/2}z_{n+1}, \nu^{-1/2}z_{n+2}, \ldots\}$ *is an orthonormal basis for the orthogonal complement* $V^{\perp}$ *of* V *in* $\mathcal{H}$. *Furthermore,* y_k *is orthogonal to* x_ℓ *for* $\ell > k \geq 0$.

Proof: Let L be the lower triangular Toeplitz matrix with first column $(1\ v_1 \ldots v_n\ 0 \ldots)^T$. Then (1.21) and (1.22) imply that

$$\langle y_k, y_\ell \rangle = e_\ell^* L T L e_k$$

From this and (1.19) it follows that

$$\begin{aligned}\langle y_k, y_\ell \rangle &= \nu e_\ell^* e_k \\ &= \nu \delta_{k,\ell}\end{aligned}$$

Therefore $\{\nu^{-1/2}y_0, \nu^{-1/2}y_1, \ldots\}$ is an orthonormal set. Next we will show that the only vector in $\mathcal{H}$ orthogonal to y_k for all $k \geq 0$ is the zero vector. This will prove that $\{\nu^{-1/2}y_0, \nu^{-1/2}y_1, \ldots\}$ is an orthonormal basis for $\mathcal{H}$. Suppose w is orthogonal to y_k for all k. Let us write

$$w = \sum_{j=0}^{\infty} a_j x_j$$

where $a = (a_0\ a_1 \ldots)^T \in \ell_2(0, \infty)$, and let $b = (b_0\ b_1 \ldots)^T$, where

$$b = L^{-1}a$$

Then for every $k \geq 0$, it follows from (1.19) and (1.21) that

$$\begin{aligned}e_k^* b &= e_k^* L^* (L^*)^{-1} L^{-1} a = e_k^* L^* \nu^{-1} T a \\ &= \nu^{-1} \langle w, y_k \rangle = 0\end{aligned}$$

This implies that b and hence a and w are zero. This completes the proof that $\{\nu^{-1/2}y_0, \nu^{-1/2}y_1, \ldots\}$ is an orthonormal basis for $\mathcal{H}$. Observe also that for $\ell > k \geq 0$,

$$\langle y_k, x_\ell \rangle = e_\ell^* T L e_k = e_\ell^* (L^*)^{-1} \nu e_k$$

Since $(L^*)^{-1}$ is upper triangular, it follows that $\langle y_k, x_\ell \rangle = 0$, so that y_k is orthogonal to x_ℓ whenever $\ell > k \geq 0$.

For $\{z_{n+1}, z_{n+2}, \ldots\}$, we observe that for $\ell \geq k \geq n+1$, z_k and z_ℓ belong to $\text{span}\{x_0, \ldots, x_\ell\}$. Let T_ℓ be the $(\ell+1) \times (\ell+1)$ principal minor of T. Then it follows from (1.21) that

$$\langle z_\ell, z_k \rangle = d^* T_\ell c \tag{1.24}$$

where

$$(1.25)\qquad c = \begin{pmatrix} 0 \\ \vdots \\ 0 \\ \bar{v}_n \\ \vdots \\ \bar{v}_1 \\ 1 \end{pmatrix}^{(\ell-n \text{ zeros})} \qquad \text{and } d = \begin{pmatrix} 0 \\ \vdots \\ 0 \\ \bar{v}_n \\ \vdots \\ \bar{v}_1 \\ 1 \\ 0 \\ \vdots \\ 0 \end{pmatrix}^{(k-n \text{ zeros})}_{(\ell-k \text{ zeros})}$$

From (1.12) we have

$$(1.26)\qquad T_\ell c = \begin{pmatrix} 0 \\ \vdots \\ 0 \\ \nu \end{pmatrix}$$

Therefore by (1.24)–(1.26)

$$\langle z_l, z_k \rangle = \nu \delta_{k,l} \qquad (k, l \geq n+1)$$

This means that $\{\nu^{-1/2} z_{n+1},\ \nu^{-1/2} z_{n+2}, \ldots\}$ is an orthonormal set.

It also follows from (1.21) and (1.26) that for all $k \leq n$ and all $\ell \geq n+1$,

$$\langle z_\ell, x_k \rangle = e_k^{(\ell)*} T_\ell c = 0$$

where $e_k^{(\ell)}$ denotes the usual k^{th} unit vector in $\mathbf{C}^{\ell+1}$. Therefore $z_\ell \in V^\perp$ for all $\ell \geq n+1$. Let x be a vector in $V^\perp$ that is orthogonal to z_k for all $k \geq n+1$. Then x is orthogonal to $x_1, \ldots, x_n$ and since

$$x_k = z_k - \sum_{j=1}^{n} \bar{v}_j x_{k-j} \qquad (k = n+1, n+2, \ldots)$$

it follows by induction that x is orthogonal to x_k for all $k \geq n+1$. Since $\text{span}\{x_0, x_1, \ldots\}$ is dense in $\mathcal{H}$, this implies that $x = 0$. We conclude that $\{\nu^{-1/2} z_{n+1}, \nu^{-1/2} z_{n+2}, \ldots\}$ is an orthonormal basis for $V^\perp$. □

Observe from (1.22) that for any $m \geq 0$,

$$\overline{\text{span}}\{y_m, y_{m+1}, \ldots\} \subseteq \overline{\text{span}}\{x_m, x_{m+1}, \ldots\}$$

The reverse inclusion can be proved as follows. By Theorem 1.5, each x_k with $k \geq m$ is orthogonal to $y_0, y_1, \ldots, y_{m-1}$. Since $\{\nu^{-1/2} y_0, \nu^{-1/2} y_1, \ldots\}$ is an orthonormal basis for $\mathcal{H}$, this implies that

$$\overline{\text{span}}\{x_m, x_{m+1}, \ldots\} \subseteq \overline{\text{span}}\{y_0, \ldots, y_{m-1}\}^\perp = \overline{\text{span}}\{y_m, y_{m+1}, \ldots\}$$

Therefore

$$\overline{\text{span}}\{y_m, y_{m+1}, \ldots\} = \overline{\text{span}}\{x_m, x_{m+1}, \ldots\} \quad (m \geq 0) \tag{1.27}$$

If we contrast (1.27) with the equality

$$\text{span}\{u_0, u_1, \ldots, u_m\} = \text{span}\{x_0, x_1, \ldots, x_m\} \quad (m \geq 0)$$

that holds when one applies the Gram-Schmidt process to $\{x_0, x_1, \ldots\}$ to produce an orthonormal basis $\{u_0, u_1, \ldots\}$, then it is natural to call $y_0, y_1, \ldots$ a *backward orthogonalization* of $x_0, x_1, \ldots$, and to call $u_0, u_1 \ldots$ a *forward orthogonalization* of $x_0, x_1, \ldots$.

Now that it has been established that $\{x_k \mid k \geq 0\}$ is a Riesz basis of $\mathcal{H}$ and that $\{\nu^{-1/2} y_k \mid k \geq 0\}$ is an orthonormal basis of $\mathcal{H}$, it is possible to show how to find an invertible bounded linear operator $A : \mathcal{H} \to \mathcal{H}$ for which $x_k = A\nu^{-1/2} y_k$. The invertible lower triangular Toeplitz matrix L with first column $(1\ v_1 \ldots\ v_n\ 0 \ldots)^T$ plays an important role. Let $M : \ell_2(0, \infty) \to \mathcal{H}$ and $N : \ell_2(0, \infty) \to \mathcal{H}$ be defined by

$$Ma = \sum_{k=0}^{\infty} a_k x_k \quad \text{and} \quad Na = \sum_{k=0}^{\infty} a_k y_k$$

for any $a = (a_0\ a_1 \ldots)^T$ in $\ell_2(0, \infty)$. Then M and N are invertible bounded linear transformations with $Me_k = x_k$ and $Ne_k = y_k$ for all $k \geq 0$. Let

$$A = \nu^{1/2} M N^{-1}.$$

Then A is an invertible bounded linear operator on $\mathcal{H}$ with

$$x_k = A\nu^{-1/2} y_k \qquad (k \geq 0)$$

Since

$$y_k = x_k + \sum_{j=1}^{n} v_j x_{j+k}$$

it follows that

$$N = M\,L$$

and hence that

$$A = \nu^{1/2} M\,L^{-1} M^{-1} = \nu^{1/2} N\,L^{-1} N^{-1}$$

Therefore if we use either M or N to identify vectors in $\mathcal{H}$ with vectors in $\ell_2(0, \infty)$, the matrix $\nu^{1/2} L^{-1}$ maps $\nu^{-1/2} y_k$ to x_k.

1.4 The Inversion Formula

Next we will use the theory we have developed to obtain the Gohberg-Semencul formula [GS] for the inverse of a positive definite Toeplitz matrix. Our proof is based on the proof given in [GL].

Theorem 1.6 *Let $T_n = (\gamma_{j-k})_{j,k=0}^n$ be a positive definite Toeplitz matrix. Then*

$$T_n^{-1} = \nu^{-1}(L_n L_n^* - \Lambda_n \Lambda_n^*)$$

where

$$L_n = \begin{pmatrix} 1 & & & 0 \\ v_1 & 1 & & \\ \vdots & & \ddots & \\ v_n & \cdots & v_1 & 1 \end{pmatrix} \text{ and } \Lambda_n = \begin{pmatrix} 0 & & & 0 \\ \bar{v}_n & 0 & & \\ \vdots & & & \\ \bar{v}_1 & \cdots & \bar{v}_n & 0 \end{pmatrix}$$

Here $v_1, \ldots, v_n$ and ν are the solutions of the equation

$$T_n \begin{pmatrix} 1 \\ v_1 \\ \vdots \\ v_n \end{pmatrix} = \begin{pmatrix} \nu \\ 0 \\ \vdots \\ 0 \end{pmatrix}$$

Proof: Let

$$a = \begin{pmatrix} a_0 \\ \vdots \\ a_n \end{pmatrix} \text{ and } b = \begin{pmatrix} b_0 \\ \vdots \\ b_n \end{pmatrix}$$

be any vectors such that

$$T_n^{-1} b = a$$

Our goal is to use Theorems 1.2, 1.4 and 1.5 to show that

$$a = \nu^{-1}(L_n L_n^* - \Lambda_n \Lambda_n^*) b$$

which will prove the theorem. Define vectors in $\ell_2(0,\infty)$ corresponding to a and b by

$$\tilde{a} = \begin{pmatrix} a_0 \\ \vdots \\ a_n \\ 0 \\ \vdots \end{pmatrix} \text{ and } \tilde{b} = \begin{pmatrix} b_0 \\ \vdots \\ b_n \\ 0 \\ \vdots \end{pmatrix} \tag{1.28}$$

Let $e_0, e_1, \ldots$ denote the usual unit vectors in $\ell_2(0,\infty)$ and let $V = \operatorname{span}\{e_0, \ldots, e_n\}$. Then T_n defines a positive definite scalar product on V if we specify that

$$\langle e_k, e_j \rangle = \gamma_{j-k} \qquad (0 \le j, k \le n)$$

By Theorem 1.2, applied with $\mathcal{X} = \ell_2(0, \infty)$ and $x_k = e_k$ for all $k \geq 0$, there is an extension of the scalar product to span$\{e_0, e_1, \ldots\}$ that is both n-informative and stationary. Let $T = (\gamma_{j-k})_{j,k\geq 0}$ be the corresponding Gram matrix. Then T contains T_n as a submatrix, so we may partition T in the form

$$T = \begin{pmatrix} T_n & P \\ Q & R \end{pmatrix}$$

Partitioning

$$\tilde{a} = \begin{pmatrix} a \\ 0 \end{pmatrix} \tag{1.29}$$

in the corresponding way, we have

$$T\tilde{a} = \begin{pmatrix} T_n & P \\ Q & R \end{pmatrix} \begin{pmatrix} a \\ 0 \end{pmatrix} = \begin{pmatrix} b \\ -c \end{pmatrix}$$

for some vector c. Since

$$\begin{pmatrix} b \\ -c \end{pmatrix} = \begin{pmatrix} b \\ 0 \end{pmatrix} + \begin{pmatrix} 0 \\ -c \end{pmatrix} = \tilde{b} - \tilde{c}$$

where

$$\tilde{c} = \begin{pmatrix} 0 \\ c \end{pmatrix}$$

we have

$$T\tilde{a} = \tilde{b} - \tilde{c}$$

and hence

$$T^{-1}\tilde{b} = \tilde{a} + T^{-1}\tilde{c} \tag{1.30}$$

Let $\mathcal{H}$ be the completion of span$\{e_0, e_1, \ldots\}$ for $\langle\, ,\, \rangle$. Then $e_0, e_1, \ldots$ is a Riesz basis for $\mathcal{H}$ and by (1.20) the mapping $x \leftrightarrow (a_0, a_1, \ldots\}^T$ allows us to identify $\mathcal{H}$ with $\ell_2(0, \infty)$ equipped with a new scalar product $\langle\, ,\, \rangle$ having Gram matrix T. By (1.29), $\tilde{a} \in V$. Furthermore, for any vector $x = \sum_{k=0}^{n} t_k e_k$ in V, (1.21) implies that $\langle T^{-1}\tilde{c}, x\rangle = (\bar{t}_0\ \bar{t}_1 \ldots \bar{t}_n\ 0 \ldots)TT^{-1}\tilde{c} = 0$ since the first $n+1$ entries in $\tilde{c}$ are all 0. Therefore $T^{-1}\tilde{c} \in V^{\perp}$. It follows that $T^{-1}\tilde{c}$ is the projection of $T^{-1}\tilde{b}$ onto $V^{\perp}$. By Theorem 1.5, $\{\nu^{-1/2}z_{n+1}, \nu^{-1/2}z_{n+2}, \ldots\}$ is an orthonormal basis for $V^{\perp}$, where

$$z_k = e_k + \sum_{j=1}^{n} \bar{v}_j e_{k-j} = S^{k-n}u \quad (k \geq n+1) \tag{1.31}$$

Here $u = (\bar{v}_n\ \bar{v}_{n-1} \ldots \bar{v}_1\ 1\ 0 \ldots)^T$ and

$$S = \begin{pmatrix} 0 & 0 & 0 & \cdots \\ 1 & 0 & 0 & \cdots \\ 0 & 1 & 0 & \cdots \\ \cdot & \cdot & \cdot & \cdots \end{pmatrix}$$

Therefore

$$T^{-1}\tilde{c} = \sum_{k=n+1}^{\infty} \langle T^{-1}\tilde{b}, \nu^{-1/2} z_k \rangle \nu^{-1/2} z_k$$

$$= \sum_{k=n+1}^{\infty} (\nu^{-1} z_k^* T\, T^{-1}\tilde{b}) z_k$$

$$= \sum_{k=n+1}^{\infty} (\nu^{-1} z_k^* \tilde{b}) z_k$$

But $z_k^* b = 0$ for $k > 2n$ by the definitions in (1.31) and (1.28). Therefore

$$T^{-1}\tilde{c} = \sum_{k=n+1}^{2n} (\nu^{-1} z_k^* \tilde{b}) z_k = \sum_{k=n+1}^{2n} \nu^{-1} z_k z_k^* \tilde{b}$$

$$= \sum_{k=n+1}^{2n} \nu^{-1} S^{k-n} u u^* (S^*)^{k-n} \tilde{b} = \sum_{k=1}^{n} \nu^{-1} S^k u u^* (S^*)^k \tilde{b}$$

By Theorem 1.4 we also have

$$T^{-1}\tilde{b} = \nu^{-1} L\, L^* \tilde{b}$$

so it follows from (1.30) that

$$\tilde{a} = T^{-1}\tilde{b} - T^{-1}\tilde{c} = \nu^{-1} \left(L\, L^* - \sum_{k=1}^{n} S^k u u^* (S^*)^k \right) \tilde{b} \tag{1.32}$$

Let

$$z = \begin{pmatrix} \bar{v} \\ \vdots \\ \bar{v}_1 \\ 1 \end{pmatrix}$$

Then Λ_n has columns $Sz, S^2 z, \ldots, S^{n+1} z$. Therefore

$$\Lambda_n = S z e_0^* + S^2 z e_1^* + \cdots + S^{n+1} z e_n^*$$

and hence

$$\Lambda_n \Lambda_n^* = \sum_{k=1}^{n+1} S^k z (S^k z)^*$$

$$= \sum_{k=1}^{n+1} S^k z z^* (S^*)^k$$

Since the first $n+1$ entries of $\tilde{a}, \tilde{b}$ and u coincide with the first $n+1$ entries of a, b and z, respectively, and since L is lower-triangular, it follows from (1.32) that

$$\begin{aligned} a &= \nu^{-1}\left(L_n L_n^* - \sum_{k=1}^{n} S^k z z^* (S^*)^k\right) b \\ &= \nu^{-1}(L_n L_n^* - \Lambda_n \Lambda_n^*) b \end{aligned}$$

Therefore

$$T_n^{-1} = \nu^{-1}(L_n L_n^* - \Lambda_n \Lambda_n^*)$$

□

2 The Matrix Case

In this section we will extend to the matrix case the results of the previous section. In doing this, we will replace vector spaces by modules over the algebra $\mathcal{M}$ of $r \times r$ complex matrices, where r is a given positive integer.

2.1 Matrix-valued Inner Products on Modules

Let $\mathcal{X}$ be a right module over the algebra $\mathcal{M}$ of all complex $r \times r$ matrices. A positive definite $r \times r$ *matrix-valued inner product* on $\mathcal{X}$ is a mapping

$$\langle\, , \,\rangle : \mathcal{X} \times \mathcal{X} \to \mathcal{M}$$

having the following properties for all x, y, z in $\mathcal{X}$ and for all $a \in M$:

1. $\langle x, x\rangle$ is positive semidefinite and $\langle x, x\rangle = 0$ only if $x = 0$.
2. $\langle y, x\rangle = \langle x, y\rangle^*$ (conjugate transpose in $\mathcal{M}$).
3. $\langle x + y, z\rangle = \langle x, z\rangle + \langle y, z\rangle$.
4. $\langle xa, y\rangle = \langle x, y\rangle a$.

Properties 2 and 4 imply that

$$\langle xa, yb\rangle = b^* \langle x, y\rangle a$$

Associated with any positive definite matrix-valued inner product $\langle\, , \,\rangle$ is the positive definite scalar product $\langle\, , \,\rangle_\tau$ defined by

$$\langle x, y\rangle_\tau = \operatorname{tr}\langle x, y\rangle \tag{2.1}$$

where tr denotes the trace. The corresponding norm defined by

$$\|x\|_\tau = \sqrt{\langle x, x\rangle_\tau}$$

determines a topology on $\mathcal{X}$. If $\mathcal{X}$ is complete for this norm, then $\mathcal{X}$ is called a *Hilbert module*.

We will use $\|a\|$ to denote the operator norm of any $r \times r$ matrix a, and we will write $\ell_2^{r\times r}(0, \infty)$ for the set of all sequences $(a_0, a_1, \ldots)^T$, frequently considered as column vectors, of $r \times r$ matrices with $\sum_{k=0}^{\infty} \|a_k\|^2 < \infty$.

For any $x \in \mathcal{X}$ and any $a \in \mathcal{M}$, we have

$$\begin{aligned}\|xa\|_\tau^2 &= \operatorname{tr}\langle xa, xa\rangle = \operatorname{tr}(a^*\langle x, x\rangle a)\\ &\le r\|a^*\langle x, x\rangle a\| \le r\|a\|^2\ \|\langle x, x\rangle\|\\ &\le r\|a\|^2 \operatorname{tr}\langle x, x\rangle \le r\|a\|^2\ \|x\|_\tau^2\end{aligned}$$

Therefore

$$\|xa\|_\tau \le \sqrt{r}\|a\|\ \|x\|_\tau \tag{2.2}$$

This implies that the mapping $(a, x) \to xa$ is continuous from $\mathcal{M} \times \mathcal{X}$ into $\mathcal{X}$.

Any right module $\mathcal{X}$ over $\mathcal{M}$ with a positive definite $r \times r$ matrix-valued inner product can be completed for the positive definite scalar product $\langle\ ,\ \rangle_\tau$ to a Hilbert space $\mathcal{H}$. The continuous mapping $(a, x) \to xa$ from $\mathcal{M} \times \mathcal{X}$ to $\mathcal{X}$ has a unique continuous extension to a mapping from $\mathcal{M} \times \mathcal{H}$ into $\mathcal{H}$. Thus $\mathcal{H}$ becomes a right module over $\mathcal{M}$. Furthermore, the matrix-valued inner product on $\mathcal{X}$, which by Proposition A.1 in [EGL] is continuous on $\mathcal{X} \times \mathcal{X}$, has a unique continuous extension to $\mathcal{H} \times \mathcal{H}$. All the properties of a positive definite matrix-valued inner product are automatically valid for the extension, except the property that $\langle x, x\rangle = 0$ only if $x = 0$. But if $\langle x, x\rangle = 0$, then $\operatorname{tr}\langle x, x\rangle = 0$ and hence $x = 0$. Thus $\mathcal{H}$ is a Hilbert module.

Now let $\langle\ ,\ \rangle$ be an $r \times r$ matrix-valued inner product on a right module $\mathcal{X}$ over $\mathcal{M}$ with a finite module basis $x_0, \ldots, x_n$. Then for any $a_0, \ldots, a_n,\ b_0, \ldots, b_n$ in $\mathcal{M}$,

$$\begin{aligned}\left\langle \sum_{k=0}^{n} x_k a_k, \sum_{j=0}^{n} x_j b_j \right\rangle &= \sum_{j,k=0}^{n} b_j^* \langle x_k, x_j\rangle a_k\\ &= b^* G(x_0, \ldots, x_n) a\end{aligned}$$

where

$$a = \begin{pmatrix} a_0\\ \vdots\\ a_n\end{pmatrix},\ b = \begin{pmatrix} b_0\\ \vdots\\ b_n\end{pmatrix}$$

and

$$G(x_0, \ldots, x_n) = (\langle x_k, x_j\rangle)_{0\le j,k\le n}$$

Thus the inner product is determined by the block matrix $G(x_0, \ldots, x_n)$, which is called the *Gram matrix* of $x_0, \ldots, x_n$ for $\langle\, , \,\rangle$ and is known to be positive semi-definite [EGL]. We can prove from the (module) linear independence of $x_0, \ldots, x_n$ that $G(x_0, \ldots, x_n)$ is positive definite as follows. Let λ be any vector in $\mathbf{C}^{\mathbf{r(n+1)}}$ and suppose $\lambda^* G(x_0, \ldots, x_n)\lambda = 0$. Let a be the $r(n+1) \times r$ matrix each of whose columns is λ, and partition a into $r \times r$ blocks:

$$a = \begin{pmatrix} a_0 \\ \vdots \\ a_n \end{pmatrix}$$

Then

$$\left\langle \sum_{k=0}^{n} x_k a_k, \sum_{k=0}^{n} x_k a_k \right\rangle = a^* G(x_0, \ldots, x_n) a = 0$$

Therefore

$$\sum_{k=0}^{n} x_k a_k = 0$$

so the (module) linear independence of $x_0, \ldots, x_n$ implies that $a = 0$ and hence $\lambda = 0$.

We can define a second $r \times r$ matrix-valued inner product $\langle\, , \,\rangle_2$ on $\mathcal{X}$ by specifying

$$\langle x_j, x_k \rangle_2 = \langle x_j, x_k \rangle^* = \langle x_k, x_j \rangle \qquad (0 \le j, k \le n) \tag{2.3}$$

and extending to all of $\mathcal{X}$, using the linearity of the inner product and the fact that $x_0, \ldots, x_n$ form a basis for $\mathcal{X}$. To see that $\langle x, x \rangle_2 \ge 0$ for all $x = \sum_{j=0}^{n} x_j a_j$, observe that

$$\begin{aligned} \langle x, x \rangle_2 &= \left\langle \sum_{j=0}^{n} x_j a_j, \sum_{j=0}^{n} x_j a_j \right\rangle_2 \\ &= \sum_{j,k=0}^{n} a_j^* \langle x_k, x_j \rangle_2 a_k \\ &= \sum_{j,k=0}^{n} a_j^* \langle x_j, x_k \rangle a_k \end{aligned}$$

Therefore

$$\langle x, x \rangle_2 = a^* G(x_0, \ldots, x_n)^T a \tag{2.4}$$

where $G(x_0, \ldots, x_n)^T$ is the block transpose of $G(x_0, \ldots, x_n)$ and

$$a = \begin{pmatrix} a_0 \\ a_1 \\ \vdots \\ a_n \end{pmatrix}$$

Observe also that

$$G(x_0, \ldots, x_n)^T = R^* G(x_0, \ldots, x_n) R \tag{2.5}$$

where R is the $(n+1) \times (n+1)$ block matrix given by

$$R = \begin{pmatrix} 0 & \cdots & 0 & 0 & I \\ 0 & \cdots & 0 & I & 0 \\ \vdots & & & \vdots & \\ I & 0 & \cdots & 0 & 0 \end{pmatrix}$$

Here I is the $r \times r$ identity matrix. Since $G(x_0, \ldots, x_n)$ is positive definite, it follows from (2.5) that $G(x_0, \ldots, x_n)^T$ is also positive definite. Thus (2.4) implies that $\langle x, x\rangle_2 \geq 0$ for all x in $\mathcal{X}$. Furthermore, if $\langle x, x\rangle_2 = 0$, then $a^* G(x_0, \ldots, x_n)^T a = 0$, so $a^{(j)*} G(x_0, \ldots, x_n)^T a^{(j)} = 0$ for $0 \leq j \leq n$, where $a^{(j)}$ is the jth column of a. It follows that $a^{(j)} = 0$ for $0 \leq j \leq n$ and hence that $a = 0$. This shows that $a_0, \ldots, a_n = 0$, so that $x = 0$. We conclude that the specification (2.3) defines a positive definite $r \times r$ matrix-valued inner product on $\mathcal{X}$.

Even though a norm has not been defined on $\mathcal{X}$, it is possible to give a definition of best approximation that reduces in the scalar case to the usual notion of best approximation. For any x in $\mathcal{X}$ and for any closed submodule $\mathcal{K}$ of $\mathcal{X}$, the *best approximation* of x in $\mathcal{K}$ for $\langle\ ,\ \rangle$ is the vector $p \in \mathcal{K}$ with the property that

$$\langle x - p, x - p\rangle \leq \langle x - w, x - w\rangle$$

for all $w \in \mathcal{K}$, where $\leq$ denotes the usual inequality for selfadjoint matrices. By Propositions A.4 and A.5 of [EGL] the best approximation of x in $\mathcal{K}$ exists and is unique, provided $\mathcal{X}$ is a Hilbert module.

2.2 *Extension of a Matrix-valued Inner Product*

Throughout the remainder of this section, $\mathcal{X}$ will be a right module over $\mathcal{M}$, and $x_0, x_1, \ldots$ will be a (module) linearly independent sequence of vectors in $\mathcal{X}$. The submodule of $\mathcal{X}$ consisting of all finite linear combinations of $x_0, x_1, \ldots$ with coefficients in $\mathcal{M}$ will be denoted by $\text{span}_{\mathcal{M}}\{x_0, x_1, \ldots\}$.

Let n be a positive integer. A positive definite $r \times r$ matrix-valued inner product on $\text{span}_{\mathcal{M}}\{x_0, x_1, \ldots\}$ is said to be *n-informative* for $x_0, x_1, \ldots$ if for every integer

$k \geq 0$, the projection of x_0 onto $\text{span}_{\mathcal{M}}\{x_1, \ldots, x_{n+k}\}$ lies in $\text{span}_{\mathcal{M}}\{x_1, \ldots, x_n\}$. (See the Appendix of [EGL] for the definition of projection in a Hilbert module.) By Proposition A.5 in [EGL], this may also be expressed by saying that the best approximation of x_0 in $\text{span}_{\mathcal{M}}\{x_1, \ldots, x_{n+k}\}$ is the same as the best approximation of x_0 in $\text{span}_{\mathcal{M}}\{x_1, \ldots, x_n\}$.

A positive definite $r \times r$ matrix-valued inner product $\langle\, ,\, \rangle$ is *stationary* for $x_0, x_1, \ldots$ if $\langle x_k, x_j \rangle$ depends only on $j - k$. If $\langle\, ,\, \rangle$ is both n-informative and stationary for $x_0, x_1, \ldots$, then for all nonnegative integers m and k, the projection of x_m onto $\text{span}_{\mathcal{M}}\{x_{m+1}, \ldots, x_{m+n+k}\}$ lies in $\text{span}_{\mathcal{M}}\{x_{m+1}, \ldots, x_{m+n}\}$.

Now suppose that n is a given positive integer and that $\langle\, ,\, \rangle$ is a positive definite $r \times r$ matrix-valued inner product on $\text{span}_{\mathcal{M}}\{x_0, \ldots, x_n\}$. Our goal is to prove that it is possible to extend $\langle\, ,\, \rangle$ to a positive definite $r \times r$ matrix-valued inner product on $\text{span}_{\mathcal{M}}\{x_0, x_1, \ldots\}$ that is both n-informative and stationary for $x_0, x_1, \ldots$. For this, we will need the following lemma.

Lemma 2.1 *Let $\mathcal{X}$ be a right $\mathcal{M}$-module and let $x_0, \ldots, x_n$ be linearly independent vectors in $\mathcal{X}$. Let $\langle\, ,\, \rangle$ be a positive definite $r \times r$ matrix-valued inner product on $\text{span}_{\mathcal{M}}\{x_0, \ldots, x_n\}$ with*

$$\langle x_k, x_j \rangle = \gamma_{j-k} \qquad (j, k = 0, 1, \ldots, n)$$

a. *The best approximation of x_0 in $\text{span}_{\mathcal{M}}\{x_1, \ldots, x_n\}$ for $\langle\, ,\, \rangle$ is the vector $-\sum_{j=1}^{n} x_j v_j$, where $v_1, \ldots, v_n$ are the unique $r \times r$ matrices that satisfy*

$$G(x_0, \ldots, x_n) \begin{pmatrix} I \\ v_1 \\ \vdots \\ v_n \end{pmatrix} = \begin{pmatrix} v \\ 0 \\ \vdots \\ 0 \end{pmatrix} \tag{2.6}$$

for some positive definite $r \times r$ matrix v.

b. *The best approximation of x_n in $\text{span}_{\mathcal{M}}\{x_0, \ldots, x_{n-1}\}$ for $\langle\, ,\, \rangle$ is $-\sum_{j=0}^{n-1} x_j u_{n-j}$, where $u_1, \ldots, u_n$ are the unique $r \times r$ matrices that satisfy*

$$G(x_0, \ldots, x_n) \begin{pmatrix} u_n \\ \vdots \\ u_1 \\ I \end{pmatrix} = \begin{pmatrix} 0 \\ \vdots \\ 0 \\ \mu \end{pmatrix} \tag{2.7}$$

for some positive definite $r \times r$ matrix μ.

c. *The best approximation of x_0 in $\text{span}_{\mathcal{M}}\{x_1, \ldots, x_n\}$ for the matrix-valued inner product $\langle\, ,\, \rangle_2$ defined by (2.2) is $-\sum_{j=1}^{n} x_j u_j$, where $u_1, \ldots, u_n$ are the matrices given by (2.7), for which $-\sum_{j=0}^{n-1} x_j u_{n-j}$ is the best approximation of x_n in $\text{span}_{\mathcal{M}}\{x_0, \ldots, x_{n-1}\}$ for $\langle\, ,\, \rangle$.*

Proof: a. For any i and j with $0 \le i \le j \le n$ let

$$\mathcal{H}_i^j = \text{span}_{\mathcal{M}}\{x_i, \ldots, x_j\}$$

By Propositions A.4 and A.5 of [EGL], the best approximations mentioned in the theorem exist and are unique. Proposition A.5 also implies that the best approximation of x_0 in $\mathcal{H}_1^n$ for $\langle\, , \,\rangle$ is the vector $-\sum_{j=1}^n x_j v_j$ for which $x_0 + \sum_{j=1}^n x_j v_j$ is orthogonal to $x_1, \ldots, x_n$ for $\langle\, , \,\rangle$, i.e., for which

$$\langle x_0, x_k \rangle + \sum_{j=1}^{n} \langle x_j, x_k \rangle v_j = 0 \qquad (1 \le k \le n) \tag{2.8}$$

This implies that

$$\gamma_k + \sum_{j=1}^{n} \gamma_{k-j} v_j = 0 \qquad (1 \le k \le n) \tag{2.9}$$

where $\gamma_{-n}, \ldots, \gamma_n$ are the block entries in the Gram matrix $G(x_0, \ldots, x_n)$. As in the Proof of Lemma 1.1, (2.9) leads to (2.6). The matrix ν is positive definite since $G(x_0, \ldots, x_n)$ is positive definite and

$$\nu = (I \; v_1^* \ldots v_n^*) G(x_0, \ldots, x_n) \begin{pmatrix} I \\ v_1 \\ \vdots \\ v_n \end{pmatrix}$$

Since (2.6) implies (2.9), which determines the unique best approximation of x_0 in $\mathcal{H}_1^n$ for $\langle\, , \,\rangle$, there can be only one positive definite matrix ν for which there are solutions $v_1, \ldots, v_n$ of (2.6). This proves part a.

b. Let $-\sum_{j=0}^{n-1} x_j w_j$ be the best approximation of x_n in $\mathcal{H}_0^{n-1}$ for $\langle\, , \,\rangle$. Then $x_n + \sum_{j=0}^{n-1} x_j w_j$ must be orthogonal to $x_0, \ldots, x_{n-1}$ for $\langle\, , \,\rangle$, i.e.

$$\langle x_n, x_k \rangle + \sum_{j=0}^{n-1} \langle x_j, x_k \rangle w_j = 0 \qquad (0 \le k \le n-1)$$

This implies that

$$\begin{pmatrix} \gamma_0 & \gamma_{-1} & \cdots & \gamma_{-n} \\ \gamma_1 & \gamma_0 & \cdots & \gamma_{-n+1} \\ \vdots & & & \\ \gamma_{n-1} & \gamma_{n-2} & \cdots & \gamma_{-1} \end{pmatrix} \begin{pmatrix} w_0 \\ \vdots \\ w_{n-1} \\ I \end{pmatrix} = \begin{pmatrix} 0 \\ \vdots \\ 0 \\ 0 \end{pmatrix}$$

Letting $\mu = \gamma_n w_0 + \cdots + \gamma_1 w_{n-1} + \gamma_0$, we obtain

$$G(x_0, \ldots, x_n) \begin{pmatrix} w_0 \\ \vdots \\ w_{n-1} \\ I \end{pmatrix} = \begin{pmatrix} 0 \\ \vdots \\ 0 \\ \mu \end{pmatrix}$$

which becomes (2.7) upon defining

$$u_k = w_{n-k} \qquad (1 \leq k \leq n)$$

Since

$$\mu = (w_0^* \ldots w_{n-1}^* I) G(x_0, \ldots, x_n) \begin{pmatrix} w_0 \\ \vdots \\ w_{n-1} \\ I \end{pmatrix}$$

it follows that μ is positive definite. The uniqueness of $u_1, \ldots, u_n$ and μ is proved as in part a.

c. Let $-\sum_{j=1}^{n} x_j s_j$ be the best approximation of x_0 in $\mathcal{H}_1^n$ for $\langle\, , \,\rangle_2$. Since the Gram matrix of $x_0, \ldots, x_n$ for $\langle\, , \,\rangle_2$ is the block transpose of the Gram matrix $G(x_0, \ldots, x_n)$ of $x_0, \ldots, x_n$ for $\langle\, , \,\rangle$, part a implies that

$$G(x_0, \ldots, x_n)^T \begin{pmatrix} I \\ s_1 \\ \vdots \\ s_n \end{pmatrix} = \begin{pmatrix} \sigma \\ 0 \\ \vdots \\ 0 \end{pmatrix} \tag{2.10}$$

for some positive definite $r \times r$ matrix σ. Let

$$R = \begin{pmatrix} 0 & & & I \\ & & \cdot & \\ & \cdot & & \\ I & & & 0 \end{pmatrix}$$

be the block matrix with entries on the off diagonal equal to the $r \times r$ identity matrix I. Then $R^2 = I$ and by (2.10),

$$RG(x_0, \ldots, x_n)^T RR \begin{pmatrix} I \\ s_1 \\ \vdots \\ s_n \end{pmatrix} = R \begin{pmatrix} \sigma \\ 0 \\ \vdots \\ 0 \end{pmatrix}$$

which yields

$$G(x_0, \ldots, x_n)\begin{pmatrix} s_n \\ \vdots \\ s_1 \\ I \end{pmatrix} = \begin{pmatrix} 0 \\ \vdots \\ 0 \\ \sigma \end{pmatrix}$$

Comparing with (2.7), we conclude that $\sigma = \mu$ and $s_k = u$ for $s \leq k \leq n$. This proves part c. □

Now suppose that $\mathcal{X}$ is a right $\mathcal{M}$-module and that $x_0, x_1, \ldots$ are (module) linearly independent vectors in $\mathcal{X}$. Let $\langle\, ,\, \rangle$ be an $r \times r$ matrix-valued inner product on $V = \mathrm{span}_{\mathcal{M}}\{x_0, \ldots, x_n\}$ for which $\langle x_k, x_j \rangle$ depends only on $j - k$. We will show that the inner product $\langle\, ,\, \rangle$ on V can be extended to $\mathrm{span}_{\mathcal{M}}\{x_0, x_1, \ldots\}$ in such a way that both $\langle\, ,\, \rangle$ and $\langle\, ,\, \rangle_2$ are n-informative and stationary for $x_0, x_1, \ldots$.

Theorem 2.2 *Let $\mathcal{X}$ be a right $\mathcal{M}$-module and let $x_0, x_1, \ldots$ be (module) linearly independent vectors in $\mathcal{X}$. Let $\langle\, ,\, \rangle$ be an $r \times r$ matrix-valued inner product on $V = \mathrm{span}_{\mathcal{M}}\{x_0, \ldots, x_n\}$ for which $\langle x_k, x_j \rangle = \gamma_{j-k}$ for $j, k = 0, \ldots, n$. Then there is a unique extension of $\langle\, ,\, \rangle$ to $\mathrm{span}_{\mathcal{M}}\{x_0, x_1, \ldots\}$ such that both $\langle\, ,\, \rangle$ and $\langle\, ,\, \rangle_2$ are stationary and n-informative for $x_0, x_1, \ldots$. This extension is determined by specifying that*

$$\langle x_k, x_j \rangle = \gamma_{j-k} \qquad (j, k = 0, 1, \ldots)$$

where $\gamma_{-n}, \ldots, \gamma_n$ are the given entries in $G(x_0, \ldots, x_n)$ and

$$\begin{aligned} \gamma^*_{-k-1} = \gamma_{k+1} &= -\gamma_k v_1 - \cdots - \gamma_{k-n+1} v_n \\ & \qquad (k = n, n+1 \ldots) \\ &= -u_1^* \gamma_k - \cdots - u_n^* \gamma_{k-n+1} \end{aligned} \tag{2.11}$$

Here $v_1, \ldots, v_n$ are the solutions of (2.6), *and $u_1, \ldots, u_n$ are the solutions of* (2.7).

Proof: For any integers j and k with $0 \leq j \leq k$ let

$$\mathcal{X}_j^k = \mathrm{span}_{\mathcal{M}}\{x_j, \ldots, x_k\}$$

The extension of $\langle\, ,\, \rangle$ will be accomplished by proving by induction that for any integer $k \geq n$ the inner product can be extended to $\mathcal{X}_0^k$ so that $\langle x_j, x_i \rangle$ depends only on $i - j$ for $0 \leq i, j \leq k$, the best approximation of x_0 in $\mathcal{X}_1^k$ for $\langle\, ,\, \rangle$ is the same as in $\mathcal{X}_1^n$, and the best approximation of x_0 in $\mathcal{X}_1^k$ for $\langle\, ,\, \rangle_2$ is the same as in $\mathcal{X}_1^n$.

For $k = n$ there is nothing to prove. For the inductive step, suppose $k \geq n$ and $\langle\, ,\, \rangle$ has been extended to $\mathcal{X}_0^k$ with the desired properties. Lemma 2.1(a) implies that in order to extend $\langle\, ,\, \rangle$ from $\mathcal{X}_0^k$ to $\mathcal{X}_0^{k+1}$ with the desired properties, we must define $\gamma_{k+1} = \langle x_0, x_{k+1}\rangle$ and $\gamma_{-k-1} = \gamma_{k+1}^*$ so that

$$\begin{pmatrix} \gamma_0 & \gamma_{-1} & \cdots & \gamma_{-k-1} \\ \gamma_1 & \gamma_0 & \cdots & \gamma_{-k} \\ \vdots & & & \\ \gamma_{k+1} & \gamma_k & \cdots & \gamma_0 \end{pmatrix} \begin{pmatrix} I \\ v_1 \\ \vdots \\ v_n \\ 0 \\ \vdots \\ 0 \end{pmatrix} = \begin{pmatrix} \nu \\ 0 \\ \vdots \\ 0 \end{pmatrix} \tag{2.12}$$

Thus we define

$$\gamma_{k+1} = -\gamma_k v_1 - \cdots - \gamma_{k-n+1} v_n$$

Similarly, Lemma 2.1(b) and (c) implies that γ_{k+1} and γ_{-k-1} must be such that

$$\begin{pmatrix} \gamma_0 & \gamma_{-1} & \cdots & \gamma_{-k-1} \\ \gamma_1 & \gamma_0 & \cdots & \gamma_{-k} \\ \vdots & & & \\ \gamma_{k+1} & \gamma_k & \cdots & \gamma_0 \end{pmatrix} \begin{pmatrix} 0 \\ \vdots \\ 0 \\ u_n \\ \vdots \\ u_1 \\ I \end{pmatrix} = \begin{pmatrix} 0 \\ \vdots \\ 0 \\ \mu \end{pmatrix} \tag{2.13}$$

which implies that γ_{-k-1} must be given by

$$\gamma_{-k-1} = -\gamma_{-k} u_1 - \cdots - \gamma_{-k+n-1} u_n$$

In order to have $\gamma_{-k-1} = \gamma_{k+1}^*$ and to verify (2.11), we must prove that

$$-\gamma_k v_1 - \cdots - \gamma_{k-n+1} v_n = -u_1^* \gamma_k - \cdots - u_n^* \gamma_{k-n+1}$$

From (2.12) and (2.13)

$$-\begin{pmatrix} \gamma_{k-n+1} \\ \vdots \\ \gamma_k \end{pmatrix} = \begin{pmatrix} \gamma_{k-n} & \cdots & \gamma_{k-2n+1} \\ \vdots & & \\ \gamma_{k-1} & \cdots & \gamma_{k-n} \end{pmatrix} \begin{pmatrix} v_1 \\ \vdots \\ v_n \end{pmatrix}$$

and

$$-\begin{pmatrix} \gamma_{-k} \\ \vdots \\ \gamma_{-k+n-1} \end{pmatrix} = \begin{pmatrix} \gamma_{-k+n} & \cdots & \gamma_{-k+1} \\ \vdots & & \\ \gamma_{-k+2n-1} & \cdots & \gamma_{-k+n} \end{pmatrix} \begin{pmatrix} u_n \\ \vdots \\ u_1 \end{pmatrix}$$

Therefore

$$
\begin{aligned}
-\gamma_k v_1 - \cdots - \gamma_{k-n+1} v_n &= -\begin{pmatrix} \gamma_{-k} \\ \vdots \\ \gamma_{-k+n-1} \end{pmatrix}^* \begin{pmatrix} v_1 \\ \vdots \\ v_n \end{pmatrix} \\
&= \begin{pmatrix} u_n \\ \vdots \\ u_1 \end{pmatrix}^* \begin{pmatrix} \gamma_{k-n} & \cdots & \gamma_{k-2n+1} \\ & \vdots & \\ \gamma_{k-1} & \cdots & \gamma_{k-n} \end{pmatrix} \begin{pmatrix} v_1 \\ \vdots \\ v_n \end{pmatrix} \\
&= -\begin{pmatrix} u_n \\ \vdots \\ u_1 \end{pmatrix}^* \begin{pmatrix} \gamma_{k-n+1} \\ \vdots \\ \gamma_k \end{pmatrix}
\end{aligned}
$$

as was to be proved. To prove that $\langle x, x\rangle \geq 0$ for any $x = \sum_{j=0}^{k+1} x_j a_j$ in $\mathcal{X}_0^{k+1}$ and that $\langle x, x\rangle = 0$ only if $x = 0$, we write

$$x = y + z$$

where

$$y = \left(x_0 + \sum_{j=1}^{n} x_j v_j\right) a_0 \ \text{ and } \ z = \sum_{j=1}^{n} x_j (a_j - v_j a_0) + \sum_{j=n+1}^{k+1} x_j a_j$$

Since $x_0 + \sum_{j=1}^{n} x_j v_j$ is orthogonal to $x_1, \ldots, x_{k+1}$, it follows that y and z are orthogonal, so

$$\langle x, x\rangle = \langle y, y\rangle + \langle z, z\rangle$$

By stationarity

$$\langle z, z\rangle = \langle w, w\rangle$$

where

$$w = \sum_{j=0}^{n-1} x_j (a_{j+1} - v_{j+1} a_0) + \sum_{j=n}^{k} x_j a_{j+1}$$

Since w and y belong to $\mathcal{X}_0^k$, the inductive hypothesis implies that $\langle w, w\rangle \geq 0$ and $\langle y, y\rangle \geq 0$. Therefore $\langle x, x\rangle \geq 0$. Furthermore, $\langle x, x\rangle = 0$ only if $\langle y, y\rangle = 0$ and $\langle w, w\rangle = 0$, which imply that $y = 0$ and $w = 0$ and hence $x = 0$. This completes the induction and hence the proof of the theorem. □

2.3 Properties of the Extended Matrix-valued Inner Product

The properties in Subsection 1.3 have matrix analogues that will be presented in this subsection.

Lemma 2.3 *Let T_n be a positive definite block Toeplitz matrix with $r \times r$ blocks and let $v_1, \ldots, v_n$ be $r \times r$ matrices satisfying*

$$T_n \begin{pmatrix} I \\ v_1 \\ \vdots \\ v_n \end{pmatrix} = \begin{pmatrix} \nu \\ 0 \\ \vdots \\ 0 \end{pmatrix}$$

for some positive definite $r \times r$ matrix ν. Then the matrix polynomial

$$p(z) = I + v_1 z + \cdots + v_n z^n$$

is invertible for all z in the closed unit disk.

Proof: Suppose that $p(\lambda)$ is not invertible and let ξ be a nonzero vector such that

$$p(\lambda)\xi = 0$$

Then $\lambda \neq 0$ and

$$\begin{aligned} (I + v_1 z + \cdots + v_n z^n)\xi &= (I + v_1 z + \cdots + v_n z^n)\xi \\ &\quad -(I + v_1 \lambda + \cdots + v_n \lambda^n)\xi \\ &= (z - \lambda)(w_1 + w_2 z + \cdots + w_n z^{n-1})\xi \end{aligned}$$

for some $r \times r$ matrices $w_1, \ldots, w_n$. Equating coefficients of like powers of z, we find that

$$b + \lambda\, c = d$$

where

$$b = \begin{pmatrix} I \\ v_1 \\ \vdots \\ v_n \end{pmatrix} \xi, \quad c = \begin{pmatrix} w_1 \\ \vdots \\ w_n \\ 0 \end{pmatrix} \xi, \quad d = \begin{pmatrix} 0 \\ w_1 \\ \vdots \\ w_n \end{pmatrix} \xi,$$

In particular,

$$\lambda w_1 \xi = -\xi \tag{2.14}$$

Because T_n is block Toeplitz, we have

$$\begin{aligned} c^* T_n c &= \xi^* (w_1^* \ldots w_n^*\ 0) T_n \begin{pmatrix} w_1 \\ \vdots \\ w_n \\ 0 \end{pmatrix} \xi \\ &= \xi^* (0\ w_1^* \ldots w_n^*) T_n \begin{pmatrix} 0 \\ w_1 \\ \vdots \\ w_n \end{pmatrix} \xi = d^* T_n d \end{aligned}$$

and by (2.6)

$$b^*T_nb \;= \xi^*(I\ v_1^* \dots v_n^*) \begin{pmatrix} \nu \\ 0 \\ \vdots \\ 0 \end{pmatrix} \xi = \xi^*\nu\xi$$

Furthermore, (2.14) and (2.7) imply that

$$\begin{aligned} \bar{\lambda}c^*T_nb \;&= \bar{\lambda}\xi^*(w_1^* \dots w_n^*\ 0) \begin{pmatrix} \nu \\ 0 \\ \vdots \\ 0 \end{pmatrix} \xi \\ &= \bar{\lambda}\ \xi^* w_1^* \nu\xi = -\xi^*\nu\xi \end{aligned}$$

Using these results and expanding the left side in the equality

$$(b^* + \bar{\lambda}c^*)T_n(b + \lambda c) = d^*T_nd$$

we obtain

$$\xi^*\nu\xi - \xi^*\nu\xi - \xi^*\nu\xi + |\lambda|^2c^*T_nc = c^*T_nc$$

which implies that

$$\xi^*\nu\xi = (|\lambda|^2 - 1)c^*T_nc$$

Since ν and T_n are positive definite and $\xi \neq 0$, it follows that $|\lambda| > 1$. Therefore $p(z)$ is invertible for all z with $|z| \leq 1$. □

Theorem 2.4 *Let $\mathcal{X}$ be a right $\mathcal{M}$-module and let $x_0, x_1, \dots$ be (module) linearly independent vectors in $\mathcal{X}$. Let $\langle\, ,\, \rangle$ be a positive definite $r \times r$ matrix-valued inner product on $\text{span}_{\mathcal{M}}\{x_0, x_1, \dots\}$ such that both $\langle\, ,\, \rangle$ and $\langle\, ,\, \rangle_2$ are n-informative and stationary for $x_0, x_1, \dots$. Then the Gram matrix $T = (\langle x_k, x_j \rangle)_{j,k\geq 0}$ defines an invertible bounded linear operator on $\ell_2^{r\times r}(0, \infty)$ and its inverse admits the factorization*

$$T^{-1} = L\nu^{-1}L^* \tag{2.15}$$

where L is the invertible block lower triangular Toeplitz matrix with first block column $(I\ v_1 \dots v_n\ 0 \dots)^T$. Here $v_1, \dots, v_n$ and ν are the solution of (2.6).

Proof: With Lemma 2.3 replacing Lemma 1.3, the Proof of Theorem 1.4 can be used almost verbatim to prove this theorem. One only has to observe that in the matrix case the equality

$$L^*TL = \nu I \tag{2.16}$$

that appeared in (1.19) implies that

$$T^{-1} = L\nu^{-1}L^*$$

□

In the next theorem we construct a Riesz basis for the Hilbert space obtained by completing $\text{span}_{\mathcal{M}}\{x_0, x_1 \ldots\}$ for the scalar product $\langle\,,\,\rangle_\tau$.

Theorem 2.5 *Let $\mathcal{X}$ be a right $\mathcal{M}$-module and let $x_0, x_1, \ldots$ be (module) linearly independent vectors in $\mathcal{X}$. Let $\langle\,,\,\rangle$ be a positive definite $r \times r$ matrix-valued inner product on $\text{span}_{\mathcal{M}}\{x_0, x_1, \ldots\}$ such that both $\langle\,,\,\rangle$ and $\langle\,,\,\rangle_2$ are n-informative and stationary for $x_0, x_1, \ldots$. Then $\{x_k E_{ij} \mid k \geq 0,\ 0 \leq i, j \leq r-1\}$ is a Riesz basis for the Hilbert space $\mathcal{H}$ obtained by completing $\text{span}_{\mathcal{M}}\{x_0, x_1, \ldots\}$ with respect to $\langle\,,\,\rangle_\tau$. Here E_{ij} is the $r \times r$ complex matrix whose only nonzero entry is 1 and is located in the (i, j) position.*

Proof: The vectors $\{x_k E_{ij} \mid k \geq 0,\ 0 \leq i, j \leq r-1\}$ span a subspace of $\mathcal{H}$ that coincides with $\text{span}_{\mathcal{M}}\{x_k \mid k \geq 0\}$ and hence is dense in $\mathcal{H}$. We will order the vectors $x_k E_{ij}$ first with respect to k, then with respect to j, and finally with respect to i. Next we will determine the Gram matrix of these vectors. For $k, \ell \geq 0$ and $0 \leq i, j, p, q \leq r-1$,

$$\begin{aligned} \operatorname{tr}\langle x_k E_{ij}, x_\ell E_{pq}\rangle &= \operatorname{tr}(E_{pq}^* \langle x_k, x_\ell\rangle E_{ij}) \\ &= \operatorname{tr}(E_{ij} E_{qp} \langle x_k, x_\ell\rangle) \\ &= \delta_{j,q} \operatorname{tr}(E_{ip} \langle x_k, x_\ell\rangle) \\ &= \delta_{j,q} \langle x_k, x_\ell\rangle_{p,i} \end{aligned}$$

where $\langle x_k, x_\ell\rangle_{p,i}$ denotes the (p, i) entry of the $r \times r$ matrix $\langle x_k, x_\ell\rangle$. Thus if we fix k and ℓ and let i, j, p, q take on the values between 0 and $r-1$, we find that the entries in the corresponding $r^2 \times r^2$ block of the Gram matrix are $\delta_{j,q}\langle x_k, x_\ell\rangle_{p,i}$. This means that the $r^2 \times r^2$ block of the Gram matrix corresponding to fixed values of k and ℓ is the block diagonal matrix

$$\begin{pmatrix} \langle x_k, x_\ell\rangle & 0 & \cdots & 0 \\ 0 & \langle x_k, x_\ell\rangle & & \vdots \\ \vdots & & \ddots & 0 \\ 0 & \cdots & 0 & \langle x_k, x_\ell\rangle \end{pmatrix} \tag{2.17}$$

From this and the fact that the Gram matrix $\{\langle x_k, x_\ell\rangle\}_{\ell,k\geq 0}$ is a bounded operator on $\ell_2^{r\times r}(0, \infty)$, it follows easily that the Gram matrix G of $\{x_k E_{ij} \mid k \geq 0,\ 0 \leq$

$i, j \leq r-1\}$ is bounded on $\ell_2^{r\times r}(0,\infty)$. Furthermore, if $\{C_{k,\ell}\}_{\ell,k\geq 0}$ is the inverse of $\{\langle x_k, x_\ell\rangle\}_{\ell,k\geq 0}$, let C be the matrix with, for any k and ℓ, the $r\times r$ block

$$\begin{pmatrix} C_{k,\ell} & 0 & \cdots & 0 \\ 0 & C_{k,\ell} & & \vdots \\ \vdots & & \ddots & 0 \\ 0 & \cdots & 0 & C_{k,\ell} \end{pmatrix}$$

in the same location in C as the block in (2.17) is in G. Then C is the inverse of G. Therefore the Gram matrix of $\{x_k E_{ij} \mid k \geq 0,\ 0 \leq i, j \leq r-1\}$ defines an invertible bounded linear operator on $\ell_2(0,\infty)$. We conclude from the same theorem of Bari used to prove Theorem 1.5 that $\{x_k E_{ij},\ k \geq 0,\ 0 \leq i, j \leq r-1\}$ is a Riesz basis of $\mathcal{H}$. (Compare with the results of A.S. Markus in Section VI.5 of [GK].) □

It follows immediately from Theorem 2.5 that every element of $\mathcal{H}$ can be expressed uniquely in the form

$$x = \sum_{k=0}^{\infty} x_k a_k \tag{2.18}$$

where $a = (a_0\ a_1 \ldots)^T \in \ell_2^{r\times r}(0,\infty)$, and for every $a \in \ell_2^{r\times r}(0,\infty)$, the series in (2.18) converges. Furthermore for all $x = \sum_{k=0}^{\infty} x_k a_k$ and $y = \sum_{k=0}^{\infty} x_k b_k$ in $\mathcal{H}$,

$$\langle x, y\rangle = b^* T a \tag{2.19}$$

A sequence $\{y_k \mid k \geq 0\}$ in a right module $\mathcal{X}$ over $\mathcal{M}$ with a positive definite $r\times r$ matrix-valued inner product is said to be *orthonormal* if $\langle y_k, y_\ell\rangle = \delta_{k,\ell} I$ $(k, \ell \geq 0)$, where I is the $r\times r$ identity matrix. An *orthonormal basis* for $\mathcal{X}$ is an orthonormal sequence $\{y_k \mid k \geq 0\}$ with the property that every vector in $\mathcal{X}$ can be expressed in the form

$$x = \sum_{k=0}^{\infty} y_k a_k$$

with convergence in the norm $\|\ \|_\tau$. In that case, $a_k = \langle x, y_k\rangle$, so the representation of each x is unique. It was proved in Proposition A.3 in [EGL] that for any orthonormal sequence $\{y_k \mid k \geq 0\}$ in $\mathcal{X}$ and for any x in $\mathcal{X}$,

$$\sum_{k=0}^{m} \langle x, y_k\rangle^* \langle x, y_k\rangle \leq \langle x, x\rangle \qquad (m \geq 0)$$

This matrix inequality may be viewed as an analogue, for modules with a matrix-valued inner product, of Bessel's Inequality. From it, one can prove that an orthonormal sequence $\{y_k \mid k \geq 0\}$ in a module $\mathcal{X}$ over $\mathcal{M}$ is an orthonormal basis if and only if the only vector in $\mathcal{X}$ that is orthogonal to all of the vectors y_k for $k \geq 0$ is zero. We will use this remark in the proof of the following theorem, which gives orthonormal bases for $\mathcal{H}$ and the orthogonal complement of $\text{span}_{\mathcal{M}}\{x_0, \ldots, x_n\}$ for the matrix-valued inner product.

Theorem 2.6 *Let $\mathcal{X}$ be a right $\mathcal{M}$-module and let $x_0, x_1, \ldots$ be (module) linearly independent vectors in $\mathcal{X}$. Let $\langle\, , \,\rangle$ be a positive definite $r \times r$ matrix-valued inner product on $\text{span}_{\mathcal{M}}\{x_0, x_1, \ldots\}$ such that both $\langle\, , \,\rangle$ and $\langle\, , \,\rangle_2$ are n-informative and stationary for $x_0, x_1, \ldots$. Let $\mathcal{H}$ be the Hilbert module obtained by completing $\text{span}_{\mathcal{M}}\{x_0, x_1, \ldots\}$ for $\langle\, , \,\rangle_\tau$. Define*

$$y_k = x_k + \sum_{j=1}^{n} x_{k+j} v_j \qquad (k = 0, 1, \ldots) \tag{2.20}$$

and

$$z_k = x_k + \sum_{j=1}^{n} x_{k-j} u_j \qquad (k = n+1, n+2, \ldots) \tag{2.21}$$

where $v_1, \ldots, v_n$ and v are the solutions of (2.6) and $u_1, \ldots, u_n$ and μ are the solutions of (2.7). Then $\{y_0 v^{-1/2}, y_1 v^{-1/2}, \ldots\}$ is an orthonormal basis for the module $\mathcal{H}$, and $\{z_{n+1}\mu^{-1/2}, z_{n+2}\mu^{-1/2}, \ldots\}$ is an orthonormal basis for the orthogonal complement of $V = \text{span}_{\mathcal{M}}\{x_0, \ldots, x_n\}$ in the module $\mathcal{H}$.

Proof: Let L be the block lower triangular Toeplitz matrix with first column $(I\ v_1 \ldots v_n\ 0 \ldots)^T$. Then (2.19) and (2.20) imply that for all k and ℓ,

$$\langle y_k, y_\ell \rangle = e_\ell^* L^* T\, L e_k$$

where for any $k \geq 0$, e_k denotes the vector in $\ell_2^{r \times r}(0, \infty)$ with all entries zero except the kth, which is the $r \times r$ identity matrix. From this and (2.16) we find that

$$\langle y_k, y_\ell \rangle = e_\ell^* v e_k = \delta_{k,\ell} v$$

Therefore $\{y_0 v^{-1/2}, y_1 v^{-1/2}, \ldots\}$ is an orthonormal set in $\mathcal{H}$. To prove that this orthonormal set is an orthonormal basis, it suffices, by the remark preceding the theorem, to prove that the only vector in $\mathcal{H}$ that is orthogonal to every y_k $(k \geq 0)$ is zero. The proof of this is analogous to the proof in the scalar case. (See the Proof of Theorem 1.5.)

Observe also that for $\ell > k \geq 0$

$$\langle y_k, x_\ell \rangle = e_\ell^* T L e_k = e_\ell^* (L^*)^{-1} v e_k$$

Since $(L^*)^{-1}$ is block upper triangular, this implies that $\langle y_k, x_\ell \rangle = 0$ and hence that y_k is orthogonal to x_ℓ whenever $\ell > k \geq 0$.

For $\{z_{n+1}\mu^{-1/2}, z_{n+2}\mu^{-1/2}, \ldots\}$ we observe that for $\ell \geq k \geq n+1$, z_k and z_ℓ belong to $\text{span}_{\mathcal{M}}\{x_0, \ldots, x_\ell\}$. Let T_ℓ be the principal minor of T with $\ell+1$ block rows and $\ell+1$ block columns. Then it follows from (2.19) that

$$\langle z_\ell, z_k\rangle = d^* T_\ell c \tag{2.22}$$

where

$$c = \begin{pmatrix} 0 \\ \vdots \\ 0 \\ u_n \\ \vdots \\ u_1 \\ I \end{pmatrix} \begin{matrix} (\ell - n \text{ zeros}) \\ \\ \\ \\ \\ \\ \\ \end{matrix} \qquad \text{and} \quad d = (S^*)^{\ell-k} c \tag{2.23}$$

Here S^* is the backward block shift matrix with $\ell+1$ block rows and $\ell+1$ block columns:

$$S^* = \begin{pmatrix} 0 & I & 0 & 0 & \cdots & 0 \\ 0 & 0 & I & 0 & \cdots & 0 \\ \vdots & & & & & \\ 0 & \cdot & \cdot & \cdot & 0 & I \\ 0 & \cdot & \cdot & \cdot & 0 & 0 \end{pmatrix}$$

By (2.13),

$$T_\ell c = \begin{pmatrix} 0 \\ \vdots \\ 0 \\ \mu \end{pmatrix} \tag{2.24}$$

From this and (2.22) it follows that

$$\langle z_k, z_\ell\rangle = \delta_{k,\ell}\mu \qquad (k, \ell \geq n+1)$$

Therefore $\{z_{n+1}\mu^{-1/2}, z_{n+2}\mu^{-1/2}, \ldots\}$ is an orthonormal set. It also follows from (2.19) and (2.24) that for all $k \leq n$ and $\ell \geq n+1$,

$$\langle z_\ell, x_k\rangle = e_k^{(\ell)*} T_\ell c = 0$$

where $e_k^{(\ell)}$ denotes the usual k^{th} unit vector in $\mathbf{C}^{\ell+1}$. Therefore $z_\ell \in V^\perp$ for all $\ell \geq n+1$. To show that $\{z_{n+1}\mu^{-1/2}, z_{n+2}\mu^{-1/2}, \ldots\}$ is an orthonormal basis for $V^\perp$, is suffices by the remark preceding the theorem to prove that the only vector in $V^\perp$ that is orthogonal to each z_k $(k \geq n+1)$ is the zero vector. Let w be any vector in $V^\perp$ that is orthogonal to z_k for all $k \geq n+1$. Then w is orthogonal to

$x_0, \ldots, x_n$ and since

$$x_k = z_k - \sum_{j=1}^{n} x_{k-j} u_k \qquad (k \geq n+1)$$

it follows by induction that w is also orthogonal to x_k for all $k \geq n+1$. Since $\text{span}_{\mathcal{M}}\{x_0, x_1, \ldots\}$ is dense in $\mathcal{H}$, this implies that $w = 0$. Therefore $\{z_{n+1}\mu^{-1/2}, z_{n+2}\mu^{-1/2}, \ldots\}$ is an orthonormal basis for $V^{\perp}$. □

As in the scalar case, it can be proved that for any $m \geq 0$,

$$\overline{\text{span}}_{\mathcal{M}}\{y_k \mid k \geq m\} = \overline{\text{span}}_{\mathcal{M}}\{x_k \mid k \geq m\}$$

Therefore $\{y_k \mid k \geq 0\}$ is called a *backward orthogonalization* of $\{x_k \mid k \geq 0\}$.

2.4 The Gohberg-Heinig Formula

In this section we will use the preceding considerations to obtain the Gohberg-Heinig formula [GH] for the inverse of a positive definite block Toeplitz matrix.

Theorem 2.7 *Let $T_n = (\gamma_{j-k})_{j,k=0}^{n}$ be a positive definite block Toeplitz matrix with $r \times r$ blocks. Then*

$$T_n^{-1} = L_n \nu^{-1} L_n^* - \Lambda_n \mu^{-1} \Lambda_n^* \tag{2.25}$$

where

$$L_n = \begin{pmatrix} I & & 0 \\ v_1 & \ddots & \\ \vdots & & \\ v_n & \cdots & v_1 \; I \end{pmatrix} \quad \text{and} \quad \Lambda_n = \begin{pmatrix} 0 & & 0 \\ u_n & \ddots & \\ \vdots & & \\ u_1 & \cdots & u_n \; 0 \end{pmatrix}$$

Here $v_1, \ldots, v_n$, and ν, and $u_1, \ldots, u_n$, and μ are the $r \times r$ matrix solutions of the equations

$$T_n \begin{pmatrix} I \\ v_1 \\ \vdots \\ v_n \end{pmatrix} = \begin{pmatrix} \nu \\ 0 \\ \vdots \\ 0 \end{pmatrix} \quad \text{and} \quad T_n \begin{pmatrix} u_n \\ \vdots \\ u_1 \\ I \end{pmatrix} = \begin{pmatrix} 0 \\ \vdots \\ 0 \\ \mu \end{pmatrix}$$

Proof: The proof will be a slight modification of the proof in the scalar case. Let

$$a = \begin{pmatrix} a_0 \\ \vdots \\ a_n \end{pmatrix} \quad \text{and} \quad b = \begin{pmatrix} b_0 \\ \vdots \\ b_n \end{pmatrix}$$

be any vectors with $r \times r$ entries such that

$$T_n^{-1} b = a$$

and define corresponding vectors in $\ell_2^{r\times r}(0, \infty)$ by

$$\tilde{a} = \begin{pmatrix} a_0 \\ \vdots \\ a_n \\ 0 \\ \vdots \end{pmatrix} \quad \text{and } \tilde{b} = \begin{pmatrix} b_0 \\ \vdots \\ b_n \\ 0 \\ \vdots \end{pmatrix} \tag{2.26}$$

For every integer $k \geq 0$, let e_k denote the vector in $\ell^{r\times r}(0, \infty)$ whose k^{th} entry is the $r \times r$ identity matrix and whose other entries are all 0. Let $V = \mathrm{span}_{\mathcal{M}}\{e_0, \ldots, e_n\}$. Then T_n defines a positive definite $r \times r$ matrix-valued inner product $\langle\, , \,\rangle$ on V if we specify

$$\langle e_k, e_j \rangle = \gamma_{j-k} \qquad (0 \leq j, k \leq n)$$

By Theorem 2.2 there is an extension of $\langle\, , \,\rangle$ to $\mathrm{span}_{\mathcal{M}}\{e_0, e_1, \ldots\}$ such that both $\langle\, , \,\rangle$ and $\langle\, , \,\rangle_2$ are n-informative and stationary for $e_0, e_1, \ldots$. Let $T = (\gamma_{j-k})_{j,k\geq 0}$ be the Gram matrix of $e_0, e_1, \ldots$ for the extended matrix-valued inner product. Since T contains T_n as a submatrix, we may partition T in the form

$$T = \begin{pmatrix} T_n & P \\ Q & R \end{pmatrix}$$

Partitioning

$$\tilde{a} = \begin{pmatrix} a \\ 0 \end{pmatrix} \tag{2.27}$$

in the corresponding way, we find that

$$T\tilde{a} = \begin{pmatrix} T_n & P \\ Q & R \end{pmatrix} \begin{pmatrix} a \\ 0 \end{pmatrix} = \begin{pmatrix} b \\ -c \end{pmatrix}$$

for some vector c. Writing

$$\begin{pmatrix} b \\ -c \end{pmatrix} = \begin{pmatrix} b \\ 0 \end{pmatrix} + \begin{pmatrix} 0 \\ -c \end{pmatrix} = \tilde{b} - \tilde{c}$$

where

$$\tilde{c} = \begin{pmatrix} 0 \\ c \end{pmatrix} \tag{2.28}$$

we have

$$T\tilde{a} = \tilde{b} - \tilde{c}$$

Therefore

$$T^{-1}\tilde{b} = \tilde{a} + T^{-1}\tilde{c} \tag{2.29}$$

Let $\mathcal{H}$ be the completion of $\operatorname{span}_{\mathcal{M}}\{e_0, e_1, \ldots\}$ for $\langle\, , \,\rangle_\tau$. Then $\mathcal{H}$ is a Hilbert module for the extended matrix-valued inner product $\langle\, , \,\rangle$. Furthermore, by the remarks following Theorem 2.5 all the elements of $\mathcal{H}$ may be expressed uniquely in the form $\sum_{k=0}^{\infty} e_k a_k$ for some sequence $(a_0\ a_1 \ldots)^T$ in $\ell_2^{r\times r}(0, \infty)$. Thus we may identifiy $\mathcal{H}$ with $\ell_2^{r\times r}(0, \infty)$ equipped with the matrix-valued inner product having Gram matrix T. As in the scalar case, (2.29) gives the decomposition of $T^{-1}\tilde{b}$ as a sum of a vector $\tilde{a}$ in V and a vector $T^{-1}\tilde{c}$ in the orthogonal complement $V^{\perp}$ of V for the matrix-valued inner product. But by Theorem 2.6, $\{z_{n+1}\mu^{-1/2}, z_{n+2}\mu^{-1/2}, \ldots\}$ is an orthonormal basis for $V^{\perp}$, where

$$z_k = e_k + \sum_{j=1}^{n} e_{k-j}u_j = S^{k-n}\tilde{z} \quad (k = n+1, n+2, \ldots) \tag{2.30}$$

with $\tilde{z} = (u_n \ldots u_1\ I\ 0 \ldots)^T$ and

$$S = \begin{pmatrix} 0 & 0 & 0 & \cdots \\ I & 0 & 0 & \cdots \\ 0 & I & 0 & \cdots \\ \cdot & \cdot & \cdot & \cdots \end{pmatrix}$$

Therefore

$$\begin{aligned} T^{-1}\tilde{c} &= \sum_{k=n+1}^{\infty} z_k\mu^{-1/2}\langle T^{-1}\tilde{b}, z_k\mu^{-1/2}\rangle \\ &= \sum_{k=n+1}^{\infty} z_k\mu^{-1/2}(\mu^{-1/2}z_k^*T\ T^{-1}\tilde{b}) \\ &= \sum_{k=n+1}^{\infty} z_k\mu^{-1}z_k^*\tilde{b} \end{aligned}$$

Since $z_k^*\tilde{b} = 0$ for $k > 2n$ by (2.30) and (2.26), we find that

$$\begin{aligned} T^{-1}\tilde{c} &= \sum_{k=n+1}^{2n} z_k\mu^{-1}z_k^*\tilde{b} \\ &= \sum_{k=1}^{n} S^k\tilde{z}\mu^{-1}\tilde{z}^*(S^*)^k\tilde{b} \end{aligned} \tag{2.31}$$

By (2.15),

$$T^{-1}\tilde{b} = L\nu^{-1}L^*\tilde{b} \tag{2.32}$$

From (2.29), (2.31) and (2.32) it follows in the same way as in the scalar case that

$$a = (L_n \nu^{-1} L_n^* - \Lambda_n \mu^{-1} \Lambda_n^*) b$$

This implies (2.25). □

References

[B] N.K. Bari, Bases in a Hilbert space, *Doklady Acad. Nauk.* **54** (1946), 383–386 (Russian).

[BEGL] A. Ben-Artzi, R.L. Ellis, I. Gohberg and D.C. Lay, The maximum distance problem and band sequences, *Linear Algebra Appl.* **87** (1987), 93–112.

[EGL] R.L. Ellis, I. Gohberg, and D.C. Lay, Infinite analogues of block Toeplitz matrices and related orthogonal functions, *Integral Equations Operator Theory* **22** (1995), 375–419.

[GGK] I.C. Gohberg, S. Goldberg and M.A. Kaashoek, Classes of Operators, *Operator Theory: Advances and Applications*, vol. 63, Birkhäuser Verlag, Basel, 1993.

[GH] I.C. Gohberg and G. Heinig, Inversion of finite-section Toeplitz matrices and consisting of elements of a non-commutative algebra, *Rev. Roum. Math. Pures et Appl.* **195** (1974), 623–663 (Russian).

[GK] I.C. Gohberg and M.G. Krein, Introduction to the Theory of Linear Nonselfadjoint Operators, *Transl. Math. Monographs*, vol. 18, *Amer. Math. Soc.*, Providence, RI, 1969.

[GL] I.C. Gohberg and H.J. Landau, *Prediction and the inverse of Toeplitz matrices*, in Approximation and Computation, R.V.M. Zahar, Ed., International Series of Numerical Mathematics, vol. 119 (1994), 219–229, Birkhäuser Verlag, Basel.

[GS] I.C. Gohberg and A.A. Semencul, On the inversion of finite Toeplitz matrices and their continuous analogs, *Mat. Issled.* **2** (1972), 201–233 (Russian).

[L] H.J. Landau, Maximum entropy and the moment problem, *Bull. Amer. Math. Soc.* **16** (1987), 47–77.

Robert L. Ellis
1301 Mathematics Building
University of Maryland
College Park, Maryland 20742-4015
USA

Israel Gohberg
Raymond and Beverly Sackler
Faculty of Exact Sciences
Tel Aviv University
Ramat Aviv, 69978
Israel

MSC 1991: 46C99, 46B15, 15A09, 47A68

Operator Theory:
Advances and Applications, Vol. 122
© 2001 Birkhäuser Verlag Basel/Switzerland

Linear Independence of Jordan Chains

K.-H. Förster and B. Nagy

Jordan chains of holomorphic operator functions are studied. It is shown that a "semi-canonical" system of Jordan chains is a linearly independent system if each single nontrivial Jordan chain is a linearly independent system. With the help of local or global linearizations of the operator function it is characterized when nontrivial Jordan chains are linearly independent systems. This is applied to give necessary and sufficient conditions for the existence of roots and spectral roots of regular holomorphic matrix functions.

1 Introduction

Jordan chains play an important role in investigations of spectral properties of holomorphic (or meromorphic) operator functions. Let $\mathcal{X}$ and $\mathcal{Y}$ denote complex Banach spaces and let $\mathcal{L}(\mathcal{X}, \mathcal{Y})$ denote the space of all bounded linear operators from $\mathcal{X}$ into $\mathcal{Y}$. For a holomorphic function $A(\cdot) : \Omega \to \mathcal{L}(\mathcal{X}, \mathcal{Y})$, where Ω is an open nonempty subset of the complex plane, and $\mu \in \Omega$ a sequence $(x_h)_{h=0}^{\ell}$ in $\mathcal{X}$ is called a Jordan chain of $A(\cdot)$ corresponding to μ if

$$\sum_{h=0}^{k} \frac{1}{h!} A^{(h)}(\mu) x_{k-h} = 0 \qquad \text{for } k = 0, 1, \ldots, \ell.$$

We call such a Jordan chain nontrivial, if $x_0 \neq 0$.

In general a nontrivial Jordan chain is not a linearly independent system of vectors, see ([Bau]; Chapter 9), ([LT], §14.3) and the example in the last section of this paper. But there are examples for which all nontrivial Jordan chains are linearly independent systems; the most important example is the function $A(\lambda) = \lambda I - A$, where I is the identity operator and $A \in \mathcal{L}(\mathcal{X}, \mathcal{Y})$, see ([Bau]; Chapter 9), ([LT], p. 230) and Example 2.3.2) below.

This paper deals with the problem, whether the nontrivial Jordan chains of a holomorphic operator (or matrix) function are linearly independent systems, and draws some consequences if the answer is yes. The paper contains three further parts: In the second part it is proved that "semi-canonical" systems of Jordan chains corresponding to the same point form a linearly independent system of vectors, if each single nontrivial Jordan chain corresponding to this point is a linearly independent system. Several examples of this situation are considered. In part three for three different types of functions it is characterized when all nontrivial Jordan chains of $A(\cdot)$ corresponding to μ up to a given length are linearly independent systems with the help of local or global linearizations of $A(\cdot)$. The last part studies the existence of roots (or spectral roots) of a regular holomorphic matrix function $A(\cdot)$, that is the existence of a matrix Z and a regular holomorphic matrix function B such that $A(\lambda) = B(\lambda)(\lambda I - Z)$ (and $B(\lambda)$ is invertible at the eigenvalues of Z, respectively); this is the case if and only if there exists a basis of nontrivial Jordan chains (with some additional properties in the case of spectral roots).

2 Linearly Independent Chains with Respect to a Sequence of Operators

Let $\ell \in \mathbb{N} \cup \{\infty\}$ and let $(A_n)_{0 \leq n < \ell}$ be a sequence of linear maps between two vector spaces $\mathcal{X}$ and $\mathcal{Y}$ (over the same field). We call a sequence $(x_k)_{k=0}^{p}$ in $\mathcal{X}$ satisfying $p < \ell$ and

$$\sum_{k=0}^{m} A_k \, x_{m-k} = 0 \quad \text{for } m = 0, 1, \ldots, p \tag{2.1}$$

a chain of length $p + 1$ with respect to the sequence $(A_n)_{0 \leq n < \ell}$. We call such a chain nontrivial if $x_0 \neq 0$. Note that you get from a chain $(x_k)_{k=0}^{p}$ a new one $(u_j)_{j=0}^{p+q}$ if $p + q < \ell$ and

$$u_j = 0 \quad \text{for } j = 0, \ldots, q - 1; \qquad u_{q+k} = x_k \quad \text{for } k = 0, \ldots, p.$$

Such chains occur in investigations of holomorphic operator functions, then they are called Jordan (or Keldysh) chains of those functions; but there are also such chains in other situations, see ([K]; §3.1).

The following theorem is the starting point of this article, it is a generalization of ([FN94]; Theorem 1.1).

Theorem 2.1 *Let $\ell \in \mathbb{N} \cup \{\infty\}$, let $(A_n)_{0 \leq n < \ell}$ a sequence of linear maps between two vector spaces $\mathcal{X}$ and $\mathcal{Y}$ (over the same field), and assume that each nontrivial chain of length less than or equal ℓ with respect to $(A_n)_{0 \leq n < \ell}$ is a linearly independent system. Let $q \in \mathbb{N}$ and for $i = 1, \ldots, q$ let $(x_k^{(i)})_{k=0}^{p_i}$ be chains of length less than or equal ℓ with respect to $(A_n)_{0 \leq n < \ell}$. If $\{x_0^{(1)}, x_0^{(2)}, \ldots, x_0^{(q)}\}$ is a linear independent system, then*

$$\{x_0^{(1)}, \ldots, x_{p_1}^{(1)}, x_0^{(2)} \ldots, x_{p_{q-1}}^{(q-1)}, x_0^{(q)} \ldots, x_{p_q}^{(q)}\}$$

is a linearly independent system.

Proof: W.l.o.g. we assume $p_1 \geq p_2 \geq \cdots \geq p_q$. For $k = 0, \ldots, p_1$ define $q_k = \max\{i : x_k^{(i)} \neq 0\}$, then $q_{p_1} \leq \cdots \leq q_1 \leq q_0 = q$. We prove by induction with respect to $r = 0, 1, \ldots, p_1$ that

$$\{x_k^{(i)} : 0 \leq k \leq r, \; 1 \leq i \leq q_k\}$$

is a linearly independent system. For $r = p_1$ we obtain the assertion of the theorem. For $r = 0$ this is just the assumption that $\{x_0^{(1)}, \dots, x_0^{(q)}\}$ is a linearly independent system.

Assume that the assertion holds for $r - 1$ with $r \leq p_1$ and that

$$\sum_{k=0}^{r} \sum_{i=1}^{q_k} \alpha_{ki}\, x_k^{(i)} = 0 \tag{2.2}$$

for some scalars α_{ki} $(0 \leq k \leq r,\ 1 \leq i \leq q_k)$. We will show that all α_{ki} vanish, thus the induction step will be proved.

Applying A_0 to (2.2) and using that $A_0\, x_0^{(i)} = 0$ and

$$A_0\, x_k^{(i)} = -\sum_{h=1}^{k} A_h\, x_{k-h}^{(i)} \quad \text{for } k = 1, \dots, p_i,$$

we obtain after a rearrangement of the sums

$$\sum_{h=1}^{r} A_h \sum_{j=h}^{r} \sum_{i=1}^{q_j} \alpha_{ji}\, x_{j-h}^{(i)} = 0. \tag{2.3}$$

Now define for $h = 0, 1, \dots, r$

$$z_{r-h} = \sum_{j=h}^{r} \sum_{i=1}^{q_j} \alpha_{ji}\, x_{j-h}^{(i)}.$$

Then $z_r = 0$ by (2.2) and

$$\sum_{k=0}^{m} A_k\, z_{m-k} = 0 \quad \text{for } m = 0, 1, \dots, r;$$

indeed for $m = r$ this is (2.3) since $z_r = 0$, and for $0 \leq m < r$ we obtain

$$\sum_{k=0}^{m} A_k\, z_{m-k} = \sum_{j=r-m}^{r} \sum_{i=1}^{q_j} \alpha_{ji} \sum_{k=0}^{j-r+m} A_k\, x_{j-r+m-k}^{(i)} = 0.$$

Thus $(z_k)_{k=0}^{r}$ is a chain with respect to $(A_n)_{0 \leq n < \ell}$. It is a linearly dependent system, since $z_r = 0$. Therefore $z_0 = \sum_{i=1}^{q_r} \alpha_{ri} x_0^{(i)} = 0$ by the assumption on chains with respect to $(A_n)_{0 \leq n < \ell}$, notice that $r \leq p_1 < \ell$. But then $\alpha_{ri} = 0$ for $i = 1, \dots, q_r$, since $\{x_0^{(1)}, \dots, x_0^{(q)}\}$ is a linearly independent system and $q_r \leq q$. From the assumption for $r - 1$ and from (2.2) we obtain that all scalars α_{ki} vanish. The proof is complete. □

Remark 2.2 If we have $\ell = \infty$ in Theorem 2.1, then its assertion is also true for finite or infinite sequences $(x_k^{(i)})_k$ which satisfy the assumptions.

Next we discuss some situations where such linearly independent sequences occur.

Examples 2.3 1) Let A and B be linear maps from a vector space $\mathcal{X}$ into a vector space $\mathcal{Y}$ and let $(\lambda_k)_{k=0}^{\infty}$ a sequence of scalars. Set $A_0 = \lambda_0 B + A$ and $A_k = \lambda_k B$ for $k > 0$. We assume that there exists a scalar $\hat{\lambda}$ such that $\hat{A} = \hat{\lambda} B + A$ is injective and A_0 is not injective. Then each nontrivial chain with respect to $(A_k)_{0 \le k < \infty}$ is a linearly independent system if and only if $\lambda_1 \neq 0$.

Proof: Let x_0 be a nontrivial vector in the kernel of A_0. If $\lambda_1 = 0$, then $(x_k)_{k=0}^{1}$ with $x_1 = x_0$ satisfies (2.1). This proves the "only if" part of the assertion.

Now we assume that $\lambda_1 \neq 0$, and let $(x_k)_{k=0}^{p}$ be a linearly dependent sequence with $x_0 \neq 0$ satisfying (2.1). W.l.o.g. we can assume that $(x_k)_{k=0}^{p-1}$ is a linearly independent system; otherwise we can find a smaller natural number p with this property. Thus

$$x_p = \sum_{k=0}^{p-1} \alpha_k x_k \quad \text{for some scalars } \alpha_k \text{ with } 0 \le k < p.$$

If we multiply (2.1) for $m = p - 1$ with α_{p-1} and subtract this from (2.1) for $m = p$, we obtain

$$0 = A_0 \left(\sum_{k=0}^{p-2} \alpha_k x_k \right) + B \left(\sum_{k=0}^{p-1} \beta_k x_k \right) = \hat{A} \left(\sum_{k=0}^{p-2} \alpha_k x_k \right) + B \left(\sum_{k=0}^{p-1} \gamma_k x_k \right)$$

with some scalars β_k and γ_k, respectively, and $\beta_{p-1} = \gamma_{p-1} = \lambda_1$. Since $(\lambda_0 B + A)x_0 = A_0 x = 0$ and $(\hat{\lambda} B + A)x_0 = \hat{A} x_0 \neq 0$, we have $\hat{\lambda}_0 := \hat{\lambda} - \lambda_0 \neq 0$ and $B x_0 = \hat{A}(\frac{1}{\hat{\lambda}_0} x)$. From (2.1) it follows easily that for $k = 0, \ldots, p$ there exist sequences $(\beta_{kj})_{j=0}^{k}$ of scalars such that

$$B x_k = \hat{A} \left(\sum_{j=0}^{k} \beta_{kj} x_j \right) \quad \text{and} \quad \beta_{kk} = \hat{\lambda}_0^{-1}.$$

Using this in the equation above and the injectivity of $\hat{A}$ we obtain that there exist scalars $\delta_0, \ldots, \delta_{p-2}$ such that

$$\lambda_1 x_{p-1} + \delta_{p-2} x_{p-2} + \cdots + \delta_0 x_0 = 0.$$

This contradicts the assumptions $\lambda_1 \neq 0$ and that $(x_k)_{k=0}^{p-1}$ is a linearly independent system. □

2) Let A and B be as above and the vector space be complex. Let Ω be an open and nonempty subset of the complex plane and let $\varphi : \Omega \to \mathbb{C}$ be a holomorphic function. Set $A_0(\lambda) = \varphi(\lambda)B + A$ and $A_k(\lambda) = \frac{1}{k!}\varphi^{(k)}(\lambda)B$ for $\lambda \in \Omega$ and $k = 0, 1, \ldots$. If $A_0(\hat{\lambda})$ is injective for some $\hat{\lambda} \in \Omega$, then the example above shows: Each chain $(x_k)_{k=0}^{p}$ with respect to $(A_k(\lambda_0))_{k=0}^{\infty}$ with $x_0 \neq 0$ is a linearly independent system if and only if $\varphi'(\lambda_0) \neq 0$.

This includes the well-known case that $\mathcal{X}$ is a Banach space, A is a linear bounded operator in $\mathcal{X}$, $B = \mathrm{Id}_{\mathcal{X}}$ and $\varphi(\lambda) = \lambda$: Let λ_0 be an eigenvalue of A, then each nontrivial Jordan chain $(x_k)_{k=0}^{p}$ of A corresponding to λ_0 where $p \in \mathbb{N} \cup \{\infty\}$, i.e.

$$x_0 \neq 0, \quad (\lambda_0 - A)x_0 = 0, \quad (A - \lambda_0)x_{k+1} = x_k \quad \text{for } k = 0, \ldots, p-1,$$

is a linearly independent system; notice that the resolvent set of A is nonempty.

Another example of this type, which attracted some interest in the literature (in the matrix case), is the operator polynomial $A(\lambda) = \lambda^2 I - \lambda I + A$, see ([LT], p. 526) and [BJW].

3) Let $A_k = 0$ for $k > 1$. If the ascent $\alpha(A_0 : A_1)$ of A_0 relative to A_1 is finite (see [BL]), then for each chain $(x_k)_{k=0}^{p}$ satisfying (2.1), i.e.

$$Ax_0 = 0 \quad \text{and} \quad A_0 x_m + A_1 x_{m-1} = 0 \quad \text{for } m = 1, 2, \ldots, p,$$

$x_0 \neq 0$ implies that $(x_k)_{k=0}^{p}$ is a linearly independent system.

Proof: Assume $(x_k)_{k=0}^{p}$ satisfies (2.1), $x_0 \neq 0$, and it is a linearly dependent system. As in Example 1) we assume w.l.o.g. that $(x_k)_{k=0}^{p-1}$ is a linearly independent system, then

$$x_p = \sum_{k=0}^{p-1} \alpha_k \, x_k \quad \text{for some scalars } \alpha_k.$$

Define recursively

$$x_{p+j} = \sum_{k=0}^{p-1} \alpha_k \, x_{k+j} \quad j = 0, 1, \ldots.$$

Then $(x_k)_{k=0}^{\infty}$ satisfies (2.1) for all $m \in \mathbb{N}$. This contradicts $\alpha(A_0 : A_1) < \infty$. □

In the next section we will characterize for a holomorphic operator function the property that nontrivial Jordan chains corresponding to an eigenvalue are linearly independent systems with help of a linearization. Then we will apply this example.

4) Let $\mathcal{X}$ be a complex Banach space, let Ω be an open and nonempty subset of the complex plane and let $A(\cdot) : \Omega \to \mathcal{L}(\mathcal{X})$ [= the space of all linear bounded

operators in $\mathcal{X}$] be a holomorphic operator valued function. If there exist $Z \in \mathcal{L}(\mathcal{X})$ and a holomorphic function $B(\cdot) : \Omega \to \mathcal{L}(\mathcal{X})$ such that

$$A(\lambda) = B(\lambda)(\lambda - Z) \quad \text{for all} \quad \lambda \in \Omega$$

and $B(\lambda)$ is invertible for all spectral values $\lambda \in \Omega$ of Z, then all nontrivial Jordan chains of A corresponding to an eigenvalue in Ω are linearly independent systems, see ([Ma]; Lemma 22.1 and 22.5) and use Example 2).

In Section 4 of this article we will discuss the existence of right (spectral) roots of a regular holomorphic matrix function in this context.

5) Let $L(\lambda) = \lambda^\ell - \lambda^{\ell-1}A_{\ell-1} - \cdots - A_0$ be a monic operator polynomial, where the coefficients A_k are positive operators in a Banach lattice. If the spectral radius of L is a pole of $L(\cdot)^{-1}$, then for each pole of $L(\cdot)^{-1}$ which is an element of the peripheral spectrum of L the nontrivial Jordan chains of L corresponding to this pole are linearly independent systems; see ([FN91]; part 4 of the theorem on p. 795, Corollary on p. 803), ([Ra]; Theorem 5.3, Corollary 5.4).

3 Jordan Chains of Holomorphic Operator Functions

Let $A(\cdot)$ be an operator function holomorphic on an open and nonempty subset Ω of the complex plane, whose values are in $\mathcal{L}(\mathcal{X}, \mathcal{Y})$, the set of all bounded linear operators between the Banach spaces $\mathcal{X}$ and $\mathcal{Y}$ over $\mathbb{C}$.

A sequence $(x_k)_{k=0}^p$ in $\mathcal{X}$ is called a Jordan chain of $A(\cdot)$ corresponding to $\mu \in \Omega$ if it is a chain with respect to the sequence $(\frac{1}{n!} A^{(n)}(\mu))_{n=0}^\infty$. In this section we will characterize the situation that all nontrivial Jordan chains of $A(\cdot)$ corresponding to a fixed $\mu \in \Omega$ are linearly independent systems with help of (local or global) linearizations or the local theory of $A(\cdot)$.

First we will consider the case that μ is a pole of $A^{-1}(\cdot)$. For the sake of simplicity we assume that $\mu = 0$ and Ω is a neighbourhood of 0. We will investigate the Jordan chains of $A(\cdot)$ corresponding to 0 with help of the linearization of the $\mathcal{Z}$-extension of $A(\cdot)$ introduced in ([Ba73]; III.4) and ([BKL]; §4). We begin by reviewing some notations and results of theses papers. The authors associate with $A(\cdot)$ a linear operator function $\mathbf{L}(\cdot)$, of the form $\mathbf{L}(\lambda) = \mathbf{T} + \lambda\mathbf{S}$ such that $\mathbf{T}$ and $\mathbf{S}$ are linear bounded operators acting between certain Banach spaces $\mathbf{X}$ and $\mathbf{Y}$; these spaces and operators are defined as follows:

Let $\mathcal{Z} = \ell^\infty(\mathcal{X}) = \{(x_i)_{i=1}^\infty : x_i \in \mathcal{X}, \sup \|x_i\| < \infty\}$ with its natural norm $\|(x_i)_{i=1}^\infty\| = \sup \|x_i\|$. Define $\mathbf{X} = \mathcal{X} \oplus \mathcal{Z}$ and $\mathbf{Y} = \mathcal{Y} \oplus \mathcal{Z}$ with norms $\|\binom{x_0}{z}\| = \max(\|x_0\|, \|z\|)$ and $\|\binom{y_0}{z}\| = \max(\|y_0\|, \|z\|)$, respectively. Then $\mathbf{X}$ and $\mathbf{Y}$ are complex Banach spaces.

Let A_n denote the n-th coefficient of the Taylor expansion of $A(\cdot)$ at 0. Choose $r > 0$ such that $A(\cdot)$ is holomorphic on an open neighbourhood of the closure of the open disc $\mathbb{D}_r = \{\lambda \in \mathbb{C} : |\lambda| < r\}$. The operators $\mathbf{T}$ and $\mathbf{S}$ are defined by

$$\mathbf{T}(x_i)_{i=0}^\infty = (A_0x_0, x_1, x_2, \ldots)^T$$

and

$$\mathbf{S}(x_i)_{i=0}^{\infty} = \left(\sum_{n=1}^{\infty} r^{n-1} A_n x_{n-1}, -\frac{1}{r} x_0, -\frac{1}{r} x_1, \dots \right)^T .$$

The operator $E(\lambda)$, given in the operator block form

$$\begin{pmatrix} I_{\mathcal{X}} & 0 & \dots & & & \\ \frac{\lambda}{r} I_{\mathcal{X}} & I_{\mathcal{X}} & 0 & \dots & & \\ \vdots & & \ddots & \ddots & & \\ \vdots & & & \ddots & 0 & \\ (\frac{\lambda}{r})^k I_{\mathcal{X}} & \dots & \dots & \dots & I_{\mathcal{X}} & 0 \\ \vdots & & & & & \ddots \end{pmatrix}, \qquad I_{\mathcal{X}} = \text{identity operator in } \mathcal{X},$$

acts as a bounded linear operator in $\mathbf{X}$ if $\lambda \in \mathbb{D}_r$; and the operator $G(\lambda)$, given in the block form

$$\begin{pmatrix} I_{\mathcal{X}} & G_{\mathcal{Z}}(\lambda) \\ 0 & I_{\mathcal{Z}} \end{pmatrix},$$

where $G_{\mathcal{Z}}(\lambda)(x_i)_{i=1}^{\infty} = \sum_{m=1}^{\infty} r^m (\sum_{j=1}^{\infty} \lambda^j A_{m+j}) x_m$, acts as a bounded linear operator in $\mathbf{Y}$ if $\lambda \in \mathbb{D}_r$; note that $G_{\mathcal{Z}}(\lambda)$ is a linear bounded operator in $\mathcal{Z}$ with norm less than $M_0 r (r_0 - r)^{-1}$ where $r_0 > r$ is such that $M_0 = \sum_{n=0}^{\infty} r_0^n \|A_n\| < \infty$. We have

$$G(\lambda)\,(A(\lambda) \oplus I_{\mathcal{Z}}) = (\mathbf{T} + \lambda \mathbf{S})\, E(\lambda), \quad \lambda \in \mathbb{D}_r.$$

Thus $\mathbf{L}(\cdot)$ and the $\mathcal{Z}$-extension of $A(\cdot)$ are equivalent at 0 in the sense of ([GKL]; 1.3).

By ([Ma]; Lemma 22.6), the Jordan chains of $G(\cdot)(A(\cdot) \oplus I_{\mathcal{Z}})$ and $A(\cdot) \oplus I_{\mathcal{Z}}$ corresponding to 0 coincide. It is easy to see that $(\binom{x_k}{z_k})_{k=0}^{p}$ is a Jordan chain of $A(\cdot) \oplus I_{\mathcal{Z}}$ corresponding to 0 if and only if $(x_k)_{k=0}^{p}$ is a Jordan chain of $A(\cdot)$ corresponding to 0 and $z_0 = z_1 = \dots = z_p = 0$. From the Proof of Theorem 4.6 in ([Ba73]; p. 83) it follows: Let $p \in \mathbb{N}$ and let $\xi_k = (x_{ki})_{i=0}^{\infty} \in \mathbf{X}$ for $k = 0, \dots, p$. Then the following statements are equivalent.

(I) $(\xi_k)_{k=0}^{p}$ is a Jordan chain of $\mathbf{L}(\cdot)$ corresponding to 0,

(II) $(x_{k,0})_{k=0}^{p}$ is a Jordan chain of $A(\cdot)$ corresponding to 0 and

$$x_{k,i} = r^{-i}\, x_{k-i} \text{ for } i = 0, 1, \dots, k; \qquad x_{k,i} = 0 \text{ for } i > k.$$

For $m \in \mathbb{N}$ we define $\mathcal{N}_m(A(\cdot), 0)$ to be the subspace of all x in $\mathcal{X}$ with the property that there exists a Jordan chain $(x_i)_{i=0}^{m-1}$ of $A(\cdot)$ corresponding to 0 such that $x = x_{m-1}$; see ([Ba73]; p. 55). By the equivalence mentioned above we obtain

$$\mathbf{P}[\mathcal{N}_m(\mathbf{L}(\cdot), 0)] = \mathcal{N}_m(A(\cdot), 0) \quad \text{for } m = 0, 1, \dots,$$

where $\mathbf{P} \in \mathcal{L}(\mathbf{X}, \mathcal{X})$ is defined by $\mathbf{P}(x_i)_{i=0}^{\infty} = x_0$.

Theorem 3.4 *Let $A(\cdot)$ be a locally holomorphic function defined on an open neighbourhood of the origin in $\mathbb{C}$ with values in $\mathcal{L}(\mathcal{X}, \mathcal{Y})$, let 0 be a pole of $A^{-1}(\cdot)$ and let $m \in \mathbb{N}$. Then the following statements are equivalent.*

(I) *Each nontrivial Jordan chain of $A(\cdot)$ corresponding to 0 of length less than or equal to m is a linearly independent system,*

(II) $\mathbf{P}$ *maps $\mathcal{N}_m(\mathbf{L}(\cdot), 0)$ bijectively onto $\mathcal{N}_m(A(\cdot), 0)$.*

Proof: Since 0 is a pole of $A^{-1}(\cdot)$, by ([Ba73]; Theorem 5.2, Theorem 4.6) the ascent $\alpha(\mathbf{T} : \mathbf{S}) = \alpha(\mathbf{L}(\cdot), 0)$ is finite. By Example 2.3, 3), the nontrivial Jordan chains of $\mathbf{L}(\cdot)$ corresponding to 0 are linearly independent systems.

Assume (I) holds. We have to show that the restriction of $\mathbf{P}$ to $\mathcal{N}_m(\mathbf{L}(\cdot), 0)$ is injective. Let $\xi \in \mathcal{N}_m(\mathbf{L}(\cdot), 0)$ and $\xi \neq 0$. There exists a Jordan chain $(\eta_k)_{k=0}^{m-1}$ of $\mathbf{L}(\cdot)$ corresponding to 0 with $\eta_{m-1} = \xi$. If $\eta_0 \neq 0$ set $\ell = 0$. If $\eta_0 = 0$, there exists a unique $\ell \in \mathbb{N}$ such that $0 < \ell < m$, $\eta_\ell \neq 0$ and $\eta_k = 0$ for $k = 0, \ldots, \ell - 1$. Then $(\xi_k)_{k=0}^{m-1-\ell}$ with $\xi_k = \eta_{k+\ell}$, $k = 0, \ldots, m-1-\ell$ is a nontrivial Jordan chain of $\mathbf{L}(\cdot)$ corresponding to 0 with $\xi_{m-1-\ell} = \xi$. From the connection between the Jordan chains of $\mathbf{L}(\cdot)$ and $A(\cdot)$ corresponding to 0 stated above, we obtain first that $\xi_0 = (x_i)_{i=0}^{\infty}$ with $x_i = 0$ for $i > 0$, therefore $\mathbf{P}\xi_0 = x_0 \neq 0$, since $\xi_0 \neq 0$, and then that $(\mathbf{P}\xi_k)_{k=0}^{m-1-\ell}$ is a Jordan chain of $A(\cdot)$ corresponding to 0 of length less than or equal to m. Assumption (I) implies $\mathbf{P}\xi = \mathbf{P}\xi_{m-1-\ell} \neq 0$. The implication (I) $\Rightarrow$ (II) is proved.

Assume (II) holds. Let $(x_k)_{k=0}^{\ell}$ be a nontrivial Jordan chain of $A(\cdot)$ corresponding to 0 with $\ell < m$. From the connection of the Jordan chains of $\mathbf{L}(\cdot)$ and $A(\cdot)$ stated above, we get a nontrivial Jordan chain $(\xi_k)_{k=0}^{\ell}$ of $\mathbf{L}(\cdot)$ corresponding to 0 with $\mathbf{P}\xi_k = x_k$ for $k = 0, 1, \ldots, \ell$. By the remark at the beginning of this proof, $(\xi_k)_{k=0}^{\ell}$ is a linearly independent system. Since the restriction of $\mathbf{P}$ to $\mathbf{N}_m(\mathbf{L}(\cdot), 0)$ is injective, $(x_k)_{k=0}^{\ell}$ is a linearly independent system. The implication (II) $\Rightarrow$ (I) is proved. □

In the case of a monic operator polynomial $A(\lambda) = \lambda^\ell I_{\mathcal{X}} + \lambda^{\ell-1} A_{\ell-1} + \cdots + A_0$ we can use the global linearization of $A(\cdot)$ given by the companion operator $\mathbf{C}_A$, see ([Ro]; p. 12 ff.). Here the connection between the Jordan chains of $\mathbf{L}$ with $\mathbf{L}(\lambda) = \lambda \mathbf{I}_{\mathcal{X}^\ell} - \mathbf{C}_A$ and $A(\cdot)$ corresponding to μ is as follows: Let $p \in \mathbb{N}$ and let $\xi_k = (x_{k,i})_{i=0}^{\ell-1} \in \mathcal{X}^\ell$ for $k = 0, \ldots, p$. Then the following statements are equivalent (see [LW]):

(I) $(\xi_k)_{k=0}^{p}$ is a Jordan chain of $\cdot\mathbf{I}_{\mathcal{X}^\ell} - \mathbf{C}_A$ corresponding to $\mu \in \mathbb{C}$,

(II) $(x_{k,0})_{k=0}^{p}$ is a Jordan chain of $A(\cdot)$ corresponding to $\mu \in \mathbb{C}$ and

$$x_{k,i} = \sum_{j=0}^{k} \binom{i}{j} \mu^{i-j} x_{k-j,0}.$$

for $i = 1, \ldots, \ell - 1$ and $k = 0, \ldots, p$.

Note that $\mathcal{N}_m((\cdot \mathbf{I}_{\mathcal{X}^\ell} - \mathbf{C}_A), \mu) = \mathcal{N}((\mu \mathbf{I}_{\mathcal{X}^\ell} - \mathbf{C}_A)^m) =$ kernel of $(\mu \mathbf{I}_{\mathcal{X}^\ell} - \mathbf{C}_A)^m$. It follows for a monic operator polynomial $A(\cdot)$ of degree ℓ

$$\mathbf{P}[\mathcal{N}((\mu \mathbf{I}_{\mathcal{X}^\ell} - \mathbf{C}_A)^m)] = \mathcal{N}_m(A(\cdot), \mu)$$

for all $\mu \in \mathbb{C}$ and $m \in \mathbb{N}$; here $\mathbf{P} \in \mathcal{L}(\mathcal{X}^\ell, \mathcal{X})$ is defined by $\mathbf{P}(x_i)_{i=0}^{\ell-1} = x_0$.

The proof of the following theorem proceeds very closely to that of Theorem 3.1; now we use Example 2.3, 2). We do not assume that μ is a pole of $A^{-1}(\cdot)$ in the next theorem.

Theorem 3.5 *Let $A(\cdot)$ be a monic operator polynomial, i.e.*

$$A(\lambda) = \lambda^\ell I_{\mathcal{X}} + \lambda^{\ell-1} A_{\ell-1} + \cdots + A_0 \quad \textit{for } \lambda \in \mathbb{C},$$

with $A_j \in \mathcal{L}(\mathcal{X})$ for $j = 0, \ldots, \ell - 1$, let $m \in \mathbb{N}$ and $\mu \in \mathbb{C}$. Then the following statements are equivalent.

(I) *Each nontrivial Jordan chain of $A(\cdot)$ corresponding to μ of length less than or equal to m is a linearly independent system,*

(II) $\mathbf{P}$ *maps $\mathcal{N}((\mu \mathbf{I}_{\mathcal{X}^\ell} - \mathbf{C}_A)^m)$ bijectively onto $\mathcal{N}_m(A(\cdot), \mu)$.*

For regular rational matrix functions we will use the results of [K] on zeros, corresponding eigenvectors and generalized eigenvectors of these functions, to obtain similar results as in the two theorems above. First we want to show that $\mathcal{N}_m(A(\cdot), \mu)$ is the image of the kernel $\mathcal{N}((\mu - T)^m)$ for a "local linearization T" of $A(\cdot)$ at μ. The following proposition generalizes the results of ([GLR]; Proposition 7.2.2) to a situation given in ([K]; Theorem 3.3.1).

Let Ω be an open and nonempty subset of the complex plane $\mathbb{C}$, and let $W : \Omega \to \mathbb{C}^{m \times m}$ be holomorphic and regular (that is, $\det W(\cdot) \not\equiv 0$ on the connected components of Ω). We assume for $\mu \in \mathbb{C}$ that

(I) $\{\lambda \in \mathbb{C} : 0 < |\lambda - \mu| < \delta\} \subset \Omega$ for some $\delta > 0$;

(II) there exists a triple (V, T, U) of matrices, which is a minimal realization of a regular rational $m \times m$ matrix function vanishing at infinity;

(III) the matrix function $W^{-1}(\cdot) - V(\cdot - T)^{-1}U$, which is defined and analytic on $\{\lambda \in \Omega : \lambda \neq \mu, \det W(\lambda) \neq 0, \lambda \text{ not an eigenvalue of } T\}$, has a holomorphic extension to μ.

Since (V, T, U) is a minimal realization of the rational matrix function $\lambda \mapsto V(\lambda - T)^{-1}U$, we know (see [GLR]; Theorem 2.7.3) that if μ is an eigenvalue of T, then the order of μ as a pole of the resolvent $\lambda \mapsto (\lambda - T)^{-1}$ is equal to the order of μ as a pole of $\lambda \mapsto V(\lambda - T)^{-1}U$. By (III), the principal part of the Laurent expansions of $W^{-1}(\cdot)$ and $V(\cdot - T)^{-1}U$ at μ coincides. It is given by

$$\sum_{-p}^{-1} (\lambda - \mu)^j \, C(T - \mu)^{-j-1} \, Q \, B,$$

where p denotes the order of μ as a pole of the resolvent of T, and Q denotes the residuum of this resolvent at μ. In the following we call a Jordan chain of $L(\lambda) = \lambda I - T$ corresponding to μ a Jordan chain of T corresponding to μ.

If $W(\cdot)$ is a rational matrix function, M.A. Kaashoek ([K]; Theorems 3.2.1 and 3.3.1) showed how to construct a triple (V, T, U) considered above. Nearly the same procedure can be applied to a regular holomorphic matrix function $A(\cdot)$ if it is defined on an open neighbourhood of the closure of Ω and the closure of Ω contains only finitely many zeros of det $A(\cdot)$, see ([GKvS]; §3).

Proposition 3.6 *Assume that the assumptions above on Ω, W, μ and (V, T, U) hold and that μ is an eigenvalue of T. Let $f(\cdot) : \Delta \to \mathbb{C}^m$ be a holomorphic function in an open neighbourhood Δ of μ with $f(\lambda) = \sum_{i=0}^{\infty}(\lambda - \mu)^i f_i$ if $|\lambda - \mu| < \delta$, $g(\lambda) = W(\lambda)f(\lambda) = \sum_{j=\ell}^{\infty}(\lambda - \mu)^j g_j$ if $0 < |\lambda - \mu| < \delta$ for some $\ell > 0$ and some positive δ, and define*

$$(*) \qquad x_h = \sum_{k=\ell}^{\infty}(T - \mu)^{k-h-1} Q_{-1} U g_j \quad \text{for } h = 0, 1, \ldots, \ell - 1.$$

Then $(x_h)_{h=0}^{\ell-1}$ is a Jordan chain of T corresponding to μ and

$$(**) \qquad f_h = V x_h \quad \text{for } h = 0, 1, \ldots, \ell - 1 .$$

Conversely: If $(x_h)_{h=0}^{\ell-1}$ is a Jordan chain of T corresponding to μ with some $\ell > 0$ then there exist functions $f(\cdot)$ and $g(\cdot)$ as above, especially (*) and (**) hold.

We have $Vx_0 \neq 0$ iff $x_0 \neq 0$, that is: $f(\cdot)$ is a right null function of W at μ of order at least ℓ iff $(x_h)_{h=0}^{\ell-1}$ is a nontrivial Jordan chain of T corresponding to μ.

Proof: By definition, $(x_h)_{h=0}^{\ell-1}$ is a Jordan chain of T corresponding to μ iff

$$(T - \mu)x_0 = 0 \text{ and } (T - \mu)x_{h+1} = x_h \qquad \text{for } h = 0, 1, \ldots, \ell - 2 .$$

From (*) the last $\ell - 1$ statements follow immediately. From the Laurent series of $W^{-1}(\cdot)$ at μ, say $W^{-1}(\lambda) = \sum_{j=-p}^{\infty}(\lambda - \mu)^j W_j$ for $0 < |\lambda - \mu| < \delta$, we obtain

$$f(\lambda) = W^{-1}(\lambda)g(\lambda) = \sum_{j=-p+\ell}^{\infty} (\lambda - \mu)^h \sum_{k=\ell}^{\infty} W_{h-k}\, g_k .$$

Notice that $W_{h-k} = 0$ if $k > h + p$, therefore the second sum is essentially finite. If $h < \ell$, then $\sum_{k=\ell}^{\infty} W_{h-k}g_k = V\sum_{k=\ell}^{\infty}(T - \mu)^{-h+k-1}g_k$. Therefore (**) holds. If $h < 0$, then we obtain $V(T - \mu)^{-h}x_0 = \sum_{k=\ell}^{\infty} W_{h-k}g_k = 0$, thus $V(T - \mu)^r(T - \mu)x_0 = 0$ for $r = 0, 1, 2, \ldots$. Since (V, T) is a null kernel pair (see[K], Theorem 1.1.5), we obtain $(T - \mu)x_0 = 0$.

Conversely, let $(x_h)_{h=0}^{\ell-1}$ be a Jordan chain for T corresponding to μ. (T, U) is a full range pair (see [K], Theorem 1.1.5). Consequently, there exists an essentially finite sequence $(g_j)_{j=\ell}^{\infty}$ of vectors in $\mathbb{C}^m$ such that

$$x_{\ell-1} = \sum_{j=\ell}^{\infty} (T-\mu)^{j-\ell}\, U g_j\,.$$

We have $Qx_{\ell-1} = x_{\ell-1}$, since $x_{\ell-1}$ is a generalized eigenvector of T corresponding to μ. Q and T commute, therefore we got (*) for $h = \ell - 1$. Since $x_h = (T-\mu)^{\ell-1+h} x_{\ell-1}$ for $h = 0, 1, \ldots, \ell-2$ we obtained (*).

Define the polynomial $g(\cdot)$ by $g(\lambda) = \sum_{j=\ell}^{\infty} (\lambda-\mu)^j g_j$; notice that only finitely many g_j are nonzero. We have to prove that $f(\cdot)$ given by $f(\lambda) = W^{-1}(\lambda)g(\lambda)$ if $0 < |\lambda-\mu| < \delta$ has a holomorphic extension at μ. If $f(\cdot)$ has the Laurent series $\sum_{i=-\infty}^{\infty} (\lambda-\mu)^i f_i$ for $0 < |\lambda-\mu| < \delta$, then from the Laurent series of $W^{-1}(\cdot)$ at μ (see above) we obtain $f_i = \sum_{k=\ell}^{\infty} W_{i-k} g_k$, $i \in \mathbb{Z}$. For $i < 0$ we have $f_i = \sum_{k=\ell}^{\infty} C(T-\mu)^{-i+k-1} QBg_k = C(T-\mu)^{-i}x_0 = 0$; therefore $f(\cdot)$ is holomorphic at μ.

For the last assertion of the proposition notice that x_0 is an eigenvector of T corresponding to μ, therefore $VT^j x_0 = \mu^j V x_0$ for $j = 0, 1, 2, \ldots$. Since (V, T) is a null kernel pair, $x_0 \neq 0$ holds iff $Vx_0 \neq 0$ holds. □

We use Proposition 3.6 to prove results on regular holomorphic matrix functions similar to Theorem 3.4 and Theorem 3.5.

Theorem 3.7 *Let Ω be an open and nonempty subset of the complex plane, let $A(\cdot) : \Omega \to \mathbb{C}^{m\times m}$ be a regular holomorphic matrix function. For $\mu \in \Omega$ and the triple (V, T, U) of matrices we assume*

(I) $\det A(\mu) = 0$;

(II) *(V, T, U) is a minimal realization of a regular rational $m \times m$ matrix function, which vanishes at infinity;*

(III) *the matrix function $A^{-1}(\cdot) - V(\cdot - T)^{-1}U$, which is defined and analytic on $\{\lambda \in \Omega : \det A(\lambda) \neq 0,\ \lambda \text{ not an eigenvalue of } T\}$, has a holomorphic extension to μ.*

Then the following statements hold.

1) *V maps $\mathcal{N}((\mu-T)^m)$ onto $\mathcal{N}_m(A(\cdot), \mu)$ for $m = 1, 2, \ldots$.*

2) *For $m = 1, 2, \ldots$ holds: All nontrivial Jordan chains of $A(\cdot)$ corresponding to μ of length less than or equal to m are linearly independent systems if and only if V maps $\mathcal{N}((\mu-T)^m)$ bijectively onto $\mathcal{N}_m(A(\cdot), \mu)$.*

Proof: A sequence $(f_h)_{h=0}^{\ell-1}$ is a Jordan chain of $A(\cdot)$ corresponding to μ iff for the polynomial $f(\cdot)$ with $f(\lambda) = \sum_{h=0}^{\ell-1} (\lambda-\mu)^h f_h$ the function $A(\cdot)f(\cdot)$ has at μ a zero of order at least ℓ. Therefore Statement 1) is a direct consequence of Proposition 3.6. In the second statement only the "only if" part needs a proof, since nontrivial Jordan chains of T are linearly independent systems; see Example 2.3.2). Assume that $Vx = 0$ for some $x \in \mathcal{N}((\mu - T)^m)$. We will prove that $x = 0$. Now $(x_h)_{h=0}^{m-1}$ with $x_h = (T-\mu)^{m-1-h}x$ is a Jordan chain of T corresponding to μ. By the second part of Proposition 3.6 and the remark at the beginning of this proof, $(Vx_h)_{h=0}^{m-1}$ is a Jordan chain of $A(\cdot)$ corresponding to μ. It is a linearly dependent system since $Vx_{m-1} = Vx = 0$. Thus $Vx_0 = 0$, but then $x_0 = 0$ as a consequence of the last statement in Proposition 3.6. That implies that $(Vx_{h+1})_{h=0}^{m-2}$ is a Jordan chain of $A(\cdot)$ corresponding to μ and it is a linearly dependent system. Thus, as above, $Vx_1 = 0$ and then $x_1 = 0$. After $m-1$ steps we obtain $x = 0$. □

In the next theorem we use the concepts and results on regular holomorphic matrix functions of [GKvS]; see also the next section.

Theorem 3.8 *Let* $A(\cdot) : \mathbb{C} \supset \Omega \to \mathbb{C}^{m\times m}$ *be a regular holomorphic matrix function. For* $\mu \in \Omega$, *let* k *be the order of* μ *as zero of* $\det A(\cdot)$ *and* p *be the order of* μ *as a pole of* $A^{-1}(\cdot)$. *Then the following statements are equivalent.*

- (I) *Each nontrivial Jordan chain of* $A(\cdot)$ *corresponding to* μ *is a linearly independent system.*
- (II) *For each Jordan pair* (X_μ, J_μ) *of* $A(\cdot)$ *at* μ *we have* $\operatorname{rank} X_\mu = k$.
- (III) *For each Jordan pair* (X_μ, J_μ) *of* $A(\cdot)$ *at* μ *the column vectors of* X_μ *form a basis of* $\mathcal{N}_p(A(\cdot), \mu)$.
- (IV) $\dim \mathcal{N}_p(A(\cdot), \mu) = k$.

Proof: By construction of a Jordan pair (X_μ, J_μ) of $A(\cdot)$ at μ,

$$X_\mu = (x_{1,0}, \dots, x_{1,k_1-1}, x_{2,0}, \dots, x_{s,0}, \dots, x_{s,k_s-1}) \in \mathbb{C}^{m\times k},$$

$k_1 \geq k_2 \geq \cdots \geq k_s > 0$, for $j = 1, 2, \dots, s$ the sequence $(x_{j,i})_{i=0}^{k_j-1}$ is a Jordan chain of $A(\cdot)$ corresponding to μ, and $x_{j,0}$ is such that for each Jordan chain $(u_i)_{i=0}^{\ell-1}$ of $A(\cdot)$ corresponding to μ with $u_0 \notin \operatorname{span}\{x_{1,0}, x_{2,0}, \dots, x_{j-1,0}\}$ we have $\ell \leq k_j$, the sequence $(x_{i,0})_{i=1}^{s}$ forms a basis of $\mathcal{N}(A(\mu))$, $p = k_1$, $k = k_1 + \cdots + k_s$ and $\operatorname{span}\{x_{1,0}, \dots, x_{s,k_s-1}\} = \operatorname{range} X_\mu$, see ([GKvS]; Section 2). With Theorem 2.1 the implication (I) ⇒ (II) follows immediately.

We will prove in the final part of this proof the following claim: Let (X_μ, J_μ) be a Jordan pair of $A(\cdot)$ at μ with $X_\mu = (x_{1,0}, \dots, x_{s,k_s-1})$ and let $(x_\ell)_{\ell=0}^{q}$ be a Jordan chain of $A(\cdot)$ corresponding to μ with $q < k_1 = p$. Then there exist scalars $\alpha_{j,i}$, $i = 0, \dots, q$, $j = 1, \dots, s_{q-i}$, such that

$$(***) \qquad x_\ell = \sum_{i=0}^{\ell} \sum_{j=1}^{s_{q-i}} \alpha_{j,i}\, x_{j,\ell-i} \qquad \ell = 0, 1, \dots, q,$$

where $s_i = \max\{j : k_j - 1 \geq i\}$ for $0 \leq i < k_1$.

This claim implies that range $X_\mu = \operatorname{span}\{x_{1,0}, \dots, x_{s,k_s-1}\} = \mathcal{N}_p(A(\cdot), \mu)$. Therefore the implications (II) $\Leftrightarrow$ (III) $\Leftrightarrow$ (IV) are evident.

For the proof of the implication (III) $\Rightarrow$ (I) let $(x_\ell)_{\ell=0}^q$ be a Jordan chain of $A(\cdot)$ corresponding to μ, which is a linearly dependent system. We have to show that the chain is trivial, i.e. $x_0 = 0$. Assume $x_0 \neq 0$, then $1 \leq q < k_1$, w.l.o.g. we can assume that $(x_i)_{i=0}^{q-1}$ is a linearly independent system. From (***) we obtain

$$\sum_{i=0}^{q} \sum_{j=1}^{s_{q-i}} \alpha_{j,i}\, x_{j,q-i} = x_q = \sum_{\ell=0}^{q-1} \beta_\ell\, x_\ell = \sum_{\ell=0}^{q-1} \sum_{i=0}^{\ell} \sum_{j=1}^{s_{q-i}} \beta_\ell\, \alpha_{j,i}\, x_{j,\ell-i}\,.$$

for some scalars β_ℓ, $0 \leq \ell < q$. Thus $\sum_{j=1}^{s_q} \alpha_{j,0}\, x_{j,q}$ belongs to $\operatorname{span}\{x_{j,i} : 0 \leq i < q,\ 1 \leq j \leq s_i\}$. But $(x_{1,0}, \dots, x_{s,k_s-1})$ is a linearly independent system by (III), thus $\alpha_{j,0} = 0$ for $j = 1, \dots, s_q$. Then (***) for $\ell = 0$ implies $x_0 = 0$. This contradicts our assumption $x_0 \neq 0$.

Proof of the Claim: Note that $1 \leq s_{k_1} \leq \cdots \leq s_i \leq \cdots \leq s_0 = s$ and $j \leq s_i \;\Leftrightarrow\; i \leq k_j - 1$. Since x_0 is in the kernel of $A(\mu)$ and $(x_{j,0})_{j=1}^s$ is a basis of this kernel, we have $x_0 = \sum_{j=1}^s \alpha_{j,0}\, x_{j,0}$ for some scalars $\alpha_{j,0}$, $j = 1, \dots, s$. Define $u_i = x_0 - \sum_{j=1}^{s_q} \alpha_{j,0}\, x_{j,i}$ for $i = 0, \dots, q$; these vectors are well-defined since $k_j - 1 \geq q$ if $j \leq s_q$. Further u_0 belongs to the kernel of $A(\mu)$ and $(u_i)_{i=0}^q$ is a Jordan chain of $A(\cdot)$ corresponding to μ. Assume that x_0 is linearly independent of $x_{1,0}, \dots, x_{s_q,0}$. Then u_0 is linearly independent of these vectors. Now $q > k_j - 1$ if $j > s_q$; this contradicts the construction of X_μ. We have proved (***) for $\ell = 0$, and proceed by induction.

Let (***) be true for $x_0, \dots, x_{\ell-1}$ with $\ell - 1 < q$. Then (note that $A_h = \frac{1}{h!} A^{(h)}(\mu)$ for $h = 0, 1, 2 \dots$)

$$\begin{aligned} 0 &= \sum_{h=0}^{\ell} A_{\ell-h}\, x_h \\ &= \sum_{h=0}^{\ell-1} A_{\ell-h} \sum_{i=0}^{h} \sum_{j=1}^{s_{q-i}} \alpha_{j,i}\, x_{j,k-i} + A_0\, x_\ell \\ &= \sum_{i=0}^{\ell-1} \sum_{h=0}^{\ell-1-i} \sum_{j=1}^{s_{q-i}} A_{\ell-i-h}\, \alpha_{j,i}\, x_{j,h} + A_0\, x_\ell \\ &= A_0 \left(x_\ell - \sum_{i=0}^{\ell-1} \sum_{j=1}^{s_{q-i}} \alpha_{j,i}\, x_{j,\ell-i} \right). \end{aligned}$$

For $h = 0, 1, \ldots, q$ define

$$u_h = x_h - \sum_{i=0}^{\ell-1} \sum_{j=1}^{s_{q-i}} \alpha_{j,i}\, x_{j,h-i}$$

with $x_{h-i} = 0$ if $h < i$, then $u_h = 0$ for $0 \leq h < \ell$, $A(\mu)u_\ell = A_0\, u_\ell = 0$, and $(u_h)_{h=0}^{q}$ is a Jordan chain of $A(\cdot)$ corresponding to μ. Thus $(v_i)_{i=0}^{q-\ell}$ is a Jordan chain of $A(\cdot)$ corresponding to μ, where $v_i = u_{\ell+i}$ for $i = 0, \ldots, q-\ell$. Therefore, by the first equation in (***) there exist $\alpha_{j,\ell}$ for $1 \leq j \leq s_{q-\ell}$ with

$$u_\ell = v_0 = \sum_{j=1}^{s_{q-\ell}} \alpha_{j,\ell}\, x_{j,0}\,.$$

This proves (***) for ℓ. The proofs of the claim and the theorem are complete. □

4 Right (spectral) Roots of a Regular Holomorphic Matrix Function

We consider a regular holomorphic matrix function $A(\cdot)$, that is, $A(\lambda)$ is a complex square (say $m \times m$) matrix, λ varies in an open and nonempty subset Ω of $\mathbb{C}$, $A(\lambda)$ depends holomorphically of λ and the scalar function $\det A(\cdot)$ does not vanish identically on any component of Ω. We use in the following the concepts and notations of [GKvS]. Further, by $\sigma(A(\cdot))$ we denote the set $\{\lambda \in \Omega : \det A(\lambda) = 0\}$; under our assumptions $\sigma(A(\cdot))$ is a discrete subset of Ω. For a matrix $Z \in \mathbb{C}^{m\times m}$ we denote, as usual, by $\sigma(Z)$ the set of eigenvalues of Z. By $J(\mu, p)$ we denote the $p \times p$ Jordan block with eigenvalue μ.

The next two theorems are in a certain sense the converse of ([Ma]; Lemma 22.1 and 22.5) in the matrix case; see Example 2.3, 4). The following theorem was proved in ([La]; Satz 2) for monic matrix polynomials; in this case the matrix Z defined below is a root of the matrix polynomial in the sense of ([Ma]; p. 114).

Theorem 4.9 *Let $A(\cdot) : \Omega \to$ be a regular holomorphic matrix function and let σ be a subset of $\sigma(A(\cdot))$. Then the following statements are equivalent.*

(I) *There exists a regular holomorphic matrix function $B(\cdot) : \Omega \to \mathbb{C}^{m\times m}$ and a matrix $Z \in \mathbb{C}^{m\times m}$ such that $\sigma(Z) = \sigma$ and $A(\lambda) = B(\lambda)(\lambda - Z)$ for all $\lambda \in \Omega$.*

(II) *There exist a basis $\{x_1, \ldots, x_m\}$ of $\mathbb{C}^m$, $\ell \in \mathbb{N}$ and $p_i \in \mathbb{N}$ for $i = 1, \ldots, \ell$ such that*

$$\{x_1, \ldots, x_m\} = \{\{x_0^{(i)}, \ldots, x_{p_i}^{(i)}\} : i = 1, \ldots, \ell\},$$

$(x_k^{(i)})_{k=0}^{p_i}$ is a Jordan chain of $A(\cdot)$ corresponding to some $\mu \in \sigma$ for $i = 1, \dots, \ell$, and for each $\mu \in \sigma$ there exists in this system at least one Jordan chain of $A(\cdot)$ corresponding to μ.

Proof: The implication (I) $\Rightarrow$ (II) follows from ([Ma]; Lemma 22.1) and the well-known fact that $\mathbb{C}^m$ has a basis of Jordan chains of Z, see ([GLR]; Theorem 2.2.4).

Assume (II) holds. For $i = 1, \dots, \ell$ let μ_i denote that point of σ to which the Jordan chain $(x_k^{(i)})_{k=0}^{p_i}$ corresponds. Let X be the row block matrix row $[X_i]_{i=1}^{\ell}$ where X_i is the $m \times (p_i + 1)$ matrix with columns $x_0^{(i)}, \dots, x_{p_i}^{(i)}$. By ([GKvS]; Lemma 2.1), the matrix function $C(\cdot)$ with

$$\begin{aligned} C(\lambda) &= A(\lambda)X(\operatorname{diag}[\lambda - J(\mu_i, p_i + 1)]_{i=1}^{\ell})^{-1} \\ &= \operatorname{row}[A(\lambda)X_i(\lambda - J(\mu_i, p_i + 1))^{-1}]_{i=1}^{\ell} \end{aligned}$$

for $\lambda \in \Omega \backslash \sigma$ has a unique holomorphic extension to Ω; we denote this extension again by $C(\cdot)$. Since X is invertible, $B(\cdot)$ with $B(\lambda) = C(\lambda)X^{-1}$ for $\lambda \in \Omega$ and $Z = X \operatorname{diag}[J(\mu_i, p_i + 1)]_{i=1}^{\ell} X^{-1}$ satisfy statement (I). □

The next theorem was proved for monic matrix polynomials in ([FN94]; Satz 1.4); in this case the matrix Z defined below is a spectral root of the matrix polynomial in the sense of ([Ma]; p. 115).

Theorem 4.10 *Let $A(\cdot) : \Omega \rightarrow$ be a regular holomorphic matrix function and let σ be a subset of $\sigma(A(\cdot))$. Then the following statements are equivalent.*

(I) *There exists a regular holomorphic matrix function $B(\cdot) : \Omega \rightarrow \mathbb{C}^{m \times m}$ and a matrix $Z \in \mathbb{C}^{m \times m}$ such that $\sigma(Z) = \sigma$, $A(\lambda) = B(\lambda)(\lambda - Z)$ for all $\lambda \in \Omega$, and $B(\lambda)$ is invertible for all $\lambda \in \sigma$.*

(II) *For all $\mu \in \sigma$ the nontrivial Jordan chains of $A(\cdot)$ corresponding to μ are linearly independent systems and $\mathbb{C}^m$ is the direct sum of the subspaces*

$$\begin{aligned} &\mathcal{N}(A(\cdot), \mu) \\ &\quad = \{x : \text{there exists a Jordan chain } (x_k)_{k=0}^{p} \text{ of } A(\cdot) \\ &\quad \text{corresponding to } \mu \text{ with } x_p = x\}, \quad \mu \in \sigma . \end{aligned}$$

Proof: The implication (I) $\Rightarrow$ (II) follows from ([Ma]; Lemma 22.5) and ([GLR], Theorem 2.2.4).

Assume (II) holds. By Theorem 3.8 for each $\mu \in \sigma$ and for each Jordan pair (X_μ, J_μ) the columns of X_μ form a basis of $\mathcal{N}(A(\cdot), \mu)$. Now the row block matrix $X = \operatorname{row}[X_\mu]_{\mu \in \sigma}$ is invertible. By ([GKvS]; Lemma 2.1), the matrix function $C(\cdot)$ with

$$C(\lambda) = A(\lambda)X(\operatorname{diag}[\lambda - J_\mu]_{\mu \in \sigma})^{-1}$$

for $\lambda \in \Omega\backslash\sigma$ has a unique holomorphic extension to Ω, which we denote again by $C(\cdot)$. Define $B(\cdot)$ by $B(\lambda) = C(\lambda)X^{-1}$ for $\lambda \in \Omega$ and $Z = X \operatorname{diag}[J_\mu]_{\mu\in\sigma} X^{-1}$. Then $A(\cdot)$ and $L(\cdot)$ with $L(\lambda) = \lambda - Z$ have the same Jordan pairs for all $\mu \in \sigma$. By ([GKvS]; Theorem 2.6), $B(\mu)$ is invertible at all $\mu \in \sigma$. Thus $B(\cdot)$ and Z satisfy statement (I). □

The following example shows the difference between the last two theorems. In Theorem 4.9 a nontrivial Jordan chain of $A(\cdot)$ corresponding to $\mu \in \sigma$ need not be a linearly independent system.

Example 4.11 Consider the monic matrix polynomial

$$\begin{aligned} A(\lambda) &= \begin{bmatrix} \lambda(\lambda-1) & -1 \\ 0 & (\lambda-1)^2 \end{bmatrix} \\ &= \left(\lambda - \begin{bmatrix} 0 & -1 \\ 0 & 1 \end{bmatrix}\right)\left(\lambda - \begin{bmatrix} 1 & 1 \\ 0 & 1 \end{bmatrix}\right). \end{aligned}$$

Then $\sigma(A(\cdot)) = \{0, 1\}$ and $\sigma\left(\begin{bmatrix} 1 & 1 \\ 0 & 1 \end{bmatrix}\right) = \{1\}$. A Jordan chain of $A(\cdot)$ corresponding to 1 is given by

$$\begin{bmatrix} 1 \\ 0 \end{bmatrix}, \begin{bmatrix} 0 \\ 1 \end{bmatrix}, \begin{bmatrix} * \\ 1 \end{bmatrix};$$

it is a linearly dependent system. But the two first vectors of this chain form a basis of $\mathbb{C}^2$. By Theorems 4.9 and 4.10, the matrix $\begin{bmatrix} 1 & 1 \\ 0 & 1 \end{bmatrix}$ is a root of $A(\cdot)$ but it is not a spectral root of $A(\cdot)$. Of course, this follows immediately from the factorization of $A(\cdot)$ given above.

References

[Ba73] H. Bart, Meromorphic Operator Valued Functions. Thesis. Vrije Universiteit, Amsterdam, 1973.

[Ba74] H. Bart, Poles of the Resolvent of an Operator Function. *Proc. Roy. Irish Acad.* **74**A (1974), 169–184.

[Bau] H. Baumgärtel, Analytic Perturbation Theory for Matrices and Operators, Operator Theory: Advances and Applications, vol. 15, Birkhäuser: Basel-Boston-Stuttgart, 1985.

[BKL] H. Bart, M.A. Kaashoek and D.C. Lay, Stability Properties of Finite Meromorphic Operator Functions. *Proc. Acad. Sci.* Amsterdam A**77** (1974), 217–259.

[BJW] G.J. Butler, C.R. Johnson and H. Wolkowicz, Nonnegative Solutions of a Quadratic Matrix Equation Arising from Comparison Theorems in Ordinary Differential Equations. *SIAM J. Alg. Disc. Meth.* **6** (1985), 47–53.

[BL] H. Bart and D.C. Lay, Poles of a Generalized Resolvent Operator. *Proc. Roy. Irish Acad.* **74**A (1974), 147–168.

[FN91] K.-H. Förster and B. Nagy, Some Properties of the Spectral Radius of a Monic Operator Polynomial with Nonnegative Compact Coefficients. *Integral Equations and Operator Theory* **14** (1991), 794–805.

[FN94] K.-H. Förster and B. Nagy, Spektraleigenschaften von Matrix- und Operatorpolynomen. *Sitzungsberichte d. Berliner Math. Gesellschaft*, 1994, 243–262.

[GKL] I. Gohberg, M.A. Kaashoek and D.C. Lay, Equivalence, Linearization, and Decomposition of Holomorphic Operator Functions. *J. Funct. Anal.* **28** (1978), 102–144.

[GKvS] I. Gohberg, M.A. Kaashoek and F. van Schagen, On the Local Theory of Regular Analytic Matrix Functions. *Linear Algebra Appl.* **182** (1993), 9–25.

[GLR] I. Gohberg, P. Lancaster and L. Rodman, Invariant Subspaces of Matrices with Applications. Wiley-Interscience: New York, 1986.

[K] M.A. Kaashoek, State-space Theory of Rational Matrix Functions and Applications, Lectures on Operator Theory and its Applications, Fields Institute Monographs, vol. 3, *Amer. Math. Soc., Providence*, 1996, 235–333.

[LW] P. Lancaster and P.N. Webber, Jordan Chains and Lambda Matrices. *Linear Algebra Appl.* **1** (1968), 563–569.

[La] H. Langer, Über Lancaster's Zerlegung von Matritzen-Scharen. *Archive Rat. Mech. Appl.* **29** (1968), 75–80.

[LT] P. Lancaster and M. Tismenetsky, The Theory of Matrices, 2nd. ed. Academic Press: New York, 1985.

[Ma] A.S. Marcus, Introduction to the Spectral Theory of Polynomial Operator Pencils. Translation of Mathematical Monograph, vol. 71, *Amer. Math. Soc., Providence*, 1988.

[Ra] R.T. Rau, On the Peripheral Spectrum of a Monic Operator Polynomial with Positive Coefficients. *Integral Equations and Operator Theory* **15** (1992), 479–495.

[Ro] L. Rodman, An Introduction to Operator Polynomials, Operator Theory: Advances and Applications, vol. 38. Birkhäuser: Basel-Boston-Berlin, 1989.

K.-H. Förster
Department of Mathematics
Technical University Berlin
Sekr. MA 6-4
Strasse des 17. Juni 135
D-10623 Berlin
Germany
e-mail: foerster@math.tu-berlin.de

B. Nagy
Department of Analysis
Institute of Mathematics
Technical University Budapest
H-1521 Budapest
Hungary
e-mail: bnagy@math.bme.hu

AMS Classification Numbers: 15A22, 47A46, 15A23

Operator Theory:
Advances and Applications, Vol. 122
© 2001 Birkhäuser Verlag Basel/Switzerland

Weighted Nevanlinna-Pick Interpolation

A.E. Frazho

This paper is dedicated to my friend and colleague Professor Kaashoek on the occasion of his sixtieth birthday

This paper uses commutant lifting techniques along with the state space method, to solve a weighted and hyper-weighted Nevanlinna-Pick interpolation problem. Necessary and sufficient conditions for the existence of a solution are given. Some specific state space formulas to calculate the corresponding interpolants are presented.

0 Introduction

By using a fixed point theorem Treil-Voberg [20] presented a generalization of the commutant lifting theorem to include the case when a certain operator is an expansion. By applying standard commutant lifting methods [2] obtained a generalization of the Treil-Voberg result, and gave some explicit formulas to compute the appropriate intertwining lifting. Starting with Glover [9] state space methods have been widely used to solve H^∞ interpolation problems; see [1, 3, 5, 8, 12, 14, 21] for further results and a history of this research. In this note we will use the commutant lifting techniques in [2, 5] along with the state space method, to solve a weighted and hyper-weighted Nevanlinna-Pick interpolation problem. Some state space solutions will be given. To be more specific, let B be an operator from $\mathcal{U}$ into $\mathcal{X}$ and Z an operator on $\mathcal{X}$. Throughout we will always assume that $\mathcal{U}$ and $\mathcal{X}$ are finite dimensional, and Z is stable, that is, all the eigenvalues of Z are in the open unit disc. The *controllability operator* corresponding to the pair $\{Z, B\}$ is the operator from $l_+^2(\mathcal{U})$ into $\mathcal{X}$ defined by

$$W = [B, ZB, Z^2B, Z^3B, \ldots]. \tag{1}$$

Because Z is stable, W is a bounded operator. Let B_1 be an operator from $\mathcal{Y}$ into $\mathcal{X}$ where $\mathcal{Y}$ is some finite dimensional space. Let W_1 be the controllability operator from $l_+^2(\mathcal{Y})$ into $\mathcal{X}$ generated by the pair $\{Z, B_1\}$, that is, W_1 is the operator defined by (1) where B_1 replaces B. Obviously, W_1 is a bounded operator. Let F be a function in $H^\infty(\mathcal{Y}, \mathcal{U})$, and let $F(\lambda) = \sum_0^\infty F_n\lambda^n$ be its power series expansion for $|\lambda| < 1$. (Throughout $H^\infty(\mathcal{E}, \mathcal{F})$ denotes the Hardy space of all uniformally bounded analytic functions in the open unit disc, whose values are linear operators from $\mathcal{E}$ into $\mathcal{F}$.) Then the *Toeplitz* operator T_F is the block matrix from $l_+^2(\mathcal{Y})$ into $l_+^2(\mathcal{U})$ whose entries are given by $(T_F)_{i,j} = F_{i-j}$ for all integers $i \geq 0$ and $j \geq 0$.

In other words,

$$(2) \qquad T_F = \begin{bmatrix} F_0 & 0 & 0 & 0 & \cdots \\ F_1 & F_0 & 0 & 0 & \cdots \\ F_2 & F_1 & F_0 & 0 & \cdots \\ F_3 & F_2 & F_1 & F_0 & \cdots \\ \vdots & \vdots & \vdots & \vdots & \vdots \end{bmatrix}.$$

The function F is called the *symbol* for the Toeplitz operator T_F. It is well known that $\|T_F\| = \|F\|_\infty$. We say that F is an *interpolant* for $\{Z, B, B_1\}$, or equivalently, for $\{W, W_1\}$ if F is a function in $H^\infty(\mathcal{Y}, \mathcal{U})$ satisfying $WT_F = W_1$.

In this note we only consider square outer functions. So, we call a function Θ *outer* if Θ is a function in $H^\infty(\mathcal{Y}, \mathcal{Y})$ and $\Theta H^2(\mathcal{Y})$ is dense in $H^2(\mathcal{Y})$, or equivalently, the range of T_Θ is dense in $l_+^2(\mathcal{Y})$. Throughout we always assume that $\mathcal{Y}$ is finite dimensional. Let Θ be an outer function in $H^\infty(\mathcal{Y}, \mathcal{Y})$. Then our weighted Nevanlinna-Pick interpolation problem is to find a function F in $H^\infty(\mathcal{Y}, \mathcal{U})$ satisfying

$$(3) \qquad WT_F = W_1 \text{ and } F(e^{\imath\omega})^* F(e^{\imath\omega}) \leq \Theta(e^{\imath\omega})^* \Theta(e^{\imath\omega})$$

for almost all ω in $[0, 2\pi)$. The set $\{Z, B, B_1, \Theta\}$ is the *data* for this interpolation problem. Here we will show that this interpolation problem has a solution if and only if there exists an operator A from $l_+^2(\mathcal{Y})$ into $l_+^2(\mathcal{U})$ satisfying $WA = W_1$ and $A^*A \leq T_\Theta^* T_\Theta$. Then we will provide an explicit state space solution when Θ is rational. Finally, it is noted that if $\Theta = \gamma I$ where $\gamma > 0$, then this weighted Nevanlinna-Pick interpolation problem reduces to the standard Nevanlinna-Pick interpolation problem discussed throughout the monograph [5]. In fact, by inverting Θ one can also convert this weighted Nevanlinna-Pick problem to the standard Nevanlinna-Pick interpolation problem.

Notice that $F(e^{\imath\omega})^* F(e^{\imath\omega}) \leq \Theta(e^{\imath\omega})^* \Theta(e^{\imath\omega})$ almost everywhere on $[0, 2\pi)$ if and only if $T_F^* T_F \leq T_\Theta^* T_\Theta$. This leads to the following hyper-weighted Nevanlinna-Pick interpolation problem, which is the main emphasis of this note: Given a positive operator Q on $l_+^2(\mathcal{Y})$, find a function F in $H^\infty(\mathcal{Y}, \mathcal{U})$ satisfying

$$(4) \qquad WT_F = W_1 \text{ and } T_F^* T_F \leq Q.$$

In this case $\{Z, B, B_1, Q\}$ is the *data* for the hyper-weighted Nevanlinna-Pick interpolation problem. If F is a function in $H^\infty(\mathcal{Y}, \mathcal{U})$ satisfying (4), then we say that F is a *solution* to the hyper-weighted Nevanlinna-Pick interpolation problem. Obviously, if $Q = T_\Theta^* T_\Theta$, then the hyper-weighted Nevanlinna-Pick interpolation problem reduces to the weighted Nevanlinna-Pick interpolation problem. Here we will follow some of the ideas in [2, 5] to solve the hyper-weighted Nevanlinna-Pick interpolation problem for certain Q. Some specific state space solutions will be given in the rational case. A simple numerical example is presented to demonstrate the difference between the solutions to the weighted and hyper-weighted interpolation problems for the same data $\{Z, B, B_1\}$. Finally, it noted that weighted interpolation problems naturally occur in H^∞ control theory; see [3, 5, 7, 8, 14, 21]. So, the hyper-weighted interpolation problem may be useful in control systems.

1 Preliminaries

In this section we will establish some notation and preliminary results. Recall that an operator C is a contraction if $\|C\| \le 1$. Let T' be a contraction on $\mathcal{H}'$. Recall that S on $\mathcal{K} \supseteq \mathcal{H}'$ is an *isometric lifting* of T' if S is an isometry, $\mathcal{H}'$ is an invariant subspace for S^* and $S^*|\mathcal{H}' = T'^*$. Let P' denote the orthogonal projection onto $\mathcal{H}'$. Then S is an isometric lifting of T' if and only if S is an isometry and $P'S = T'P'$. Let T be an isometry on $\mathcal{H}$. Let A be an operator intertwining T with T', that is, assume that A is an operator from $\mathcal{H}$ into $\mathcal{H}'$ satisfying $T'A = AT$. We say that Γ is an *intertwining lifting* of A with respect to S, if Γ is an operator from $\mathcal{H}$ into $\mathcal{K}$ satisfying $S\Gamma = \Gamma T$ and $P'\Gamma = A$. If A is a contraction, and S is an isometric lifting of T', then the Sz.-Nagy-Foias commutant lifting theorem shows that there exists a contractive intertwining lifting Γ of A; see [4, 5, 18, 19]. The set of all contractive liftings of A is parameterized by the closed unit ball in some $H^\infty(\mathcal{E}, \mathcal{F})$; see [4, 5] for further details. Let us begin with the following result taken form [2].

Theorem 1 ([2]). *Let Q be a positive operator on $\mathcal{H}$ and T an operator on $\mathcal{H}$ satisfying $Q \le T^*QT$. Let A be an operator intertwining T with the contraction T' on $\mathcal{H}'$. Finally, let S be an isometric lifting of T'. If $A^*A \le Q$, then there exists an intertwining lifting Γ of A satisfying $\Gamma^*\Gamma \le Q$.*

Recall that an operator T is a unilateral shift on $l^2_+(\mathcal{Y})$, if T can be identified with the block matrix on $l^2_+(\mathcal{Y})$ with the identity I immediately below the main diagonal and zeros elsewhere. In other words, T is a unilateral shift if $T = T_{\lambda I}$ where $T_{\lambda I}$ is the Toeplitz operator on $l^2_+(\mathcal{Y})$ whose symbol is $F(\lambda) = \lambda I$.

Theorem 2 *Let $\{Z, B, B_1, Q\}$ be the data for the hyper-weighted Nevanlinna-Pick interpolation problem where Q is a positive operator on $l^2_+(\mathcal{Y})$. Assume that $Q \le T^*QT$ where T is the unilateral shift on $l^2_+(\mathcal{Y})$. Let W (respectively W_1) be the controllability operator generated by the pair $\{Z, B\}$ (respectively $\{Z, B_1\}$). Let $\mathcal{H}'$ be the range of W^*. Then there exists a solution to the hyper-weighted Nevanlinna-Pick interpolation problem if and only if there exists an operator A from $l^2_+(\mathcal{Y})$ into $\mathcal{H}'$ satisfying $WA = W_1$ and $A^*A \le Q$.*

Proof: Assume that F is a solution to the hyper-weighted Nevanlinna-Pick interpolation problem, that is, assume that (4) holds. Let A be the operator from $l^2_+(\mathcal{Y})$ into $\mathcal{H}'$ defined by $A = P'T_F$ where P' is the orthogonal projection onto $\mathcal{H}'$. Since $WT_F = W_1$, it follows that $WA = W_1$. Using $T_F^*T_F \le Q$, we have $A^*A \le Q$.

On the other hand, assume that A is an operator from $l^2_+(\mathcal{Y})$ into $\mathcal{H}'$ satisfying $WA = W_1$ and $A^*A \le Q$. Let S be the unilateral shift on $l^2_+(\mathcal{U})$, and let T' the compression of S to $\mathcal{H}'$, that is, $T' = P'S|\mathcal{H}'$. Notice that $WS = ZW$. Thus, $S^*W^* = W^*Z^*$, and it follows that $\mathcal{H}'$ is an invariant subspace for S^*. Hence, S is an isometric lifting for T'. In particular, $P'S = T'P'$. We claim that A intertwines T with T'. Using $WA = W_1$ and $W_1T = ZW_1$, we obtain

$$WAT = W_1T = ZW_1 = ZWA = WSA = WP'SA = WT'A. \tag{5}$$

Because $W|\mathcal{H}'$ is one to one, we have $T'A = AT$. By employing Theorem 1, there exists an operator Γ from $l^2_+(\mathcal{Y})$ into $l^2_+(\mathcal{U})$ satisfying the following three conditions $\Gamma T = S\Gamma$ and $P'\Gamma = A$ and $\Gamma^*\Gamma \le Q$. Since Γ intertwines two unilateral shifts, there exists a function F in $H^\infty(\mathcal{Y}, \mathcal{U})$ satisfying $\Gamma = T_F$. Obviously, $T_F^*T_F \le Q$. Moreover, using $P'\Gamma = A$, we have $WT_F = WP'\Gamma = WA = W_1$. Therefore (4) holds and this F is a solution to the hyper-weighted Nevanlinna-Pick interpolation problem. This completes the proof. □

By setting $Q = T_\Theta^*T_\Theta$ in the above theorem, we readily obtain the following result.

Corollary 1 *Let Θ be an outer function in $H^\infty(\mathcal{Y}, \mathcal{Y})$. Then there exists a solution to the weighted Nevanlinna-Pick interpolation problem with data $\{Z, B, B_1, \Theta\}$ if and only if there exists an operator A from $l^2_+(\mathcal{Y})$ into $\mathcal{H}'$ satisfying $WA = W_1$ and $A^*A \le T_\Theta^*T_\Theta$.*

Let Φ be a function in $L^\infty(\mathcal{Y}, \mathcal{Y})$ and $\Phi = \sum \Phi_n e^{\imath n\omega}$ be its Fourier series expansion. Then the Toeplitz operator generated by Φ is the block matrix on $l^2_+(\mathcal{Y})$ defined by

$$T_\Phi := \begin{bmatrix} \Phi_0 & \Phi_{-1} & \Phi_{-2} & \cdots \\ \Phi_1 & \Phi_0 & \Phi_{-1} & \cdots \\ \Phi_2 & \Phi_1 & \Phi_0 & \cdots \\ \vdots & \vdots & \vdots & \vdots \end{bmatrix}. \tag{6}$$

If T is the unilateral shift on $l^2_+(\mathcal{Y})$, then $T^*T_\Phi T = T_\Phi$. Recall that Υ is a *Toeplitz* operator on $l^2_+(\mathcal{Y})$, if $T^*\Upsilon T = \Upsilon$. It is well known that Υ is a Toeplitz operator if and only if $\Upsilon = T_\Phi$ where Φ is a function in $L^\infty(\mathcal{Y}, \mathcal{Y})$. In this case, Φ is called the *symbol* for Υ. Recall that H is a *Hankel* operator form $l^2_+(\mathcal{Y})$ into $l^2_+(\mathcal{E})$, if $S_\mathcal{E}^*H = HT$ where $S_\mathcal{E}$ is the unilateral shift on $l^2_+(\mathcal{E})$. Therefore, H is a Hankel operator if and only if H admits a block matrix decomposition of the form

$$H = \begin{bmatrix} H_1 & H_2 & H_3 & \cdots \\ H_2 & H_3 & H_4 & \cdots \\ H_3 & H_4 & H_5 & \cdots \\ \vdots & \vdots & \vdots & \vdots \end{bmatrix}, \tag{7}$$

where $\{H_j\}_1^\infty$ are a sequence of operators from $\mathcal{Y}$ into $\mathcal{E}$. Clearly, the sequence $\{H_j\}_1^\infty$ and the Hankel operator H uniquely determine each other. So, we say that

H in (7) is the Hankel operator generated by the sequence $\{H_j\}_1^\infty$. The following result was obtained in collaboration with C. Foias in connection with some other research.

Proposition 1 *Let T be the unilateral shift on $l_+^2(\mathcal{Y})$, and Q a positive operator on $l_+^2(\mathcal{Y})$. Then $T^*QT \geq Q$ if and only if Q admits a decomposition of the form*

$$Q = \Upsilon - H^*H \tag{8}$$

where Υ is a positive Toeplitz operator on $l_+^2(\mathcal{Y})$ and H is a Hankel operator from $l_+^2(\mathcal{Y})$ into some $l_+^2(\mathcal{E})$ space.

Proof: If the decomposition in (8) holds, then using $T^*\Upsilon T = \Upsilon$ and $S_\mathcal{E}^*H = HT$, we obtain

$$T^*QT = T^*(\Upsilon - H^*H)T = \Upsilon - H^*S_\mathcal{E}S_\mathcal{E}^*H \geq \Upsilon - H^*H = Q.$$

Hence, $T^*QT \geq Q$.

Now assume that $T^*QT \geq Q$. Then $T^{n+1*}QT^{n+1} \geq T^{n*}QT^n$ for all integers $n \geq 0$. So, $T^{n*}QT^n$ forms an increasing sequence of operators. For h in $l_+^2(\mathcal{Y})$, we have

$$(T^{n*}QT^nh, h) = (QT^nh, T^nh) \leq \|Q\| \, \|T^nh\|^2 = \|Q\| \, \|h\|^2.$$

Hence, the sequence $T^{n*}QT^n$ is uniformly bounded by $\|Q\|I$. Because, $T^{n*}QT^n$ is an increasing sequence of operators, it follows that $T^{n*}QT^n$ strongly converges to a positive operator Υ on $l_+^2(\mathcal{Y})$. Since $T^{n*}QT^n$ and $T^{n+1*}QT^{n+1}$ have the same limit, $T^*\Upsilon T = \Upsilon$. In other words, Υ is a positive Toeplitz operator.

Because $T^*QT - Q$ is positive, there exists an operator R from $l_+^2(\mathcal{Y})$ into some space $\mathcal{E}$ satisfying $T^*QT - Q = R^*R$. The operator R admits a block matrix decomposition of the form:

$$R = [H_1, H_2, H_3, H_4, \dots]. \tag{9}$$

Obviously, $Q = T^*QT - R^*R$. By recursively substituting $T^*QT - R^*R$ for Q, we obtain

$$\begin{aligned} Q &= T^*QT - R^*R = T^{2*}QT^2 - T^*R^*RT - R^*R \\ &= T^{n+1*}QT^{n+1} - \sum_{k=0}^{n} T^{k*}R^*RT^k. \end{aligned}$$

By taking the limit with h in $l_+^2(\mathcal{Y})$, we have

$$(Qh, h) = (\Upsilon h, h) - \sum_{k=0}^{\infty} \|RT^kh\|^2 . \tag{10}$$

In particular, the sum $\sum_0^\infty \|RT^k h\|^2$ is finite for all h in $l_+^2(\mathcal{Y})$. Let H be the Hankel operator generated by $\{H_j\}_1^\infty$, that is, let H be the operator defined in (7). Using this it follows that for h in $l_+^2(\mathcal{Y})$,

$$\|Hh\|^2 = \sum_{k=0}^{\infty} \|RT^k h\|^2 .$$

Substituting this into (10) yields $(Qh, h) = (\Upsilon h, h) - \|Hh\|^2$. Therefore, H is a bounded Hankel operator and (8) holds. This completes the proof. □

A Commutant Lifting Proof of Theorem 2: Following some of the ideas in [2], let us obtain a Proof of Theorem 2 directly from the commutant lifting theorem. As before, let $\{Z, B, B_1, Q\}$ be the data for the hyper-weighted Nevanlinna-Pick interpolation problem where Q is a positive operator on $l_+^2(\mathcal{Y})$. Assume that $Q \leq T^*QT$ where T is the unilateral shift on $l_+^2(\mathcal{Y})$. If F is a solution to the hyper-weighted Nevanlinna-Pick interpolation problem, then (4) holds. Let A be the operator from $l_+^2(\mathcal{Y})$ into $\mathcal{H}'$ defined by $A = P'T_F$. Using $WT_F = W_1$, it follows that $WA = W_1$. Moreover, $A^*A \leq T_F^*T_F \leq Q$.

Now assume that there exists an operator A from $l_+^2(\mathcal{Y})$ into $\mathcal{H}'$ satisfying $WA = W_1$ and $A^*A \leq Q$. According to the previous proposition, Q admits a decomposition of the form $Q = \Upsilon - H^*H$ where Υ is a positive Toeplitz operator and H is a Hankel operator from $l_+^2(\mathcal{Y})$ into $l_+^2(\mathcal{E})$. This readily implies that $A^*A + H^*H \leq \Upsilon$. Let $\Upsilon^{1/2}$ be the positive square root of Υ, and $\mathcal{H}$ be the closure of the range of $\Upsilon^{1/2}$. Notice that for all h in $l_+^2(\mathcal{Y})$, we have

$$\|\Upsilon^{1/2}Th\|^2 = (\Upsilon Th, Th) = (\Upsilon h, h) = \|\Upsilon^{1/2}h\|^2.$$

So, there exists an isometry U on $\mathcal{H}$ satisfying $U\Upsilon^{1/2} = \Upsilon^{1/2}T$. Let $\tilde{A}$ be the operator from $l_+^2(\mathcal{Y})$ into $\tilde{\mathcal{H}}' := \mathcal{H}' \oplus l_+^2(\mathcal{E})$ defined by

$$\tilde{A} := \begin{bmatrix} A \\ H \end{bmatrix}. \tag{11}$$

Let $S_{\mathcal{E}}$ be the unilateral shift on $l_+^2(\mathcal{E})$. As before, let S be the unilateral shift on $l_+^2(\mathcal{U})$, and let T' be the compression of S to $\mathcal{H}'$. Recall that S is an isometric lifting for T'. Moreover, (5) shows that A intertwines T with T'. Using $T'A = AT$, it follows that $\tilde{A}$ intertwines T with $T' \oplus S_{\mathcal{E}}^*$. Since $A^*A + H^*H \leq \Upsilon^{1/2}\Upsilon^{1/2}$, there exists a contraction C form $\mathcal{H}$ into $\tilde{\mathcal{H}}'$ satisfying $\tilde{A} = C\Upsilon^{1/2}$. Using the fact that $\tilde{A}$ intertwines T with $T' \oplus S_{\mathcal{E}}^*$ and $\Upsilon^{1/2}$ intertwines T with U, it follows that C intertwines U with $T' \oplus S_{\mathcal{E}}^*$. Let V on $\mathcal{K}$ be an isometric lifting of $S_{\mathcal{E}}^*$. In fact, one can take V to be the backward shift operator on $\mathcal{K} = l^2(\mathcal{E})$, that is, V admits a block matrix representation on $l^2(\mathcal{E})$ with I immediately above the main diagonal and zero's elsewhere. Then $S \oplus V$ is an isometric lifting of $T' \oplus S_{\mathcal{E}}^*$. By applying the commutant lifting theorem to C, there exists a contractive intertwining lifting

Γ of C satisfying $(S \oplus V)\Gamma = \Gamma U$ and $\tilde{P}\Gamma = C$ where $\tilde{P}$ is the orthogonal projection onto $\tilde{\mathcal{H}}'$. Notice that Γ is a contraction from $\mathcal{H}$ into $l_+^2(\mathcal{U}) \oplus \mathcal{K}$, and $\tilde{A} = \tilde{P}\Gamma\,\Upsilon^{1/2}$.

Let Γ_+ be the operator form $\mathcal{H}$ into $l_+^2(\mathcal{U})$ defined by $\Gamma_+ = P_+\Gamma$ where P_+ is orthogonal projection onto $l_+^2(\mathcal{U})$. Since Γ intertwines U with $(S \oplus V)$, it follows that Γ_+ intertwines U with S. Hence, $\Gamma_+\Upsilon^{1/2} = P_+\Gamma\,\Upsilon^{1/2}$ intertwines the unilateral shift T with the unilateral shift S. So, there exists a function F in $H^\infty(\mathcal{Y}, \mathcal{U})$ satisfying $T_F = \Gamma_+\Upsilon^{1/2}$. We claim that F is the solution to the hyper-weighted Nevanlinna-Pick interpolation problem. To see this, notice that $P' = P'P_+$ and $\tilde{A} = \tilde{P}\Gamma\,\Upsilon^{1/2}$, yields

$$\begin{aligned} WT_F &= WP'\Gamma_+\Upsilon^{1/2} = WP'\Gamma\,\Upsilon^{1/2} = WP'\tilde{P}\Gamma\,\Upsilon^{1/2} \\ &= WP'\tilde{A} = WA = W_1. \end{aligned}$$

Thus, $WT_F = W_1$. Let P_{2+} be the orthogonal projection onto $l_+^2(\mathcal{E})$. Observe that

$$H = P_{2+}\tilde{A} = P_{2+}\tilde{P}\Gamma\,\Upsilon^{1/2} = P_{2+}\Gamma\,\Upsilon^{1/2}.$$

So, $H = P_{2+}\Gamma\,\Upsilon^{1/2}$. Since $T_F = P_+\Gamma\,\Upsilon^{1/2}$, it follows that $\Gamma\,\Upsilon^{1/2}$ admits a matrix representation of the form $\Gamma\,\Upsilon^{1/2} = [T_F, H, R]^{tr}$ where R is an operator from $l_+^2(\mathcal{Y})$ into $\mathcal{K} \ominus l_+^2(\mathcal{E})$. (The transpose is denoted by *tr*.) Because Γ is a contraction, we have

$$T_F^*T_F + H^*H \leq \Upsilon^{1/2}\Gamma^*\Gamma\,\Upsilon^{1/2} \leq \Upsilon.$$

Hence, (4) holds. Therefore, F is a solution to the hyper-weighted Nevanlinna-Pick interpolation problem. This completes the alternate Proof of Theorem 2.

Remark 1 There is not necessarily a one to one correspondence between the set of all solutions F to the hyper-weighted Nevanlinna-Pick interpolation problem and the set of all contractive intertwining liftings Γ of C. To see this consider the hyper-weighted Nevanlinna-Pick interpolation problem with data $\{Z, B, B_1, I - H^*H\}$. Moreover, assume that there exists an operator A from $l_+^2(\mathcal{Y})$ into $\mathcal{H}'$ satisfying $WA = W_1$. Obviously, $\Upsilon^{1/2} = I$ and $\mathcal{H} = l_+^2(\mathcal{Y})$. Moreover, $C = \tilde{A} = [A, H]^{tr}$. Hence, there exists a solution to this hyper-weighted Nevanlinna-Pick interpolation problem if and only if $\tilde{A}$ is a contraction. In this case, if Γ is any contractive intertwining lifting of $\tilde{A}$, then $\Gamma_+ = P_+\Gamma$ intertwines T with S. So, there exists a function F in $H^\infty(\mathcal{Y}, \mathcal{U})$ satisfying $\Gamma_+ = T_F$, and F is a solution to this hyper-weighted Nevanlinna-Pick interpolation problem. Therefore, every contractive intertwining lifting Γ of $\tilde{A}$ determines a solution F to the hyper-weighted Nevanlinna-Pick interpolation problem with data $\{Z, B, B_1, I - H^*H\}$. However, the same solution F can correspond to infinitely many contractive intertwining liftings Γ of $\tilde{A}$. For example, let $B = 1$, $Z = 0$ and $B_1 = 0$. Let H be any Hankel matrix on l_+^2 satisfying $\|H\| < 1$. Clearly, $F = 0$ is a solution to this hyper-weighted Nevanlinna-Pick problem. In this case, $A = 0$ and $\tilde{A} = [0, H]^{tr}$. Since

$\|H\| < 1$, the solution to the Nehari interpolation problem shows that there are infinitely many contractions Γ_2 from l_+^2 into $\mathcal{K}$ intertwining T with V and satisfying $P_{2+}\Gamma_2 = H$; see [1, 4, 5, 11, 13]. Obviously, $\Gamma = [0, \Gamma_2]^{tr}$ is a contractive intertwining lifting of $\tilde{A}$ and $\Gamma_+ = 0$. So, there are infinitely contractive intertwining lifting Γ of $\tilde{A}$ satisfying $\Gamma_+ = 0$. Therefore, there is not necessarily a one to one correspondence between the set of all contractive intertwining lifting Γ of C and the set of all solutions to the hyper-weighted Nevanlinna-Pick interpolation problem.

2 Some State Space Existence Results

In this section we will use state space methods to establish some necessary and sufficient conditions for the existence of a solution to the hyper-weighted Nevanlinna-Pick interpolation problem. First let us establish some notation and classical results from linear systems theory. For further results on state space and linear systems; see [1, 5, 15, 16]. As before, let W be the controllability operator in (1) generated by the pair $\{Z, B\}$. Let P be the positive operator on $\mathcal{X}$ defined by $P = WW^*$. This P is called the *controllability Gramian* for the pair $\{Z, B\}$. Because Z is stable,

$$P = \sum_{j=0}^{\infty} Z^j BB^* Z^{*j}. \tag{12}$$

The sum converges in the operator topology. Moreover, it is well known that P is the unique solution to the following Lyapunov equation

$$P = ZPZ^* + BB^*. \tag{13}$$

Recall that the pair $\{Z, B\}$ *controllable*, if the range of W is onto all of $\mathcal{X}$. Therefore $\{Z, B\}$ is controllable if and only if P is strictly positive, that is, there exists a scalar $\delta > 0$ such that $P \geq \delta I$. In other words, $\{Z, B\}$ is controllable if and only if its controllability Gramian P is invertible.

Let $\{H_j\}_1^\infty$ be a sequence of operators from $\mathcal{Y}$ into $\mathcal{E}$. Throughout it is always assumed that both $\mathcal{Y}$ and $\mathcal{E}$ are finite dimensional. We say that $\{Z_h, B_h, C_h\}$ is a *realization* of $\{H_j\}_1^\infty$, if

$$H_j = C_h Z_h^{j-1} B_h \qquad (\text{for } j = 1, 2, 3, \ldots). \tag{14}$$

Here Z_h is an operator on $\mathcal{X}_h$, while B_h maps $\mathcal{Y}$ into $\mathcal{X}_h$ and C_h maps $\mathcal{X}_h$ into $\mathcal{E}$. We say that $\{Z_h, B_h, C_h\}$ is a realization for the Hankel operator H if H is the Hankel operator in (7) determined by $H_j = C_h Z_h^{j-1} B_h$ for all integers $j \geq 1$. The realization $\{Z_h, B_h, C_h\}$ is finite dimensional, if the state $\mathcal{X}_h$ is finite dimensional. Let H be the Hankel operator from $l_+^2(\mathcal{Y})$ into $l_+^2(\mathcal{E})$ in (7) generated by $\{H_j\}_1^\infty$. Then H has finite rank if and only if H admits a finite dimensional realization $\{Z_h, B_h, C_h\}$. Furthermore, H is a bounded, finite rank Hankel operator if and only if H admits a finite dimensional, stable realization. In this note we only consider finite dimensional realizations or finite rank operator Hankel operators. Finally, recall that the pair $\{C_h, Z_h\}$ is *observable* if the pair $\{Z_h^*, C_h^*\}$ is controllable. The realization $\{Z_h, B_h, C_h\}$ is called *minimal* if $\{Z_h, B_h\}$ is controllable and $\{C_h, Z_h\}$ is observable. Moreover, all minimal realizations are similar. In particular, if the rank of H is finite, then H is a bounded operator if and only if its minimal realization is stable.

Let $\{Z_h, B_h, C_h\}$ be a finite dimensional, stable, realization for the Hankel operator H in (7). Let W_o be the *observability operator* from $\mathcal{X}_h$ into $l_+^2(\mathcal{E})$ defined by

(15) $$W_o = \begin{bmatrix} C_h \\ C_h Z_h \\ C_h Z_h^2 \\ \vdots \end{bmatrix}.$$

Because Z_h is stable, W_o is a bounded operator. The *observability Gramian* P_o generated by the pair $\{C_h, Z_h\}$ is the operator on $\mathcal{X}_h$ defined by $P_o = W_o^* W_o$. Notice that P_o is the controllability Gramian for the pair $\{Z_h^*, C_h^*\}$. Moreover, P_o is the unique solution to the Lyapunov equation

(16) $$P_o = Z_h^* P_o Z_h + C_h^* C_h.$$

Let W_h be the controllability operator from $l_+^2(\mathcal{Y})$ into $\mathcal{X}_h$ generated by the pair $\{Z_h, B_h\}$. Then using (14), it follows that H admits a factorization of the form $H = W_o W_h$. In other words, if H is the Hankel operator generated by the realization $\{Z_h, B_h, C_h\}$, then $H = W_o W_h$. For further results on linear systems theory see [15, 16]. The previous notation is used in the following result.

Theorem 3 *Let* $\{Z, B, B_1, \Upsilon - H^*H\}$ *be the data for the hyper-weighted Nevanlinna-Pick interpolation problem where the pair* $\{Z, B\}$ *is controllable,* Υ *is a positive Toeplitz operator and* H *is a finite rank Hankel operator. Let* $\{Z_h, B_h, C_h\}$ *be a stable, realization for* H*. Finally, let* P *and* P_o *be respectively the unique solutions to the Lyapunov equations* (13) *and* (16)*. Then there exists a solution to this hyper-weighted Nevanlinna-Pick interpolation problem if and only if*

(17) $$W_1^* P^{-1} W_1 + W_h^* P_o W_h \le \Upsilon .$$

Proof: Let A be the operator from $l_+^2(\mathcal{Y})$ into $\mathcal{H}'$ defined by

(18) $$A = W^* P^{-1} W_1 .$$

Using $P = WW^*$, it follows that $WA = W_1$. According to Theorem 2, there exists a solution to this hyper-weighted Nevanlinna-Pick interpolation problem if and only if $A^*A \le Q$ where $Q = \Upsilon - H^*H$. Notice that $A^*A = W_1^* P^{-1} W_1$. So, there exists a solution to this hyper-weighted Nevanlinna-Pick interpolation problem if and only if $W_1^* P^{-1} W_1 + H^* H \leq \Upsilon$. Recall that H admits a factorization of the form $H = W_o W_h$ and $P_o = W_o^* W_o$. Hence, $H^* H = W_h^* P_o W_h$. Therefore, there exists a solution to this hyper-weighted Nevanlinna-Pick interpolation problem if and only if (17) holds. This completes the proof. □

By choosing $\Upsilon = T_\Theta^* T_\Theta$ and $H = 0$ in the previous theorem, we readily obtain the following result.

Corollary 2 *Let Θ be an outer function in $H^\infty(\mathcal{Y}, \mathcal{Y})$ and assume that the pair $\{Z, B\}$ is controllable. Then there exists a solution to the weighted Nevanlinna-Pick interpolation problem with data $\{Z, B, B_1, \Theta\}$ if and only if*

$$W_1^* P^{-1} W_1 \leq T_\Theta^* T_\Theta . \tag{19}$$

3 The Outer Spectral Factor Case

In this section we will solve the hyper-weighted Nevanlinna-Pick interpolation problem when Υ admits an outer spectral factor. Let Υ be a positive Toeplitz operator on $l_+^2(\mathcal{Y})$. Then we say that Θ is an *outer spectral factor* for Υ if Θ is an outer function in $H^\infty(\mathcal{Y}, \mathcal{Y})$ satisfying $T_\Theta^* T_\Theta = \Upsilon$. The outer spectral factor Θ is unique up to a unitary constant, that is, if Ψ is another outer spectral factor for Υ, then $\Psi = U\Theta$ where U is a unitary operator on $\mathcal{Y}$; see Proposition V.4.2 in [19]. Finally, it is noted that not all positive Toeplitz operator on $l_+^2(\mathcal{Y})$ admit an outer spectral factor. However, if Υ is invertible, then Υ admits an outer spectral factor.

Now assume that Θ in $H^\infty(\mathcal{Y}, \mathcal{Y})$ is the outer spectral factor for the Toeplitz operator Υ. Throughout $\Omega(\lambda)$ is the inverse of $\Theta(\lambda)$ in the open unit disc. We claim that Ω is analytic in the open unit disc. To see this, let $\det[\Theta(\lambda)]$ be the determinant of any matrix representation of $\Theta(\lambda)$. Because Θ is outer, $\det[\Theta(\lambda)]$ is outer; see Corollary 6.3 page 220 in [19]. In particular, $\det[\Theta]$ has no zeros in the open unit disc. Clearly, $\Omega(\lambda) = \operatorname{adj}(\Theta(\lambda))/\det[\Theta(\lambda)]$, where $\operatorname{adj}(\Theta)$ be the algebraic adjoint of Θ. Since $\operatorname{adj}(\Theta)$ is in $H^\infty(\mathcal{Y}, \mathcal{Y})$ and $\det[\Theta(\lambda)]$ is outer, this implies that $\Omega(\lambda)$ is analytic in the open unit disc.

Because Ω is analytic in the open unit disc, Ω admits a power series expansion of the form

$$\Theta(\lambda)^{-1} := \Omega(\lambda) = \sum_{n=0}^{\infty} \Omega_n \lambda^n \qquad (\text{for } |\lambda| < 1) . \tag{20}$$

In particular, this implies that $\limsup \|\Omega_n\|^{1/n} \leq 1$. Furthermore, let T_Ω be the block Toeplitz matrix generated by $\{\Omega_n\}_0^\infty$, that is, let T_Ω be the lower triangular matrix given by (2) where Ω replaces F and Ω_n replaces F_n. By matching coefficients of λ^n in $\Omega(\lambda)\Theta(\lambda) = I$ $\Theta(\lambda)\Omega(\lambda) = I$, it follows that T_Ω is the algebraic inverse of T_Θ, that is, $T_\Omega T_\Theta = I$ and $T_\Theta T_\Omega = I$. Moreover, T_Ω is a bounded operator on $l_+^2(\mathcal{Y})$ if and only if Θ^{-1} is a function in $H^\infty(\mathcal{Y}, \mathcal{Y})$.

Let $\{Z, B, B_1, Q\}$ be the data for the hyper-weighted Nevanlinna-Pick interpolation problem where the pair $\{Z, B\}$ is controllable and $T^*QT \geq Q$. As before, T is the unilateral shift on $l_+^2(\mathcal{Y})$. Recall that Q admits a decomposition of the form $Q = \Upsilon - H^*H$ where H is a Hankel operator. Assume that Θ in $H^\infty(\mathcal{Y}, \mathcal{Y})$ is an outer spectral factorization for Υ and $\{Z_h$ on $\mathcal{X}_h, B_h, C_h\}$ is a stable, finite dimensional realization for H. Let $\tilde{B}$ be the operator from $\mathcal{Y}$ into $\mathcal{X}$ and $\tilde{B}_h$ the operator from $\mathcal{Y}$ into $\mathcal{X}_h$ defined by

$$\tilde{B} := \sum_{n=0}^{\infty} Z^n B_1 \Omega_n \text{ and } \tilde{B}_h := \sum_{n=0}^{\infty} Z_h^n B_h \Omega_n . \tag{21}$$

Because Z and Z_h are stable, and $\limsup \|\Omega_n\|^{1/n} \leq 1$, the operators $\tilde{B}$ and $\tilde{B}_h$ are well defined. The following result uses the operators $\tilde{B}$ and $\tilde{B}_h$ to solve the hyper-weighted Nevanlinna-Pick interpolation problem.

Theorem 4 *Let Θ be an outer function in $H^\infty(\mathcal{Y}, \mathcal{Y})$. Let $\{Z_h, B_h, C_h\}$ be a stable, finite dimensional realization for the Hankel operator H, and let $\tilde{H}$ be the Hankel operator determined by the realization $\{Z_h, \tilde{B}_h, C_h\}$. Then F is a solution to the hyper-weighted Nevanlinna-Pick interpolation problem with data $\{Z, B, B_1, T_\Theta^* T_\Theta - H^*H\}$ if and only if F admits a factorization of the form $F = G\Theta$ and G is a solution to the hyper-weighted Nevanlinna-Pick interpolation problem with data $\{Z, B, \tilde{B}, I - \tilde{H}^*\tilde{H}\}$.*

Proof: As before, let W, W_1 and W_h be the controllability operators generated by the pairs $\{Z, B\}$,$\{Z, B_1\}$ and $\{Z_h, B_h\}$, respectively, while W_o is the observability operator generated by the pair $\{C_h, Z_h\}$. Recall also that $H = W_o W_h$. Equation (21) implies that $W_1 T_\Omega = \tilde{W}$ and $W_h T_\Omega = \tilde{W}_h$ where $\tilde{W}$ and $\tilde{W}_h$ are the controllability operators generated by the pairs $\{Z, \tilde{B}\}$ and $\{Z_h, \tilde{B}_h\}$, respectively. Since T_Ω is the algebraic inverse of T_Θ, we also have $W_1 = \tilde{W} T_\Theta$ and $W_h = \tilde{W}_h T_\Theta$. Using $H = W_o W_h$, it also follows that $\tilde{H} = W_o \tilde{W}_h = H T_\Omega$. In other words, $\tilde{H} T_\Theta = H$. Finally, it is noted that because Z and Z_h are stable and $\limsup \|\Omega_n\|^{1/n} \leq 1$, all the previous calculations involving T_Ω are well defined.

Assume that F is a solution to the hyper-weighted Nevanlinna-Pick interpolation problem with data $\{Z, B, B_1, T_\Theta^* T_\Theta - H^*H\}$, that is,

$$W T_F = W_1 \text{ and } T_F^* T_F + H^*H \leq T_\Theta^* T_\Theta . \tag{22}$$

In particular, $T_F^* T_F \leq T_\Theta^* T_\Theta$. This implies that there exists a contraction C from the closure of the range of T_Θ into $l_+^2(\mathcal{U})$ satisfying $T_F = C\, T_\Theta$. Because Θ is outer, the range of T_Θ is dense in $l_+^2(\mathcal{Y})$. So, C maps $l_+^2(\mathcal{Y})$ into $l_+^2(\mathcal{U})$. As before, let T be the unilateral shift on $l_+^2(\mathcal{Y})$ and S the unilateral shift on $l_+^2(\mathcal{U})$. Using $T_F = C\, T_\Theta$, we have

$$C\, T T_\Theta = C\, T_\Theta T = T_F T = S T_F = SC\, T_\Theta .$$

Since the range of T_Θ is dense in $l_+^2(\mathcal{Y})$, it follows that $SC = CT$. Therefore, there exists a function G in $H^\infty(\mathcal{Y}, \mathcal{U})$ satisfying $C = T_G$. Hence, $T_F = T_G T_\Theta$, and F admits a factorization of the form $F = G\Theta$.

We claim that G is a solution to the hyper-weighted Nevanlinna-Pick interpolation problem with data $\{Z, B, \tilde{B}, I - \tilde{H}^*\tilde{H}\}$, that is, G satisfies

$$(23) \qquad W T_G = \tilde{W} \text{ and } T_G^* T_G + \tilde{H}^*\tilde{H} \leq I\,.$$

Using $WT_F = W_1$ along with $W_1 T_\Omega = \tilde{W}$, we have $\tilde{W} T_\Theta = W_1 = WT_F = WT_G T_\Theta$. Because the range of T_Θ is dense in $l_+^2(\mathcal{Y})$, we have $WT_G = \tilde{W}$. Recall that $H = \tilde{H} T_\Theta$. Since F is a solution to the original hyper-weighted Nevanlinna-Pick interpolation problem, the inequality in (22) yields

$$T_\Theta^* T_G^* T_G T_\Theta + T_\Theta^* \tilde{H}^* \tilde{H} T_\Theta \leq T_\Theta^* T_\Theta.$$

This readily implies that $T_G^* T_G + \tilde{H}^*\tilde{H} \leq I$. Therefore, (23) holds and G is a solution to hyper-weighted Nevanlinna-Pick interpolation problem with data $\{Z, B, \tilde{B}, I - \tilde{H}^*\tilde{H}\}$.

Now assume that G is a solution to the hyper-weighted Nevanlinna-Pick interpolation problem with data $\{Z, B, \tilde{B}, I - \tilde{H}^*\tilde{H}\}$. In other words, assume that (23) holds. Then we claim that $F = G\Theta$ is a solution to the original hyper-weighted Nevanlinna-Pick problem, that is, (22) holds. Using $WT_G = \tilde{W}$, we obtain

$$WT_F = WT_G T_\Theta = \tilde{W} T_\Theta = W_1\,.$$

Hence $WT_F = W_1$, and the first equation in (22) holds. The inequality $T_G^* T_G + \tilde{H}^*\tilde{H} \leq I$ along with $\tilde{H} T_\Theta = H$, yields

$$T_F^* T_F + H^* H = T_\Theta^* T_G^* T_G T_\Theta + T_\Theta^* \tilde{H}^* \tilde{H} T_\Theta \leq T_\Theta^* T_\Theta\,.$$

Therefore, (22) holds and $F = G\Theta$ is a solution to the original hyper-weighted Nevanlinna-Pick problem. This completes the proof. □

By applying the previous theorem to the weighted Nevanlinna-Pick problem, we obtain the following result.

Corollary 3 *Let Θ be an outer function in $H^\infty(\mathcal{Y}, \mathcal{Y})$. Then F is a solution to the weighted Nevanlinna-Pick interpolation problem with data $\{Z, B, B_1, \Theta\}$ if and only if F admits a factorization of the form $F = G\Theta$ and G is a solution to the weighted Nevanlinna-Pick interpolation problem with data $\{Z, B, \tilde{B}, I\}$.*

To condense some notation, let Z_2 be the block diagonal matrix on $\mathcal{X} \oplus \mathcal{X}_h$, and B_2 the block matrix from $\mathcal{Y}$ into $\mathcal{X} \oplus \mathcal{X}_h$, respectively defined by

$$(24) \qquad Z_2 = \begin{bmatrix} Z & 0 \\ 0 & Z_h \end{bmatrix} \text{ and } B_2 = \begin{bmatrix} \tilde{B} \\ \tilde{B}_h \end{bmatrix}.$$

Throughout P_2 on $\mathcal{X} \oplus \mathcal{X}_h$ is the controllability Gramian generated by the pair $\{Z_2, B_2\}$. In other words, P_2 is the unique solution to the following Lyapunov equation

$$P_2 = Z_2 P_2 Z_2^* + B_2 B_2^* . \tag{25}$$

This notation is used in the following result.

Corollary 4 *Let Θ be an outer function in $H^\infty(\mathcal{Y}, \mathcal{Y})$, and $\{Z, B\}$ be a controllable pair. Let $\{Z_h, B_h, C_h\}$ be a stable realization for the Hankel operator H. Finally, let J be the block diagonal matrix on $\mathcal{X} \oplus \mathcal{X}_h$ defined by $J = P^{-1} \oplus P_o$. Then there exists solution to the hyper-weighted Nevanlinna-Pick interpolation problem with data $\{Z, B, B_1, T_\Theta^* T_\Theta - H^* H\}$ if and only if*

$$r_{spec}(J P_2) \leq 1 . \tag{26}$$

Proof: Let $\tilde{W}$ and $\tilde{W}_h$ be the controllability operators generated by the pairs $\{Z, \tilde{B}\}$ and $\{Z_h, \tilde{B}_h\}$, respectively. Theorem 3 shows that there exists a solution to the hyper-weighted Nevanlinna-Pick interpolation problem with data $\{Z, B, \tilde{B}, I - \tilde{H}^* \tilde{H}\}$ if and only if

$$E := \tilde{W}^* P^{-1} \tilde{W} + \tilde{W}_h^* P_o \tilde{W}_h \leq I . \tag{27}$$

By consulting Theorem 4, we see that there exists a solution to the hyper-weighted Nevanlinna-Pick problem with data $\{Z, B, B_1, T_\Theta^* T_\Theta - H^* H\}$ if and only if (27) holds. Notice that

$$E = [\tilde{W}^*, \tilde{W}_h^*] \begin{bmatrix} P^{-1} & 0 \\ 0 & P_o \end{bmatrix} \begin{bmatrix} \tilde{W} \\ \tilde{W}_h \end{bmatrix} .$$

Hence, (27) holds if and only if $E \leq I$. So, there exists a solution to hyper-weighted Nevanlinna-Pick interpolation problem if and only if $r_{spec}(E) \leq 1$. Recall that if M and N are two operators acting between the appropriate spaces, then $r_{spec}(MN) = r_{spec}(NM)$; see equation (3) on page 38 in [10]. Thus, E and

$$\begin{bmatrix} P^{-1} & 0 \\ 0 & P_o \end{bmatrix} \begin{bmatrix} \tilde{W} \\ \tilde{W}_h \end{bmatrix} [\tilde{W}^*, \tilde{W}_h^*]$$

have the same spectral radius. Notice that $[\tilde{W}, \tilde{W}_h]^{tr}$ is the controllability operator from $l_+^2(\mathcal{Y})$ into $\mathcal{X} \oplus \mathcal{X}_h$ generated by the pair $\{Z_2, B_2\}$. This implies that controllability Gramian P_2 generated by the pair $\{Z_2, B_2\}$ is given by

$$P_2 = \begin{bmatrix} \tilde{W} \\ \tilde{W}_h \end{bmatrix} [\tilde{W}^*, \tilde{W}_h^*].$$

Hence, $J P_2$ and E have the same spectral radius. In particular, $r_{spec}(E) \leq 1$ if and only if (26) holds. This completes the proof. □

Remark 2 Let us note that $I - \tilde{W}^* P^{-1} \tilde{W} - \tilde{W}_h^* P_o \tilde{W}_h$ is positive and invertible if and only if $r_{spec}(J P_2) < 1$. To see this simply notice that $I - E$ is positive and invertible if and only if $r_{spec}(E) < 1$. The previous proof shows that E and $J P_2$ have the same spectral radius. Therefore, $I - E$ is positive and invertible if and only if $r_{spec}(J P_2) < 1$.

If $H = 0$, then we can set $\mathcal{X}_h = \{0\}$. In this case, $P_2 = \tilde{P}$ where $\tilde{P}$ is the controllability Gramian determined by the pair $\{Z, \tilde{B}\}$, and $J = P^{-1}$. So, when $H = 0$, the above corollary reduces to the following result.

Corollary 5 *Assume that Θ is an outer function, the pair $\{Z, B\}$ is controllable, and $\tilde{P}$ is the controllability Gramian determined by the pair $\{Z, \tilde{B}\}$. Then there exists a solution to weighted Nevanlinna-Pick interpolation problem with data $\{Z, B, B_1, \Theta\}$ if and only if $r_{spec}(P^{-1}\tilde{P}) \le 1$.*

Finally, it is noted that if $\Theta = \gamma I$ where γ is a positive constant, then the above corollary reduces to Theorem I.4.1 in [5].

Example 1 Let us apply Corollaries 3 and 5 to a weighted version of the classical Nevanlinna-Pick interpolation problem, which naturally arises in H^∞ control theory; see [3]. To this end, let $\{\alpha_j\}_1^n$ be a set of distinct complex numbers in the open unit disc, and $\{\tilde{b}_j\}_1^n$ be a set of complex numbers. Recall that the classical Nevanlinna-Pick interpolation problem is to find (if possible) a function g in H^∞ satisfying $g(\alpha_j) = \tilde{b}_j$ for all $j = 1, 2, \ldots, n$ and $\|g\|_\infty \le 1$. Now let $\{\gamma_j\}_1^n$ be a set of complex numbers, and let Θ be a scalar valued outer function in H^∞. Then a weighted version of the classical Nevanlinna-Pick interpolation problem is to find (if possible) a function f in H^∞ satisfying

$$f(\alpha_j) = \gamma_j \text{ for all } j = 1, 2, \ldots, n \text{ and } |f(e^{\imath\omega})| \le |\Theta(e^{\imath\omega})| \tag{28}$$

for almost all ω in $[0, 2\pi)$. To solve this weighted interpolation problem, let $\{\tilde{b}_j\}_1^n$ be the complex numbers defined by

$$\tilde{b}_j = \gamma_j / \Theta(\alpha_j) \qquad (\text{for } j = 1, 2, 3, \ldots, n). \tag{29}$$

Because Θ is outer, and $\{\alpha_j\}_1^n$ are in the open disc, $\Theta(\alpha_j)$ is nonzero for all j. Therefore, the complex numbers $\{\tilde{b}_j\}_1^n$ in (29) are well defined.

We claim that f is a function in H^∞ satisfying (28) if and only if f admits a factorization of the form $f = g\Theta$ and g is a solution to the classical Nevanlinna-Pick interpolation problem with data $\{\alpha_j\}_1^n$ and $\{\tilde{b}_j\}_1^n$ defined in (29).

To prove this, we simply convert these interpolation problems to the framework of our weighted Nevanlinna-Pick interpolation problem. To this end, let Z be the diagonal matrix on C^n defined by $Z = \text{diag}\{\alpha_1, \alpha_2, \ldots, \alpha_n\}$. Obviously, Z is stable. Let B be the column vector in C^n defined by taking the transpose of $[1, 1, \ldots, 1]$, while B_1 is the transpose of $[\gamma_1, \gamma_2, \ldots, \gamma_n]$. Let W and W_1

be the controllability operators generated by $\{Z, B\}$ and $\{Z, B_1\}$, respectively. Notice that $WT_f = W_1$ if and only if $f(\alpha_j) = \gamma_j$ for all $j = 1, 2, \ldots, n$. Therefore, f is a function in H^∞, satisfying (28) if and only if f is a solution to the weighted Nevanlinna-Pick problem with data $\{Z, B, B_1, \Theta\}$. To convert this weighted Nevanlinna-Pick problem to the classical Nevanlinna-Pick interpolation problem, let $\Omega(\lambda) = 1/\Theta(\lambda)$. Then for any α in the open unit disc, we have

$$\sum_{k=0}^{\infty} \alpha^k \Omega_k = \Omega(\alpha) = 1/\Theta(\alpha).$$

This readily implies that the operator $\tilde{B}$ in (21) is given by

$$\tilde{B} = [\gamma_1/\Theta(\alpha_1), \gamma_2/\Theta(\alpha_2), \ldots, \gamma_n/\Theta(\alpha_n)]^{tr},$$

where tr denotes the transpose. Hence, $\tilde{B}$ is the transpose of $[\tilde{b}_1, \tilde{b}_2 \ldots, \tilde{b}_n]$ where $\{\tilde{b}_j\}_1^n$ are defined in (29). Corollary 3 shows that f is a function in H^∞ satisfying (28) if and only if f admits a factorization of the form $f = g\Theta$ and g is a solution to the weighted Nevanlinna-Pick interpolation problem with data $\{Z, B, \tilde{B}, I\}$. However, the weighted Nevanlinna-Pick interpolation problem with data $\{Z, B, \tilde{B}, I\}$ is precisely the classical Nevanlinna-Pick interpolation problem with data $\{\alpha_j\}_1^n$ and $\{\tilde{b}_j\}_1^n$. Hence, f is a function in H^∞ satisfying (28) if and only if f admits a factorization of the form $f = g\Theta$ and g is a solution to the classical Nevanlinna-Pick problem with data $\{\alpha_j\}_1^n$ and $\{\gamma_j/\Theta(\alpha_j)\}_1^n$. This verifies our claim.

Because $\{\alpha_j\}_1^n$ are distinct, the controllability Gramian $P = WW^*$ is positive and invertible. Let $\tilde{P}$ be the controllability Gramian generated by the pair $\{Z, \tilde{B}\}$. Since Z is a diagonal matrix, it well known and easy to verify that P and $\tilde{P}$ are the Pick matrices whose entries are given by

$$P_{i,j} = \frac{1}{1 - \alpha_i \alpha_j^*} \qquad \text{and} \qquad \tilde{P}_{i,j} = \frac{\tilde{b}_i \tilde{b}_j^*}{1 - \alpha_i \alpha_j^*},$$

for $i, j = 1, 2, \ldots, n$. According to Corollary 5, there exists a function f in H^∞ solving the weighted Nevanlinna-Pick interpolation problem with data $\{\alpha_j\}_1^n$ and $\{\gamma_j\}_1^n$, that is, (28) holds if and only if $r_{spec}(P^{-1}\tilde{P}) \leq 1$.

4 A State Space Solution

In this section we will provide a state space solution to the hyper-weighted Nevanlinna-Pick interpolation problem. To begin, recall that Z_2 is the block diagonal matrix on $\mathcal{X} \oplus \mathcal{X}_h$, and B_2 is the block matrix from $\mathcal{Y}$ into $\mathcal{X} \oplus \mathcal{X}_h$ defined in (24), while P_2 is the controllability Gramian in (25) generated by the pair $\{Z_2, B_2\}$. Finally, recall that P and P_o are respectively the controllability Gramians determined by the pairs $\{Z, B\}$ and $\{Z_h^*, C_h^*\}$.

Theorem 5 *Consider the hyper-weighted Nevanlinna-Pick interpolation problem corresponding to the data* $\{Z, B, B_1, T_\Theta^* T_\Theta - H^* H\}$, *where the pair* $\{Z, B\}$ *is controllable, and* Θ *is an outer function. Assume that* $\{Z_h, B_h, C_h\}$ *is a stable, observable realization for the Hankel operator* H, *and* $r_{spec}(\Lambda^{-1} P_2) < 1$, *where* $\Lambda = P \oplus P_o^{-1}$. *Finally, let* M *and* N *be the operator valued analytic functions defined by*

$$
\begin{aligned}
N &= [B^*(I - \lambda Z^*)^{-1}, 0](\Lambda - P_2)^{-1} B_2 \\
M &= I + B_2^*(I - \lambda Z_2^*)^{-1}(\Lambda - P_2)^{-1} B_2.
\end{aligned} \tag{30}
$$

Then M *is an outer function. Moreover, if* $G(\lambda) = N(\lambda)M(\lambda)^{-1}$, *then* G *is a function in* $H^\infty(\mathcal{Y}, \mathcal{U})$. *Furthermore,* $F(\lambda) = G(\lambda)\Theta(\lambda)$ *is a solution to this hyper-weighted Nevanlinna-Pick interpolation problem.*

The Fourier transform from $l_+^2(\mathcal{E})$ onto $H^2(\mathcal{E})$ is denoted by $\mathcal{F}_\mathcal{E}$. In other words, if $g = (g_n)_0^\infty$ is a vector in $l_+^2(\mathcal{E})$, then $\mathcal{F}_\mathcal{E} g = \sum_0^\infty g_n \lambda^n$. Finally, let Π_0 be the operator form $l_+^2(\mathcal{Y})$ onto $\mathcal{Y}$ defined by $y_0 = \Pi_0 (y_n)_0^\infty$, that is, Π_0 picks out the first component of the vector $(y_n)_0^\infty$. We will use the following result taken form [2] to prove the previous theorem.

Theorem 6 ([2]). *Let* Q *be a positive operator on* $l_+^2(\mathcal{Y})$ *and* T *the unilateral shift on* $l_+^2(\mathcal{Y})$ *satisfying* $T^* Q T \geq Q$. *Let* S *be the unilateral shift on* $l_+^2(\mathcal{U})$. *Let* $\mathcal{H}'$ *be an invariant subspace for* S^* *and* T' *the contraction on* $\mathcal{H}'$ *defined by* $T'^* = S^* | \mathcal{H}'$. *Let* A *be an operator intertwining* T *with* T' *and assume that* $Q - A^* A$ *is strictly positive. Finally, let* N *and* M *be the operator valued analytic functions defined by*

$$
\begin{aligned}
N(\lambda) y &= (\mathcal{F}_\mathcal{U} A (Q - A^* A)^{-1} \Pi_0^* y)(\lambda) \qquad (y \in \mathcal{Y}) \\
M(\lambda) y &= (\mathcal{F}_\mathcal{Y} (Q - A^* A)^{-1} \Pi_0^* y)(\lambda) \qquad (y \in \mathcal{Y}).
\end{aligned} \tag{31}
$$

Then M *is an outer function. Moreover, if* $G(\lambda) = N(\lambda)M(\lambda)^{-1}$, *then* G *is a function in* $H^\infty(\mathcal{Y}, \mathcal{U})$. *Furthermore,* T_G *is an intertwining lifting of* A *satisfying* $T_G^* T_G \leq Q$.

Assume that in the previous theorem, $Q = \gamma^2 I$ where $\gamma > 0$. Then obviously $T^* Q T = Q$. Clearly, $Q - A^* A$ is strictly positive if and only if $\|A\| < \gamma$. In this case, Theorem 6 reduces to the formula for the central solution for the commutant lifting theorem with tolerance γ presented in Theorem IV.4.1 of [5]; see also [6]. In other words, if $\|A\| < \gamma$, and M and N are the operator valued functions defined in (31), then $F(\lambda) = N(\lambda)M(\lambda)^{-1}$ is a well defined function in $H^\infty(\mathcal{Y}, \mathcal{U})$. Furthermore, T_F is an intertwining lifting of A satisfying $\|F\|_\infty \leq \gamma$. Finally, it is noted that one can also use Proposition 1 and modify the Proof of Theorem IV.4.1 in [5], to obtain a Proof of Theorem 6.

Proof of Theorem 5: As before, let $\tilde{B}$ and $\tilde{B}_h$ be the operators computed by (21), and $\tilde{H}$ the Hankel operator determined by the realization $\{Z_h, \tilde{B}_h, C_h\}$. Let us use the above theorem to construct a solution G to the hyper-weighted Nevanlinna-Pick problem with data $\{Z, B, \tilde{B}, I-\tilde{H}^*\tilde{H}\}$. Then according to Theorem 4, the function $G\Theta$ is a solution to our original hyper-weighted Nevanlinna-Pick problem. To this end, let A be the operator from $l_+^2(\mathcal{Y})$ into $l_+^2(\mathcal{U})$ defined by $A = W^*P^{-1}\tilde{W}$ where $\tilde{W}$ is the controllability operator generated by the pair $\{Z, \tilde{B}\}$. Using $P = WW^*$, it follows that $WA = \tilde{W}$. Notice that $A^*A = \tilde{W}^*P^{-1}\tilde{W}$. Since $\tilde{H} = W_o\tilde{W}_h$ and $P_o = W_o^*W_o$, we obtain $\tilde{H}^*\tilde{H} = \tilde{W}_h^*P_o\tilde{W}_h$. So, by taking $Q = I - \tilde{H}^*\tilde{H}$, we have

$$Q - A^*A = I - \tilde{W}^*P^{-1}\tilde{W} - \tilde{W}_h^*P_o\tilde{W}_h\,. \tag{32}$$

According to Theorem 3, there exists a solution to the hyper-weighted Nevanlinna-Pick problem with data $\{Z, B, \tilde{B}, I - \tilde{H}^*\tilde{H}\}$ if and only if $I - \tilde{W}^*P^{-1}\tilde{W} - \tilde{W}_h^*P_o\tilde{W}_h$ is positive. Since, $J = \Lambda^{-1}$ and $r_{spec}(\Lambda^{-1}P_2) < 1$, Remark 2 shows that $I - \tilde{W}^*P^{-1}\tilde{W} - \tilde{W}_h^*P_o\tilde{W}_h$ is positive and invertible. Thus, $Q - A^*A$ is also positive and invertible. Recall that T is the unilateral shift on $l_+^2(\mathcal{Y})$, and S is the unilateral shift on $l_+^2(\mathcal{U})$. As before, let $\mathcal{H}'$ be the range of W^*. Using $S^*W^* = W^*Z^*$, it follows that $\mathcal{H}'$ is an invariant subspace for S^*. Let T' be the contraction on $\mathcal{H}'$ defined by $T'^* = S^*|\mathcal{H}'$. Then S is an isometric lifting of T'. By replacing W_1 by $\tilde{W}$ in (5), we see that $T'A = AT$. So, if $G = NM^{-1}$ where M and N are computed by (31), then Theorem 6 shows that T_G is an intertwining lifting of A satisfying $T_G^*T_G \leq Q$. Hence, G is a solution to the hyper-weighted Nevanlinna-Pick problem with data $\{Z, B, \tilde{B}, I - \tilde{H}^*\tilde{H}\}$.

To apply the above theorem we must invert the operator $Q - A^*A$. To this end, recall that if R, X and Y are operators acting between the appropriate spaces, then we have the following well known identity; see page 656 in [15]

$$(I - XR^{-1}Y)^{-1} = I + X(R - YX)^{-1}Y\,. \tag{33}$$

Let W_2 be the controllability Gramian from $l_+^2(\mathcal{Y})$ into $\mathcal{X} \oplus \mathcal{X}_h$ generated by the pair $\{Z_2, B_2\}$. Recall that $W_2 = [\tilde{W}, \tilde{W}_h]^{tr}$. Using $P_2 = W_2W_2^*$ in (32) along with (33), we obtain

$$\begin{aligned}(Q - A^*A)^{-1} &= (I - W_2^*\Lambda^{-1}W_2)^{-1} = I + W_2^*(\Lambda - W_2W_2^*)^{-1}W_2 \\ &= I + W_2^*(\Lambda - P_2)^{-1}W_2.\end{aligned} \tag{34}$$

Notice that $\Lambda - P_2$ is positive and invertible. To see this, observe that

$$r_{spec}(\Lambda^{-1/2}P_2\Lambda^{-1/2}) = r_{spec}(\Lambda^{-1}P_2) < 1\,.$$

This implies that $I - \Lambda^{-1/2}P_2\Lambda^{-1/2}$ is positive and invertible. Therefore, $\Lambda - P_2$ is positive and invertible. The Fourier transform of W_2^* is given by

$$\mathcal{F}_{\mathcal{Y}}\, W_2^* = \sum_{n=0}^{\infty} \lambda^n B_2^* Z_2^{*n} = B_2^*(I - \lambda Z_2^*)^{-1}\,.$$

Using this along with $\mathcal{F}_{\mathcal{Y}}\Pi_0^* = I$ and $W_2\Pi_0^* = B_2$ in (34), yields

$$\mathcal{F}_{\mathcal{Y}}(Q - A^*A)^{-1}\Pi_0^* = \mathcal{F}_{\mathcal{U}}[\Pi_0^* + W_2^*(\Lambda - P_2)^{-1}W_2\Pi_0^*] = M,$$

where M is the function defined in the second equation of (30). Because $W_2 = [\tilde{W}, \tilde{W}_h]^{tr}$ and $A = W^*P^{-1}\tilde{W}$, we have the identity $A = \left[W^*, 0\right]\Lambda^{-1}W_2$. Using this in (34), gives

$$\begin{aligned}
&A(Q - A^*A)^{-1}\Pi_0^* \\
&\quad = [W^*, 0]\Lambda^{-1}W_2[I + W_2^*(\Lambda - P_2)^{-1}W_2]\Pi_0^* \\
&\quad = [W^*, 0]\Lambda^{-1}[I + P_2(\Lambda - P_2)^{-1}]W_2\Pi_0^* \\
&\quad = [W^*, 0][\Lambda^{-1}(\Lambda - P_2) + \Lambda^{-1}P_2](\Lambda - P_2)^{-1}B_2 \\
&\quad = [W^*, 0](\Lambda - P_2)^{-1}B_2 .
\end{aligned}$$

Notice that the Fourier transform $\mathcal{F}_{\mathcal{U}}W^* = B^*(I - \lambda Z^*)^{-1}$. Using this in the previous equation, we arrive at

$$\mathcal{F}_{\mathcal{U}}\,A(Q - A^*A)^{-1}\Pi_0^* = N(\lambda),$$

where N is the function defined in (30). Theorem 6 shows that $G = NM^{-1}$ is a solution to the hyper-weighted Nevanlinna-Pick problem with data $\{Z, B, \tilde{B}, I - \tilde{H}^*\tilde{H}\}$. By Theorem 4, the function $G\Theta$ is a solution to our original hyper-weighted Nevanlinna-Pick problem. This completes the proof.

Corollary 6 *Consider the hyper-weighted Nevanlinna-Pick interpolation problem corresponding to the data $\{Z, B, B_1, T_\Theta^*T_\Theta - H^*H\}$, where the pair $\{Z, B\}$ is controllable, and Θ is an outer function. Assume that $\{Z_h, B_h, C_h\}$ is a stable, observable realization for the Hankel operator H, and $r_{spec}(\Lambda^{-1}P_2) < 1$. Let G be the function defined by*

$$G(\lambda) = \hat{C}(I - \lambda\hat{Z})^{-1}B_2, \tag{35}$$

where $\hat{C}$ form $\mathcal{X} \oplus \mathcal{X}_h$ into $\mathcal{U}$ and $\hat{Z}$ on $\mathcal{X} \oplus \mathcal{X}_h$ are the operators defined by

$$\begin{aligned}
\hat{C} &= [B^*, 0](\Lambda - Z_2P_2Z_2^*)^{-1} \text{ and} \\
\hat{Z} &= (\Lambda - P_2)Z_2^*(\Lambda - Z_2P_2Z_2^*)^{-1}.
\end{aligned} \tag{36}$$

Then $F = G\Theta$ is a solution to this hyper-weighted Nevanlinna-Pick interpolation problem.

Proof: To prove this result it is sufficient to show that the function G in (35) is given by $G = NM^{-1}$ where N and M are defined in (30) of Theorem 5. To this end, notice that

$$N(\lambda) = [B^*, 0](I - \lambda Z_2^*)^{-1}(\Lambda - P_2)^{-1}B_2 .$$

So, if $R = (\Lambda - P_2)(I - \lambda Z_2^*)$, then the function $G = NM^{-1}$ is given by $G = [B^*, 0]G_2$ where G_2 is defined by

$$G_2 = R^{-1}B_2(I + B_2^* R^{-1} B_2)^{-1}.$$

By employing (33), we obtain

$$\begin{aligned} G_2 &= R^{-1}B_2[I - B_2^*(R + B_2B_2^*)^{-1}B_2] \\ &= R^{-1}[I - B_2B_2^*(R + B_2B_2^*)^{-1}]B_2 \\ &= R^{-1}[(R + B_2B_2^*) - B_2B_2^*](R + B_2B_2^*)^{-1}B_2 \\ &= (R + B_2B_2^*)^{-1}B_2 \\ &= [\Lambda - P_2 + B_2B_2^* - \lambda(\Lambda - P_2)Z_2^*]^{-1}B_2 \\ &= (\Lambda - P_2 + B_2B_2^*)^{-1}[I - \lambda(\Lambda - P_2) \\ &\quad Z_2^*(\Lambda - P_2 + B_2B_2^*)^{-1}]^{-1}B_2. \end{aligned}$$

Recall that $\Lambda - P_2$ is positive and invertible. Hence, $\Lambda - P_2 + B_2B_2^*$ is also positive and invertible. Using $P_2 = Z_2P_2Z_2^* + B_2B_2^*$ and $G = [B^*, 0]G_2$, yields the formula for G in (35). This completes the proof. □

If $H = 0$ in the previous corollary, then we can take $\mathcal{X}_h = \{0\}$. Hence, $\Lambda = P$. Moreover, $Z_2 = Z$ and $B_2 = \tilde{B}$ while $P_2 = \tilde{P}$ the controllability Gramian generated by the pair $\{Z, \tilde{B}\}$. Using this in Corollary 6 readily yields the following result.

Corollary 7 *Consider the weighted Nevanlinna-Pick interpolation problem corresponding to the data $\{Z, B, B_1, \Theta\}$, where the pair $\{Z, B\}$ is controllable, and Θ is an outer function. Let $\tilde{P}$ be the controllability Gramian generated by the pair $\{Z, \tilde{B}\}$, and assume that $r_{spec}(P^{-1}\tilde{P}) < 1$. Let G be the function defined by*

$$G(\lambda) = \tilde{C}(I - \lambda\tilde{Z})^{-1}\tilde{B}, \tag{37}$$

where $\tilde{C}$ form $\mathcal{X}$ into $\mathcal{U}$ and $\tilde{Z}$ on $\mathcal{X}$ are the operators defined by

$$\tilde{C} = B^*(P - Z\tilde{P}Z^*)^{-1} \text{ and } \tilde{Z} = (P - \tilde{P})Z^*(P - Z\tilde{P}Z^*)^{-1}. \tag{38}$$

Then $F = G\Theta$ is a solution to the weighted Nevanlinna-Pick interpolation problem with data $\{Z, B, B_1, \Theta\}$.

Finally, it is noted that if $\Theta = \gamma I$ where γ is a positive constant, then $\tilde{B} = \gamma^{-1}B_1$. In this case, the previous corollary reduces to Theorem V.1.2 in [5].

5 A State Space Computation for $\tilde{B}$ and $\tilde{B}_h$

In this section we will show how one can use state space techniques to compute the operators $\tilde{B}$ and $\tilde{B}_h$ in (21) when Θ is a rational outer function. For completeness, let us recall some elementary facts concerning outer functions. First notice that a scalar valued rational function is outer if and only if all its poles are in $\{\lambda : |\lambda| > 1\}$ and all its zeros are in $\{\lambda : |\lambda| \geq 1\}$. It is well known that a function Θ is outer if and only if Θ is in $H^\infty(\mathcal{Y}, \mathcal{Y})$ and $\det[\Theta]$ is an outer function in H^∞; see Corollary 6.3 page 220 in [19]. We say that Θ is an *invertible* outer function if Θ is an outer function, its inverse Θ^{-1} exists and is also in $H^\infty(\mathcal{Y}, \mathcal{Y})$. So, Θ is an invertible outer function if and only if both Θ and its inverse Θ^{-1} are functions in $H^\infty(\mathcal{Y}, \mathcal{Y})$. A scalar valued rational function is an invertible outer function if and only if all its poles and zeros are in $\{\lambda : |\lambda| > 1\}$. An outer function Θ is an invertible outer function if and only if its determinant is an invertible outer function. To see this, assume that $\det[\Theta]$ is an invertible outer function. Let adj (Θ) be the algebraic adjoint of Θ. Because Θ is in $H^\infty(\mathcal{Y}, \mathcal{Y})$, the function adj (Θ) is also in $H^\infty(\mathcal{Y}, \mathcal{Y})$. Since $1/\det[\Theta]$ is a function in H^∞, the inverse $\Theta^{-1} = \text{adj}\,(\Theta)/\det[\Theta]$ is in $H^\infty(\mathcal{Y}, \mathcal{Y})$. Therefore, Θ is an invertible outer function. On the other hand, if Θ is an invertible outer function, then $1/\det[\Theta] = \det[\Theta^{-1}]$ is in H^∞, and thus $\det[\Theta]$ is an invertible outer function. This proves our claim.

Recall that $\{Z, B, C, D\}$ is a realization for a transfer function Φ if

$$\Phi(\lambda) = D + \lambda C(I - \lambda Z)^{-1} B\,. \tag{39}$$

Here Z is an operator on $\mathcal{X}$ and B is an operator from $\mathcal{U}$ into $\mathcal{X}$, while C maps $\mathcal{X}$ into $\mathcal{Y}$ and D is an operator from $\mathcal{U}$ into $\mathcal{Y}$. Notice that Φ is analytic in some neighborhood of the origin. Throughout we assume that the state space $\mathcal{X}$ is finite dimensional. It is well known that a function Φ admits a finite dimensional realization if and only if Φ is a rational function which is analytic in some neighborhood of the origin. For further results on finite dimensional realization theory see [15, 16].

If $\{Z_o, B_o, C_o, D_o\}$ is any realization for an outer function Θ in $H^\infty(\mathcal{Y}, \mathcal{Y})$, then D_o is invertible. To see this simply notice that $\Theta(0) = D_o$. Because Θ is outer, $\det[\Theta]$ is also an outer function. In particular, $\det[\Theta(0)] = \det[D_o]$ is nonzero. Therefore, D_o is invertible. The following result uses realization theory to determine when a rational function Θ is outer.

Lemma 1 *Let $\{Z_o$ on $\mathcal{X}_o, B_o, C_o, D_o\}$ be a stable realization for a rational function Θ in $H^\infty(\mathcal{Y}, \mathcal{Y})$, and assume that D_o is invertible. Then the inverse of Θ is given by*

$$\Theta(\lambda)^{-1} = D_o^{-1} - \lambda D_o^{-1} C_o (I - \lambda Z^\sharp)^{-1} B_o D_o^{-1}, \tag{40}$$

where $Z^\sharp = Z_o - B_o D_o^{-1} C_o$. Furthermore, Θ is an outer function, respectively an invertible outer function, if and only if $r_{spec}(Z^\sharp) \leq 1$, respectively $r_{spec}(Z^\sharp) < 1$.

Proof: For completeness a proof is given. If D_o is invertible, then it is well known that the inverse of Θ is given by (40); see Lemma III.1.2 in [5] or A.21 in the Appendix of [15]. In fact, this follows from (33) with $R = I - \lambda Z_o$ and $X = -\lambda D_o^{-1} C_o$ and $Y = B_o$. We claim that

$$\det[\Theta(\lambda)] = \det[D_o] \frac{\det[I - \lambda Z^\sharp]}{\det[I - \lambda Z_o]}. \tag{41}$$

To see this, recall that if M and N are two operators acting between the appropriate finite dimensional spaces, then $\det[I + MN] = \det[I + NM]$. Using this along with the fact that $\{Z_o, B_o, C_o, D_o\}$ is a finite dimensional realization of Θ, we have

$$\begin{aligned}
\det[\Theta(\lambda)] &= \det[D_o + \lambda C_o(I - \lambda Z_o)^{-1} B_o] \\
&= \det[D_o] \det[I + \lambda D_o^{-1} C_o (I - \lambda Z_o)^{-1} B_o] \\
&= \det[D_o] \det[I + \lambda (I - \lambda Z_o)^{-1} B_o D_o^{-1} C_o] \\
&= \det[D_o] \det[(I - \lambda Z_o)^{-1} (I - \lambda Z_o + \lambda B_o D_o^{-1} C_o)] \\
&= \det[D_o] \frac{\det[I - \lambda Z^\sharp]}{\det[I - \lambda Z_o]}. [3pt]
\end{aligned}$$

Hence (41) holds.

Notice that if V is any operator on $\mathcal{X}_o$, then

$$\det[I - \lambda V] = \prod_{j=1}^{n} (1 - \lambda \lambda_j)$$

where $\{\lambda_1, \lambda_2, \ldots, \lambda_n\}$ are the eigenvalues for V and n is the dimension of $\mathcal{X}_o$. In particular, λ_o is a nonzero eigenvalue of V if and only if $1/\lambda_o$ is a root of $\det[I - \lambda V]$, and $\lambda_o = 0$ is an eigenvalue of V if and only if the degree of the polynomial $\det[I - \lambda V]$ is strictly less than n. Therefore, the zeros of $\det[I - \lambda V]$ are contained in $\{\lambda : |\lambda| \geq 1\}$, respectively $\{\lambda : |\lambda| > 1\}$ if and only if $r_{spec}(V) \leq 1$, respectively $r_{spec}(V) < 1$.

Because $\{Z_o, B_o, C_o, D_o\}$ is a stable realization, all the zeros of $\det[I - \lambda Z_o]$ are contained in $\{\lambda : |\lambda| > 1\}$. By consulting (41), it follows that $r_{spec}(Z^\sharp) \leq 1$, respectively $r_{spec}(Z^\sharp) < 1$ if and only if $\det[\Theta]$ is outer, respectively an invertible outer function in H^∞. Therefore, Θ is outer, respectively an invertible outer function, if and only if $r_{spec}(Z^\sharp) \leq 1$, respectively $r_{spec}(Z^\sharp) < 1$. This completes the proof. □

The following result uses realization theory to present an algorithm to compute the operators $\tilde{B}$ and $\tilde{B}_h$ defined in (21). This may be useful when implementing Theorems 4, 5, Corollary 4 or any of the previous results involving the operators $\tilde{B}$ and $\tilde{B}_h$.

Proposition 2 *Let $\{Z_o$ on $\mathcal{X}_o, B_o, C_o, D_o\}$ be a stable realization for a rational outer function Θ in $H^\infty(\mathcal{Y}, \mathcal{Y})$. Let Z on $\mathcal{X}$ and Z_h on $\mathcal{X}_h$ be two finite dimensional stable operators. Let B_1 be an operator from $\mathcal{Y}$ into $\mathcal{X}$ and B_h be an operator from $\mathcal{Y}$ into $\mathcal{X}_h$. Finally, let R and R_h be the unique solution to the following Stein equations:*

$$R = ZRZ^\sharp + B_1 D_o^{-1} C_o \text{ and } R_h = Z_h R_h Z^\sharp + B_h D_o^{-1} C_o, \tag{42}$$

where $Z^\sharp = Z_o - B_o D_o^{-1} C_o$. Then the operators $\tilde{B}$ and $\tilde{B}_h$ defined in (21) can be computed by

$$\tilde{B} = B_1 D_o^{-1} - ZRB_o D_o^{-1} \text{ and } \tilde{B}_h = B_h D_o^{-1} - ZR_h B_o D_o^{-1}. \tag{43}$$

Proof: If L is any operator on $\mathcal{X}_o$, then $(I - \lambda L)^{-1}$ admits a power series of the form $\sum_0^\infty \lambda^n L^n$ in some neighborhood of the origin. Applying this to the expression for the inverse of Θ in (40) shows that Θ^{-1} admits a power series of the form

$$\Theta(\lambda)^{-1} = \sum_{n=0}^{\infty} \Omega_n \lambda^n = D_o^{-1} - \sum_{n=1}^{\infty} \lambda^n D_o^{-1} C_o Z^{\sharp n-1} B_o D_o^{-1}. \tag{44}$$

By matching like coefficients of λ^n, we readily arrive at the following result

$$\Omega_0 = D_o^{-1} \text{ and } \Omega_n = -D_o^{-1} C_o (Z^\sharp)^{n-1} B_o D_o^{-1} \quad (\text{for } n = 1, 2, 3, \ldots). \tag{45}$$

Substituting this into the definition for $\tilde{B}$ in (21), we obtain

$$\tilde{B} = B_1 D_o^{-1} - Z \sum_{n=0}^{\infty} Z^n B_1 D_o^{-1} C_o Z^{\sharp n} B_o D_o^{-1}. \tag{46}$$

Because Θ is outer, Lemma 1 shows that $r_{spec}(Z^\sharp) \leq 1$. Let R be the operator from $\mathcal{X}_o$ into $\mathcal{X}$ defined by

$$R = \sum_{n=0}^{\infty} Z^n B_1 D_o^{-1} C_o Z^{\sharp n}.$$

Since Z is stable and $r_{spec}(Z^\sharp) \leq 1$, this sum converges in the operator topology. Hence, the operator R is well defined. Substituting this into (46) shows that $\tilde{B}$ is given by the first equation in (43). Moreover, R is the unique solution to the Stein equation $R = ZRZ^\sharp + B_1 D_o^{-1} C_o$. This shows that one can compute $\tilde{B}$ by first solving the Stein equation for R in (42). Then $\tilde{B}$ is computed from the first equation in (43). By replacing Z by Z_h and B_1 by B_h in the above analysis, we see that $\tilde{B}_h$ is computed from the second equation in (43), where R_h is the solution to the Stein equation in (42). This completes the proof. □

6 The Case when $TQT^* \leq Q$

In this section we will use the previous theory to solve a special hyper-weighted Nevanlinna-Pick interpolation problem. Then we will present a simple numerical example to demonstrate the difference between the hyper-weighted and weighted Nevanlinna-Pick problem for the same data. Let us first establish some notation. Let Φ be a function in $L^\infty(\mathcal{Y}, \mathcal{E})$ and $\Phi = \sum \Phi_n e^{\imath n\omega}$ be its Fourier series expansion. Then the *Laurent* operator L_Φ is the block matrix from $l^2(\mathcal{Y})$ into $l^2(\mathcal{E})$ whose entries are given by $(L_\Phi)_{i,j} = \Phi_{i-j}$ for all integers i and j. The function Φ is the *symbol* for L_Φ. Using this notation, it follows that the Toeplitz operator T_Φ is the operator from $l^2_+(\mathcal{Y})$ into $l^2_+(\mathcal{E})$ given by $T_\Phi = P_+ L_\Phi | l^2_+(\mathcal{Y})$ where P_+ is the orthogonal projection onto $l^2_+(\mathcal{E})$. Let $l^2_-(\mathcal{E})$ be the orthogonal complement of $l^2_+(\mathcal{E})$ in $l^2(\mathcal{E})$, and let P_- be the orthogonal projection onto $l^2_-(\mathcal{E})$. Let H_Φ be the Hankel operator from $l^2_+(\mathcal{Y})$ into $l^2_-(\mathcal{E})$ defined by $H_\Phi h = P_- \Phi h$ where h is in $l^2_+(\mathcal{Y})$. Let H be the Hankel operator from $l^2_+(\mathcal{Y})$ into $l^2_+(\mathcal{E})$ defined in (7) where the operators H_n are replaced by Φ_{-n} for all integers $n \geq 1$. Then H is a matrix representation for H_Φ. In particular, $H^*H = H^*_\Phi H_\Phi$.

As before, let T be the unilateral shift on $l^2_+(\mathcal{Y})$, and Q a positive operator on $l^2_+(\mathcal{Y})$. Proposition V.5.1 in [19] shows that $TQT^* \leq Q$ if and only if Q admits a factorization of the form $Q = T^*_\Psi T_\Psi$ where $\Psi(1/\lambda)$ is a function in some $H^\infty(\mathcal{Y}, \mathcal{E})$ space, that is, $\Psi(e^{-\imath\omega})$ is a function in $H^\infty(\mathcal{Y}, \mathcal{E})$. If Q is a positive operator on $l^2_+(\mathcal{Y})$ satisfying $TQT^* \leq Q$, then using $T^*T = I$, it follows that $Q \leq T^*QT$. According to Proposition 1, the operator Q also admits a decomposition of the form $Q = \Upsilon - H^*H$ where Υ is a positive Toeplitz operator and H is a Hankel operator. To obtain this decomposition directly from the factorization $Q = T^*_\Psi T_\Psi$, notice that

$$\|T_\Psi h\|^2 = \|L_\Psi h\|^2 - \|H_\Psi h\|^2 \qquad (h \in l^2_+(\mathcal{Y}))\,. \tag{47}$$

Let $\sum_0^\infty e^{-\imath n\omega}\Psi_n$ be the Fourier series expansion for Ψ, and H be the Hankel operator defined in (7) where the operators H_n are replaced by Ψ_n for all integers $n \geq 1$. Then H is a matrix representation for H_Ψ, and $H^*H = H^*_\Psi H_\Psi$. Let Υ be the Toeplitz operator on $l^2_+(\mathcal{Y})$ defined by $\Upsilon = P_+ L^*_\Psi L_\Psi | l^2_+(\mathcal{Y})$ where P_+ is the orthogonal projection onto $l^2_+(\mathcal{Y})$. Notice that $(\Upsilon h, h) = \|L_\Psi h\|^2$ for all h in $l^2_+(\mathcal{Y})$. Using this in (47) shows that $T^*_\Psi T_\Psi = \Upsilon - H^*H$, which is precisely the decomposition for $Q = T^*_\Psi T_\Psi$ obtained in Proposition 1.

In this section, we will use our previous results to solve the hyper-weighted Nevanlinna-Pick interpolation problem with data $\{Z, B, B_1, Q\}$ where $Q = T^*_\Psi T_\Psi$ and $\Psi(1/\lambda)$ is a function in $H^\infty(\mathcal{Y}, \mathcal{Y})$. Then we will provide a simple numerical example to show that these interpolation problems may be useful in applications. Using (47) it follows that F is a solution to this hyper-weighted Nevanlinna-Pick interpolation problem if and only if F is a function in $H^\infty(\mathcal{Y},\mathcal{U})$ satisfying

$$WT_F = W_1 \text{ and } \|T_F h\|^2 + \|H_\Psi h\|^2 \le \|L_\Psi h\|^2 \tag{48}$$
$$(\text{for all } h \in l^2(\mathcal{Y})).$$

Now let Θ be an outer function in $H^\infty(\mathcal{Y},\mathcal{Y})$, satisfying $\Theta^*\Theta = \Psi^*\Psi$ almost everywhere on the unit circle. This Θ is called the outer spectral factor for $\Psi^*\Psi$. Propositions V.4.2 and V.7.1 in [19] show that there always exists an outer spectral factor for $\Psi^*\Psi$. Moreover, the outer spectral factor is unique up to a unitary constant on the left. Hence, $\Upsilon = P_+ L_\Psi^* L_\Psi | l_+^2(\mathcal{Y}) = T_\Theta^* T_\Theta$. Thus, F is a solution to the hyper-weighted Nevanlinna-Pick interpolation problem with data $\{Z, B, B_1, T_\Psi^* T_\Psi\}$ if and only if F is a solution to the hyper-weighted Nevanlinna-Pick interpolation problem with data $\{Z, B, B_1, T_\Theta^* T_\Theta - H^* H\}$. In particular, if F is a solution to the hyper-weighted Nevanlinna-Pick interpolation problem with data $\{Z, B, B_1, T_\Psi^* T_\Psi\}$, then $F(e^{\imath\omega})^* F(e^{\imath\omega}) \le \Theta(e^{\imath\omega})^*\Theta(e^{\imath\omega})$. Therefore, this F is also a solution to the weighted Nevanlinna-Pick interpolation problem with data $\{Z, B, B_1, \Theta\}$.

For the moment assume that $\mathcal{Y} = C^1$ and $\Psi(1/\lambda)$ is a function in H^∞. If $\Psi(1/\lambda)$ is not an outer function, then the hyper-weighted Nevanlinna-Pick interpolation problem with data $\{Z, B, B_1, T_\Psi^* T_\Psi\}$ is trivial. To be precise, if $\Psi(1/\lambda)$ is not an outer function, then the hyper-weighted Nevanlinna-Pick problem has a solution if and only if $B_1 = 0$. In this case, $F = 0$ is a solution to this problem. To see this, assume that $\Psi(1/\lambda)$ is not an outer function. Then Ψ^* admits an inner-outer factorization of the form $\Psi^* = \Psi_i \Psi_o$ where Ψ_i is a non-trivial inner function and Ψ_o is an outer function. According to the Beurling-Lax-Halmos Theorem, the space $l_+^2 \ominus L_{\Psi_i} l_+^2$ is nonzero. Let h be any nonzero vector in $l_+^2 \ominus L_{\Psi_i} l_+^2$. Then $L_{\Psi_i}^* h$ is orthogonal to l_+^2, that is, $L_{\Psi_i}^* h$ is a vector in l_-^2. Because l_-^2 is an invariant subspace for $L_{\Psi_o}^*$, we have $H_\Psi h = P_- L_{\Psi_o}^* L_{\Psi_i}^* h = L_\Psi h$. Using this h shows that the second constraint in (48) holds if and only if $T_F h = 0$. Therefore, $F\hat{h} = 0$ where $\hat{h}$ is the Fourier transform of h, that is, $\hat{h} = \sum h_n e^{\imath n\omega}$ where $h = (h_n)_0^\infty$. Because h is a non-zero vector in l_+^2, its Fourier transform $\hat{h}$ is a nonzero function in H^2. So, $F\hat{h} = 0$, implies that $F = 0$. If $F = 0$, then obviously $W_1 = WT_F$ must also be zero, and thus $B_1 = 0$. Therefore, if $\Psi(1/\lambda)$ is not an outer function, then the hyper-weighted Nevanlinna-Pick interpolation problem with data $\{Z, B, B_1, T_\Psi^* T_\Psi\}$ has a solution if and only if $B_1 = 0$. Finally, it is noted that the example given below shows that this hyper-weighted Nevanlinna-Pick interpolation problem can have a nontrivial solution when $\Psi(1/\lambda)$ is an outer function.

Now let us proceed to obtain a solution to this hyper-weighted Nevanlinna-Pick interpolation problem in the rational case. To this end, let $\{Z_h \text{ on } \mathcal{X}_h, B_h, C_h, D_h\}$ be a minimal realization for a rational outer function $\Psi(1/\lambda)$ in $H^\infty(\mathcal{Y},\mathcal{Y})$. In other words, the pair $\{Z_h, B_h\}$ is controllable, $\{C_h\, Z_h\}$ is observable and

$$\Psi(\lambda) = D_h + C_h(\lambda I - Z_h)^{-1} B_h\,. \tag{49}$$

In this case, $\Psi(\lambda)$ admits a power series expansion of the form

$$\Psi(\lambda) = \sum_{n=0}^{\infty} \lambda^{-n} \Psi_n = D_h + \sum_{n=1}^{\infty} \lambda^{-n} C_h Z_h^{n-1} B_h \,. \tag{50}$$

Hence, $\Psi_n = C_h Z_h^{n-1} B_h$ for all integers $n \geq 1$. Now let H be the Hankel operator in (7) generated by the realization $\{Z_h, B_h, C_h\}$. Then this H is precisely a matrix representation for the Hankel operator H_Ψ. In particular, $H_\Psi^* H_\Psi = H^* H$.

As before, let Θ in $H^\infty(\mathcal{Y}, \mathcal{Y})$ be the outer spectral factor for $\Psi^*\Psi$, that is, Θ is outer and $\Theta^*\Theta = \Psi^*\Psi$. If $\mathcal{Y} = C^1$, then $\Theta = \Psi^*$. In this case, $\{Z_h^*, C_h^*, B_h^*, D_h^*\}$ is a minimal realization for Θ, that is,

$$\Theta(\lambda) = D_h^* + \lambda B_h^* (I - \lambda Z_h^*)^{-1} C_h^* \,. \tag{51}$$

If the dimension of $\mathcal{Y}$ is greater then or equal to two, then one can use standard state space techniques to compute the outer spectral factor Θ for $\Psi^*\Psi$; for example see Theorem 2.1 in the Appendix in [5]. In this case, Θ will have a state space realization of the form $\{Z_o, B_o, C_o, D_o\}$ where $Z_o = Z_h^*$.

Consider the hyper-weighted Nevanlinna-Pick problem with data $\{Z, B, B_1, T_\Psi^* T_\Psi\}$ where $\{Z, B\}$ is controllable and $\Psi(1/\lambda)$ is a rational outer function in $H^\infty(\mathcal{Y}, \mathcal{Y})$. To solve this hyper-weighted Nevanlinna-Pick interpolation problem, let $\{Z_h, B_h, C_h, D_h\}$ be a minimal realization for $\Psi(1/\lambda)$. Let H be the Hankel matrix determined by the realization $\{Z_h, B_h, C_h\}$. Then compute the minimal realization $\{Z_o, B_o, C_o, D_o\}$ for Θ, the outer spectral factor of $\Psi^*\Psi$. According to our previous analysis F is a solution to the hyper-weighted Nevanlinna-Pick interpolation problem with data $\{Z, B, B_1, T_\Psi^* T_\Psi\}$ if and only if F is a solution to the hyper-weighted Nevanlinna-Pick interpolation problem with data $\{Z, B, B_1, T_\Theta^* T_\Theta - H^* H\}$. By consulting Corollary 4, we see that there exists a solution to the hyper-weighted Nevanlinna-Pick interpolation problem with data $\{Z, B, B_1, T_\Psi^* T_\Psi\}$ if and only if $r_{spec}(J P_2) \leq 1$ where $J = P^{-1} \oplus P_o$ and P_2 is the solution to the Lyapunov equation in (25) with $\{Z_2, B_2\}$ defined in (24). Moreover, if $r_{spec}(J P_2) < 1$, then the function $F = G\Theta$ defined in Theorem 5 or Corollary 6 provides a solution to this hyper-weighted Nevanlinna-Pick interpolation problem.

Example 2 Assume that Θ is a scalar valued outer function in H^∞ and set $\Psi = \Theta^*$. Then $\Psi(1/\lambda)$ is also a scalar valued outer function, and $|\Theta|^2 = |\Psi|^2$ almost everywhere on the unit circle. In other words, Θ is the outer spectral factor for $\Psi^*\Psi$. Notice that if F is a solution to the hyper-weighted Nevanlinna-Pick interpolation problem with data $\{Z, B, B_1, T_\Psi^* T_\Psi\}$, then F is also a solution to the weighted Nevanlinna-Pick interpolation problem with data $\{Z, B, B_1, \Theta\}$. In this example, we will compare the solution to the hyper-weighted Nevanlinna-Pick interpolation problem with data $\{Z, B, B_1, T_\Psi^* T_\Psi\}$ computed in Theorem 5, or equivalently, Corollary 6, to the solution of the weighted Nevanlinna-Pick interpolation problem

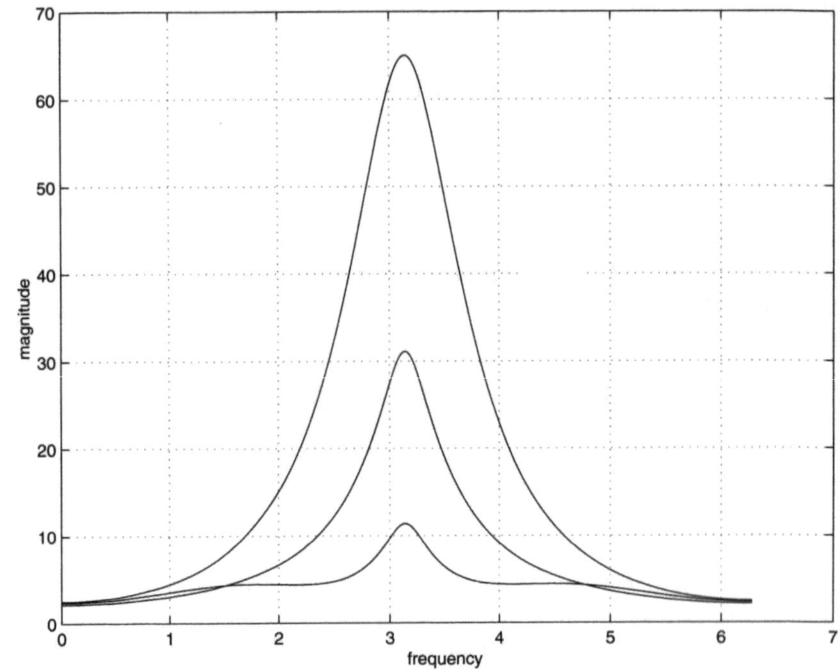

Figure 1

with data $\{Z, B, B_1, \Theta\}$ computed according to Corollary 7. To this end, consider the system $\{Z, B, B_1\}$ defined by

$$(52) \qquad Z = \begin{bmatrix} 0 & 0 & 0 \\ 0 & 1/2 & 0 \\ 0 & 0 & -4/5 \end{bmatrix}, \quad B = \begin{bmatrix} 1 \\ 1 \\ 1 \end{bmatrix}, \text{ and } B_1 = \begin{bmatrix} 1 \\ 2 \\ 3 \end{bmatrix}.$$

Let Θ be the rational outer function given by

$$(53) \qquad \Theta(\lambda) = \frac{10(10-2\lambda)(10-3\lambda)}{(10+4\lambda)(10+6\lambda)}.$$

In this case, F is a function in H^∞ satisfying $WT_F = W_1$ if and only F is a function in H^∞ satisfying the interpolating conditions $F(0) = 1$ and $F(1/2) = 2$ and $F(-4/5) = 3$. By applying the Lyapunov functions in Matlab, we obtained $r_{spec}(\tilde{P}P^{-1}) = 0.5122$. So, there exists a solution to this weighted Nevanlinna-Pick problem. Let F_o be the solution to the weighted Nevanlinna-Pick interpolation problem with data $\{Z, B, B_1, \Theta\}$ computed according to Corollary 7. By implementing standard Matlab computing techniques, we discovered that F_o is given by

$$(54)\qquad F_o(\lambda) = \frac{1 + 4.7718\lambda + 2.9588\lambda^2 - 2.4510\lambda^3 + 0.3321\lambda^4}{1 + 1.5531\lambda + 0.6100\lambda^2 - 0.0504\lambda^3 - 0.0439\lambda^4}.$$

Now let Ψ be the rational function defined by $\Psi(\lambda) = \Theta(1/\lambda)$. Obviously, Θ is the outer spectral factor for $\Psi^*\Psi$. By applying the Lyapunov functions in Matlab, we computed $r_{spec}(JP_2) = 0.988$. So, there exists a solution to the hyper-weighted Nevanlinna-Pick problem with data $\{Z, B, B_1, T_\Psi^* T_\Psi\}$. Let F be the solution to this hyper-weighted Nevanlinna-Pick interpolation problem given in Theorem 5, or equivalently, Corollary 6. By applying standard Matlab computing methods, we discovered that

$$(55)\qquad F(\lambda) = \frac{1 + 3.0803\lambda + 2.3887\lambda^2 - 1.8446\lambda^3 + 0.2471\lambda^4}{1 + 0.8374\lambda + 0.1252\lambda^2 + 0.0846\lambda^3 + 0.0079\lambda^4}.$$

Figure 1 presents a plot of $|\Theta(e^{\imath\omega})|$ and $|F_o(e^{\imath\omega})|$ and $|F(e^{\imath\omega})|$, as the frequency ω varies form zero to 2π. In this plot we have

$$|F(e^{\imath\omega})| \leq |F_o(e^{\imath\omega})| \leq |\Theta(e^{\imath\omega})|$$

for $1.8 \leq \omega \leq 4.5$, that is, the graph at top of Figure 1 is $|\Theta(e^{\imath\omega})|$, while the graph in the middle corresponds to $|F_o(e^{\imath\omega})|$, and the graph at the bottom corresponds to $|F(e^{\imath\omega})|$. Clearly $\|F\|_\infty < \|F_o\|_\infty$. This may be useful in applications. For example, one can convert a H^∞ control problem to a Nevanlinna-Pick problem; see [1, 3, 5, 8, 14]. In particular, obtaining an interpolant with a small H^∞ norm solving the Nevanlinna-Pick problem is useful in constructing robust controllers. Since $\|F\|_\infty < \|F_o\|_\infty$, the solution to the hyper-weighted Nevanlinna-Pick interpolation problem may be useful in H^∞ control design. Finally, notice that both F_o and F have the same McMillan degree. So, in a H^∞ control application, the solution F to the hyper-weighted Nevanlinna-Pick problem did not increase the McMillan degree of the controller.

References

[1] J.A. Ball, I. Gohberg and L. Rodman, *Interpolation For Rational Matrix Functions*, Operator Theory: Advances and Applications **45**, Birkhäuser Verlag, Basel, 1990.

[2] A. Biswas, C. Foias and A.E. Frazho, Weighted commutant lifting, Submitted.

[3] J.C. Doyle, B.A. Francis and A. Tannenbaum, *Feedback Control Theory*, MacMillan, New York, 1991.

[4] C. Foias and A.E. Frazho, *The Commutant Lifting Approach to Interpolation Problems*, Operator Theory Advances and Applications **44**, Birkhäuser Verlag, Basel, 1990.

[5] C. Foias, A.E. Frazho, I. Gohberg and M.A. Kaashoek, *Metric Constrained Interpolation, Commutant Lifting and Systems*, Operator Theory: Advances and Applications **100**, Birkhäuser Verlag, Basel, 1998.

[6] C. Foias, A.E. Frazho and W.S. Li, The exact H^2 estimate for the central H^∞ interpolant, in: *New aspects in interpolation and completion theories* (Ed. I Gohberg) Operator Theory Advances and Applications **64**, Birkhäuser-Verlag, Basel, 1993, 119–156.

[7] C. Foias, H. Ozbay and A.R. Tannenbaum, *Robust Control of Infinite Dimensional Systems*, Springer-Verlag, London, 1996.

[8] B.A. Francis, *A Course in Control Theory*, Lecture Notes in Control and Information Sciences, Springer-Verlag, New York, 1987.

[9] K. Glover, All optimal Hankel-norm approximations of linear multivariable systems and their error bounds, *International J. Control* **39** (1984), 1115–1193.

[10] I. Gohberg, S. Goldberg and M.A. Kaashoek, *Classes of Linear Operators I*, **49**, Birkhäuser-Verlag, Basel, 1990.

[11] I. Gohberg, S. Goldberg and M.A. Kaashoek, *Classes of Linear Operators II*, **63**, Birkhäuser-Verlag, Basel, 1993.

[12] I. Gohberg, M.A. Kaashoek and F. Van Schagen, Rational contractive and unitary interpolants in realized form, *Integral Equations and Operator Theory* **11** (1988), 105–127.

[13] I. Gohberg, M.A. Kaashoek and H. J. Woerdeman, The band method for positive and strictly contractive extension problems: an alternative version and new applications, *Integral Equations and Operator Theory* **12** (1989), 343–382.

[14] M. Green and D. Limebeer, *Linear Robust Control*, Prentice Hall, Inc., 1995.

[15] T. Kailath, *Linear Systems*, Englewood Cliffs: Prentice Hall, New Jersey, 1980.

[16] W.J. Rugh, *Linear System Theory*, Prentice Hall, Inc., New Jersey, 1993.

[17] D. Sarason, Generalized interpolating in H^∞, *Trans. American Math. Soc.* **127** (1967), 179–203.

[18] B. Sz.-Nagy and C. Foias, Dilatation des commutants d'opérateurs, *C. R. Acad. Sci. Paris série A* **266** (1968), 493–495.

[19] B. Sz.-Nagy and C. Foias, *Harmonic Analysis of Operators on Hilbert Space*, North-Holland Publishing Co., Amsterdam, 1970.

[20] S. Treil and A. Volberg, A fixed point approach to Nehari's problem and its applications, *Operator Theory: Advances and Applications,* vol 71, Birkhäuser, 1994, 165–186.

[21] K. Zhou, J.C. Doyle and K. Glover, *Robust and Optimal Control*, Prentice Hall, Inc., 1996.

School of Aeronautics and Astronautics
Purdue University
West Lafayette
IN 47907
U.S.A.

MSC 1992: Primary, 47A20, 47A57, 93B36.

Operator Theory:
Advances and Applications, Vol. 122
© 2001 Birkhäuser Verlag Basel/Switzerland

Effects of Small Delays on Stability and Control

Jack K. Hale and Sjoerd M. Verduyn Lunel

Dedicated to Rien Kaashoek on the 60th anniversary of his birth

Stabilization and control of partial differential equations through the application of forces on the boundary turns out to be very important in applications. When the boundary forces are applied with no delays there is a rather complete theory. In applications, however, it is very likely that time delays will occur when applying the boundary forces. So it is of vital importance to understand the sensitivity with respect to small time delays and problems of this type have attracted a lot of attention, especially in the engineering literature. In this paper we shall explain the underlying mechanisms and the important role difference equations and almost periodic functions are playing in these results.

1 Introduction

In the implementation of any feedback control system, it is very likely that time delays will occur. It is therefore of vital importance to understand the sensitivity of the control system to the introduction of small delays in the loop. For some systems, small delays can lead to destabilization. One of the earliest papers to address this topic using frequency domain techniques was [2]. A more recent and more complete discussion using this approach is [22, 23, 24, 25]. At approximately the same time, studies were made directly on the characteristic equation associated with a neutral delay differential equations to determine robustness of stability with respect to variations in the delays [1, 14, 17, 27, 28].

It is our aim to present an attempt for a unifying theory that explains the underlying mechanisms and the role difference equations are playing in these results. A source of confusion in the literature is that sometimes small delays have no effect and sometimes dramatic changes occur. The following simple examples are characteristic for the general theory.

Example 1.1 Consider the linear control problem

$$\dot{x}(t) = Ax(t) + Bu(t),$$

where A and B are matrices, the state x and the control u are vector-valued functions. Suppose that the system is (exponentially) stabilizable, i.e., if we close the

---

The research was supported by the N.S.F under DMS-970483 and by an N.W.O. pionier-grant under 600-61-410.

loop by $u(t) = Fx(t)$ for some matrix F, then the differential equation

$$\dot{x}(t) = (A + BF)x(t) \tag{1.1}$$

is exponentially stable. In other words, the roots of the characteristic polynomial of (1.1)

$$f(z) = \det(zI - (A + BF)) \tag{1.2}$$

are all in the left half plane $\{z \in \mathbb{C} : \operatorname{Re} z < 0\}$ of the complex plane.

A time delay in the feedback loop leads to a differential difference equation

$$\dot{x}(t) = Ax(t) + BFx(t - \epsilon). \tag{1.3}$$

This equation is exponentially stable if and only if the roots of the exponential polynomial

$$g_\epsilon(z) = \det\left(zI - A - BFe^{-\epsilon z}\right) \tag{1.4}$$

are in the left half plane $\{z \in \mathbb{C} : \operatorname{Re} z < 0\}$ (see [12, 15] for the general theory). It is a fact that there exists an ϵ_0 such that for every $\epsilon < \epsilon_0$ the differential delay equation (1.3) is exponentially stable if and only if the differential equation (1.1) is exponentially stable. The proof of this fact is a consequence of the Rouché theorem and holds for arbitrary linear delay equations with delays at most ϵ. Since the argument of proof plays an important role in the sequel, we shall outline the main idea of the proof.

From general estimates, see [12, Section I.4], it follows that g_ϵ has finitely many roots in any right half plane $\operatorname{Re} z > -\delta$, $\delta > 0$. In addition, there exists a contour Γ such that the roots of g_ϵ with real part larger than $-\delta$ are inside Γ and

$$\min_{z\in\Gamma} |f(z)| > 0. \tag{1.5}$$

Furthermore, the functions f and g_ϵ defined in (1.2) and (1.4) can be expanded as follows

$$f(z) = z^n + a_1 z^{n-1} + \cdots + a_n$$

and

$$g_\epsilon(z) = z^n + b_1(\epsilon, z)z^{n-1} + \cdots + b_n(\epsilon, z),$$

where the functions $b_j(\epsilon, \cdot)$, $j = 1, \ldots, n$, are analytic, uniformly bounded in the right half plane $\operatorname{Re} z > -\delta$, and

$$b_j(\epsilon, z) \to a_j \quad \text{as } \epsilon \to 0 \quad \text{uniformly on } \Gamma. \tag{1.6}$$

Estimate (1.6) implies that $|f(z) - g_\epsilon(z)|$ can be made arbitrary small on Γ. Hence, it follows from (1.5) that

$$|f(z) - g_\epsilon(z)| < |f(z)| \quad \text{for } z \in \Gamma.$$

So Rouché's theorem [4] implies that f and g_ϵ have the same number of roots in the interior of Γ.

Thus there exists an $\epsilon_0 > 0$ such that for $\epsilon < \epsilon_0$ the functions f and g_ϵ have the same number of zeros in the closed right half plane $\mathrm{Re}\, z \geq 0$ of the complex plane and we obtain that for $\epsilon < \epsilon_0$ system (1.1) is exponentially stable if and only if system (1.3) is exponentially stable and small time delays in the feedback *do not matter*.

The next example is an illustration of the fact that small delays *do matter*.

Example 1.2 Consider the difference equation

$$x(t) - a_1 x(t-h) = 0, \tag{1.7}$$

where $a_1 \in \mathbb{R}$ and $h > 0$. The roots of the characteristic function

$$h(z) = 1 - a_1 e^{-zh} \tag{1.8}$$

determine the stability of (1.7). So, the system is not exponentially stable if and only if $|a_1| \geq 1$. Given $a_1 \geq 1$, we can apply a feedback control to stabilize (1.7)

$$x(t) - a_1 x(t-h) - f x(t-h) = 0. \tag{1.9}$$

This system is exponentially stable if and only if $|a_1 + f| < 1$. Suppose now that the feedback control cannot be applied instantaneously and that there is a small time delay in the feedback:

$$x(t) - a_1 x(t-h) - f x(t-(h+\epsilon)) = 0. \tag{1.10}$$

We claim that, although (1.10) is exponentially stable for $\epsilon = 0$, there is a sequence of positive real numbers ϵ_j tending to zero so that (1.10) is exponentially unstable. To prove the claim, we have to show that the characteristic function associated with (1.10)

$$h_\epsilon(z) = 1 - a_1 e^{-zh} - f e^{-z(h+\epsilon)} \tag{1.11}$$

has roots in the right half plane $\mathrm{Re}\, z > 0$.

To prove this, we compare the roots of h_ϵ with the roots of

$$g(z) = 1 - (|a_1| + |f|) e^{-zh} \tag{1.12}$$

using the Rouché theorem. Take Γ to be a rectangle with corners it, $H + it$, $H + i(t+l)$ and $i(t+l)$, where $H > h^{-1} \log(|a_1| + |f|)$ and t is fixed so that g has roots in the interior of Γ and such that $g(z) > 0$ on Γ (this is always possible, see the discussion after Theorem 3.2). So, to prove that h_ϵ has roots in the open right half plane, it suffices to show that there exists a sequence of ϵ_j such that

$$|h_{\epsilon_j}(z) - g(z)| = |f| e^{-\mathrm{Re}\, zh} |e^{-\epsilon_j z} - 1| \tag{1.13}$$

can be made arbitrary small on Γ. This follows from a direct computation.

It is illustrative to derive this result directly in a specific case. Take $h = 1$, $a_1 = 5/4$ and $f = -3(\sqrt{2})^{-1}$. So (1.7) is unstable and (1.9) is exponentially stable. Let us now consider (1.10). The characteristic function h_ϵ is given by

$$h_\epsilon(z) = 1 - \frac{5}{4}e^{-z} + \frac{3}{\sqrt{2}}e^{-z(1+\epsilon)}. \tag{1.14}$$

In order to show that (1.14) has roots in the right half plane Re $z > 0$, we solve

$$e^{-z} = \frac{1}{2} \quad \text{and} \quad e^{-z(1+\epsilon)} = -\frac{1}{4\sqrt{2}}$$

or

$$\begin{aligned} z &= \log 2 + 2k\pi i \\ z(1+\epsilon) &= \frac{1}{2}\log 32 + (2l+1)\pi i \end{aligned}$$

for integers k and l. From these equations we can solve ϵ explicitly. We would like to choose k and l such that ϵ is real and positive. A direct computation shows that this is indeed possible if $k = 2l + 1$.

Examples 2.1 and 2.2 are illustrative for the general theory. Sometimes small delays do matter and sometimes they don't. In this paper we shall describe some of the results in this area and present a number of examples. In case of stabilization and control of partial differential equations through the application of forces on the boundary, we will also suggest some other types of control for which there will be robustness of stability with respect to the introduction of delays. Some of the work we shall present is work in progress and we hope to return to these questions in a sequel paper.

The organisation of this paper is as follows. In Section 2 we shall describe the abstract setting of the problem using the theory of strongly continuous semigroups. In Section 3 we shall discuss robustness of stability with respect to delays for difference equations and present a new result motivated by work from [24]. In Section 4, we recall the close connection between neutral functional differential equations and difference equations and state a consequence of the theory from Section 3. Finally, in Section 5 we shall explain why difference equations appear in the study of stabilization and control of partial differential equations through the application of forces on the boundary.

2 The Abstract Setting of the Problem

Let X be a complex Banach space, and let $A(X \to X)$ be a densely defined linear operator acting on X. The domain $\mathcal{D}(A)$ of A is a linear subspace of X and $A : \mathcal{D}(A) \to X$ is a linear operator. Consider the abstract Cauchy problem

$$\dot{u}(t) = Au, \quad u(0) = u_0 \in X. \tag{2.1}$$

If A is the generator of a $\mathcal{C}_0$-semigroup $T(t)$, then the unique solution to (2.1) is given by $u(t) = T(t)u_0$. The asymptotic behaviour of the solutions is determined by the spectrum $\sigma(T(1))$ of the operator $T(1)$. If $\sigma(T(1))$ is inside the unit circle in the complex plane, then each orbit $\{T(t)u_0, t \geq 0\}$ will approach zero exponentially and uniformly. If there is a point in $\sigma(T(1))$ outside the unit circle, then there is an unbounded orbit and we have instability.

Let $B : X \to X$ be a bounded linear operator and consider the perturbed equation

$$\dot{u}(t) = Au + Bu, \quad u(0) = u_0 \in X. \tag{2.2}$$

The solution to (2.2) is given by the variation-of-constants formula

$$u(t) = T(t)u_0 + \int_0^t T(t-s)Bu(s)\,ds, \quad t \geq 0.$$

In turns out that a fundamental role in the study of the preservation of stability of (2.1) with respect to perturbations B is played by the behaviour of the essential spectrum $\sigma_e(T)$ of $T = T(1)$ under perturbations.

Although there are different definitions for the essential spectrum in the literature all of them have the same radius (see [10, pp. 80–81]). According to Schechter, one can define the essential spectrum as the part of the spectrum of T that cannot be removed by compact perturbations

$$\sigma_e(T) \stackrel{\text{def}}{=} \bigcap_{K \text{ compact}} \sigma(T+K). \tag{2.3}$$

Since any nonzero point in the spectrum of a compact operator is an eigenvalue of finite type, we have that nonzero points λ with the property that the generalized eigenspace at λ,

$$\mathcal{M}_\lambda = \bigcup_{k \geq 1} \mathcal{N}((\lambda I - T)^k),$$

is infinite dimensional, belong to the essential spectrum of T.

Essential spectrum arises naturally for semigroups, even if the generator A of a semigroup $T(t)$ has compact resolvent, i.e., for $z \in \rho(A)$ the operator $(zI - A)^{-1}$ is compact. For example, if $\rho + i2k\pi \in \sigma_p(A)$ for every k, then there exist independent eigenvectors v_k,

$$Av_k = (\rho + i2k\pi)v_k,$$

and, for every k, we have that

$$T(1)v_k = e^\rho v_k, \quad k = 1, 2, \ldots.$$

This implies that e^ρ belongs to the essential spectrum of $T(1)$.

To further explain the role of the essential spectrum we recall the notion of hyperbolicity. A $\mathcal{C}_0$-semigroup $(T(t))_{t\geq 0}$ on a Banach space X is called *hyperbolic* when the space X decomposes into $X = X_- \oplus X_+$ such that $T(t)X_\pm \subset X_\pm$,

$$T_-(t) : X_- \to X_-, \qquad T_-(t)x = T(t)x$$

extends to a $\mathcal{C}_0$-group on X_- over $-\infty < t < \infty$, and there are positive constants K, α, β such that

$$\begin{aligned} \|T(t)\Pi x\| &\leq Ke^{\beta t}\|\Pi x\|, \qquad t \leq 0, \\ \|T(t)(I-\Pi)x\| &\leq Ke^{-\alpha t}\|(I-\Pi)x\|, \qquad t \geq 0. \end{aligned}$$

Here $\Pi : X \to X$ denotes the projection of X along X_+ onto X_-.

It is clear that $(T(t))_{t\geq 0}$ is hyperbolic if and only if there is an open annulus containing the circle $\{z \in \mathbb{C} : |z| = 1\}$ in the resolvent set of $T(1)$. A closely related concept is that of an exponentially dichotomous operator (see [3]). Let $P : X \to X$ be a (bounded) projection of X commuting with A, i.e., P maps $\mathcal{D}(A)$ into itself and $PAx = APx$ for all $x \in \mathcal{D}(A)$. Set $X_- = \mathcal{R}(P)$ and $X_+ = \mathcal{N}(P)$. Then

$$X = X_- \oplus X_+ \tag{2.4}$$

and this decomposition *reduces* A. By this we mean

$$\mathcal{D}(A) = [\mathcal{D}(A) \cap X_-] \oplus [\mathcal{D}(A) \cap X_+] \tag{2.5}$$

and A maps $\mathcal{D}(A) \cap X_-$ into X_- and $\mathcal{D}(A) \cap X_+$ into X_+. So with respect to the decomposition (2.4) and (2.5) the operator A has the matrix representation

$$A = \begin{pmatrix} A_- & 0 \\ 0 & A_+ \end{pmatrix}. \tag{2.6}$$

Here $A_\pm = A\big|_{X_\pm} (X_\pm \to X_\pm)$ is the restriction of A to $X_\pm$. The domain $\mathcal{D}(A_\pm)$ of $X_\pm$ is $\mathcal{D}(A) \cap X_\pm$, so (2.5) can be rewritten as $\mathcal{D}(A) = \mathcal{D}(A_-) \oplus \mathcal{D}(A_+)$.

The operator A is said to be *exponentially dichotomous* if the operators $-A_-$ and A_+ in (2.6) are generators of $\mathcal{C}_0$-semigroups $T_-(-t)$ and $T_+(t)$, respectively, such that there are positive constants K, α, β

$$\|T_-(t)\| \leq Ke^{\beta t}, \quad t \leq 0, \tag{2.7}$$

$$\|T_+(t)\| \leq Ke^{-\alpha t}, \quad t \geq 0. \tag{2.8}$$

We call A of *exponential type* $-\omega$, where $\omega = \min\{\alpha, \beta\}$.

In case of an exponential dichotomous operator the projection P turns out to be unique and is called the *separating* projection for A. Furthermore, by the Hille-Yosida theorem, the spectrum $\sigma(A_-)$ of A_- is contained in the closed half plane Re $z \geq \omega$, whereas $\sigma(A_+)$ is a subset of Re $z \leq -\omega$. In particular, the strip $\{z \in \mathbb{C} : |\mathrm{Re}\, z| < \omega\}$ is contained in the resolvent set $\rho(A)$ of A and the resolvent of A is bounded in this strip. This justifies the use of the term "separating projection" for P. One can prove the following integral representation for P

$$Px = \frac{-1}{2\pi i}\int_{h-i\infty}^{h+i\infty} R(z, A)A^2 x \frac{dz}{z^2}, \qquad x \in \mathcal{D}(A^2), \tag{2.9}$$

where h is any real number between 0 and ω (see [3]). Since $\mathcal{D}(A^2)$ is dense in X, the bounded operator P is completely determined by (2.9). Finally, we recall a simple observation from [19].

Theorem 2.1 *If $A(X \to X)$ is the generator of a C_0-semigroup $(T(t))_{t\geq 0}$, then the following statements are equivalent:*

(i) *A is exponentially dichotomous;*

(ii) *$T(t)$ is hyperbolic;*

(iii) *There is an annulus $\{z \in \mathbb{C} : 1 - \delta < |z| < 1 + \delta,\ \delta > 0\}$ contained in the resolvent set of $T(1)$.*

Furthermore, in that case the separating projection of A is equal to the Riesz projection of $T(1)$ relative to the part of the spectrum outside the unit disk.

We now return to the exponential stabilization of (2.1) using a perturbation B as in (2.2). If the radius of the essential spectrum of $T(1)$ is less than one and

$$\sigma(T(1)) \cap \{z \in \mathbb{C} : |z| = 1\} = \emptyset,$$

then the semigroup $T(t)$ is hyperbolic and the space X_+ is finite dimensional. So the exponential stabilization of $T(t)$ reduces to a finite dimensional problem:

$$\begin{pmatrix} \dot{u}_-(t) \\ \dot{u}_+(t) \end{pmatrix} = \begin{pmatrix} A_- & 0 \\ 0 & A_+ \end{pmatrix}\begin{pmatrix} u_-(t) \\ u_+(t) \end{pmatrix} + \begin{pmatrix} 0 & 0 \\ 0 & B_+ \end{pmatrix}\begin{pmatrix} u_-(t) \\ u_+(t) \end{pmatrix} \tag{2.10}$$

and this problem is robust with respect to small time delays (see Example 1.1).

If on the other hand, the semigroup is hyperbolic and the radius of the essential spectrum of $T(1)$ is greater than one, then the space X_+ is infinite dimensional and the exponential stabilization of $T(t)$ is an infinite dimensional problem. And for the exponential stabilization of system (2.2) one also has to change the radius of the essential spectrum. Therefore the perturbation B cannot be compact and such perturbations turn out, in general, not to be robust with respect to small time delays. In the next sections we shall discuss several examples according to this framework.

To conclude this section, we formulate the robustness with respect to time delays more general as continuity with respect to parameters. Suppose that τ is a parameter varying in a metric space S and $T_\tau(t) :\to X$ is a C_0-semigroup of linear transformations on a Banach space X which is continuous in τ; that is, $T_\tau(t)x$ is continuous in $(\tau, t, x) \in S \times [0, \infty) \times X$. Let $r_\sigma(T_\tau(1))$ be the spectral radius of $T_\tau(1)$. In general, eigenvalues of $T_\tau(1)$ can be chosen to be continuous functions of τ. However, the spectral radius $r_\sigma(T_\tau(1))$ is not necessarily continuous with respect to τ. In general, the spectral radius $r_\sigma(T_\tau(1))$ is only lower semicontinuous (see [13, Theorem II.4.1]). In order to be continuous, further conditions are needed. For example, the spectral radius $r_\sigma(T_\tau(1))$ will be continuous at a point τ_0 if there exists a $\epsilon > 0$ such that the difference operators

$$T_\tau(1) - T_{\tau_0}(1), \quad |\tau - \tau_0| < \epsilon \tag{2.11}$$

are compact (see [13, Theorem II.4.2]).

If the operators in (2.11) are only bounded, then there is the possibility of a large shift in the spectral radius if we vary τ. For example, if $r_\sigma(T_\tau(1)) < 1$, and the perturbation is only bounded, it is possible to have $r_\sigma(T_{\tau_j}(1)) > 1$ for a sequence of $\tau_j \to \tau$ and the semigroup $T_{\tau_j}(t)$ will be unstable for each j (see the examples in Section 3). In many of the results that have been given in the literature for which stability is destroyed by the introduction of a small delay, this instability is due to the fact that the operators in (2.11) are only bounded.

3 Difference Equations

In this section, we discuss how variations in parameters affect the property of difference equations being stable. We shall only be concerned with difference operators $D = D(r, A)$ of the form

$$D(r, A)\varphi = \varphi(0) - \sum_{j=1}^{M} A_j\varphi(-r_j), \tag{3.1}$$

where each A_k is an $n \times n$ constant matrix, each $r_k > 0$ is a constant, $A = (A_1, \dots, A_M)$, $r = (r_1, \dots, r_M)$ and $h = \max_{j=1,\dots,M} r_j$.

In a problem, both the A_k and the r_k are generally parameters that are not known exactly. It is important, therefore, to know the effect that small changes in the parameters will have on the stability of the zero solution of the equation

$$D(r, A)x_t = 0,$$

where $x_t(\theta) = x(t+\theta)$, $-h \le \theta \le 0$. This difference equation can, equivalently, be written as follows

$$x(t) - \sum_{j=1}^{M} A_j x(t - r_j) = 0. \tag{3.2}$$

To use the results from Section 2, we shall first formulate (3.2) as an abstract Cauchy problem on a Banach space. Let $C = C([-h, 0]; \mathbb{R}^n)$ be the space of continuous functions from $[-h, 0]$ to $\mathbb{R}^n$ with the usual supremum topology. If we define the closed subspace $C_D \subset C$ as follows

$$C_D = \left\{ \varphi \in C : \varphi(0) - \sum_{j=1}^{M} A_j \varphi(-r_j) = 0 \right\},$$

then for any $\varphi \in C_D$, there is a unique solution $x(t; \varphi)$ of (3.2) which is continuous for $t \geq -h$ and coincides with φ on $[-h, 0]$. This implies that translation along the solutions of (3.2)

$$T_{D(r,A)}(t)\varphi = x_t(\cdot; \varphi) \tag{3.3}$$

defines a C_0-semigroup on C_D with infinitesimal generator

$$\mathcal{D}(B_{D(r,A)}) = \{\varphi \in C_D : \dot{\varphi} \in C,\ D(r, A)\dot{\varphi} = 0\}, \quad B\varphi = \frac{d\varphi}{d\theta} \tag{3.4}$$

So (3.2) can be written as an abstract Cauchy problem

$$\dot{u}(t) = Bu(t),$$

where B is defined by (3.4).

The spectrum of B consists of point spectrum only and is given by the roots of the determinant of the characteristic matrix

$$\Delta_{D(r,A)}(z) = I - \sum_{j=1}^{M} A_j e^{-zr_j}. \tag{3.5}$$

Let $r_\sigma(T_{D(r,A)}(1))$ denote the spectral radius of $T_{D(r,A)}(1)$ and let $r_e(T_{D(r,A)}(1))$ denote the radius of the essential spectrum of $T_{D(r,A)}(1)$.

The results in Theorems 3.1 and 3.2 can be found in [17], see also [1, 15, 14, 18, 27] for related results and generalizations.

Theorem 3.1 *Let $T_{D(r,A)}(t)$ be the C_0-semigroup associated with* (3.2). *The spectral radius of $T_{D(r,A)}(1)$ equals the radius of the essential spectrum of $T_{D(r,A)}(1)$ and*

$$r_\sigma(T_{D(r,A)}(1)) = e^\alpha, \tag{3.6}$$

where $\alpha = \alpha(r, A)$ is defined by

$$\alpha = \sup\{\operatorname{Re}\lambda : \det \Delta_{D(r,A)}(\lambda) = 0\}$$

and $\Delta_{D(r,A)}$ is given by (3.5). *Furthermore, the spectral radius of $T_{D(r,A)}(1)$ is continuous in A but only lower semicontinuous in r.*

Next we study the dependence of the spectral radius of $T_{D(r,A)}(1)$ upon the vector-valued parameters r and A. Variations in A lead to compact perturbations of the semigroup $T_{D(r,A)}(1)$, whereas variations in r lead to bounded perturbations which are not compact. So, according to the results from Section 2, it is this fact which is the source of the strange behaviour that can occur from variations in the delays.

In the scalar case we write $A_j = a_j$ for $j = 1, \dots, M$ and $a = (a_1, \dots, a_M)$ and have the following detailed result.

Theorem 3.2 *Let $T_{D(r,a)}(t)$ be the C_0-semigroup associated with (3.2) in the scalar case. The spectral radius of $T_{D(r,a)}(1)$ is continuous at a point r_0 if the components of r_0 are rationally independent. Furthermore, if the components of r are rationally independent, then*

$$r_\sigma(T_{D(r,a)}(1)) = e^{\rho_0}, \tag{3.7}$$

where ρ_0 is the solution to the equation

$$\sum_{j=1}^{M} |a_j| e^{-\rho_0 r_j} = 1. \tag{3.8}$$

The proof of (3.7) follows from basic properties of almost periodic functions and Kronecker's theorem (see, for example [5]). In particular, we shall use the following result for almost periodic functions analytic in a strip (see [20, p. 268]). There exists a number N such that, for all real t, the function $\Delta_{D(r,a)}(z) = 1 - \sum_{j=1}^{M} a_j e^{-r_j z}$ has no more than N zeros in the box

$$\{z \in \mathbb{C} : \alpha \le \operatorname{Re} z \le \beta,\ t \le \operatorname{Im} z \le t+1\}.$$

Furthermore, for every $\delta > 0$, there exists an $m = m(\delta)$ such that if z is at distance at least δ from every zero of $\Delta_{D(r,a)}$ then

$$|\Delta_{D(r,a)}(z)| \ge m(\delta). \tag{3.9}$$

If $\Delta_{(r,a)}(\rho + i\tau) = 0$, then

$$\sum_{j=1}^{M} a_j e^{-\rho r_j} e^{-i\tau r_j} = 1 \tag{3.10}$$

and hence

$$1 \le \sum_{j=1}^{M} |a_j| e^{-\rho r_j}. \tag{3.11}$$

The larger ρ, the smaller the right hand side of (3.11). Thus to find the solution of

$$\Delta_{D(r,a)}(\rho + i\tau) = 0$$

with the largest real part, we would like to choose τ such that approximately

$$e^{-i\tau r_j} = \operatorname{sign}(a_j) \quad j = 1, \ldots, M.$$

In other words, we would like to solve

$$|\tau r_j - \theta_j| < \delta \quad (\mathrm{mod}\ 2\pi),$$

where δ is sufficiently small and θ_j are given. Kronecker's theorem tells us that this is possible if the components of r are rationally independent, i.e., $\sum_{j=1}^{M} m_j r_j = 0$, $m_j \in \mathbf{Z}$, implies $m_j = 0$ for $j = 1, \ldots, M$. Together with (3.9), this proves (3.7).

Corollary 3.3 *Exponential stability of the difference equation* (3.2) *is preserved under small perturbations in the delays if and only if*

$$\sum_{j=1}^{M} |a_j| < 1. \tag{3.12}$$

The proof of this corollary is an easy consequence of Theorem 3.2. If the components of r are rationally independent, then exponential stability of (3.2) is equivalent to having $\rho_0 < 0$ in (3.8). Also, from the definition of ρ_0, we see that $\rho_0 < 0$ if and only if (3.12) is satisfied. Since the rationals are dense in $\mathbb{R}$, we obtain the result.

Example 3.1 (Destruction of stability with small changes in delays). Suppose that $0 < r_1 \le r_2 \le h$, are constants and consider the difference equation

$$x(t) + \frac{1}{2}x(t - r_1) + \frac{1}{2}x(t - r_2) = 0. \tag{3.13}$$

Observe from (3.11) that, if $\Delta_{(r_1,r_2)}(\rho + i\nu) = 0$, then $\rho \le 0$. Since $a = (-1/2, -1/2)$ is fixed, we suppress the dependence of the semigroup, characteristic equation, etc. upon a and we shall compute the spectral radius of $T_{(r_1,r_2)}(t)$, the C_0-semigroup associated with (3.13) for various values of (r_1, r_2).

For $(r_1, r_2) = (1, 2)$, the function $\Delta_{(1,2)}(z)$ is quadratic in e^{-z} and it is easy to show that the zeros are on the line

$$\operatorname{Re} z = -\log \sqrt{2}.$$

Thus,

$$r_\sigma(T_{(1,2)}(1)) = \frac{1}{\sqrt{2}} < 1,$$

which implies that $T_{(1,2)}(t)$ is uniformly exponentially stable.

For $(r_1, r_2) = (1 - \frac{1}{2k+1}, 2)$, where $k \geq 0$ is an odd integer. It is easy to verify that the equation $\Delta_{(1-1/(2k+1),2)}(z) = 0$ has solutions

$$\lambda_k = i\left(k + \frac{1}{2}\right)\pi.$$

The corresponding eigenfunctions are solutions of (3.13) which are periodic of period $2\pi(\lambda_k)^{-1}$. This implies that

$$r_\sigma(T_{(1-\frac{1}{2k+1},2)}(1)) = 1 \quad \text{for all odd integers } k.$$

Finally, for (r_1, r_2) rationally independent we can apply Theorem 3.2 to conclude that

$$r_\sigma(T_{(r_1,r_2)}(1)) = 1.$$

Example 3.2 (Feedback control can be sensitive to delay). Consider the system

$$x(t) - a_1x(t - r_1) - a_2x(t - r_2) = 0, \tag{3.14}$$

where r_1 and r_2 are rationally independent. According to Corollary 3.3, this equation is not exponentially stable if and only if

$$|a_1| + |a_2| \geq 1.$$

Given $|a_1| + |a_2| \geq 1$, we can apply a feedback control to stabilize (3.14)

$$x(t) - (a_1 + f_1)x(t - r_1) - (a_2 + f_2)x(t - r_2) = 0. \tag{3.15}$$

This system is exponentially stable if and only if

$$|a_1 + f_1| + |a_2 + f_2| < 1.$$

Suppose now that the feedback control cannot be applied instantaneously and that there is a small time delay in the feedback:

$$\begin{aligned} x(t) \;&-\; a_1x(t - r_1) - a_2x(t - r_2) \\ &-\; f_1x(t - r_1 - \epsilon_1) - f_2x(t - r_2 - \epsilon_2) = 0. \end{aligned} \tag{3.16}$$

We claim that, although (3.16) is exponentially stable for $\epsilon_1 = \epsilon_2 = 0$, there is a sequence $(\epsilon_1^j, \epsilon_2^j)$ tending to zero so that (3.16) is exponentially unstable. To prove the claim, we choose ϵ_1 and ϵ_2 such that $r_1, r_2, r_1 + \epsilon_1$ and $r_2 + \epsilon_2$ are rationally independent. From Corollary 3.3 it follows that (3.16) is always exponentially unstable, since

$$|a_1| + |f_1| + |a_2| + |f_2| > |a_1| + |a_2| \geq 1.$$

Example 3.3 (Large delays help for unstable systems). Suppose that $r > 0$ is an irrational number, a, b are scalars with $a \neq 0$ and $|b| > 1$. If we consider the equation

$$x(t) - ax(t-1) + bx(t-r) = 0 \tag{3.17}$$

then characteristic function is given by

$$h_\epsilon(z) = 1 - ae^{-z} + be^{-rz}.$$

The number $\rho_0 = \rho_0(r)$ in (3.8) of Theorem 3.2, satisfying

$$1 = |a|e^{-\rho_0} + |b|e^{-\rho_0 r},$$

is always positive for all $r > 0$. This implies that (3.17) is unstable for every value of the delay r.

It is easy to see that $\rho_0(r) \to \infty$ as $r \to 0$. Therefore, there are solutions of the difference equation with arbitrarily large rate of growth as $r \to 0$. Furthermore $\rho_0'(r) < 0$ for all $r > 0$. This implies that the maximal exponential growth rate of the solutions of (3.17) is a decreasing function of the delay r. So, even though the system is unstable, it is less unstable for larger values of the delay. If $r \to \infty$, then there is a $\gamma \geq 0$ such that $\rho_0(r) \to \gamma$.

If $\gamma > 0$, then $|a| > 1$ and $e^{-\gamma} = |a|^{-1}$. In this case, the exponential growth rate of the solutions of the difference equation exceeds a certain value γ.

If $\gamma = 0$, then $|a| \leq 1$ and for $r \to \infty$, we have $\rho_0(r)r \to c$, where

$$e^{-c} = (1 - |a|)/|b|.$$

In this case, the growth rate of the solutions of the difference equation can be made as small as we want by taking the delay τ large enough. If this were a control problem, then the system cannot be stabilized in a practical way with delays (b is the control variable). However, using a large delay in the control is better than using a small delay.

Example 3.4 In this example we shall analyse the zeros of the characteristic equation

$$\Delta_r(z) = 1 + ke^{-zr} - ke^{-2a}e^{-z(2+r)} + e^{-2a}e^{-2z}, \tag{3.18}$$

where $a \geq 0, k > 0$ are constants.

If $r = 0$, then $\Delta_0(z) = 0$ if and only if

$$e^{-2(z+a)} = \frac{k+1}{k-1}.$$

Therefore

$$\text{Re } z + a = -\frac{1}{2}\log\left|\frac{k+1}{k-1}\right| < 0 \tag{3.19}$$

for every $k > 0$. So, if $r = 0$, then the zeros of $\Delta_0(z)$ are in the left half plane $\{z \in \mathbb{C} : \text{Re } z < 0$ for every $k > 0$.

If $r > 0$, then $\Delta_r(z) = 0$ if and only if

$$ke^{-zr} = \frac{e^{-2a}e^{-2z} + 1}{e^{-2a}e^{-2z} - 1}. \tag{3.20}$$

Since

$$\left|\frac{e^{-2a}e^{-2z} + 1}{e^{-2a}e^{-2z} - 1}\right| \geq \frac{1 - e^{-2a}e^{-2\text{Re } z}}{1 + e^{-2a}e^{-2\text{Re } z}},$$

we obtain that the zeros of $\Delta_r(z)$ are in the left half plane $\{z \in \mathbb{C} : \text{Re } z < 0\}$ for every

$$k \leq \frac{1 - e^{-2a}}{1 + e^{-2a}}.$$

Next we consider the case that r is irrational. To find the supremum of the real parts of the zeros of $\Delta_r(z)$, $\rho_0 = \sup\{\text{Re } \lambda : \Delta_r(\lambda) = 0\}$, we apply Theorem 3.2. So we have to solve, see (3.8), the following equation for ρ_0

$$1 = ke^{-\rho_0 r} + ke^{-2a}e^{-\rho_0(2+r)} + e^{-2a}e^{-2\rho_0}.$$

Or, equivalently,

$$ke^{-\rho_0 r} = \frac{1 - e^{-2(\rho_0 + a)}}{1 + e^{-2(\rho_0 + a)}}. \tag{3.21}$$

As a consequence, for any $\rho_0 > 0$, if $k > k_0(\rho_0)$, then we can find an $r = r(\rho_0, k) > 0$ such that (3.21) is satisfied and $\Delta_r(z)$ has zeros in the right half plane $\{z \in \mathbb{C} : \text{Re } z > 0\}$. In particular, if $k > 1$, this implies that we can choose a ρ_0 as large as we want and then choose an r sufficiently small so that (3.21) is satisfied.

As in the previous example, we can analyze the dependence of $\rho_0(r)$ on r as $r \to \infty$. We see that $\rho_0'(r) < 0$ for all r, $\rho_0(r) \to 0$ as $r \to \infty$, $\rho_0(r)r \to c$, $c \geq 0$, where $e^{-c} = k_0/k$ provided that $k \geq k_0$. As a consequence, we can obtain slow exponential growth for the delay r large.

We continue this section with a characterization of those matrix-valued equations that preserve stability when small perturbations are made in the delays.

Definition 3.1 Let $\mathbb{R}_{>0}$ be the set of positive real numbers. Equation (3.2) is said to be stable locally in the delays if there is an open neighborhood $I(r) \subset \mathbb{R}^M_{>0}$ of r such that (3.2) is stable for each $\hat{r} \in I(r)$. If (3.2) is stable for each $r \in \mathbb{R}^M_{>0}$ then it is said to be stable globally in the delays.

The following result [15, Theorem 9.6.1] describes an interesting result for systems of difference equations.

Theorem 3.4 *Consider equation* (3.2) *in the matrix case*

$$x(t) - \sum_{j=1}^{M} A_j x(t - r_j) = 0.$$

The following statements are equivalent:

(i) *For some fixed* $r \in \mathbb{R}^M_{>0}$ *with components that are rationally independent* (3.2) *is exponentially stable.*

(ii) *If*

$$\rho(A) \stackrel{\text{def}}{=} \sup \left\{ r_\sigma \left(\sum_{k=1}^{M} A_k e^{i\theta_k} \right) : \theta_k \in [0, 2\pi], \right. \tag{3.22}$$
$$\left. k = 1, 2, \ldots, M \right\},$$

then $\rho(A) < 1$.

(iii) *Equation* (3.2) *is stable locally in the delays.*

(iv) *Equation* (3.2) *is stable globally in the delays.*

The Proof of Theorem 3.4 follows in principle the same arguments as given in the discussion after Theorem 3.2. Since in the sequel, we shall also need the slightly more general result that, if $\rho(A) > 1$ then (3.2) exponentially unstable, we prove (i) $\Rightarrow$ (ii). If (ii) does not hold, then there exist θ_k^*, $k = 1, 2, \ldots, M$, and $\gamma_0 \geq 1$ such that

$$r_\sigma \left(\sum_{k=1}^{M} A_k e^{i\theta_k^*} \right) = \gamma_0.$$

Define the matrix-valued function

$$F(\mu) = \sum_{k=1}^{M} A_k e^{-\mu r_k + i\theta_k^*}, \quad \mu \in \mathbb{R}. \tag{3.23}$$

Since $r_k > 0$, $r_\sigma(F(\mu)) \to 0$ as $\mu \to \infty$. Since $r_\sigma(F(0)) = \gamma_0 \geq 1$, there are $\mu_0 \geq 0$ and $\nu_0 \in [0, 2\pi]$, such that $r_\sigma(F(\mu_0)) = 1$ and

$$0 = \det\left[I - e^{-i\nu_0} F(\mu_0)\right] = \det\left[I - \sum_{k=1}^{M} A_k e^{-\mu_0 r_k} e^{i(\theta_k^* - \nu_0)} \right] \tag{3.24}$$

Since the r_k are rationally independent, it follows from Kronecker's theorem that there is a sequence $\{t_n\}$ of real numbers such that

$$\lim_{n\to\infty}(t_n r_1, \ldots, t_n r_N) = (\nu_0 - \theta_1^*, \ldots, \nu_0 - \theta_M^*) \quad \mathrm{mod}\ 2\pi.$$

We may also assume that the sequence $\{t_n\}$ is such that the sequence

$$e^{\mu_0 + it_n} \to e^{\lambda_0} \quad \text{as } n \to \infty,$$

where Re $\lambda_0 = \mu_0$. Therefore, using Equation (3.24), we have

$$0 = \det\left[I - \sum_{k=1}^{M} A_k e^{-(\mu_0 + it_n)r_k} e^{i(\theta_k - \nu_0 + t_n r_k)}\right]$$

and so, as $n \to \infty$,

$$0 = \det\left[I - \sum_{k=1}^{M} A_k e^{-\lambda_0 r_k}\right].$$

This contradicts the fact that (3.2) is stable since Re $\lambda_0 = \mu_0 \geq 0$ and completes the proof that (i) implies (ii) in Theorem 3.2.

Note that by construction, we have $\mu_0 > 0$ if and only if $\gamma_0 > 1$. This implies that the characteristic function has roots with positive real part and the corresponding equation is exponentially unstable. This shows the following lemma which we will use in the sequel.

Lemma 3.5 *If the components of r are rationally independent, then the equation*

$$x(t) - \sum_{j=1}^{M} A_j x(t - r_j) = 0$$

is exponentially unstable if and only if

$$\sup\left\{r_\sigma\left(\sum_{k=1}^{M} A_k e^{i\theta_k}\right) : \theta_k \in [0, 2\pi],\ k = 1, 2, \ldots, M\right\} > 1.$$

In [25], there is a general result saying that stabilizing an unstable difference equation by feedback is sensitive to variations in the delays. Consider the following linear control system with time delays

$$x(t) - \sum_{k=1}^{M} A_k x(t - r_k) - \sum_{k=1}^{M} B_k u(t - r_k) = 0 \tag{3.25}$$

where $x \in \mathbb{R}^n$ and each A_k is an $n \times n$ matrix. Let $r_e(T_D(t))$ be the radius of the essential spectrum of the semigroup $T_D(t)$ generated by

$$x(t) - \sum_{k=1}^{M} A_k x(t - r_k) = 0. \tag{3.26}$$

If $r_e(T_D(1)) \geq 1$, then the solutions of (3.26) in C_D are not uniformly exponentially stable. Suppose that system (3.25) is (exponentially) stabilizable, i.e, if we close the loop by $u(t) = Fx(t)$ for some matrix F, then the difference equation

$$x(t) - \sum_{k=1}^{M} (A_k + B_k F) x(t - r_k) = 0 \tag{3.27}$$

is exponentially stable. Let $r_e(T_{D,F}(t))$ be the radius of the essential spectrum of the semigroup $T_{D,F}(t)$ generated by the solutions of (3.27) in $C_{D,F} = \{\varphi \in C([-R, 0], \mathbb{R}^n) : \varphi(0) - \sum_{k=1}^{M} (A_k - B_k F)\varphi_k(-r_k) = 0\}$.

In the sequel, we shall assume that the delays r_k in (3.26) are rationally independent. This is not an unreasonable assumption from the point of view of applications, since any stability assertion that is made about the equation should be insensitive to small changes in the r_k.

From the practical point of view the control $u(t) = Fx(t)$ should also be insensitive to a small time change at which it is implemented; that is, the family of difference equations

$$v(t) - \sum_{k=1}^{M} A_k x(t - r_k) - \sum_{k=1}^{M} B_k F x(t - r_k - \epsilon_k) = 0, \tag{3.28}$$

where $|\epsilon_k|$, $k = 1, \ldots, M$, is small, should also be stable. However, the following result shows that this is almost never the case, compare [25].

Theorem 3.6 *Suppose that the time delays in* (3.26) *are rationally independent and that* (3.26) *is not uniformly exponentially stable (exponentially unstable). If the control system* (3.25) *is stabilizable using a feedback control* $u(t) = Fx(t)$, *then there exists a dense subset* E *of* $\mathbb{R}$ *such that for every* $\epsilon_k \in E$, $k = 1, \ldots, M$, *the closed-loop control system* (3.28) *is not uniformly exponentially stable (exponentially unstable).*

Proof: If (3.26) is not uniformly exponentially stable and the delays r_k are rationally independent, then it follows from Theorem 3.4 that there exist θ_k^*, $k = 1, \ldots, M$, such that the spectral radius of the matrix

$$G = \sum_{k=1}^{M} A_k e^{i\theta_k^*}$$

is greater or equal to one, i.e., $r_\sigma(G) \geq 1$. If (3.26) is exponentially unstable, then it follows from Lemma 3.5 that $r_\sigma(G) > 1$.

If the control $u(t)$ is not applied instantaneously but with a time delay ϵ, i.e., $u(t) = Fx(t-\epsilon)$, then the closed-loop system becomes (3.28). Choose a set $E \subset \mathbb{R}$ such that for $\epsilon_k \in E$, $k = 1, \ldots, M$, the reals $r_1, r_2, \ldots, r_M, r_1+\epsilon_1, r_2+\epsilon_2, \ldots, r_M+\epsilon_M$ are rationally independent. We claim that for every $\epsilon_k \in E$, $k = 1, \ldots, M$, equation (3.28) is not uniformly exponentially stable (exponentially unstable). To prove this claim we apply Theorem 3.4 and Lemma 3.5. So it suffices to prove that

$$\sup\left\{ r\left(\sum_{k=1}^{M} A_k e^{i\theta_k} + \sum_{k=1}^{M} B_k F e^{i\psi_k}\right) : \theta_k, \psi_k \in [0, 2\pi], \right. \tag{3.29}$$
$$\left. k = 1, 2, \ldots, M \right\} \geq r_\sigma(G).$$

To prove (3.29) we restrict ourselves to the following choices of θ_k and ψ_k. Take $\theta_k = \theta_k^*$, $\psi_k = \theta + \theta_k^*$ and define

$$H = \sum_{k=1}^{M} B_k F e^{i\theta_k^*}. \tag{3.30}$$

Therefore to prove (3.29), it suffices to show that

$$\sup\{r_\sigma(G + e^{i\theta} H) : 0 \leq \theta \leq 2\pi\} \geq r_\sigma(G).$$

This inequality turns out to be a general fact for matrices and we state it as a separate lemma. □

Lemma 3.7 *If G and H are square matrices, then*

$$\sup\{r_\sigma(G + e^{i\theta} H) : 0 \leq \theta \leq 2\pi\} \geq r_\sigma(G). \tag{3.31}$$

Proof: To arrive at a contradiction we assume that (3.31) is false and

$$\sup\{r_\sigma(G + e^{i\theta} H) : 0 \leq \theta \leq 2\pi\} < r_\sigma(G).$$

From Gelfand's formula for the spectral radius of a matrix M

$$r_\sigma(M) = \lim_{n\to\infty} \|M^n\|^{\frac{1}{n}},$$

we derive that there exists a $\gamma_0 < r_\sigma(G)$ such that for any δ sufficiently small, there exists an n_0 such that,

$$\|(G + e^{i\theta} H)^n\| \leq (\gamma_0 + \delta)^n \quad \text{for all } n \geq n_0 \text{ and } \theta \in [0, 2\pi]. \tag{3.32}$$

If we choose n odd and $\theta_0 = 2\pi/n$, then we have the following identities

$$\sum_{j=1}^{n}(G \pm e^{j\theta_0 i} H)^n = nG^n \pm nH^n. \tag{3.33}$$

Using (3.32) and (3.33) we obtain the following estimates

$$\begin{aligned} n\|G^n - H^n\| &\leq \sum_{j=1}^{n} \|(G + e^{(j\theta_0+\pi)i} H)^n\| \\ &\leq *n(\gamma_0 + \delta)^n \end{aligned}$$

and

$$\begin{aligned} n\|G^n + H^n\| &\leq \sum_{j=1}^{n} \|(G + e^{j\theta_0 i} H)^n\| \\ &\leq n(\gamma_0 + \delta)^n. \end{aligned}$$

Next we combine these estimates to obtain the following estimate

$$\begin{aligned} 2\|G^n\| &= \|G^n - H^n + G^n + H^n\| \\ &\leq \|G^n - H^n\| + \|G^n + H^n\| \\ &\leq 2(\gamma_0 + \delta)^n. \end{aligned} \tag{3.34}$$

Thus for every δ there exists a n_0 such that for every $n \geq n_0$

$$\|G^n\|^{\frac{1}{n}} \leq \gamma_0 + \delta,$$

but this yields $r_\sigma(G) \leq \gamma_0$. A contradiction to the assumption that $\gamma_0 < r_\sigma(G)$. This completes the proof of the lemma. □

To complete the Proof of Theorem 3.6, note that Lemma 3.7 implies that the supremum in (3.29) is bigger or equal to $r_\sigma(G)$. This completes the Proof of Theorem 3.6.

Note that in the Proof of Theorem 3.6 we actually did not use the fact that system (3.25) is stabilizable using a feedback control.

4 Neutral Delay Differential Equations

In this section, we present some results on neutral delay differential equations which show that the difference operator associated with the neutral part determines whether stabilization by feedback control is a finite or infinite dimensional problem.

We have seen that, if it is necessary to have an infinite dimensional control function to stabilize an unstable system, then it will be unstable under small variations in the delays. Also, for the system to be controllable by a finite dimensional control and, if it is required that the stabilization be insensitive to small variations in the delays, then the difference operator for the neutral part must by stable for all values of the delays.

We consider a scalar neutral differential difference equation

$$\frac{d}{dt}\left[x(t) - \sum_{j=1}^{M} a_j x(t - r_j)\right] = \sum_{j=0}^{M} b_j x(t - r_j), \tag{4.1}$$

where $r_0 = 1$, $0 < r_j \leq h$, a_j, b_j, $j = 1, \ldots, M$ are constants. Without loss of generality, we can assume that $r_1 \leq r_2 \leq \cdots \leq r_M$. Let $r = (r_1, \ldots, r_M)$, $a = (a_1, \ldots, a_M)$ and $b = (b_0, b_1, \ldots, b_M)$. Translation along the solutions of (4.1) defines a C_0-semigroup $T_r(t) \equiv T_{(r,a,b)}(t)$ on the state space $C = C([-r, 0], \mathbb{R}]$. Also, let $T_{D(r,a)}(t)$ be the C_0-semigroup on $C_D = \{\varphi \in C : \varphi(0) - \sum_{j=1}^{M} a_j \varphi(-r_j) = 0\}$ defined by the difference equation

$$x(t) - \sum_{j=1}^{M} a_j x(t - r_j) = 0. \tag{4.2}$$

We have the following result (see [15] for historical references and [19] for related results).

Theorem 4.1 (i) $r_\sigma(T_{(r,a,b)}(1)) = e^{\beta}$, *where* $\beta = \beta(r, a, b)$ *is defined by*

$$\beta = \sup\{\mathrm{Re}\,\lambda : \Delta_{(r,a,b)}(\lambda) = 0\}, \tag{4.3}$$

where Δ *denotes the characteristic function of* (4.1)

$$\Delta_{(r,a,b)}(z) = z\left[1 - \sum_{j=1}^{M} a_j e^{-zr_j}\right] - \sum_{j=0}^{M} b_j e^{-zr_j}.$$

(ii) $r_e(T_{(r,a,b)}(1)) = e^{\alpha}$, *where* $\alpha = \alpha(r, a)$ *is defined by*

$$\alpha = \sup\{\mathrm{Re}\,\lambda : \Delta_{(r,a)}(\lambda) = 0\}, \tag{4.4}$$

$$\Delta_{(r,a)}(z) = 1 - \sum_{j=1}^{M} a_j e^{-zr_j}.$$

From this result, it is clear that the eigenvalues of the difference equation (4.2); that is, the zeros of $\Delta_{(r,a)}(\lambda) = 0$, play a fundamental role in the discussion of the asymptotic behavior of the solutions of (4.1) when we subject the equation to perturbations in the delays. If we apply a control function which consists of a delayed feedback, then it will be finite dimensional if $r_\sigma(T_{D(r,a)}(1)) < 1$ and it will be infinite dimensional if $r_\sigma(T_{D(r,a)}(1)) \geq 1$. Furthermore, if we want the stabilization to be independent of the delay, then we must have $r_\sigma(T_{D(r,a)}(1)) < 1$ for all values of the delays. For scalar equations (see Corollary 3.2), this implies that

$$\sum_{j=1}^{M} |a_j| < 1. \tag{4.5}$$

If we consider matrix equations, then similar remarks hold except that we will not have the simple criterion (4.5). See Theorem 3.4 for details.

5 Delayed Boundary Control in a Hyperbolic Equation

In recent years, there has been considerable effort devoted to the problem of stabilization and control of PDE through the application of forces on the boundary. The mathematical theory is very complete in the situation when the boundary forces are applied with no delays in time (see [21] for a review). Small delays in the boundary forces can lead to a destabilization of the system (see [6, 7, 8, 9, 11, 16, 22, 23, 24, 25]).

In this section, we consider the one dimensional linear wave equation on a compact interval with zero boundary conditions at one end fixed, and strain at the other end, being controlled by the velocity with a small delay. Various methods have been used to show that a stable system can be destabilized even with small delays. We show how this problem can be reduced to a discussion of a difference equation with three delays.

Consider the solution $w : [0, 1] \times [0, \infty) \to \mathbb{R}$ of the linear wave equation

$$w_{tt} + 2aw_t - w_{xx} + bw = 0, \quad 0 < x < 1, \quad t > 0, \tag{5.1}$$

with initial data $w(x, 0) = w_0(x)$ and $w_t(x, 0) = w_1(x)$ and with the boundary conditions

$$w(0, t) = 0, \quad w_x(1, t) = -kw_t(1, t - r), \tag{5.2}$$

where $a \geq 0, k > 0, r \geq 0$ and b are constants. System (5.1)–(5.2) corresponds to a boundary stabilization problem where the control function is $u(1, t) = kw_t(1, t - r)$.

If $r = 0$, system (5.1)–(5.2) generates a C_0-semigroup $S_{0,k}(t)$ on the Banach space $X = H^1(0, 1) \times L^2(0, 1)$. In the space X, the control function $kw_t(1, t)$ is a bounded perturbation but not compact. Therefore, it might be possible to shift the radius of the essential spectrum of $S_{0,k}(1)$ by varying k. In fact, even if $a = 0$, it is easy to see that $r_e(S_{0,k}(1)) < 1$ for any $k > 0$. Therefore, any instability in the equation is due to finitely many eigenvalues of $S_{0,k}(1)$ with positive real part. These unstable modes are easily controlled by a finite dimensional control (see Section 2).

If we introduce a time delay r in the problem then the situation becomes more complicated. For the space of initial data, let us choose an $h > 0$ and define

$$X_h = C([-h,0]; H^1(0,1)) \times C([-h,0]; L^2(0,1)).$$

For any $0 \le r \le h$, system (5.1)–(5.2) with a time delay generates a $\mathcal{C}_0$-semigroup $S_{r,k}(t)$ on the state space X_h. In the space X_h, the control function $kw_t(1, t-r)$ is a bounded perturbation but not compact. Therefore, the radius of the essential spectrum of $S_{r,k}(1)$ may change with r for fixed k.

Next we study the dependence of the spectral radius of the essential spectrum of $S_{r,k}(1)$ upon the parameters k and r. By definition $r_e(S_{r,k}(1))$ does not change when we apply a compact perturbation to $S_{r,k}(1)$, and it is possible to discard such terms whenever they appear.

If we replace $w(x,t)$ by $e^{-at}w(x,t)$ and use the fact that, for any constant c, a term cw corresponds to a compact perturbation, then we see that

$$r_e(S_{r,k}(1)) = e^{-a} r_e(T_{r,k}(1)),$$

where $T_{r,k}(t)$ is the semigroup generated by the wave equation

(5.3) $$w_{tt} - w_{xx} = 0, \quad 0 < x < 1, \quad t > 0,$$

with the boundary conditions

(5.4) $$w(0,t) = 0, \quad w_x(1,t) = -ke^{ar} w_t(1, t-r).$$

We can now easily compute the eigenvalues of (5.3)–(5.4). If we let $w(x,t) = e^{\lambda t} v(x)$, then it is easy to see that $\lambda \neq 0$ and that λ must satisfy the equation

$$e^{\lambda} + e^{-\lambda} = -ke^{ar}(e^{\lambda} - e^{-\lambda})e^{-\lambda r}.$$

If we let $z = \lambda - a$, we obtain

(5.5) $$1 + ke^{-zr} - ke^{-2a}e^{-z(2+r)} + e^{-2a}e^{-2z} = 0.$$

From our derivations above, $r_e(S_{r,k}(1))e^{\rho}$, where ρ is determined by the supremum of the real parts of the solutions of (5.5). If this supremum is negative, then there are only finitely many independent solutions which are exponentially unbounded as $t \to \infty$.

The roots of equation (5.5) have been studied in Example 3.4 and we have the following theorem.

Theorem 5.1 *Consider the boundary stabilization problem* (5.1)–(5.2). *For* $a \geq 0$ *and* b *fixed, the dependence of the solutions upon* k *and* r *is as follows.*

(i) *If* $r = 0$, $k > 0$, *then the problem* (5.1)–(5.2) *is uniformly exponential stabilizable.*

(ii) *If* $r > 0$ *and* $0 < k < (1 - e^{-2a})/(1 + e^{-2a})$, *then the problem* (5.1)–(5.2) *is still uniformly exponential stabilizable.*

(iii) *If* $k > (1 - e^{-2a})/(1 + e^{-2a})$ *then there exists a dense set of* $r > 0$ *such that the problem* (5.1)–(5.2) *is exponentially unstable.*

(iv) *If* $k > 1$, *then we can choose a* ρ *as large as we want and then choose* r *sufficiently small so that there is a solution of problem* (5.1)–(5.2) *which becomes unbounded at the rate* $e^{\rho t}$ *as* $t \to \infty$.

As a consequence of this theorem, it follows that problem (5.1)–(5.2) could have been stabilized with a control which involves no delay and then there are arbitrarily small delays in the control which lead to destabilization. These remarks are contained in [7, 8] with a somewhat different proof.

It is not surprising that the radius of the essential spectrum of the semigroup generated by (5.1)–(5.2) is determined by a difference equation. In fact, any linear wave equation in one space dimension (even with nonlinear boundary conditions) is equivalent to a neutral delay differential equation (see [15] for references). For such equations, we have seen in Section 4 that the radius of the essential spectrum is determined by the difference equation obtained from the neutral part. For these applications, it becomes important to develop methods to compute the radius of the essential spectrum. The paper [31] gives an interesting discussion of how to determine the radius of the essential spectrum of the semigroup generated by more general hyperbolic systems. In work in progress we shall return to these issues.

The above analysis shows that boundary stabilization in the wave equation with no delay leads to very undesirable behavior when the control is implemented with a small delay. The main conclusion of our work is that one should try to avoid to stabilize systems with infinitely many unstable modes. Instead one should put more effort to develop models that put less emphasis on the high frequency modes (the eigenvalues and eigenvectors corresponding with large imaginary part). It has been noted in [29, 30] that such an approach is indeed possible.

We shall conclude this paper with two illustrative examples that are characteristic for the direction in which we hope to further develop the theory.

Example 5.1 Consider the strongly damped wave equation

$$(5.6) \qquad w_{tt} + 2aw_t - w_{xx} - cw_{xxt} + bw = 0, \quad 0 < x < 1, \quad t > 0,$$

with the boundary conditions

$$(5.7) \qquad w(0,t) = 0, \quad w_x(1,t) + cw_{xt}(1,t) = -kw_t(1,t-r),$$

where $a \geq 0, c > 0, k > 0, h \geq 0, b$ are constants. Equation (5.6) contains both viscous damping if $a > 0$ and internal damping which takes into account the rate of change of strain. The boundary condition (5.7) at $x = 1$ corresponds to a boundary stabilization problem with the control function $kw_t(1, t - r)$ acting on a natural function of the strain.

If $r > 0$ and $0 \leq \sigma \leq \beta < 1$, then we can define the fractional power spaces X^σ and X^β associated with the differential operator $\partial^2/\partial x^2$ with homogeneous Dirichlet boundary conditions at $x = 0$ and homogeneous Neumann boundary conditions at $x = 1$. Next we define the state space to be

$$Y_{h\sigma\beta} = C([-h, 0], X^\beta) \times C([-h, 0], X^\sigma).$$

For $r \in [0, h]$, it is possible to follow the same type of argument as in [26] to show that (5.6)–(5.7) defines a C_0-semigroup $\tilde{T}_{r,c,k}(t)$ on $Y_{h\frac{1}{2}\frac{1}{2}}$. Furthermore, in this setting the control function $-kw_t(1, t)$ leads to a compact perturbation. Therefore

$$r_e(\tilde{T}_{0,c,k}(1)) = r_e(T_{0,c,k}(1)),$$

where $T_{0,c,k}(t)$ denotes the semigroup generated by the equation

$$w_{tt} - w_{xx} - cw_{xxt} = 0, \quad 0 < x < 1, \quad t > 0, \tag{5.8}$$

with the boundary conditions

$$w(0, t) = 0, \quad w_x(1, t) + cw_{xt}(1, t) = 0. \tag{5.9}$$

The eigenvalues of (5.8)–(5.9) can be computed directly. First we find the solutions (λ, v), $v \neq 0$, of the auxiliary problem

$$v_{xx} = -\lambda v, \quad 0 < x < 1, \quad v(0) = 0, \quad v_x(1) = 0. \tag{5.10}$$

These are $\lambda_k = (2k + 1)^2\pi^2$ and $v_k = \sin(2k + 1)\pi$, $k = 0, 1, 2, \ldots$. The eigenvalues μ of (5.8)–(5.9) are now the solutions of the equation

$$\mu^2 + c\lambda_k\mu + \lambda_k = 0. \tag{5.11}$$

One can show that the solutions of (5.11) are in the left half of the complex plane and that the only accumulation point is $-1/c$. The points not accumulating at $-1/c$ approach $-\infty$ as $k \to \infty$. This computation shows that

$$r_e(T_{0,c,k}(1)) = e^{-1/c} < 1.$$

Thus, as a consequence, if one chooses the control $-kw_t(1, t)$ in such a way that problem (5.8)–(5.9) is uniformly exponentially stable, then it will remain exponentially stable even when the control is implemented with a small delay r (see Example 1.1). Note that the implementation of the control with a large delay perhaps could lead to instability, but it will occur on a finite dimension subspace since it will be the result of a finite number of eigenvalues crossing the imaginary axis.

In the next example, we consider the Euler-Bernoulli beam equation (see [11]).

Example 5.2 Consider the equation

$$w_{tt} + w_{xxxx} = 0 \quad 0 < x < 1, \tag{5.12}$$

with the boundary conditions

$$w(0,t) = w_x(0,t) = 0 \tag{5.13}$$

and

$$w_{xxx}(1,t) = 0, \quad w_{xx}(1,t) = -kw_{tx}(1,t-r). \tag{5.14}$$

In the state space for which this problem is well posed, the term $-kw_{tx}(1,t)$ is a noncompact perturbation. If $r = 0$, the system is stabilized for any $k > 0$. On the other hand, there are sequences $r_j \to 0$ as $j \to \infty$ such that the system is unstable (see, for example, [7, 11, 23]). Now if we change the boundary condition (5.12) to

$$w_{xxx}(1,t) = kw_t(1,t-r), \quad w_{xx}(1,t) = 0, \tag{5.15}$$

then the system is stabilized even for small delays (see [11]). The basic reason for this is that the system is stabilized for $r = 0$ if $k > 0$ and the control term $kw_t(1,t)$ is a compact perturbation and therefore does not affect the radius of the essential spectrum of the semigroup generated by (5.10), (5.11) and (5.13).

Finally we consider the beam equation with an internal damping (see [32])

$$w_{tt} + cw_{txxxx} + w_{xxxx} = 0 \qquad 0 < x < 1, \tag{5.16}$$

with the boundary conditions

$$w(0,t) = w_x(0,t) = 0 \tag{5.17}$$

and

$$w_{xxx}(1,t) = 0, \quad w_{xx}(1,t) + cw_{txx}(1,t) = -kw_{tx}(1,t-r). \tag{5.18}$$

In this case, the control term $w_{tx}(1,t)$ is a compact perturbation. Therefore, stabilization for $r = 0$ leads to stabilization for small delays.

References

[1] C.E. Avellar and J.K. Hale, "On the zeros of exponential polynomials", *J. Math. Anal. Appl.* **73** (1980), 434–452.

[2] J.F. Barman, F.M. Callier and C.A. Desoer, "L^2-stability and L^2-instability of linear time-invariant distributed feedback systems perturbed by a small delay in the loop", *IEEE Trans. Automat. Control* **18** (1973), 479–484.

[3] H. Bart, I. Gohberg and M.A. Kaashoek, *Minimal Factorization of Matrix and Operator Functions*, Birkhäuser Verlag, Basel, 1979.

[4] J.B. Conway, *Functions of One Complex Variable (2nd ed.)*, Springer-Verlag, New York, 1978.

[5] C. Corduneanu, *Almost Periodic Functions*, Interscience Publishers, 1968.

[6] R. Datko, "Not all feedback stabilized hyperbolic systems are robust with respect to small time delays in their feedbacks", *SIAM J. Control and Optimization* **26** (1988), 697–713.

[7] R. Datko, "Two questions concerning the boundary control of certain elastic systems", *J. Differential Equations* **92** (1991), 27–44.

[8] R. Datko, "A paradigm of ill-posedness with respect to time delays", *IEEE Trans. Automat. Control* **43** (1998), 964–967.

[9] R. Datko, J. Lagnese and M.P. Polis, "An example of the effect of time delays in boundary feedback stabilization of wave equations", *SIAM J. Control and Optimization* **24** (1986), 152–156.

[10] K. Deimling, *Nonlinear Functional Analysis*, Springer-Verlag, New York, 1985.

[11] W. Desch and R.L. Wheeler, "Destabilization due to delay in one dimensional feedback", *Int. Series Num. Math.* **91** (1989), 61–83.

[12] O. Diekmann, S.A. van Gils, S.M. Verduyn Lunel and H.O. Walther, *Delay Equations: Complex, Functional, and Nonlinear Analysis*, Springer-Verlag, New York, 1995.

[13] I. Gohberg, S. Goldberg and M.A. Kaashoek, *Classes of Linear Operators I*, Birkhäuser Verlag, Basel, 1990.

[14] J.K. Hale, "Parametric stability in difference equations", *Bol. Un. Mat. Ital. (4)* **11** (1975), Suppl. Fasc. 3, 209–214.

[15] J.K. Hale and S.M. Verduyn Lunel, *An Introduction to Functional Differential Equations*, Springer-Verlag, New York, 1993.

[16] K.B. Hannsgen, Y. Renardy and R.L. Wheeler, "Effectiveness and robustness with respect to time delays of boundary feedback stabilization in one-dimensional viscoelasticity", *SIAM J. Control and Optimization* **26** (1988), 1200–1234.

[17] D. Henry, "Linear autonomous neutral functional differential equations", *J. Differential Equations* **15** (1974), 106–128.

[18] D. Henry, "Topics in Analysis", *Pub. Mat. UAB* **31** (1987), 29–84.

[19] M.A. Kaashoek and S.M. Verduyn Lunel, "An integrability condition on the resolvent for hyperbolicity of the semigroup", *J. Differential Equations* **112** (1994), 374–406.

[20] B.JA. Levin, Distribution of Zeros of Entire Function, *Amer. Math. Soc., Providence*, 1972.

[21] J.L. Lions, "Exact controllability, stabilization and perturbations for distributed parameter systems", *SIAM Review* **30** (1988), 1–68.

[22] H. Logemann, "Destabilizing effects of small time delays on feedback-controlled descriptor systems", *Linear Algebra Appl.* **272** (1998), 131–153.

[23] H. Logemann, R. Rebarber and G. Weiss, "Conditions for robustness and nonrobustness of the stability of feedback systems with respect to small delays in the feedback loop", *SIAM J. Control and Optimization* **34** (1996), 572–600.

[24] H. Logemann and R. Rebarber, "The effect of small time-delays on the closed-loop stability of boundary control systems", *Math. Control Signals Systems* **9** (1996), 123–151.

[25] H. Logemann and S. Townley, "The effect of small delays in the feedback loop on the stability of neutral systems", *Systems and Control Letters* **27** (1996), 267–274.

[26] P. Massatt, "Limiting behavior for strongly damped nonlinear wave equations", *J. Differential Equations* **48** (1983), 334–349.

[27] W.R. Melvin, "Stability properties of functional differential equations", *J. Math. Anal. Appl.* **48** (1974), 749–763.

[28] C.J. Moreno, "The zeros of exponential polynomials, I.", *Comp. Math.* **26** (1973), 69–78.

[29] Ö. Morgül, "On stabilization and stability robustness against small delays of some damped wave equations", *IEEE Trans. Automat. Control* **40** (1995), 1626–1630.

[30] Ö. Morgül, "Stabilization and disturbance rejection for the wave equation", *IEEE Trans. Automat. Control* **43** (1998), 89–95.

[31] A.F. Neves, H.S. Ribeiro and O. Lopes, "On the spectrum of evolution operators generated by hyperbolic systems", *J. Functional Analysis* **67** (1985), 320–344.

[32] Y. Sakawa, "Feedback control of second-order evolution equations with damping", *SIAM J. Control Optimization* **22** (1984), 343–361.

Jack K. Hale
School of Mathematics
Georgia Institute of Technology
Atlanta, 30332-0160 U.S.A

Sjoerd M. Verduyn Lunel
Faculteit Exacte Wetenschappen
Vrije Universiteit
de Boelelaan 1081
1081 HV Amsterdam
The Netherlands

1991 *Mathematics Subject Classification.* Primary 34K15, 39A10, 34K35, 93B52

Operator Theory:
Advances and Applications, Vol. 122
© 2001 Birkhäuser Verlag Basel/Switzerland

Generalized Bezoutian, Factorization of Rational Matrix Functions and Matrix Quadratic Equations

*Irina Karelin and Leonid Lerer**

To Rien Kaashoek with affection and admiration

Various notions of generalized Bezoutian for families of rational matrix functions are introduced and studied. In particular, it turns out that using these Bezoutians one can describe that precise nature of the well-known correspondence between solution of matrix quadratic equations (descrete and continuous) and factorization of certain associated rational matrix functions.

0 Introduction

In this paper we deal with connections between solutions of matrix quadratic equations and factorizations of certain associated rational matrix functions.

We consider matrix quadratic equations of two types. The first one is of the form

$$A_1X - XA_2 = -XWX + Q \tag{0.1}$$

where A_1, A_2, W, Q are given complex matrices of sizes $p \times p$, $q \times q$, $q \times p$, $p \times q$, respectively, and X is the unknown $p \times q$ matrix.

The second type of quadratic equations that are considered in this paper is of the form

$$X = VXA - VX\Psi(D + \Phi X\Psi)^{-1}\Phi XA + Q, \tag{0.2}$$

where V, A, Ψ, D, Φ, Q are given complex matrices of sizes $p \times p$, $q \times q$, $q \times s$, $s \times s$, $s \times p$, $p \times q$, respectively. and X is again the unknown $p \times q$ matrix.

The equations (0.1) and (0.2) are non-symmetric counterparts of the wellknown Continuous and Discrete Algebraic Riccati Equations, respectively, which are obtained as particular cases of (0.1) and (0.2) by imposing certain symmetry conditions on the coefficients involved. In this case, as a rule, one is interested in hermition solutions X. There is an enormous number of works devoted to the

*The research of this author was partially supported by the USA-Israel Binational Science Foundation (BSF) Grant no. 9400271 and by the fund for promotion of research at the Technion-Israel Institute of Technology.

study of Riccati Equations among which we mention only the recent monograph [LaR] and the paper [IW], where one can find the main elements of the theory and further references.

The present paper deals with the general quadratic equations (0.1) and (0.2). The results and techniques of the paper will be applied to the study of the problems arising in the theory of Riccati Equations in another publication.

To describe the results of this paper consider first the equation (0.1) and assume that (W, A_1) is observable, while (A_2, W) is controllable. Write a rank decomposition of W: $W = \Psi\Phi$ and introduce the Popov function

$$G_0(\lambda) = I + \Phi(\lambda I - A_1)^{-1}Q(\lambda I - A_2)^{-1}\Psi.$$

Also, for arbitrary feedback matrices C and K of appropriate sizes introduce the function

$$G_{C,K}(\lambda) = I - [K \;\; \Phi]\left(\lambda I - \begin{bmatrix} A_2 - \Psi K & 0 \\ Q + CK & A_1 - C\Phi \end{bmatrix}\right)^{-1}\begin{bmatrix} \Psi \\ C \end{bmatrix}. \tag{0.3}$$

In this paper we show that there is a one-to-one correspondence between the set of solutions of equation (0.1) and the set of factorizations

$$G_{C,K}(\lambda) = G^{(1)}_{C,K}(\lambda)G^{(2)}_{C,K}(\lambda) \tag{0.4}$$

(with arbitrary (fixed) C and K) where $G^{(i)}_{C,K}$ $(i = 1, 2)$ are of a certain specific form, whereas this correspondence is provided by the generalized Bezoutian operator as introduced and studied in [GL]. A similar result is proved for the equation (0.2). In this case the functions $G_{C,K}(\lambda)$ and $G^{(i)}_{C,K}(\lambda)$ $(i = 1, 2)$ are replaced by certain appropriate anologues $\hat{G}_{C,K}(\lambda)$ and $\hat{G}^{(i)}_{C,K}(\lambda)$ $(i = 1, 2)$, respectively, and the role of the generalized Bezoutian is taken over by the generalized T-Bezoutian which is introduced and studied in this paper.

We note that the fact that there is a bijective correspondence between the solution of (0.1) and factorizations (0.4) was established earlier in [M1] (see also [M2]) by different methods and under some additional (technical) assumtions (e.g. minimality of factorization (0.4)). Also, the corresponding results for the equation (0.2) were proved in [M3] in the symmetric framework. The main contribution of the present paper is the discovery of the precise nature of this correspondence which is provided by the generalized Bezoutian and the generalized T-Bezoution.

We also note that for the homogeneous equation (0.1) (i.e. when $Q = 0$) results of similar nature have been obtained in [L], [LR2], [LR3], [LRa]. In this case the role of the generalized Bezoutian is played by the Bezoutian of a quadruple of polynomial and rational matrix functions as introduced in [LT] and [LR1], respectively.

It is our pleasure to thank A. Ran for useful discussions.

1 Bezoutian of Rational Matrix Functions and Matrix Quadratic Equations

In this section we'll discuss the connection between matrix quadratic equation and the notion of generalized Bezoutian of rational matrix functions as introduced in [GL]. This notion of Bezoutian is based on realization of rational matrix functions.

A realization of a rational matrix function $W(\lambda)$ is, by definition, a representation of the form

$$W(\lambda) = D + C(\lambda I - A)^{-1}B \tag{1.1}$$

where D, C, A and B are constant (i.e., independent on λ) matrices. The realization (1.1) is called *minimal* if the size of the matrix A is minimal among all realizations of $W(\lambda)$. The realization (1.1) is called *controllable* if the pair of matrices (A, B) is *controllable*, i.e., for some integer $l \geq 1$ the matrix

$$row[A^{j-1}B]_{j=1}^{l} := [B, AB, \ldots, A^{l-1}B]$$

is right invertible. Similarly, (1.1) is called *observable* if the pair (C, A) is *observable*, i.e., the matrix

$$col[CA^{j-1}]_{j=1}^{r} := \begin{bmatrix} C \\ CA \\ \vdots \\ CA^{r-1} \end{bmatrix}$$

is left invertible for some integer $r \geq 1$. Recall (see e.g. [S]) that a realization is minimal if and only if it is controllable and observable.

Let $L(\lambda)$ be $n \times r$ rational matrix function analytic at ∞ which is represented as

$$L(\lambda) = L(\infty) + \Phi(\lambda I - V)^{-1}C_L \tag{1.2}$$

where $L(\infty)$, Φ, V, C_L are $n \times r$, $n \times p$, $p \times p$, $p \times r$ matrices, respectively, and the pair (Φ, V) is observable. Let $M(\lambda)$ be $r \times n$ rational matrix function analytic at infinity with realization

$$M(\lambda) = M(\infty) + K_L(\lambda I - A)^{-1}\Psi \tag{1.3}$$

where $M(\infty)$, K_L, A, Ψ are $r \times n$, $r \times q$, $q \times q$, $q \times n$ matrices, respectively, and the pair (A, Ψ) is controllable. Assume that $L(\lambda)$ and $M(\lambda)$ satisfy the equality

$$L(\lambda)M(\lambda) = 0 \tag{1.4}$$

for any $\lambda \in \Omega$, where $\Omega = \{\lambda : \lambda \in \rho(A) \cap \rho(V)\}$.

Theorem 1.1 ([GL]). *There exists a unique matrix B such that the equation*

$$\frac{L(\lambda)M(\mu)}{\lambda-\mu} = \Phi(\lambda I - V)^{-1}B(\mu I - A)^{-1}\Psi \tag{1.5}$$

holds true for all $(\lambda, \mu) \in C^2$ *for which both sides of this equality make sense.*

The matrix B from Theorem 1.1 will be called the *generalized Bezoutian* associated with the equality (1.4) and the realizations (1.2) and (1.3).

Now consider the matrix quadratic equation

$$A_1X - XA_2 = -XWX + Q \tag{1.6}$$

for the unknown $p \times q$ matrix X, where A_1, A_2, W, Q are given $p \times p$, $q \times q$, $q \times p$, $p \times q$ complex matrices, respectively, the pair (W, A_1) is observable and the pair (A_2, W) is controllable. We shall establish connections between the quadratic equation (1.6) and the generalized Bezoutian generated by certain rational matrix functions which will be introduced now.

First write a rank decomposition of $W: W = \Psi\Phi$, where $\Psi \in C^{q\times n}$, $\Phi \in C^{n\times p}$ and note that the observability of (W, A_1) and controllability of (A_2, W) imply that (Φ, A_1) is observable and (A_2, Ψ) is controllable. Now define the Popov function

$$G_0(\lambda) = I + \Phi(\lambda I - A_1)^{-1}Q(\lambda I - A_2)^{-1}\Psi. \tag{1.7}$$

Next for arbitrary feedback matrices of appropriate sizes C and K define

$$U = A_1 - C\Phi, \ V = A_2 - \Psi K.$$

A well known result in system theory (see e.g. [GKvS], [S]) yields that (Φ, U) and (V, Ψ) are observable and controllable, respectively. We associate with the above matrices the rational matrix function

$$G_{C,K}(\lambda) = I - [K \ \ \Phi]\left(\lambda I - \begin{bmatrix} V & 0 \\ Q+CK & U \end{bmatrix}\right)^{-1}\begin{bmatrix} \Psi \\ C \end{bmatrix}.$$

We claim that

$$G_{C,K}(\lambda) = (I - \Phi(\lambda I - U)^{-1}C)G_0(\lambda)(I - K(\lambda I - V)^{-1}\Psi).$$

Indeed,

$$\begin{aligned}
&(I - \Phi(\lambda I - U)^{-1}C)G_0(\lambda)(I - K(\lambda I - V)^{-1}\Psi) \\
&= I - \Phi(\lambda I - U)^{-1}C + \Phi(\lambda I - A_1)^{-1}Q(\lambda I - A_2)^{-1}\Psi \\
&\quad -\Phi(\lambda I - U)^{-1}C\Phi(\lambda I - A_1)^{-1}Q(\lambda I - A_2)^{-1}\Psi
\end{aligned}$$

$$+\Phi(\lambda I - U)^{-1}CK(\lambda I - V)^{-1}\Psi - K(\lambda I - V)^{-1}\Psi$$

$$-\Phi(\lambda I - A_1)^{-1}Q(\lambda I - A_2)^{-1}\Psi K(\lambda I - V)^{-1}\Psi$$

$$+\Phi(\lambda I - U)^{-1}C\Phi(\lambda I - A_1)^{-1}Q(\lambda I - A_2)^{-1}\Psi K(\lambda I - V)^{-1}\Psi$$

$$= I - \Phi(\lambda I - U)^{-1}C - K(\lambda I - V)^{-1}\Psi$$

$$+\Phi(\lambda I - U)^{-1}(CK + Q)(\lambda I - V)^{-1}\Psi$$

$$= I - [K \ \Phi]\left(\lambda I - \begin{bmatrix} V & 0 \\ Q+CK & U \end{bmatrix}\right)^{-1}\begin{bmatrix} \Psi \\ C \end{bmatrix}.$$

Now decompose $Q = PY$ for some $P \in C^{p\times r}$, $Y \in C^{r\times q}$ and put

$$L_1(\lambda) = I - \Phi(\lambda I - U)^{-1}C, \tag{1.8}$$

$$P(\lambda) = \Phi(\lambda I - U)^{-1}P, \tag{1.9}$$

$$L_2(\lambda) = I - K(\lambda I - V)^{-1}\Psi, \tag{1.10}$$

$$Y(\lambda) = Y(\lambda I - V)^{-1}\Psi. \tag{1.11}$$

Then from the above computations we have

$$G_{C,K}(\lambda) = L_1(\lambda)L_2(\lambda) + P(\lambda)Y(\lambda). \tag{1.12}$$

Indeed, from (1.8)–(1.12)

$$G_{C,K}(\lambda) = I - \Phi(\lambda I - U)^{-1}C - K(\lambda I - V)^{-1}$$

$$+\Phi(\lambda I - U)^{-1}(Q + CK)(\lambda I - V)^{-1}\Psi$$

$$= I - [K \ \Phi]\left(\lambda I - \begin{bmatrix} V & 0 \\ Q+CK & U \end{bmatrix}\right)^{-1}\begin{bmatrix} \Psi \\ C \end{bmatrix}.$$

Suppose that $G_{C,K}(\lambda)$ can be factored into a product of two rational matrix functions

$$G_{C,K}(\lambda) = R_1(\lambda)R_2(\lambda) \tag{1.13}$$

such that $R_1(\lambda)$ and $R_2(\lambda)$ have the realizations

$$R_1(\lambda) = I - \Phi(\lambda I - U)^{-1}C_1, \tag{1.14}$$

$$R_2(\lambda) = I - K_1(\lambda I - V)^{-1}\Psi \tag{1.15}$$

with some C_1 and K_1.

Then we can rewrite the equation (1.13) as

$$L(\lambda)M(\lambda) = 0, \tag{1.16}$$

where

$$
\begin{aligned}
L(\lambda) &= [\,L_1(\lambda)P(\lambda)R_1(\lambda)\,] \\
&= [I \;\; 0 \;\; I] - \Phi(\lambda I - U)^{-1}[C \;\; P \;\; C_1],
\end{aligned}
\tag{1.17}
$$

$$
M(\lambda) = \begin{bmatrix} L_2(\lambda) \\ Y(\lambda) \\ -R_2(\lambda) \end{bmatrix} = \begin{bmatrix} I \\ 0 \\ -I \end{bmatrix} - \begin{bmatrix} K \\ Y \\ -K_1 \end{bmatrix} (\lambda I - V)^{-1}\Psi. \tag{1.18}
$$

Let B denote the generalized Bezoutian of the rational matrix functions L and M associated with the equality (1.16) and the realizations (1.18) and (1.18) as defined by (1.5). Sometimes we say that B is the generalized Bezoutian of the sextet $(L_1, P, R_1; L_2, Y, -R_2)$ associated with the equality

$$
L_1(\lambda)L_2(\lambda) + P(\lambda)Y(\lambda) - R_1(\lambda)R_2(\lambda) = 0 \tag{1.19}
$$

and the realizations (1.8), (1.9), (1.14); (1.10), (1.11), (1.15). The theorem below states that there exists one-to-one correspondence between the solutions of the equation (1.6) and the factorizations of the type (1.13). This correspondence is provided by the generalized Bezoutian.

Theorem 1.2 *With notations as above assume that a factorization* (1.13) *of* $G_{C,K}(\lambda)$ *holds true. Then the generalized Bezoutian B of $L(\lambda)$ and $M(\lambda)$ associated with the equality* (1.16) *and the realizations* (1.18), (1.18) *is a solution of the equation* (1.6).

Conversely, if X is a solution of equation (1.6) *define*

$$
\begin{aligned}
R_1(\lambda) &= L_1(\lambda) + \Phi(\lambda I - U)^{-1}X\Psi \\
&= I - \Phi(\lambda I - U)^{-1}(C - X\Psi),
\end{aligned}
\tag{1.20}
$$

$$
\begin{aligned}
R_2(\lambda) &= L_2(\lambda) - \Phi X(\lambda I - V)^{-1}\Psi \\
&= I - (K + \Phi X)(\lambda I - V)^{-1}\Psi.
\end{aligned}
\tag{1.21}
$$

Then $R_1(\lambda)$, $R_2(\lambda)$ satisfy (1.13) *and this solution X of the equation* (1.6) *is the generalized Bezoutian of the sextet $(L_1, P, R_1; L_2, Y, -R_2)$ associated with* (1.19) *and the realizations* (1.8), (1.9), (1.20); (1.10), (1.11), (1.21).

Such correspondence between the solutions of the equation (1.6) *and the factorizations of the type* (1.13)–(1.15) *is bijective.*

Proof: We start with the converse part of the theorem. Assume that the matrix X is a solution of (1.6). Multiply (1.6) from the left by $\Phi(\lambda I - U) = \Lambda(\lambda)$ and from the right by $(\mu I - V)\Psi = \Delta(\mu)$:

$$
0 = \Lambda(\lambda)(A_1X - XA_2 + X\Psi\Phi X - PY)\Delta(\mu). \tag{1.22}
$$

Now compute

$$(1.23)\qquad \begin{aligned} A_2\Delta(\mu) &= (V+\Psi K)\Delta(\mu) = (-\mu I + V + \mu I + \Psi K)\Delta(\mu) \\ &= -\Psi + \mu\Delta(\mu) + \Psi K\Delta(\mu) = -\Psi L_2(\mu) + \mu\Delta(\mu). \end{aligned}$$

In a similar way we obtain that

$$(1.24)\qquad \Lambda(\lambda)A_1 = -L_1(\lambda)\Phi + \lambda\Lambda(\lambda).$$

So (1.22) becomes

$$(1.25)\qquad \begin{aligned} 0 &= -L_1(\lambda)\Phi X\Delta(\mu) + \lambda\Lambda(\lambda)X\Delta(\mu) + \Lambda(\lambda)X\Psi L_2(\mu) \\ &\quad -\Lambda(\lambda)X\mu\Delta(\mu) + \Lambda(\lambda)(X\Psi\Phi X - PY)\Delta(\mu) \\ &= -L_1(\lambda)X_2(\mu) + (\lambda-\mu)X(\lambda,\mu) \\ &\quad +X_1(\lambda)L_2(\mu) + X_2(\lambda)X_1(\mu) - P(\lambda)Y(\mu) \\ &= -L_1(\lambda)L_2(\mu) - P(\lambda)Y(\mu) + (L_1(\lambda) \\ &\quad +X_1(\lambda))(L_2(\mu) - X_2(\mu)) + (\lambda-\mu)X(\lambda,\mu) \end{aligned}$$

where $X_1(\lambda) = \Lambda(\lambda)X\Psi$, $X_2(\mu) = \Phi X\Delta(\mu)$, $X(\lambda,\mu) = \Lambda(\lambda)X\Delta(\mu)$. Setting $\lambda = \mu$ in (1.25) we have that

$$L_1(\lambda)L_2(\lambda) + P(\lambda)Y(\lambda) = R_1(\lambda)R_2(\lambda)$$

for

$$R_1(\lambda) = L_1(\lambda) + X_1(\lambda) = I - \Phi(\lambda I - U)^{-1}(C - X\Psi),$$

$$R_2(\lambda) = L_2(\lambda) - X_2(\lambda) = I - (K + \Phi X)(\lambda I - V)^{-1}\Psi.$$

So we have a desired factorization of $G_{C,K}(\lambda)$ and from (1.25) we see that X is the generalized Bezoutian of the sextet $(L_1, P, R_1; L_2, Y, -R_2)$ associated with the equality (1.19) and the realizations (1.8), (1.9), (1.20) (1.10), (1.11), (1.21).

To prove the direct part of the theorem let $G_{C,K}(\lambda) = R_1(\lambda)R_2(\lambda)$ where $R_1(\lambda)$ and $R_2(\lambda)$ have the realizations (1.14), (1.15). We have to show that the Bezoutian B of the sextet of functions $(L_1, P, R_1; L_2, Y, -R_2)$ associated with the equality (1.19) and the realizations (1.8), (1.9), (1.14); (1.10), (1.11), (1.15) is a solution of the equation (1.6). Indeed, denote $\Omega = A_1B - BA_2$ and compute the expression $\Lambda(\lambda)\Omega\Delta(\mu)$. Using the formulas (1.23), (1.24) we have

$$\Lambda(\lambda)\Omega\Delta(\mu) = (\lambda-\mu)B(\lambda,\mu) - L_1(\lambda)B_2(\mu) + B_1(\lambda)L_2(\mu),$$

where $B(\lambda,\mu) = \Lambda(\lambda)B\Delta(\mu)$, $B_1(\lambda) = \Lambda(\lambda)B\Psi$ and $B_2(\mu) = \Phi B\Delta(\mu)$. Since B is the Bezoutian of the sextet of functions $(L_1, P, R_1; L_2, Y, -R_2)$ associated with the equality (1.19) and the realizations (1.8), (1.9), (1.14); (1.10), (1.11), (1.15) we can write that

$$(\lambda - \mu)B(\lambda, \mu) = L_1(\lambda)L_2(\mu) + P(\lambda)Y(\mu) - R_1(\lambda)R_2(\mu),$$

and hence

$$\begin{aligned} \Lambda(\lambda)\Omega\Delta(\mu) &= L_1(\lambda)L_2(\mu) + P(\lambda)Y(\mu) - R_1(\lambda)R_2(\mu) \\ &\quad -L_1(\lambda)B_2(\mu) + B_1(\lambda)L_2(\mu). \end{aligned} \tag{1.26}$$

Assuming that $\lambda \to \infty$ in the last equation we obtain

$$R_2(\mu) = L_2(\mu) - B_2(\mu). \tag{1.27}$$

In a similar way when $\mu \to \infty$ we have

$$R_1(\lambda) = L_1(\lambda) + B_1(\lambda). \tag{1.28}$$

Substituting (1.27) and (1.28) into (1.27) we obtain the equation

$$\Lambda(\lambda)\Omega\Delta(\mu) = P(\lambda)Y(\mu) + B_1(\lambda)B_2(\mu)$$

which holds true for any λ, μ for which both sides of this equation are well defined. Since (Φ, U) is observable and (V, Ψ) is controllable the last equation yields

$$A_1B - BA_2 = Q - B\Psi\Phi B$$

which proves the direct part of the theorem.

To prove the last statement of the theorem assume that for a certain solution X of the equation (1.6) there exists a factorization $G_{C,K}(\lambda) = Z_1(\lambda)Z_2(\lambda)$ where $Z_1(\lambda)$ and $Z_2(\lambda)$ are of the form (1.14) and (1.15), respectively. Then from the direct part of the theorem $X = B(L_1, P, Z_1; L_2, Y, -Z_2)$. On the other hand from the converse part of the theorem we know that $G_{C,K}(\lambda) = R_1(\lambda)R_2(\lambda)$, where $R_1(\lambda)$ and $R_2(\lambda)$ are given by (1.20) and (1.21), respectively, and $X = B(L_1, P, R_1; L_2, Y, -R_2)$. We claim that $Z_1(\lambda) = R_1(\lambda)$ and $Z_2(\lambda) = R_2(\lambda)$. Indeed, the equality $B(L_1, P, R_1; L_2, Y, -R_2) = B(L_1, P, Z_1; L_2, Y, -Z_2)$ implies

$$R_1(\lambda)R_2(\mu) = Z_1(\lambda)Z_2(\mu).$$

Now let $\lambda \to \infty$, then we see that $R_2(\mu) = Z_2(\mu)$. In a similar way as $\mu \to \infty$ we have that $R_1(\lambda) = Z_1(\lambda)$ which completes the proof of the theorem. □

2 Generalized T-Bezoutian of Rational Matrix Functions

In this section we introduce a slightly different notion of a Bezoutian which is more suitable for the study of "discrete" quadratic equations (0.2). In the case of a quadruple of matrix polynomials and special realizations a similar notion was considered previously in various publications (see e.g. [GSh], [HR]) and was referred in these works to as a *T-Bezoutian,* since it is related to inversion of Toeplitz matrices. We will keep the term "*T-Bezoutian*" in the more general situation, considered in the present paper.

Let $L(\lambda)$ be $n \times r$ matrix function that is analytic at the origin and is given by the realization

$$L(\lambda) = L(0) + \lambda\Phi(I - \lambda V)^{-1}C_L \tag{2.1}$$

where $L(0)$, Φ, V, C_L are $n \times r$, $n \times p$, $p \times p$, $p \times \mu$ matrices, respectively, and the pair (Φ, V) is observable. Let $M(\lambda)$ be $r \times n$ matrix function analytic at ∞ which is represented as

$$M(\lambda) = M(\infty) + K_M(\lambda I - A)^{-1}\Psi \tag{2.2}$$

where $M(0)$, K_M, A, Ψ are $r \times n, r \times q, q \times q, q \times n$ matrices, respectively, such that the pair (A, Ψ) is controllable. Assume that $L(\lambda)$, $M(\lambda)$ satisfy the equation

$$L(\lambda)M(\lambda) = 0 \tag{2.3}$$

for $\lambda \in \Omega$, where $\Omega = \{\lambda : \lambda^{-1} \in \rho(V) \text{ and } \lambda \in \rho(A)\}$.

Theorem 2.1 *There exists a unique matrix T such that the equation*

$$\frac{L(\lambda)M(\mu^{-1})}{\lambda\mu - 1} = \Phi(I - \lambda V)^{-1}T(I - \mu A)^{-1}\Psi \tag{2.4}$$

holds true for all $(\lambda, \mu) \in C^2$ for which both sides of this equality make sence.

Proof: Using (2.3) we can write that

$$\frac{L(\lambda)M(\mu^{-1})}{\lambda\mu - 1} = \frac{L(\lambda)(M(\mu^{-1}) - M(\lambda))}{\lambda\mu - 1}.$$

Substituting the realization (2.2) we obtain

$$\frac{M(\mu^{-1}) - M(\lambda)}{\lambda\mu - 1} = \frac{K_M(\lambda I - A)^{-1}(\lambda - \mu^{-1})(\mu^{-1}I - A)^{-1}\Psi}{\lambda\mu - 1}$$

and hence

$$\frac{L(\lambda)M(\mu^{-1})}{\lambda\mu - 1} = L(\lambda)K_M(\lambda I - A)^{-1}(I - \mu A)^{-1}\Psi. \tag{2.5}$$

Denote

$$D(\lambda) = L(\lambda)K_M(\lambda I - A)^{-1}. \tag{2.6}$$

We claim that $D(\lambda)$ is analytic at the origin. Indeed, suppose that $D(\lambda)$ has a pole at $\lambda = 0$ and write the Laurent expansion at $\lambda = 0$:

$$D(\lambda) = \sum_{j=-k}^{\infty} C_j \lambda^j. \tag{2.7}$$

From the equation (2.3) we see that $D(\lambda)\Psi$ is analytic at 0. Then clearly $C_j \Psi = 0$ for $j > 0$. From the residue formula we have that

$$C_{-1} = \frac{1}{2\pi i} \int_\Gamma L(\lambda)K_M(\lambda I - A)^{-1} d\lambda,$$

where Γ is a sufficiently small closed contour surrounding 0. Now compute the coefficient C_{-2} in the Laurent series of $D(\lambda)$:

$$\begin{aligned} C_{-2} &= \frac{1}{2\pi i} \int_\Gamma \lambda L(\lambda)K_M(\lambda I - A)^{-1} d\lambda \\ &= \frac{1}{2\pi i} \int_\Gamma L(\lambda)K_M(\lambda I - A)^{-1}(\lambda I - A + A) d\lambda \\ &= \frac{1}{2\pi i} \Bigg(\int_\Gamma L(\lambda)K_M d\lambda \\ &\quad + \left(\int_\Gamma L(\lambda)K_M(\lambda I - A)^{-1} d\lambda \right) A \Bigg). \end{aligned}$$

Since $L(\lambda)K_M$ is analytic at 0 we obtain that $\int_\Gamma L(\lambda)K_M d\lambda = 0$ and $C_{-2} = C_{-1}A$. In a similar way we obtain that

$$C_{-j} = C_{-1}A^{j-1} \tag{2.8}$$

for any $j > 0$. Since $D(\lambda)\Psi$ is analytic we have $C_{-j}\Psi = 0$ for $j > 0$ and rewriting (2.8) we obtain that $C_{-1}\Psi = 0, C_{-1}A\Psi = 0, \ldots, C_{-1}A^p\Psi = 0$ for p sufficiently large. Since the pair (A, Ψ) is controllable, the matrix $[\Psi \;\; A\Psi \;\; \ldots \;\; A^p\Psi]$ is right invertible for p sufficiently large, which yields $C_{-1} = 0$. It follows from (2.8) that $C_{-j} = 0$ for any $j > 0$, i.e. the function $D(\lambda)$ is analytic at $\lambda = 0$ and (2.7) becomes

$$D(\lambda) = \sum_{j=0}^{\infty} C_j \lambda^j. \tag{2.9}$$

On the other hand we have

$$\frac{L(\lambda)M(\mu^{-1})}{\lambda\mu - 1} = \frac{(L(\lambda) - L(\mu^{-1}))M(\mu^{-1})}{\lambda\mu - 1} \tag{2.10}$$
$$= \Phi(I - \lambda V)^{-1}(\mu I - V)^{-1}C_L M(\mu^{-1}).$$

Denote

$$E(\mu) = (\mu I - V)^{-1}C_L M(\mu^{-1}). \tag{2.11}$$

In a similar manner as for $D(\lambda)$ one shows that $E(\mu)$ is analytic in a neighbourhood of $\lambda = 0$. Write its expansion into powers of μ: $E(\mu) = \sum_{j=0}^{\infty} E_j\mu^j$. From (2.5) and (2.11) we have

$$\Phi(I - \lambda V)^{-1}E(\mu) = D(\lambda)(I - \mu A)^{-1}\Psi.$$

Expanding both sides of the above equation into powers of λ in neighbourhood of $\lambda = 0$ we compute that

$$\Phi V^{i-1}E(\mu) = D_i(I - \mu A)^{-1}\Psi, (i = 0, 1, \ldots). \tag{2.12}$$

Since the pair (Φ, V) is observable we can choose an integer k such that matrix $\Delta = col\{\Phi V^i\}_{i=0}^k$: $C^p \to R$ is left invertible. Introduce the matrix $\Gamma = col\{D_i\}_{i=0}^k$, $\Gamma : C^q \to R$, where $R = \{col\{x_i\}_{i=0}^k, x_i \in C^n\}$. We can rewrite (2.12) as

$$\Delta E(\mu) = \Gamma(I - \mu A)^{-1}\Psi. \tag{2.13}$$

Introduce

$$T = \Delta^{(-1)}\Gamma, \tag{2.14}$$

where $\Delta^{(-1)}$ stands for a left inverse of Δ. From (2.13), (2.14) we conclude that

$$E(\mu) = T(I - \mu A)^{-1}$$

and from (2.5), (2.11) it follows that

$$\frac{L(\lambda)M(\mu^{-1})}{\lambda\mu - 1} = \Phi(I - \lambda V)^{-1}T(I - \mu A)^{-1}\Psi.$$

It remains to prove that T is unique. Indeed, let T_1 and T_2 be two operators satisfying (2.4). Then we have that

$$\Phi(I - \lambda V)^{-1}(T_1 - T_2)(I - \mu A)^{-1}\Psi = 0$$

and since (Φ, V) is observable and (A, Ψ) is controllable, it follows that $T_1 = T_2$, and the theorem is proved. □

The linear transformation T in (2.4) will be called the *generalized T-Bezoutian* associated with the equality (2.3) and the realizations (2.1), (2.2).

3 Discrete Quadratic Equation and Factorizations of Rational Matrix Functions

In this section we show that the generalized T-Bezoutian serves as an adequate link between "discrete" quadratic equations and factorizatian of certain rational matrix functions associated with such equations.

In what follows we preserve the notations of the previous section. Let $L_1(\lambda)$, $P(\lambda)$ be two $n \times n$ matrix functions which are analytic at 0 of which the realizations (in the neighbourhood of 0) have the form

$$L_1(\lambda) = (I - \lambda\Phi(I - \lambda V)^{-1}C_{L_1})D_1, \tag{3.1}$$

$$P(\lambda) = \lambda\Phi(I - \lambda V)^{-1}P, \tag{3.2}$$

where D_1 is invertible. Let $L_2(\lambda)$, $Y(\lambda)$ be two $n \times n$ matrix functions that are analytic at ∞ with realizations of form

$$L_2(\lambda) = I - K_{L_2}(\lambda I - A)^{-1}\Psi, \tag{3.3}$$

$$Y(\lambda) = Y(\lambda I - A)^{-1}\Psi. \tag{3.4}$$

As before the pairs of matrices (Φ, V) and (A, Ψ) are assumed to be observable and controllable, respectively. Introduce the matrix function

$$G(\lambda) = L_1(\lambda)L_2(\lambda) + P(\lambda)Y(\lambda)$$

and consider factorizations of $G(\lambda)$ into a product of two functions

$$G(\lambda) = R_1(\lambda)R_2(\lambda), \tag{3.5}$$

where R_2 is analytic at 0, R_2 is analytic at ∞,

$$R_1(\lambda) = (I - \lambda\Phi(I - \lambda V)^{-1}C_{R_1})D_2, \tag{3.6}$$

for invertible D_2 and

$$R_2(\lambda) = I - K_{R_2}(\lambda I - A)^{-1}\Psi. \tag{3.7}$$

We can rewrite equation (3.5) in the following form

$$L_1(\lambda)L_2(\lambda) + P(\lambda)Y(\lambda) = R_1(\lambda)R_2(\lambda). \tag{3.8}$$

Let T be the generalized T-Bezoution defined by (2.4) for $L = [L_1, P, R_1]$ and $M = col[L_2, Y, -R_2]$ associated with the equality

$$L(\lambda)M(\lambda) = 0 \tag{3.9}$$

and the realizations

$$L(\lambda) = \begin{bmatrix} D_1 & 0 & D_1 \end{bmatrix} - \lambda\Phi(I - \lambda V)^{-1} \begin{bmatrix} C_{L_1}D_1 & -P & C_{R_1}D_1 \end{bmatrix} \tag{3.10}$$

and

$$M(\mu) = \begin{bmatrix} I \\ 0 \\ -I \end{bmatrix} - \begin{bmatrix} K_{L_2} \\ -Y \\ -K_{R_2} \end{bmatrix} (\mu I - A)^{-1}\Psi, \tag{3.11}$$

i.e.

$$\frac{L(\lambda)M(\mu^{-1})}{\lambda\mu - 1} = \Phi(I - \lambda V)^{-1}T(I - \mu A)^{-1}\Psi.$$

We also say that T is *the generalized T-Bezoutian of the sextet of functions* $(L_1, P, R_1; L_2, Y, -R_2)$ associated with equation (3.8) and realizations (3.1)–(3.7) and in this case (2.4) can be rewritten as

$$\begin{aligned} &\frac{L_1(\lambda)L_2(\mu^{-1}) + P(\lambda)Y(\mu^{-1}) - R_1(\lambda)R_2(\mu^{-1})}{\lambda\mu - 1} \\ &= \Phi(I - \lambda V)^{-1}T(I - \mu A)^{-1}\Psi. \end{aligned} \tag{3.12}$$

Theorem 3.1 *The generalized T-Bezoutian T of functions $(L_1, P, R_1; L_2, Y, -R_2)$ associated with equations* (3.8) *and realizations* (3.1)–(3.7) *satisfies the matrix quadratic equation*

$$T = V^{\times}TA^{\times} - V^{\times}T\Psi(D_1 + \Phi T\Psi)^{-1}\Phi TA^{\times} + PY, \tag{3.13}$$

where $V^{\times} = V + C_{L_1}\Phi$, $A^{\times} = A + \Psi K_{L_2}$.

Proof: First rewrite (3.13) in the following form

$$\begin{aligned} &(\lambda\mu - 1)\Phi(I - \lambda V)^{-1}T(I - \mu A)^{-1}\Psi \\ &\quad = (D_1 - D_2) - \lambda\Phi(I - \lambda V)^{-1}(C_{L_1}D_1 - C_{R_1}D_2) - \mu(D_1K_{L_2} \\ &\quad -D_2K_{R_2})(I - \mu A)^{-1}\Psi + \lambda\mu\Phi(I - \lambda V)^{-1}(C_{L_1}D_1K_{L_2} + PY \\ &\quad -C_{R_1}D_2K_{R_2})(I - \mu A)^{-1}\Psi. \end{aligned} \tag{3.14}$$

Developing both sides of this equation at $\lambda = \mu = 0$ and equating the coefficients of $\lambda^0\mu^0, \lambda^0\mu^j, \lambda^j\mu^0$ $(j = 1, 2, \ldots,)$ on both sides we have $D_1 - D_2 = -\Phi T\Psi$, $\Phi TA^j\Psi = (D_1K_{L_2} - D_2K_{R_2})A^{j-1}\Psi$ for $j = 1, 2, \ldots, \Phi V^i$

$T\Psi = \Phi V^{i-1}(C_{L_1}D_1 - C_{R_1}D_2)$ for $i = 1, 2, \dots,$. Since D_2 is invertible and the pairs (Φ, V), (A, Ψ) are observable and controllable, respectively, we can rewrite the above equations as

$$D_2 = D_1 + \Phi T\Psi, \tag{3.15}$$

$$K_{R2} = (D_1 + \Phi T\Psi)^{-1}(D_1 K_{L_2} - \Phi T A), \tag{3.16}$$

$$C_{R_1} = (C_{L_1}D_1 - VT\Psi)(D_1 + \Phi T\Psi)^{-1}. \tag{3.17}$$

Now using (3.15)–(3.17) we can rewrite (3.15) as

$$\begin{aligned}(\lambda\mu - 1)&\Phi(I - \lambda V)^{-1}T(I - \mu A)^{-1}\Psi \\ &= -\Phi T\Psi - \lambda\Phi(I - \lambda V)^{-1}VT\Psi \\ &\quad -\mu\Phi TA(I - \mu A)^{-1}\Psi + \lambda\mu(I - \lambda V)^{-1}(C_{L_1}D_1 KL_2 \\ &\quad +PY - (C_{L_1}D_1 - VT\Psi)(D_1 + \Phi T\Psi)^{-1}(D_1 K_{L_2} - \Phi TA)).\end{aligned} \tag{3.18}$$

Computations show that

$$\begin{aligned}(I - \lambda V)&((\lambda\mu - 1)(I - \lambda V)^{-1}T(I - \mu A)^{-1} + T \\ &\quad +\lambda(I - \lambda V)^{-1}VT + \mu TA(I - \mu A)^{-1})(I - \mu A) \\ &= (\lambda\mu - I)T + (I - \lambda V)T(I - \mu A) \\ &\quad +\lambda VT(I - \mu A) + \mu(I - \lambda V)TA = \lambda\mu(T - VTA).\end{aligned} \tag{3.19}$$

Using (3.19) rewrite (3.18) in the following way

$$\begin{aligned}\lambda\mu\Phi&(I - \lambda V)^{-1}(T - VTA)(I - \mu A)^{-1}\Psi \\ &= \lambda\mu\Phi(I - \lambda V)^{-1}(C_{L_1}D_1 KL_2 \\ &\quad +PY - (C_{L_1}D_1 - VT\Psi)(D_1 + \Phi T\Psi)^{-1} \\ &\quad (D_1 K_{L_2} - \Phi TA))(I - \mu A)^{-1}\Psi.\end{aligned}$$

Note that the last equation holds true for any λ, μ such that $\lambda^{-1} \in \rho(V)$ and $\mu^{-1} \in \rho(A)$, and hence T satisfies the equation

$$\begin{aligned}T - VTA &= C_{L_1}D_1 KL_2 + PY - (C_{L_1}D_1 - VT\Psi) \\ &\quad (D_1 + \Phi T\Psi)^{-1}(D_1 K_{L_2} - \Phi TA).\end{aligned} \tag{3.20}$$

The third term in the right hand part of (3.20) can be transformed as follows

$$(C_{L_1}D_1 - VT\Psi)(D_1 + \Phi T\Psi)^{-1}(D_1K_{L_2} - \Phi TA)$$
$$= (C_{L_1}D_1 - VT\Psi + C_{L_1}\Phi T\Psi - C_{L_1}\Phi T\Psi)$$
$$(D_1 + \Phi T\Psi)^{-1}(D_1K_{L_2} - \Phi TA - \Phi T\Psi K_{L_2} + \Phi T\Psi K_{L_2})$$
$$= (C_{L_1}(D_1 + \Phi T\Psi) - V^{\times}T\Psi)(D_1 + \Phi T\Psi)^{-1}$$
$$((D_1 + \Phi T\Psi)K_{L_2} - \Phi TA^{\times})$$
$$= V^{\times}T\Psi(D_1 + \Phi T\Psi)^{-1}\Phi TA - C^{L_1}\Phi TA^{\times}$$
$$-V^{\times}T\Psi K_{L_2} + C_{L_1}(D_1 + \Phi T\Psi)K_{L_2}.$$

We need also the following computation

$$VTA + C_{L_1}\Phi TA^{\times} + V^{\times}T\Psi K_{L_2}$$
$$= (V^{\times} - C_{L_1}\Phi)T(A^{\times} - \Psi K_{L_2})$$
$$+C_{L_1}\Phi TA^{\times} + V^{\times}T\Psi K_{L_2}$$
$$= V^{\times}TA^{\times} + C_{L_1}\Phi T\Psi K_{L_2}.$$

Now (3.20) can be rewritten as

$$T = V^{\times}TA^{\times} - V^{\times}T\Psi(D_1 + \Phi T\Psi)^{-1}\Phi TA^{\times}$$
$$+C_{L_1}(\Phi T\Psi - (D_1 + \Phi T\Psi) + D_1)K_{L_2} + Q,$$

where $Q = PY$. The theorem is proved. □

Now consider a general discrete quadratic equation

$$X = \hat{V}X\hat{A} - \hat{V}X\Psi(D + \Phi X\Psi)^{-1}\Phi X\hat{A} + Q. \tag{3.21}$$

Choose some feedback matrices C and K of appropriate sizes and introduce the matrices $V = \hat{V} - C\Phi$, $A = \hat{A} - \Psi K$. Also factor $Q = PY$, where $P \in C^{p\times n}$, $Q \in C^{n\times q}$. Denote

$$L_1 = D - \lambda\Phi(I - \lambda V)^{-1}CD, \tag{3.22}$$

$$P(\lambda) = \lambda\Phi(I - \lambda V)^{-1}P \tag{3.23}$$

and observe that $L_1(\lambda)$ and $P(\lambda)$ are analytic at $\lambda = 0$. Also denote

$$L_2(\lambda) = I - K(\lambda - A)^{-1}\Psi, \tag{3.24}$$

$$Y(\lambda) = Y(\lambda - A)^{-1}\Psi \tag{3.25}$$

and observe that $L_2(\lambda)$ and $Y(\lambda)$ are analytic at infinity. Now introduce the matrix function

$$G_{C,K}(\lambda) := L_1(\lambda)L_2(\lambda) + P(\lambda)Y(\lambda).$$

Remark that $G_{C,K}$ depends on the choice of C, K.

Theorem 3.2 *Any solution X of* (3.21) *generates a factorization of $G_{C,K}(\lambda)$ of the type*

$$G_{C,K}(\lambda) := L_1(\lambda)L_2(\lambda) + P(\lambda)Y(\lambda) = R_1(\lambda)R_2(\lambda), \tag{3.26}$$

where $R_1(\lambda)$, $R_2(\lambda)$ are rational matrix functions, such that $R_1(\lambda)$ is analytic at 0 and $R_2(\lambda)$ is analytic at ∞ with the following realizations

$$R_1(\lambda) = (D + \Phi X\Psi) - \lambda\Phi(I - \lambda V)^{-1}(CD - VX\Psi), \tag{3.27}$$

$$R_2(\lambda) = I - (D + \Phi X\Psi)^{-1}(DK - \Phi XA)(\lambda I - A)^{-1}\Psi. \tag{3.28}$$

The solution X coincides with the generalized T-Bezoutian of $(L_1, P, R_1; L_2, Y, -R_2)$ associated with realizations (3.22), (3.23), (3.27); (3.24), (3.25), (3.28) *and equation* (3.26).

Conversely, let $G_{C,K} = R_1(\lambda)R_2(\lambda)$ be a factorization of $G_{C,K}(\lambda)$ into a product of the rational matrix functions $R_1(\lambda)$ and $R_2(\lambda)$ such that $R_1(\lambda)$ is analytic at 0, $R_2(\lambda)$ is analytic at infinity with the realizations

$$R_1(\lambda) = (I - \lambda\Phi(I - \lambda V)^{-1}C_{R_1})D_2, \tag{3.29}$$

$$R_2(\lambda) = I - K_{R_2}(\lambda - A)^{-1}\Psi \tag{3.30}$$

for some matrices C_{R_1}, K_{R_2}, D_2. Then the generalized T-Bezoutian T of the sextet of functions $(L_1, P, R_1; L_2, Y, -R_2)$ associated with equality (3.26) *and the realizations* ((3.22), (3.23), (3.29); (3.24), (3.25), (3.30)) *is the solution of the equation* (3.21).

The correspondence between solutions of the equation (3.21) *and such factorizations of $G(\lambda)$ is one-to-one.*

Proof: First transform the equation (3.21) into

$$\begin{aligned} X \;=\; & VXA - (VX\Psi - CD)(D + \Phi X\Psi)^{-1} \\ & (\Phi XA - DK) + Q + CDK. \end{aligned} \tag{3.31}$$

Next, compute using (3.31)

$$\begin{aligned} R_1(\lambda)R_2(\lambda) &= (D+\Phi X\Psi) - \lambda\Phi(I-\lambda V)^{-1}(CD - VX\Psi) \\ &\quad -(DK - \Phi XA)(\lambda I - A)^{-1}\Psi + \lambda\Phi(I-\lambda V)^{-1} \\ &\quad (CD - VX\Psi)(D+\Phi X\Psi)^{-1}(DK - \Phi A)(\lambda I - A)^{-1}\Psi \\ &= (D+\Phi X\Psi) - \lambda\Phi(I-\lambda V)^{-1}(CD - VX\Psi) \\ &\quad -(DK - \Phi XA)(\lambda I - A)^{-1}\Psi + \lambda\Phi(I-\lambda V)^{-1} \\ &\quad (VXA - X + Q + CDK)(\lambda I - A)^{-1}\Psi. \end{aligned}$$

Simple computations show that

$$\begin{aligned} &\Phi X\Psi + \lambda\Phi(I-\lambda V)^{-1}VX\Psi + \Phi XA(\lambda I - A)^{-1} \\ &\quad \Psi + \lambda\Phi(I-\lambda V)^{-1}(VXA - X)(\lambda I - A)^{-1}\Psi \\ &\quad = \Phi(I-\lambda V)^{-1}((I-\lambda V)X(\lambda I - A) + \lambda VX(\lambda I - A) \\ &\quad +(I-\lambda V)XA + \lambda(VXA - X))(\lambda I - A)^{-1}\Psi = 0 \end{aligned}$$

and hence

$$\begin{aligned} R_1(\lambda)R_2(\lambda) &= D - \lambda\Phi(I-\lambda V)^{-1}CD - DK(\lambda I - A)^{-1}\Psi \\ &\quad +\lambda\Phi(I-\lambda V)^{-1}(Q + CDK)(\lambda I - A)^{-1}\Psi \\ &= G_{C,K}(\lambda). \end{aligned}$$

which proves the factorization (3.26).

Now we show that the solution X of (3.21) generating the factorization (3.26) of $G(\lambda)$ is the generalized T-Bezoutian of the sextet $(L_1, P, R_1; L_2, Y, -R_2)$ associated with the equation (3.26) and realizations (3.22), (3.23), (3.27); (3.24), (3.25), (3.28). Indeed,

$$\begin{aligned} &L_1(\lambda)L_2(\mu^{-1}) + P(\lambda)Y(\mu^{-1}) - R_1(\lambda)R_2(\mu^{-1}) \\ &\quad = D - \Phi\lambda(I-\lambda V)^{-1}CD - \mu DK \\ &\qquad (I-\mu A)^{-1}\Psi + \lambda\mu\Phi \\ &\qquad (I-\lambda V)^{-1}(CDK + Q)(I-\mu A)^{-1} \\ &\qquad -((D+\Phi X\Psi) - \lambda\Phi(I-\lambda V)^{-1} \\ &\qquad (CD - VX\Psi) - \mu(DK - \Phi XA)(I-\mu A)^{-1}\Psi \\ &\qquad +\lambda\mu\Phi(I-\lambda V)^{-1}(CD - VX\Psi)(D+\Phi X\Psi)^{-1}(DK - \Phi XA)) \end{aligned}$$

$$= -\Phi X\Psi - \lambda\Phi(I - \lambda V)^{-1} V X\Psi - \mu\Phi X A(I - \mu A)^{-1}$$

$$\Psi + \lambda\mu\Phi(I - \lambda V)^{-1}(VXA - X)(I - \mu A)^{-1}\Psi$$

$$= (\lambda\mu - 1)X,$$

i.e. X is the generalized T-Bezoutian of $(L_1, P, R_1; L_2, Y, -R_2)$ associated with the equation (3.26) and realizations (3.22), (3.23), (3.27); (3.24), (3.25), (3.28).

The converse part of this theorem follows easily from the Theorem 3.1. □

References

[GKvS] I. Gohberg, M.A. Kaashoek and F. van Schagen, Rational matrix operator functions with prescribed singularities, *Integral Equations Operator Theory* **5** (1982), 673–717.

[GS] I. Gohberg and T. Shalom, On Bezoutians of nonsquare matrix polynomials and inversion of matrices with non-square blocks, *Linear Algebra Appl.* **137/138** (1990), 249–323.

[GL] G. Gomez and L. Lerer, Generalized Bezoutian for analytic operator functions and inversion of structured operators, *System and Networks: Mathematical Theory and Applications (U. Helmke, R. Mennicken, J. Saures, Eds.)*, Academie Verlag, 1994, 691–696.

[HR] G. Heining and K. Rost, Algebraic Methods for Toeplitz-like Matrices and Operators, *OT13, Birkhauser Verlag*, 1984.

[IW] V. Ionescu and M. Weiss, Continuous and discrete-time Riccati theory: A Popov-function approach. *Linear Algebra Appl.* **193** (1993), 173–209.

[LaR] P. Lancaster and L. Rodman, Algebraic Riccati Equations, *Oxford University Press, Oxford*, 1995.

[L] L. Lerer, The matrix quadratic equations and factorization of matrix polynomials, *Operator Theory: Advanced and Applications* **40** (1989), 279–324.

[LRa] L. Lerer and A.C.M. Ran, J-pseudo-spectral and J-inner-pseudo-outer factorizations for matrix polynomials, *Integral Equations and Operator Theory* **29** (1997), 23–51.

[LR1] L. Lerer and L. Rodman, Bezoutians of rational matrix functions, *Journal of Functional Analysis* **141** (1996), 1–36.

[LR2] L. Lerer and L. Rodman, Symmetric factorizations and locations of zeroes of rational matrix functions, *Linear and Multilinear Algebra* **40** (1996), 259–281.

[LR3] L. Lerer and L. Rodman, Bezoutians of rational matrix functions, matrix equations and factorizations, *Linear Algebra Apl.*, to appear.

[LT] L. Lerer and M. Tismenetsky, The Bezoutian and the eigenvalue separation problem, *Integral Equations and Operator Theory* **5** (1982), 386–445.

[M1] B.P. Molinari, Equivalence relations for the algebraic Riccati equation, *SIAM J. Control* **11** (1973), 272–285.

[M2] B.P. Molinari, The stabilizing solution of the algebraic Riccati equation, *SIAM J. Control* **11** (1973), 262–271.

[M3] B.P. Molinari, The stabilizing solution of the discrete algebraic Riccati equation, *IEEE Trans. AC*, June 1975, 396–399.

[S] E.D. Sontag, Mathematical Control Theory: Deterministic Finite Dimenstional System, *Springer Verlag*, 1990.

I. Karelin, L. Lerer
Department of Mathematics
Technion-Israel Institute of Technology
Haifa 32000
Israel
email: kirina@techunix.technion.ac.il
email: llerer@techunix.technion.ac.il

MSC Primary 47A68, 15A24; Secondary 47A56

Operator Theory:
Advances and Applications, Vol. 122
© 2001 Birkhäuser Verlag Basel/Switzerland

A Note on Factorization of Analytic Matrix Functions

P. Lancaster and A. Markus[1]

Dedicated to Rien Kaashoek on the occasion of his 60th birthday, with respect and affection

Let $L(\lambda)$ be an analytic matrix function, let F be a bounded and isolated part of its numerical range, and Γ be a closed contour of regular points of $L(\lambda)$ with F inside Γ. Defining $c(F) = \text{ind}_\Gamma(L(\lambda)f, f)$, $f \neq 0$, we give new proofs of the existence of spectral divisors of $L(\lambda)$ with respect to Γ in the cases that either $c(F) = 1$ or Γ is a circle and, in both cases, there is no completeness hypothesis. Other results on the existence of Γ-spectral divisors (using completeness) are extended from the case of matrix polynomials to that of analytic matrix functions.

1 Introduction

Let $L(\lambda)$ denote an $n \times n$ matrix-valued function which is analytic in the domain G of $\mathbb{C}$. For brevity, we call $L(\lambda)$ an analytic matrix function, or a.m.f. *Eigenvalues* of $L(\lambda)$ are points in G for which $\det L(\lambda) = 0$, and the *spectrum* of $L(\lambda)$, written $\sigma(L)$, is the set of all its eigenvalues. If $\det L(\lambda_0) \neq 0$, λ_0 is said to be a *regular* point of $L(\lambda)$. Γ will denote a closed contour (possibly consisting of several simple contours) of regular points of $L(\lambda)$ and is the positive oriented boundary of a bounded domain Γ_+. A monic matrix polynomial $L_1(\lambda) = \sum_{j=0}^{k} \lambda^j A_j$, $A_k = I$, is a (right) divisor of a.m.f. $L(\lambda)$ if there is an a.m.f. $L_2(\lambda)$ such that $L(\lambda) = L_2(\lambda)L_1(\lambda)$. If also $\sigma(L)\bigcap\Gamma_+ = \sigma(L_1)$ and $\sigma(L_2)\bigcap\Gamma_+ = \phi$, then $L_1(\lambda)$ is said to be a Γ-*spectral divisor* of $L(\lambda)$.

The *numerical range* of a.m.f. $L(\lambda)$ is the set

$$W(L) := \{\lambda \in \mathbb{C} \; : \; (L(\lambda)f, \; f) = 0, f \neq 0\}$$

and we note that $\sigma(L) \subset W(L)$. Let F be a bounded and isolated part of $W(L)$, i.e. $\text{dist}(F, W(L)\backslash F) > 0$. The contour Γ will generally be chosen so that $F \subset \Gamma_+$ and $W(L)\backslash F$ is outside Γ. Then for some $f \neq 0$ define

$$c(F) := \text{ind}_\Gamma(L(\lambda)f, \; f),$$

[1]The authors were supported in part by grants from, respectively, the Natural Science and Engineering Research Council of Canada, and the Israel Science Foundation founded by the Israel Academy of Sciences and Humanities.

and, by continuity, it is clear that $c(F)$ is independent of the choice of f. We study the following question: Given an isolated part F of $W(L)$ as above, when does $c(F) = k$ imply the existence of a Γ-spectral divisor of degree k?

This question has quite a long history. An account of related results for matrix polynomials appears in Chapter 4 of [4]. In that work, and elsewhere, answers to our question involve a completeness property of eigenvectors and generalized eigenvectors. However, an unexpected answer of a different kind was discovered by Markus and Matsaev [8] (see also Sections 26 and 27 of [6]), and shows that completeness can be avoided in two cases: (a) $c(F) = 1$; (b) there is a *circular* contour Γ having the required properties with respect to F.

Our main objective is the presentation of a new, and more accessible line of argument to establish the Markus-Matsaev (and related) results for matrix functions (the original results were established for analytic functions with values in the bounded linear operators on a Hilbert space). We remark also that in [8] and [6] it was assumed that Γ_+ was simply connected. The case of an arbitary bounded domain Γ_+ was considered in [9] for the case of matrix polynomials $L(\lambda)$. The proof in [9], as in [8] and [6], is based on some results about Wiener-Hopf factorization. Apart from the new proof, additions made here to the existing theory are relatively minor.

For the new proof we need some machinery which captures all the spectral properties of an a.m.f. at an eigenvalue. This has been developed in a sequence of works including (in this order) [4], [5], [1], and [3]. We use the last of these references as a starting point. An essential idea is that of a *Jordan pair* (X_i, J_i) associated with an eigenvalue λ_i of algebraic multiplicity m_i of a.m.f. $L(\lambda)$. Here, J_i has Jordan canonical form, size m_i, and the single eigenvalue λ_i, and X_i is an $n \times m_i$ matrix whose columns are a canonical set of Jordan chains of $L(\lambda)$ at λ_i. In short, the pair (X_i, J_i) encapsulates complete information about partial multiplicities (see below) and canonical chains of (right) eigenvectors and generalized eigenvectors of $L(\lambda)$ at λ_i. Furthermore, (X_i, J_i) is a zero-kernel pair, i.e. $\bigcap_{r=0}^{n-1} \mathrm{Ker}(X_i J_i^r) = \{0\}$.

It is shown in Theorem 3.1 of [3] that there is a unique $m_i \times n$ matrix Y_i such that $X_i(\lambda I - J_i)^{-1} Y_i$ is the principal part of the Laurent expansion for $L^{-1}(\lambda)$ at λ_i and (J_i, Y_i) is a full-range pair, i.e. $\sum_{j=0}^{n-1} \mathrm{Im}(J_i^j Y_i) = \mathbb{C}^n$. Then (X_i, J_i, Y_i) is known as a *canonical triple* at λ_i. We remark also that (Y_i^T, J_i^T) is permutationally similar to a Jordan pair for $L^T(\lambda)$ at λ_i.

If $\sigma(L) \bigcap \Gamma_+ = \{\lambda_1, \ldots, \lambda_r\}$, then a *$\Gamma$-canonical triple*, (X, J, Y) for $L(\lambda)$ is defined by

$$X = [X_1 \ X_2, \ldots, X_r], \quad J = \mathrm{diag}[J_1, \ldots, J_r], \quad Y = \begin{bmatrix} Y_1 \\ \vdots \\ Y_r \end{bmatrix}.$$

Also, (X, J) is said to be a *Γ-canonical pair.*

If the total algebraic multiplicity of eigenvalues in Γ_+ is kn, there is an $l \geq k$ such that

$$\begin{bmatrix} X \\ XJ \\ \vdots \\ XJ^{l-1} \end{bmatrix}$$

has rank kn (see p. 114 of [5]). The pair (X, J) is then said to be *k-fold complete* if $l = k$. This definition is equivalent to that introduced by M.V. Keldysh in 1951 (see [6] for this definition and historical references). Clearly, such a pair is k-fold complete if and only if any similar pair, $(XT, T^{-1}JT)$, is k-fold complete.

2 Preliminary Results

Lemma 1 *The pair (X, J) is a Γ-canonical pair for $L(\lambda)$ if and only if $(X, J+\alpha I)$ is a canonical pair for $L(\lambda-\alpha)$ with respect to the shifted contour $\Gamma_1 = \{\lambda - \alpha : \lambda \in \Gamma\}$.*

The easy verification is left to the reader.

Lemma 2 *The total algebraic multiplicity of eigenvalues inside Γ_+ is kn if and only if $c(F) = k$.*

Proof: Lemma 6 of [2] gives

$$\text{ind}_\Gamma \text{det} L(\lambda) = n\, \text{ind}_\Gamma (L(\lambda)g, g), \quad g \neq 0.$$

But it is well known (see p. 93 of [5], for example) that the left side of this equality is the total algebraic multiplicity of eigenvalues in Γ_+. □

The following easy proposition is also needed subsequently, and we give a proof for the convenience of the reader.

Lemma 3 *If a function $\phi(\lambda)$ is analytic in $\Gamma_+ \bigcup \Gamma$ with the possible exception of a finite number of poles in Γ_+, and if $\int_\Gamma \lambda^m \phi(\lambda)\, d\lambda = 0$ for any integer $m \geq 0$, then there are no poles in Γ_+.*

Proof: Argue by contradiction and suppose that

$$\phi(\lambda) = \sum_{j=1}^{s} \sum_{k=1}^{m_j} \frac{a_{jk}}{(\lambda - \lambda_j)^k} + \psi(\lambda)$$

where $s \geq 1$, $a_{j m_j} \neq 0$ $(j = 1, 2, \ldots, s)$, and $\psi(\lambda)$ is analytic in $\Gamma_+ \bigcup \Gamma$. Let

$$P(\lambda) = (\lambda - \lambda_1)^{m_1 - 1} \prod_{j=2}^{s} (\lambda - \lambda_j)^{m_j}.$$

Then

$$\int_\Gamma P(\lambda)\phi(\lambda)\,d\lambda = \int_\Gamma \frac{a_{1m_1}}{\lambda-\lambda_1}\,d\lambda = 2\pi i a_{1m_1} \neq 0$$

and, as $P(\lambda)$ is a polynomial, this contradicts the hypothesis. □

The next lemma has prototypes in the literature. See, for example, Section 18.7 of [6]. We give a relatively transparent proof that will be useful in the sequel.

Lemma 4 *Let $c(F) = k$. Then Γ-canonical pairs are k-fold complete if and only if all functions of the form $(L^{-1}(\lambda)f,\ h(\bar{\lambda}))$ are analytic inside Γ for all $f \in \mathbb{C}^n$ and some vector polynomial $h(\lambda) = \sum_{j=0}^{k-1}\lambda_j h_j$ (in $\mathbb{C}^n$) only when $h(\lambda) \equiv 0$.*

Proof: Let (X, J) be a Γ-canonical pair for $L(\lambda)$. Then there is a Γ-canonical triple (X, J, Y) and, for $j = 0, 1, 2, \ldots,$

$$\frac{1}{2\pi i}\int_\Gamma \lambda^j L^{-1}(\lambda)d\lambda = \frac{1}{2\pi i}X\int_\Gamma \lambda^j(\lambda I - J)^{-1}d\lambda Y = XJ^jY.$$

Hence, for any integer $m \geq 0$,

$$\begin{aligned}\int_\Gamma \lambda^m(L^{-1}(\lambda)f,\ h(\bar{\lambda}))d\lambda &= \int_\Gamma \sum_{j=0}^{k-1}\lambda^{j+m}(L^{-1}(\lambda)f, h_j)d\lambda \\ &= \sum_{j=0}^{k-1}(XJ^{j+m}Yf, h_j), \qquad (1)\\ &= [h_0^*\ h_1^*\ \ldots\ h_{k-1}^*]\begin{bmatrix} X \\ XJ \\ \vdots \\ XJ^{k-1}\end{bmatrix} J^mYf.\end{aligned}$$

Recall that there is an integer $l \geq k$ such that $[Y\ \ JY\ \ \ldots\ J^{l-1}Y]$ has full rank, and consider the $kn \times ln$ block matrix:

$$M_{k,l} = [XJ^{i+j-2}Y]_{i,j=1}^{k,l} = \begin{bmatrix} X \\ XJ \\ \vdots \\ XJ^{k-1}\end{bmatrix}[Y\ \ JY\ \ \ldots\ J^{l-1}Y]. \qquad (2)$$

If $(L^{-1}(\lambda)f,\ h(\bar{\lambda}))$ is analytic inside Γ then all the integrals of equation (1) vanish and we obtain

$$[h_0^*\ h_1^*\ \ldots\ h_{k-1}^*]M_{k,l} = 0. \qquad (3)$$

This equation implies that $h_0 = h_1 = \cdots = h_{k-1} = 0$ if $M_{k,l}$ has full rank, kn. But this is equivalent to the k-fold completeness of (X, J).

Conversely, if (3) is satisfied for some vectors $h_0, \ldots, h_{k-1}$, then it follows from equation (1) that the conditions of Lemma 3 are satisfied for the function $\phi(\lambda) = (L^{-1}(\lambda)f, h(\bar{\lambda}))$, and hence $\phi(\lambda)$ is analytic in Γ_+. Hence, by the condition of the lemma, $h(\lambda) \equiv 0$, i.e. the equation (3) has only the trivial solution. Therefore $M_{k,l}$ has rank kn and the pair (X, J) is k-fold complete. □

The next lemma is Lemma 2.1 of [7]. For clarity, we include the short proof.

Lemma 5 *If Γ contains F, a bounded and isolated part of $W(L)$, and if $f(\lambda)$ is a non-vanishing continuous vector function on Γ with values in $\mathbb{C}^n$, then*

$$\mathrm{ind}_\Gamma(L(\lambda)f(\lambda), f(\lambda)) = c(F).$$

Proof: Since, for $k > 2$, the unit sphere in $\mathbb{R}^k$ is simply connected, every continuous mapping f from Γ to $\mathbb{C}^n \backslash \{0\}$ is homotopic to a constant map. Hence $(L(\lambda)f(\lambda), f(\lambda))$ is homotopic to $(L(\lambda)g, g)$, $g \neq 0$. □

3 Spectral Divisors and the Case $c(F) = 1$

We can make a natural generalization of Theorem 4.2 of [4]:

Theorem 1 *Let $L(\lambda)$ be an a.m.f. on G and let Γ contain F, a bounded and isolated part of $W(L)$. Then $L(\lambda)$ has a spectral right divisor of degree k with respect to Γ if and only if $c(F) = k$ and Γ-canonical pairs are k-fold complete.*

Proof: If $L(\lambda)$ has a Γ-spectral right divisor $L_1(\lambda)$ of degree k, then the total algebraic multiplicity of eigenvalues in Γ_+ is the same as that for $L(\lambda)$, namely kn. Hence, by Lemma 2, $c(F) = k$. By Corollary 1.5 of [5], the Jordan pairs of $L_1(\lambda)$ are also Γ-canonical pairs for $L(\lambda)$. Hence they are k-fold complete.

Conversely, if $c(F) = k$ and a Γ-canonical pair (X, J) of $L(\lambda)$ is k-fold complete then, by definition, the matrix $\mathrm{col}(XJ^i)_{i=0}^{k-1}$ is nonsingular. Hence the equation

$$A_0X + A_1XJ + \cdots + A_{k-1}XJ^{k-1} = -XJ^k$$

has a unique solution

$$[A_0\ A_1\ \ldots\ A_{k-1}] = -XJ^k[\mathrm{col}(XJ^i)_{i=0}^{k-1}]^{-1},$$

and, by Theorem 1.23 of [4], the monic matrix polynomial

$$L_1(\lambda) = \lambda^k I + \sum_{i=0}^{k-1} A_i$$

has the Jordan pair (X, J). Hence, by Corollary 1.5 of [5] (or Theorem 2.6 of [3]), $L_1(\lambda)$ is a Γ-spectral right divisor of $L(\lambda)$. □

We also obtain the natural generalization of Theorem 4.11 of [4]:

Corollary 1 *The following are equivalent:*

1. *The a.m.f.* $L(\lambda)$ *has a* Γ*-spectral right divisor of degree* k *and also a* Γ*-spectral left divisor of degree* k.
2. $c(F) = k$ *and the matrix* $M_{k,k}$ *of equation* (2) *is nonsingular.*
3. $c(F) = k$ *and any* Γ*-canonial triple* (X, J, Y) *has the properties that* (X, J) *and* (Y^T, J^T) *are* k*-fold complete.*

Proof: The theorem gives the equivalence of items 1. and 2. The equivalence of items 2. and 3. follows from the factorization of equation (2) (with $l = k$). □

We note that, as in Chapter 4 of [4], it is possible to write down explicit formulae for the divisors in Theorem 1 and Corollary 1.

The following theorem and Theorem 4 below are our main results. They give sufficient conditions for the existence of right divisors of analytic matrix functions *without* the necessity of checking completeness conditions.

Theorem 2 *If* $c(F) = 1$, *then* $L(\lambda)$ *has a linear* Γ*-spectral right divisor.*

Proof: When $k = 1$, k-fold completeness reduces to completeness of the columns of a matrix X of a Γ-canonical pair (X, J). But this is established in Theorem 3.2 of [7]. The result now follows from Theorem 1. □

4 When Γ is a Circle

Theorem 3 *If* Γ *is a circle and* $c(F) = k$, *then* Γ*-canonical pairs are* k*-fold complete.*

Proof: Observe first that, by Lemma 2, the total algebraic multiplicity of eigenvalues in Γ_+ is kn. If Γ is a circle then, in view of Lemma 1, we can assume that the center of Γ is at the origin. Let Γ have radius $r > 0$.

Now suppose that there exists a Γ-canonical pair (X, J) which is not k-fold complete. Then, by Lemma 4, there exist $h_0, h_1, \ldots, h_{k-1} \in \mathbb{C}^n$, not all zero, such that, for any $f \in \mathbb{C}^n$, the function

$$F(\lambda) = \sum_{j=0}^{k-1} \lambda^j (L^{-1}(\lambda) f, h_j) = (L^{-1}(\lambda) f, h(\bar{\lambda}))$$

is analytic inside Γ. Define

$$\phi(\lambda) = (L^{-1}(\lambda) h(\bar{\lambda}), h(\bar{\lambda})). \tag{4}$$

Then

$$\phi(\lambda) = \sum_{j=0}^{k-1} \bar{\lambda}^j (L^{-1}(\lambda)h_j, h(\bar{\lambda}))$$

and since $\bar{\lambda} = r^2/\lambda$ on Γ,

$$\phi(\lambda) = \sum_{j=0}^{k-1} \frac{r^{2j}}{\lambda^j} (L^{-1}(\lambda)h_j, h(\bar{\lambda})), \qquad \lambda \in \Gamma.$$

By hypothesis, $(L^{-1}(\lambda)h_j,\ h(\bar{\lambda}))$ is analytic inside and on Γ for each j, so ϕ is analytic inside Γ except, possibly, for a pole at the origin whose order does not exceed $k-1$. Hence $\mathrm{ind}_\Gamma \phi(\lambda) \geq -(k-1)$.

On the other hand, writing $L^{-1}(\lambda)h(\bar{\lambda}) = g(\lambda)$, we have $h(\bar{\lambda}) = L(\lambda)g(\lambda)$ and, from the definition (4), $\phi(\lambda) = (g(\lambda),\ L(\lambda)g(\lambda))$, or

$$\overline{\phi(\lambda)} = (L(\lambda)g(\lambda),\ g(\lambda)).$$

The vector polynomial $h(\lambda)$ vanishes at only finitely many points so we can adjust the radius r of Γ in such a way that $h(\bar{\lambda}) \neq 0$ on Γ, and hence $g(\lambda) \neq 0$ on Γ. Now apply Lemma 5 to get $\mathrm{ind}_\Gamma \overline{\phi(\lambda)} = c(F) = k$. But this implies $\mathrm{ind}_\Gamma \phi(\lambda) = -k$, a contradiction. □

Theorems 1 and 3 immediately yield:

Theorem 4 *If Γ is a circle and $c(F) = k$ then $L(\lambda)$ has a Γ-spectral divisor of degree k.*

References

[1] J.A. Ball, I. Gohberg and L. Rodman, Interpolation of rational matrix functions, *Operator Theory: Adv. & Applic.* **45** Birkhäuser, Basel, 1990.

[2] I. Krupnik, A. Markus and V. Matsaev, Factorization of matrix functions and characteristic properties of the circle, *Integral Equations and Operator Theory* **17** (1993), 554–566.

[3] I. Gohberg, M.A. Kaashoek and F. Van Schagen, On the local theory of regular analytic matrix functions, *Linear Algebra and its Applications* **182** (1993), 9–25.

[4] I. Gohberg, P. Lancaster and L. Rodman, *Matrix Polynomials*, Academic Press, New York, 1982

[5] I. Gohberg and L. Rodman, Analytic matrix functions with prescribed local data, *Journal d'Analyse Mathématique* **40** (1981), 90–128.

[6] A. Markus, Introduction to Spectral Theory of Polynomial Operator Pencils, *Transl. of Math. Monographs, Amer. Math. Society, Providence*, 1988.

[7] A. Markus, J. Maroulas and P. Psarrakos, Spectral properties of a matrix polynomial connected with a component of its numerical range, *Operator Theory: Adv. & Applic.* **106** (1998), 305–308.

[8] A. Markus and V. Matsaev, On the spectral theory of holomorphic operator-valued functions in Hilbert space, *Funct. Anal. Appl.* **9** (1975), 76–77.

[9] A. Markus and L. Rodman, Some results on numerical ranges and factorization of matrix polynomials, *Linear and Multilinear Algebra* **42** (1997), 169–185.

P. Lancaster
Department of Mathematics and Statistics
University of Calgary
Calgary, AB T2N 1N4
Canada

A. Markus
Department of Mathematics
Ben Gurion University
of the Negev
Beer Sheva
Israel

Operator Theory:
Advances and Applications, Vol. 122
© 2001 Birkhäuser Verlag Basel/Switzerland

Diagonalization of Certain Block Operator Matrices and Applications to Dirac Operators

Heinz Langer and Christiane Tretter[1]

To our friend and colleague Rien Kaashoek on the occasion of his sixtieth birthday

In this paper several classes of self-adjoint and non-self-adjoint block operator matrices with unbounded entries are studied. The main results concern the existence of maximal spectral invariant subspaces which correspond to the right and the left half planes and their representation by means of angular operators. Sometimes this yields a diagonalization of the block operator matrix under consideration. Applications to abstract Dirac operators with supersymmetry and to Dirac operators with a potential are given.

0 Introduction

In this paper we consider closable block operator matrices of the form

$$\widetilde{A}_0 = \begin{pmatrix} A & B \\ B^* & D \end{pmatrix} \tag{0.1}$$

with possibly unbounded entries $A,\ B,\ D$ in a product Hilbert space $\widetilde{\mathcal{H}} = \mathcal{H} \oplus \widehat{\mathcal{H}}$. It is always supposed that $\sigma(A)$ is in the right half plane and that $\sigma(D)$ is in the left half plane. We find conditions which assure that the closure $\widetilde{A}$ of $\widetilde{A}_0$ has a maximal spectral invariant subspace $\widetilde{\mathcal{L}}_+$ corresponding to the right half plane and a maximal spectral invariant subspace $\widetilde{\mathcal{L}}_-$ corresponding to the left half plane, such that $\widetilde{\mathcal{L}}_+$ is 'supported' by the first component $\mathcal{H}$ and $\widetilde{\mathcal{L}}_-$ is 'supported' by the second component $\widehat{\mathcal{H}}$. The latter means that these subspaces admit representations

$$\widetilde{\mathcal{L}}_+ = \left\{ \begin{pmatrix} x \\ K_+ x \end{pmatrix} : x \in \mathcal{H} \right\}, \quad \widetilde{\mathcal{L}}_- = \left\{ \begin{pmatrix} K_- \widehat{x} \\ \widehat{x} \end{pmatrix} : \widehat{x} \in \widehat{\mathcal{H}} \right\} \tag{0.2}$$

[1] H. Langer was supported by the "Fonds zur Förderung der wissenschaftlichen Forschung" of Austria, Project P 12176-MAT. C. Tretter gratefully acknowledges the support of the German Research Foundation, DFG, Grant TR 368/4–1.

with so-called angular operators K_+ and K_-. It turns out that these angular operators are even contractions. In some cases a diagonalization

$$\begin{aligned}&\begin{pmatrix} I & K_- \\ K_+ & I \end{pmatrix}^{-1} \begin{pmatrix} A & B \\ B^* & D \end{pmatrix} \begin{pmatrix} I & K_- \\ K_+ & I \end{pmatrix} \\ &= \begin{pmatrix} A + BK_+ & 0 \\ 0 & D + B^*K_- \end{pmatrix}\end{aligned} \tag{0.3}$$

of the block operator matrix (0.1) follows.

The main tools which we use in the proofs are a result of H. Bart, I.C. Gohberg and M.A. Kaashoek (see [BGK], [GGK]) about the separation of the spectrum of an operator at infinity, and a theorem about accretive operators in Krein spaces which goes back to I.S. Iokhvidov, M.G. Krein and T.Ya. Azizov (see [IK], [AI]). The present paper contains, as special cases, some results of [AL] and [LT][2]. The new feature compared with these papers is the use of an indefinite inner product. With respect to this inner product the block operator matrices are accretive, and this yields the angular operator representations (0.2) of the invariant subspaces with contractions K_+, K_-. In [AL] and [LT] the proofs of the corresponding statements were more complicated, and the statements sometimes less complete. E.g. the results of the present paper imply that also in [LT], Theorem 4.3, the angular operators $K_\pm$ are contractions.

In the first section below the above mentioned main tools are formulated and adapted to our situation. The second section contains some auxiliary results about block operator matrices. In the third section we consider a self-adjoint block operator matrix and prove the angular operator representation of its spectral subspaces corresponding to the positive and the negative half axis. As an example, abstract Dirac operators with supersymmetry are considered. In this case the angular operators K_+ and $K_- = -K_+^*$ can be calculated explicitly, and the transformation (0.3) coincides with the well-known Foldy-Wouthuysen transformation from quantum mechanics. The fourth section contains results for two non-self-adjoint cases, namely in Theorem 4.1 the diagonal operators A and D are dominating in the sense that they determine the domain of $\widetilde{A}$, in Theorem 4.3 the off diagonal operators B and B^* are dominating. The latter situation appears also in the self-adjoint case for the Dirac operator. Finally, in the last section the results of Section 3 are applied to the Dirac operator with a potential with respect to two different decompositions of the underlying Hilbert space $L_2(\mathbb{R}^3)^4$, namely the natural 2×2 decomposition and the decomposition induced by the positive and negative spectral subspaces of the free Dirac operator.

Angular operator representations of the subspaces $\widetilde{\mathcal{L}}_\pm$ were also obtained in the papers [ALMS], [MS], and [M]. But e.g. in [ALMS] and [MS] the assumptions

---

[2] We use this occasion to point out that in the paper [LT], Theorem 4.3, (iii) and (iv) the operators $I + K_\pm K_\pm^*$ have to be replaced by $I - K_\pm K_\pm^*$.

about the domains of the entries in (0.1) are such that their results cannot be applied to the Dirac operator.

The block operator matrices considered in Section 4 are exponentially dichotomous (see, e.g., [BGK], [KVL]), therefore they are infinitesimal generators of hyperbolic semigroups. A study of semigroups generated by block operator matrices can be found in [E].

1 Basic Propositions

1.1 In the following $\mathbb{C}_+$ denotes the open right and $\mathbb{C}_-$ the open left half plane. The first theorem is an easy consequence of Theorem XV.3.1 of [GGK].

Theorem 1.1 *Let S be a closed densely defined operator in a Banach space $\mathcal{X}$ such that for some $h > 0$ the following conditions are satisfied:*

(i) $\{z \in \mathbb{C} : -h < \Re(z) < h\} \subset \rho(S)$ *and* $\sup_{-h<\Re(z)<h} \|(S-z)^{-1}\| < \infty$.

(ii) $\lim_{\eta\to\infty} \sup\{\|(S-z)^{-1}\| : z = \xi \pm \mathrm{i}\eta, 0 \le \xi < h\} = 0$.

(iii) *The integral* $\dfrac{1}{\pi \mathrm{i}} \displaystyle\int'_{\mathrm{i}\mathbb{R}} (S-z)^{-1}\mathrm{d}z$*, exists in the strong operator topology, where the prime denotes the Cauchy principal value at infinity,*

Then

$$\frac{1}{\pi \mathrm{i}} \int'_{\mathrm{i}\mathbb{R}} (S-z)^{-1}\mathrm{d}z = P_+ - P_- \tag{1.1}$$

where P_+, P_- are projections in $\mathcal{X}$ such that $P_+ + P_- = I$, for their ranges $\mathcal{X}_\pm := P_\pm\mathcal{X}$ the sets $\mathcal{X}_\pm \cap \mathcal{D}(S)$ are dense in $\mathcal{X}_\pm$ and

$$\begin{aligned} &\mathcal{X}_\pm \cap \mathcal{D}(S) = P_\pm\mathcal{D}(S), \quad S(\mathcal{X}_\pm \cap \mathcal{D}(S)) = \mathcal{X}_\pm, \\ &\sigma(S|\mathcal{X}_\pm) = \sigma(S) \cap \mathbb{C}_\pm. \end{aligned} \tag{1.2}$$

Proof: Choose α such that $0 < \alpha < h$. Since for $\nu > 0$ we have

$$\begin{aligned} \frac{1}{2\pi}\int_{-\nu}^{\nu} (S-\mathrm{i}t)^{-1}\mathrm{d}t &= \frac{1}{2\pi}\int_{-\nu}^{\nu} (S-(\alpha+\mathrm{i}t))^{-1}\mathrm{d}t \\ &\quad -\frac{1}{2\pi \mathrm{i}}\int_{0}^{\alpha} (S-(\mathrm{i}\nu+s))^{-1}\mathrm{d}s \\ &\quad -\frac{1}{2\pi \mathrm{i}}\int_{\alpha}^{0} (S-(-\mathrm{i}\nu+s))^{-1}\mathrm{d}s, \end{aligned}$$

the assumptions (ii) and (iii) imply that also the integral $\frac{1}{2\pi \mathrm{i}}\int'_{\alpha+\mathrm{i}\mathbb{R}}(S-z)^{-1}\mathrm{d}z$ exists in the strong operator topology and that it coincides with $\frac{1}{2\pi \mathrm{i}}\int'_{\mathrm{i}\mathbb{R}}(S-z)^{-1}\mathrm{d}z$.

Further, the relation

$$z^{-2}(S-z)^{-1}S^2 = (S-z)^{-1} + z^{-1} + z^{-2}S$$

yields for $x \in \mathcal{D}(S^2)$

$$\begin{aligned}
&\frac{1}{2\pi \mathrm{i}} \int_{\alpha+\mathrm{i}\mathbb{R}} z^{-2}(S-z)^{-1}S^2 x \mathrm{d}z \\
&= \frac{1}{2\pi \mathrm{i}} \int'_{\alpha+\mathrm{i}\mathbb{R}} (S-z)^{-1} x \mathrm{d}z \\
&+ \frac{1}{2\pi \mathrm{i}} \int'_{\alpha+\mathrm{i}\mathbb{R}} z^{-1} \mathrm{d}z\, x + \frac{1}{2\pi \mathrm{i}} \int_{\alpha+\mathrm{i}\mathbb{R}} z^{-2} \mathrm{d}z\, Sx.
\end{aligned}$$

Here the last integral vanishes, and the second last integral equals $x/2$. Therefore the integral on the left hand side defines a bounded operator in $\mathcal{X}$ and according to [GGK], Theorem XV.3.1, this is a projection P_+. It follows that

$$P_+ = \frac{1}{2\pi \mathrm{i}} \int'_{\mathrm{i}\mathbb{R}} (S-z)^{-1} \mathrm{d}z + \frac{1}{2}(P_+ + P_-)$$

with $P_- = I - P_+$, and hence the relation (1.1) holds. The other statements are easily obtained from [GGK], Theorem XV.3.1. □

Possibly the assumption (ii) in Theorem 1.1 is superfluous. However, it will always be satisfied in the examples we have in mind and it allowed us to refer to [GGK], Theorem XV.3.1, for the essential part of the proof.

The next theorem was proved by K.Veselič [V] in the more special situation of the Klein-Gordon equation.

Theorem 1.2 *Suppose that the operator S fulfills the conditions* (i), (ii), (iii) *of Theorem* 1.1 *and that, additionally, with some $\gamma > 0$,*

(iv) $$\|(S-\mathrm{i}t)^{-1}\| \leq \frac{\gamma}{1+|t|}, \quad t \in \mathbb{R}.$$

If T is a bounded operator in $\mathcal{X}$ such that for some $h' > 0$ the strip $\{z \in \mathbb{C} : -h' < \Re(z) < h'\}$ belongs to $\rho(S+T)$, then for the operator $S+T$ the relation

$$\frac{1}{\pi \mathrm{i}} \int'_{\mathrm{i}\mathbb{R}} (S+T-z)^{-1} \mathrm{d}z = P'_+ - P'_-$$

holds with two projections $P'_\pm$ which have the same properties with respect to $S+T$ as the projections $P_\pm$ in Theorem 1.1 *have with respect to S.*

Proof: We only have to show that the operator $S+T$ satisfies the same conditions as S (with possibly another h). By assumption this is true for the first condition in (i). For the second condition in (i) and for (ii) we use the relation

$$(S+T-z)^{-1} = (S-z)^{-1}(I+T(S-z)^{-1})^{-1}, \quad z \in \rho(S) \cap \rho(S+T),$$

which also shows that

$$\|(S+T-\mathrm{i}t)^{-1}\| \le \frac{\gamma'}{1+|t|}, \qquad t \in \mathbb{R}, \tag{1.3}$$

with some $\gamma' > 0$. Now (iii) follows from the relation

$$\frac{1}{\pi \mathrm{i}} \int'_{\mathrm{i}\mathbb{R}} (S+T-z)^{-1} \mathrm{d}z = \frac{1}{\pi \mathrm{i}} \int'_{\mathrm{i}\mathbb{R}} (S-z)^{-1} \mathrm{d}z$$
$$- \frac{1}{\pi \mathrm{i}} \int'_{\mathrm{i}\mathbb{R}} (S+T-z)^{-1} T (S-z)^{-1} \mathrm{d}z,$$

since the last integral exists even in the uniform operator topology because of the assumption (iv) and the relation (1.3). □

A similar reasoning can be applied if the perturbation T is unbounded. The proof of the following theorem uses the same method as the Proof of Theorem 1.2 and is left to the reader.

Theorem 1.3 *Suppose that the operator S fulfills the conditions* (i), (ii), (iii) *of Theorem* 1.1. *Let T be an operator in $\mathcal{X}$ such that $\mathcal{D}(S) \subset \mathcal{D}(T)$, and that with positive numbers γ, γ' and η the inequalities*

$$\|T(S-\mathrm{i}t)^{-1}\| \le \frac{\gamma}{1+|t|^{\eta}}, \qquad \|(S+T-\mathrm{i}t)^{-1}\| \le \frac{\gamma'}{1+|t|}, \qquad t \in \mathbb{R}, \tag{v}$$

are satisfied. Then the operator $S+T$ is closed. If for some $h' > 0$ the strip $\{z \in \mathbb{C}: -h' < \Re(z) < h'\}$ belongs to $\rho(S+T)$, then the relation

$$\frac{1}{\pi \mathrm{i}} \int'_{\mathrm{i}\mathbb{R}} (S+T-z)^{-1} \mathrm{d}z = P'_+ - P'_-$$

holds with two projections $P'_\pm$ which have the same properties with respect to $S+T$ as the projections $P_\pm$ in Theorem 1.1 *have with respect to S.*

We mention that if the inequality (iv) is satisfied, then the second inequality in (v) is a consequence of the first one.

1.2 We need a result about accretive operators in Krein spaces. It is a slight extension of a theorem of T.Ya. Azizov, see [AI], Theorem II.2.21, for dilating operators it was proved by I.S. Iohkvidov and M.G. Krein in [IK], Theorem 3.7, see also [IKL], Theorem 11.1. For the following definitions and simple facts about Krein spaces we refer to [AI].

Let $(\mathcal{K}, [\cdot,\cdot])$ be a Krein space. Recall that this means that $\mathcal{K}$ is a direct sum

$$\mathcal{K} = \mathcal{K}_+ \dot{+} \mathcal{K}_- \tag{1.4}$$

where $(\mathcal{K}_\pm, \pm[\cdot,\cdot])$ are Hilbert spaces and for $x_\pm \in \mathcal{K}_\pm$ we have $[x_+, x_-] = 0$. It follows that if $x = x_+ + x_-$ and $y = y_+ + y_-$ are the decompositions of the elements $x, y \in \mathcal{K}$ according to (1.4), then

$$(x, y) := [x_+, y_+] - [x_-, y_-]$$

defines a Hilbert space inner product on $\mathcal{K}$. If in this Hilbert space Q_+ denotes the orthogonal projection onto the first and Q_- denotes the orthogonal projection onto the second component of $\mathcal{K}$ in (1.4) and $J := Q_+ - Q_-$, then $[x, y] = (Jx, y)$, $x, y \in \mathcal{K}$. A subspace $\mathcal{L}$ of the Krein space $\mathcal{K}$ is called *nonnegative* (*positive, uniformly positive*, respectively), if $[x, x] \geq 0$ (> 0 or $\geq \gamma \|x\|^2$ with some $\gamma > 0$, respectively) for all $x \in \mathcal{L}$, $x \neq 0$. A nonnegative subspace $\mathcal{L}$ is called a *maximal nonnegative subspace* if it is not properly contained in another nonnegative subspace. For a maximal nonnegative subspace $\mathcal{L}$ of the Krein space $\mathcal{K}$ there exists a contraction K from $\mathcal{K}_+$ into $\mathcal{K}_-$ such that

$$\mathcal{L}_+ = \left\{ \begin{pmatrix} x \\ Kx \end{pmatrix} : x \in \mathcal{H} \right\}.$$

If $\mathcal{L}$ is positive, then the contraction K is strict (that is $\|Kx_+\| < \|x_+\|$ for all $x_+ \in \mathcal{K}_+$, $x_+ \neq 0$), if $\mathcal{L}$ is uniformly positive, then $\|K\| < 1$.

A closed densely defined linear operator S in $\mathcal{K}$ is called *accretive* (*strictly accretive*, respectively) if

$$\Re\,[Sx, x] \geq 0 \quad (> 0, \text{ respectively}), \qquad x \in \mathcal{D}(S), x \neq 0,$$

and *uniformly accretive* if there exists a $\beta > 0$ such that

$$\Re\,[Sx, x] \geq \beta\,\|x\|^2, \quad x \in \mathcal{D}(S). \tag{1.5}$$

In the following the expressions under the integrals may also have a singularity at zero. In this case also at zero the integral is to be understood in the sense of the Cauchy principal value.

Theorem 1.4 *Let S be an accretive operator in the Krein space $\mathcal{K}$ such that $\mathrm{i}\mathbb{R}\backslash\{0\} \subset \rho(S)$ and that the integral $\frac{1}{\pi \mathrm{i}} \int'_{\mathrm{i}\mathbb{R}} (S-z)^{-1}\mathrm{d}z$ exists in the strong operator topology and is the difference of two complementary projections $P_\pm$:*

$$\frac{1}{\pi \mathrm{i}} \int'_{\mathrm{i}\mathbb{R}} (S - z)^{-1}\mathrm{d}z = P_+ - P_-, \qquad P_+ + P_- = I.$$

Then $\mathcal{L}_+ := P_+\mathcal{K}$ is a maximal nonnegative subspace of $\mathcal{K}$ and $\mathcal{L}_- := P_-\mathcal{K}$ is a maximal nonpositive subspace of $\mathcal{K}$. If S is strictly accretive (bounded and uniformly accretive, respectively), then $\mathcal{L}_+$ is a positive (uniformly positive, respectively) subspace and $\mathcal{L}_-$ is a negative (uniformly negative, respectively) subspace.

Proof: We consider an element $x \in \mathcal{L}_+$. Then

$$\begin{aligned}[x,x] &= [P_+x,x] = \Re\,[P_+x,x] = \Re\,[(P_+ - P_-)x,x] \\ &= \Re\left(\frac{1}{\pi \mathrm{i}}\int'_{\mathrm{i}\mathbb{R}} [(S-z)^{-1}x,x]\mathrm{d}z\right) \\ &= \frac{1}{\pi}\int'^{+\infty}_{-\infty} \Re\left([S(S-\mathrm{i}t)^{-1}x,(S-\mathrm{i}t)^{-1})x]\right)\mathrm{d}t.\end{aligned}$$

Evidently, for $x \neq 0$ the last integral is nonnegative (positive, respectively) if S is accretive (strictly accretive, respectively) in $\mathcal{K}$. Therefore $\mathcal{L}_+$ is a nonnegative (positive, respectively) subspace of $\mathcal{K}$, and, analogously, $\mathcal{L}_-$ is a nonpositive (negative, respectively) subspace of $\mathcal{K}$. If S is uniformly accretive and bounded, we consider an arbitrary nonempty open bounded interval Δ with a positive distance from zero. Then with β from (1.5) the integral in the last relation is greater or equal than

$$\frac{1}{\pi}\int_\Delta \beta\|(S-\mathrm{i}t)^{-1}x\|^2\,dt \geq \frac{\beta}{\pi}\,l(\Delta)\left(\max_{t\in\Delta}\|(S-\mathrm{i}t)\|\right)^{-2}\|x\|^2.$$

That the subspace $\mathcal{L}_+$ ($\mathcal{L}_-$, respectively) is maximal nonnegative (maximal nonpositive, respectively) follows from the decomposition $\mathcal{K} = \mathcal{L}_+ \dotplus \mathcal{L}_-$ and [AI], I.1.25°. □

Corollary 1.5 *With the decomposition* (1.4) *there exist contractions* $K_+ \in L(\mathcal{K}_+, \mathcal{K}_-)$ *and* $K_- \in L(\mathcal{K}_-, \mathcal{K}_+)$ *such that*

$$\mathcal{L}_+ = \left\{\begin{pmatrix} x_+ \\ K_+x_+\end{pmatrix} : x_+ \in \mathcal{K}_+\right\}, \qquad \mathcal{L}_- = \left\{\begin{pmatrix} K_-x_- \\ x_-\end{pmatrix} : x_- \in \mathcal{K}_-\right\}.$$

If S is strictly accretive (*bounded and uniformly accretive, respectively*), *then K_+ and K_- are strict contractions* (*uniform contractions, respectively*).

2 Block Operator Matrices

2.1 In this section we consider the Hilbert space

$$\widetilde{\mathcal{H}} = \mathcal{H} \oplus \widehat{\mathcal{H}}, \tag{2.1}$$

which is the orthogonal sum of the two Hilbert spaces $\mathcal{H}$ and $\widehat{\mathcal{H}}$, and in $\widetilde{\mathcal{H}}$ the operator

$$\widetilde{A}_0 := \begin{pmatrix} A & B \\ B^* & D\end{pmatrix}. \tag{2.2}$$

The operators A, B and D are supposed to be densely defined and closed, the operator $\widetilde{A}_0$ is defined on its natural domain $(\mathcal{D}(A) \cap \mathcal{D}(B^*)) \oplus (\mathcal{D}(B) \cap \mathcal{D}(D))$, and we always assume that this domain is dense in $\widetilde{\mathcal{H}}$ and that $\widetilde{A}_0$ is also closable. The closure of $\widetilde{A}_0$ is denoted by $\widetilde{A}$, and (2.2) is called a *block operator matrix representation* of $\widetilde{A}_0$.

It is clear that any bounded operator $\widetilde{A}$ in $\widetilde{\mathcal{H}}$ for a given decomposition (2.1) of the Hilbert space $\widetilde{\mathcal{H}}$ admits a block operator matrix representation. This is not true, in general, for an unbounded operator $\widetilde{A}$. Consider e.g. in the space $L^2([-1, 1])$ the operator $\widetilde{A}$ given by the second derivative f'' with boundary conditions $f(-1) = f(1) = 0$. It does not admit a block operator matrix representation with respect to the decomposition $L^2([-1, 1]) = L^2([-1, 0]) \oplus L^2([0, 1])$. However, its inverse $\widetilde{A}^{-1}$, which is a bounded integral operator, does admit such a representation.

It is our aim to find conditions such that for the closure $\widetilde{A}$ of the block operator matrix $\widetilde{A}_0$ besides the decomposition (2.1) a decomposition

$$\widetilde{\mathcal{H}} = \widetilde{\mathcal{L}}_+ \dot{+} \widetilde{\mathcal{L}}_- \tag{2.3}$$

with $\widetilde{A}$-invariant subspaces $\widetilde{\mathcal{L}}_\pm$ holds, where $\sigma(\widetilde{A}|\widetilde{\mathcal{L}}_\pm) = \sigma(\widetilde{A}) \cap \mathbb{C}_\pm$, and that these subspaces admit representations

$$\widetilde{\mathcal{L}}_+ = \left\{ \begin{pmatrix} x \\ K_+x \end{pmatrix} : x \in \mathcal{H} \right\} \tag{2.4}$$

with a contraction K_+ from $\mathcal{H}$ into $\widehat{\mathcal{H}}$, and

$$\widetilde{\mathcal{L}}_- = \left\{ \begin{pmatrix} K_-\widehat{x} \\ \widehat{x} \end{pmatrix} : \widehat{x} \in \widehat{\mathcal{H}} \right\} \tag{2.5}$$

with a contraction K_- from $\widehat{\mathcal{H}}$ into $\mathcal{H}$. Here $\widetilde{A}$-invariance of these subspaces $\widetilde{\mathcal{L}}_\pm$ means that with the projections $\widetilde{P}_\pm$ onto $\widetilde{\mathcal{L}}_\pm$ according to the decomposition (2.3) the following relations hold:

$$\widetilde{\mathcal{L}}_\pm \cap \mathcal{D}(\widetilde{A}) = \widetilde{P}_\pm \mathcal{D}(\widetilde{A}), \quad \widetilde{A}(\widetilde{\mathcal{L}}_\pm \cap \mathcal{D}(\widetilde{A})) \subset \widetilde{\mathcal{L}}_\pm. \tag{2.6}$$

The subspaces $\widetilde{\mathcal{L}}_\pm$ of the decomposition (2.3) are in general not orthogonal to each other in $\widetilde{\mathcal{H}}$.

In order to find a decomposition (2.3), Theorem 1.4 about accretive operators in Krein spaces is used. To this end, in the Hilbert space $\widetilde{\mathcal{H}}$ besides the inner product

$$(\widetilde{x}, \widetilde{y}) = (x, y) + (\widehat{x}, \widehat{y}), \quad \widetilde{x} = x + \widehat{x},\ \widetilde{y} = y + \widehat{y},\ x, y \in \mathcal{H},\ \widehat{x}, \widehat{y} \in \widehat{\mathcal{H}},$$

also the indefinite inner product

$$[\widetilde{x}, \widetilde{y}] := (x, y) - (\widehat{x}, \widehat{y}), \quad \widetilde{x} = x + \widehat{x},\ \widetilde{y} = y + \widehat{y},\ x, y \in \mathcal{H},\ \widehat{x}, \widehat{y} \in \widehat{\mathcal{H}},$$

is considered. Then $\widetilde{\mathcal{H}}$ becomes a Krein space with fundamental symmetry

$$\widetilde{J} := \begin{pmatrix} I & 0 \\ 0 & -I \end{pmatrix} = \widetilde{Q}_+ - \widetilde{Q}_-, \quad \widetilde{Q}_+ := \begin{pmatrix} I & 0 \\ 0 & 0 \end{pmatrix}, \quad \widetilde{Q}_- := \begin{pmatrix} 0 & 0 \\ 0 & I \end{pmatrix},$$

with respect to the decomposition (2.1). We denote this Krein space by $\widetilde{\mathcal{K}}$. The elements $\widetilde{x}$ of $\widetilde{\mathcal{H}}$ or $\widetilde{\mathcal{K}}$ will often be written as vectors $x = \begin{pmatrix} x \\ \widehat{x} \end{pmatrix} = \begin{pmatrix} x_+ \\ x_- \end{pmatrix}$.

2.2 We formulate some general statements about block operator matrices of the form (2.2) for which the spectra of A and D are 'separated'.

Theorem 2.1 *Suppose that there exist positive numbers α, δ such that*

$$\begin{aligned} \Re\,(Ax, x) &\geq \alpha\,\|x\|^2, & x &\in \mathcal{D}(A) \cap \mathcal{D}(B^*), \\ \Re\,(D\widehat{x}, \widehat{x}) &\leq -\delta\,\|\widehat{x}\|^2, & \widehat{x} &\in \mathcal{D}(B) \cap \mathcal{D}(D). \end{aligned} \tag{2.7}$$

Then the strip $\{z \in \mathbb{C} : -\delta < \Re(z) < \alpha\}$ belongs to the set $r(\widetilde{A})$ of points of regular type of $\widetilde{A}$.

Proof: Let $z \in \mathbb{C}$ be such that $-\delta < \Re(z) < \alpha$. The theorem is proved if we show that for a sequence $(\widetilde{x}_n)_0^\infty$ of elements of $\mathcal{D}(\widetilde{A}_0)$ such that $(\widetilde{A}_0 - z)\widetilde{x}_n \to 0$ it follows that $\|\widetilde{x}_n\| \to 0$ if $n \to \infty$.

Assume that $(\widetilde{x}_n)_0^\infty$ does not tend to 0 for $n \to \infty$. Then there exists a subsequence $(\widetilde{x}_{n_k})_{k=0}^\infty$ such that $\|\widetilde{x}_{n_k}\| \geq M$, $k = 0, 1, \ldots,$ with $M > 0$, and it follows that $(\widetilde{A}_0 - z)(\widetilde{x}_{n_k}/\|\widetilde{x}_{n_k}\|) \to 0$, $n \to \infty$. Therefore without loss of generality we can assume that for the original sequence $(\widetilde{x}_n)_0^\infty$ we have $\|\widetilde{x}_n\| = 1$. With $\widetilde{x}_n = \begin{pmatrix} x_n \\ \widehat{x}_n \end{pmatrix}$ we obtain

$$(A - z)x_n + B\widehat{x}_n \to 0, \quad B^*x_n + (D - z)\widehat{x}_n \to 0, \qquad n \to \infty,$$

which implies

$$\begin{aligned} &((A - z)x_n, x_n) + (B\widehat{x}_n, x_n) \to 0, \\ &(B^*x_n, \widehat{x}_n) + ((D - z)\widehat{x}_n, \widehat{x}_n) \to 0, \qquad n \to \infty. \end{aligned}$$

If we form the difference of these relations it follows that

$$\begin{aligned} &((A - z)x_n, x_n) - ((D - z)\widehat{x}_n, \widehat{x}_n) + (B\widehat{x}_n, x_n) \\ &-\overline{(B\widehat{x}_n, x_n)} \to 0, \qquad n \to \infty, \end{aligned}$$

whence

$$\Re\,(Ax_n, x_n) - \Re(z)\|x_n\|^2 - \Re\,(D\widehat{x}_n, \widehat{x}_n) + \Re(z)\|\widehat{x}_n\|^2 \to 0, \qquad n \to \infty,$$

a contradiction because of the assumptions (2.7) and $\|\widetilde{x}_n\| = 1$. Hence $\|\widetilde{x}_n\| \to 0$ for $n \to \infty$. □

For a linear operator T in the Hilbert space $\mathcal{H}$ its numerical range is denoted by $W(T)$:

$$W(T) := \{(Tx, x) : x \in \mathcal{D}(T),\ \|x\| = 1\}.$$

In the sequel we always suppose that A and $-D$ are m-sectorial operators (see [K]), that is

$$W(A) \subset \{z \in \mathbb{C} : |\arg z| \leq \Theta_A\}, \tag{2.8}$$

$$W(D) \subset \{z \in \mathbb{C} : |\arg z| \geq \pi - \Theta_D\} \tag{2.9}$$

for some $\Theta_A, \Theta_D \in [0, \frac{\pi}{2})$ and $\mathbb{C}_- \subset \rho(A)$, $\mathbb{C}_+ \subset \rho(D)$.

We need the following simple statement about m-sectorial operators.

Lemma 2.2 *Suppose that T is m-sectorial in the Hilbert space $\mathcal{H}$. Then for a sequence $(x_n) \subset \mathcal{D}(T)$ the relations*

$$x_n \to x_0 \text{ and } (Tx_n, x_n) \to 0, \qquad n \to \infty, \tag{2.10}$$

imply $x_0 \in \mathcal{D}(T)$ and $Tx_0 = 0$.

Proof: Consider on $\mathcal{D}(T)$ the sesquilinear form $\mathbf{t}[x, y] := (Tx, y)$, and let $\Theta_T \in [0, \frac{\pi}{2})$ be its semiangle (i.e. $W(T) \subset \{z \in \mathbb{C} : |\arg z| \leq \Theta_T\}$). According to [K], Theorem VI.1.27, it is closable; denote its closure by $\widetilde{\mathbf{t}}$. The assumptions (2.8) and (2.10) imply $x_n \xrightarrow{\mathbf{t}} x_0$ if $n \to \infty$ (we recall that this means $\|x_n - x_0\| \to 0$ and $\mathbf{t}[x_n - x_m, x_n - x_m] \to 0$ if $n, m \to \infty$). Indeed, if $\mathbf{h}$ denotes the form $\Re(\mathbf{t})$, then (2.10) implies $\mathbf{h}[x_n, x_n] \to 0$, hence $\mathbf{h}[x_n - x_m, x_n - x_m] \to 0$ and

$$|\mathbf{t}[x_n - x_m, x_n - x_m]| \leq (1 + \tan\Theta_T)\mathbf{h}[x_n - x_m, x_n - x_m] \to 0.$$

For an arbitrary sequence $(u_n) \in \mathcal{D}(T)$ such that $u_n \xrightarrow{\mathbf{t}} u$ we obtain

$$\begin{aligned} |\mathbf{t}[x_n, u_n]| &\leq (1 + \tan\Theta_T)(\mathbf{h}[x_n, x_n])^{1/2}(\mathbf{h}[u_n, u_n])^{1/2} \\ &\leq (1 + \tan\Theta_T)|\mathbf{t}[x_n, x_n]|^{1/2}(\mathbf{h}[u_n, u_n])^{1/2} \longrightarrow 0, \end{aligned}$$

hence by [K], Theorem VI.1.2, $\mathbf{t}[x_0, u] = 0$ for all $u \in \mathcal{D}(T)$. Since $\mathcal{D}(T)$ is a core of $\widetilde{\mathbf{t}}$, the First Representation Theorem ([K], Theorem VI.2.1) yields $x_0 \in \mathcal{D}(T)$ and $Tx_0 = 0$. □

We say that for the closure $\widetilde{A}$ of the block operator matrix $\widetilde{A}_0$ the kernel $\ker\widetilde{A}$ has the *splitting property* if

$$\begin{pmatrix} x_0 \\ \widehat{x}_0 \end{pmatrix} \in \ker\widetilde{A} \implies \begin{pmatrix} x_0 \\ 0 \end{pmatrix} \in \ker\widetilde{A}, \begin{pmatrix} 0 \\ \widehat{x}_0 \end{pmatrix} \in \ker\widetilde{A}. \tag{2.11}$$

Theorem 2.3 *Suppose that A and $-D$ are m-sectorial and that B is closed. Then the relation $\widetilde{A}\widetilde{x}_0 = 0$ with $\widetilde{x}_0 = \begin{pmatrix} x_0 \\ \widehat{x}_0 \end{pmatrix} \in \mathcal{D}(\widetilde{A})$ implies*

$$x_0 \in \mathcal{D}(A), \; Ax_0 = 0, \quad \widehat{x}_0 \in \mathcal{D}(D), \; D\widehat{x}_0 = 0.$$

If, additionally, at least one of the operators $A,\ B,\ D$ *is bounded, then* $\ker \widetilde{A}$ *has the splitting property, if* B *is bounded, or if* A *and* D *are bounded, then*

$$\ker \widetilde{A} = \ker \widetilde{A}^* = \left\{ \begin{pmatrix} x_0 \\ \widehat{x}_0 \end{pmatrix} \in \mathcal{D}(\widetilde{A}_0) : \ Ax_0 = B\widehat{x}_0 = 0, \ D\widehat{x}_0 = B^* x_0 = 0 \right\}.$$

Proof: There exists a sequence $(\widetilde{x}_n) \subset \mathcal{D}(\widetilde{A}_0)$, $\widetilde{x}_n = \begin{pmatrix} x_n \\ \widehat{x}_n \end{pmatrix}$ with $x_n \in \mathcal{D}(A) \cap \mathcal{D}(B^*)$, $\widehat{x}_n \in \mathcal{D}(B) \cap \mathcal{D}(D)$, such that $x_n \to x_0$, $\widehat{x}_n \to \widehat{x}_0$ and

$$(2.12) \qquad Ax_n + B\widehat{x}_n \longrightarrow 0, \quad B^* x_n + D\widehat{x}_n \longrightarrow 0, \qquad n \to \infty.$$

It follows that $(Ax_n, x_n) + (B\widehat{x}_n, x_n) \to 0$, $(B^* x_n, \widehat{x}_n) + (D\widehat{x}_n, \widehat{x}_n) \to 0$ for $n \to \infty$ and hence $(Ax_n, x_n) - \overline{(D\widehat{x}_n, \widehat{x}_n)} \to 0$. This implies $(Ax_n, x_n) \to 0$, $(D\widehat{x}_n, \widehat{x}_n) \to 0$ for $n \to \infty$, and the first statement follows from Lemma 2.2.

If A is bounded, then the first relation in (2.12) yields $B\widehat{x}_n \to 0$ and hence $\widehat{x}_0 \in \mathcal{D}(B)$, $B\widehat{x}_0 = 0$. It follows that $\begin{pmatrix} 0 \\ \widehat{x}_0 \end{pmatrix} \in \ker \widetilde{A}$ and hence also $\begin{pmatrix} x_0 \\ 0 \end{pmatrix} \in \ker \widetilde{A}$. If D is bounded, we get $x_0 \in \mathcal{D}(B^*)$, $B^* x_0 = 0$. If B is bounded, then $B\widehat{x}_n \to B\widehat{x}_0$ and from the first relation in (2.12) we get again $B\widehat{x}_0 = 0$, and, analogously, $B^* x_0 = 0$. It remains to observe that the assumptions (2.8) and (2.9) imply that $\ker A = \ker A^*$, $\ker D = \ker D^*$, and that

$$\widetilde{A}_0^* = \begin{pmatrix} A^* & B \\ B^* & D^* \end{pmatrix}$$

if B is bounded, or if A and D are bounded. □

3 The Self-adjoint Case

3.1 Now we consider the case that the operator $\widetilde{A}$ is self-adjoint, that is, that the operator $\widetilde{A}_0$ is essentially self-adjoint. Then, evidently, the components $A|_{\mathcal{D}(A)\cap\mathcal{D}(B^*)}$ and $D|_{\mathcal{D}(B)\cap\mathcal{D}(D)}$ are symmetric. We recall that the operators $A,\ B$, and D are assumed to be closed.

The following theorem contains, as a special case, the corresponding result of [AL] which was proved there without making use of a Krein space structure for the case that $\widetilde{A}_0$ is self-adjoint, $\sigma(D) < \sigma(A)$ and bounded B.

Theorem 3.1 *Suppose that the operator* $\widetilde{A}$ *is self-adjoint in the Hilbert space* $\widetilde{\mathcal{H}}$, $0 \notin \sigma_p(\widetilde{A})$, *and that the relations*

$$(3.1) \qquad \begin{gathered} (Ax, x) \geq 0, \quad (D\widehat{x}, \widehat{x}) \leq 0, \\ x \in \mathcal{D}(A) \cap \mathcal{D}(B^*), \ \widehat{x} \in \mathcal{D}(B) \cap \mathcal{D}(D), \end{gathered}$$

hold. Then with the spectral subspaces $\widetilde{\mathcal{L}}_+$ and $\widetilde{\mathcal{L}}_-$ of $\widetilde{A}$ corresponding to $[0, +\infty)$ *and* $(-\infty, 0]$, *respectively, the Hilbert space $\widetilde{\mathcal{H}}$ admits a decomposition $\widetilde{\mathcal{H}} = \widetilde{\mathcal{L}}_+ \dot{+} \widetilde{\mathcal{L}}_-$, and these subspaces can be represented as*

$$\widetilde{\mathcal{L}}_+ = \left\{ \begin{pmatrix} x \\ Kx \end{pmatrix} : x \in \mathcal{H} \right\}, \qquad \widetilde{\mathcal{L}}_- = \left\{ \begin{pmatrix} -K^*\widehat{x} \\ \widehat{x} \end{pmatrix} : \widehat{x} \in \widehat{\mathcal{H}} \right\}$$

with a contraction K from $\mathcal{H}$ into $\widehat{\mathcal{H}}$. If A and D are bounded or if B is bounded, and if the inequalities in (3.1) *are strengthened to*

$$\begin{aligned} &(Ax, x) > 0, \quad (D\widehat{x}, \widehat{x}) < 0, \quad x \in \mathcal{D}(A) \cap \mathcal{D}(B^*), \\ &\widehat{x} \in \mathcal{D}(B) \cap \mathcal{D}(D),\ x,\ \widehat{x} \neq 0, \end{aligned} \tag{3.2}$$

then the contraction K is strict; if the operator B is bounded and the inequalities in (3.1) *are further strengthened to*

$$\begin{aligned} &(Ax, x) \geq \alpha\, \|x\|^2, \quad (D\widehat{x}, \widehat{x}) \leq -\delta\, \|\widehat{x}\|^2, \\ &x \in \mathcal{D}(A),\ \widehat{x} \in \mathcal{D}(D), \end{aligned} \tag{3.3}$$

with positive numbers α, δ, then the contraction K is uniform.

Proof: It is well known that

$$\frac{1}{\pi \mathrm{i}} \int_{\mathrm{i}\mathbb{R}}' (\widetilde{A} - z)^{-1}\, \mathrm{d}z = \widetilde{P}_+ - \widetilde{P}_-,$$

where $\widetilde{P}_+$ and $\widetilde{P}_-$ are the orthogonal projections onto the spectral subspaces $\widetilde{\mathcal{L}}_+$ and $\widetilde{\mathcal{L}}_-$, respectively. If we observe the relations $\widetilde{J}\widetilde{A} = \overline{\widetilde{J}\widetilde{A}_0}$ and

$$\Re(\widetilde{J}\widetilde{A}_0\widetilde{x}, \widetilde{x}) = (Ax, x) - (D\widehat{x}, \widehat{x}) \geq 0, \quad \widetilde{x} = \begin{pmatrix} x \\ \widehat{x} \end{pmatrix} \in \mathcal{D}(\widetilde{A}_0),$$

the first and second assertion follow immediately from Theorem 1.4 and Corollary 1.5. If the operator B is bounded, $\widetilde{A}_0$ is closed and the self-adjointness of $\widetilde{A} = \widetilde{A}_0$ implies that A and D are self-adjoint. As in the Proof of Theorem 1.4 we use the relation

$$[\widetilde{x}, \widetilde{x}] = \frac{1}{\pi} \int_{\mathbb{R}}' \Re[\widetilde{A}(\widetilde{A} - \mathrm{i}t)^{-1}\widetilde{x}, (\widetilde{A} - \mathrm{i}t)^{-1}\widetilde{x}]\, \mathrm{d}t, \qquad \widetilde{x} \in \widetilde{\mathcal{L}}_+,$$

and estimate this integral from below.
With the Schur complements $S_A(z) := A - z - B(D - z)^{-1}B^*$ and $S_D(z) := D - z - B^*(A - z)^{-1}B$ for $z = \mathrm{i}t$ the integrand becomes

$$\begin{aligned} &(AS_A(\mathrm{i}t)^{-1}(x - B(D - \mathrm{i}t)^{-1}\widehat{x}), S_A(\mathrm{i}t)^{-1}(x - B(D - \mathrm{i}t)^{-1}\widehat{x}) \\ &-(DS_D(\mathrm{i}t)^{-1}(\widehat{x} - B^*(A - \mathrm{i}t)^{-1}x), \\ &S_D(\mathrm{i}t)^{-1}(\widehat{x} - B^*(A - \mathrm{i}t)^{-1}x). \end{aligned} \tag{3.4}$$

If we integrate this expression from $-\infty$ to $+\infty$, the integral will be not smaller than the integral from some t_0 to $+\infty$, the t_0 (> 0) will be chosen appropriately later. The first term in (3.4) can be written as $(Ay(t), y(t))$ where

$$
\begin{aligned}
y(t) &= ((A - \mathrm{i}t)^{-1} + (A - \mathrm{i}t)^{-1} B (D - \mathrm{i}t)^{-1} \\
&\qquad B^* S_A(\mathrm{i}t)^{-1})(x - B(D - \mathrm{i}t)^{-1}\widehat{x}) \\
&= (A - \mathrm{i}t)^{-1} x - (A - \mathrm{i}t)^{-1} B (D - \mathrm{i}t)^{-1} \widehat{x} + \cdots,
\end{aligned} \tag{3.5}
$$

and the second one as $-(Dv(t), v(t))$ with

$$
\begin{aligned}
v(t) &= ((D - \mathrm{i}t)^{-1} + (D - \mathrm{i}t)^{-1} B^* (A - \mathrm{i}t)^{-1} \\
&\qquad B S_D(\mathrm{i}t)^{-1})(\widehat{x} - B^*(A - \mathrm{i}t)^{-1} x) \\
&= (D - \mathrm{i}t)^{-1} \widehat{x} - (D - \mathrm{i}t)^{-1} B^* (A - \mathrm{i}t)^{-1} x + \cdots.
\end{aligned} \tag{3.6}
$$

We start with the 'leading terms' of the integrals:

$$
\begin{aligned}
a(t_0, x) &:= \int_{t_0}^{+\infty} (A(A - \mathrm{i}t)^{-1} x, (A - \mathrm{i}t)^{-1} x) \\
\mathrm{d}t &\geq \left(\frac{\pi}{2} - \operatorname{Arc\,tan} \frac{t_0}{\alpha}\right) \|x\|^2 =: \gamma_1^\alpha(t_0)\, \|x\|^2,
\end{aligned} \tag{3.7}
$$

$$
\begin{aligned}
d(t_0, \widehat{x}) &:= -\int_{t_0}^{+\infty} (D(D - \mathrm{i}t)^{-1} \widehat{x}, (D - \mathrm{i}t)^{-1} \widehat{x}) \\
\mathrm{d}t &\geq \left(\frac{\pi}{2} - \operatorname{Arc\,tan} \frac{t_0}{\delta}\right) \|\widehat{x}\|^2 =: \gamma_1^\delta(t_0)\, \|\widehat{x}\|^2.
\end{aligned} \tag{3.8}
$$

Observe that e.g.

$$
\frac{\pi}{2} - \operatorname{Arc\,tan} \frac{t}{\alpha} = \mathrm{O}(t^{-1}), \quad t \to +\infty.
$$

Further, for the first mixed terms from the last expressions in (3.5) we get

$$
\begin{aligned}
&\left| \int_{t_0}^{+\infty} (A(A - \mathrm{i}t)^{-1} x, (A - \mathrm{i}t)^{-1} B (D - \mathrm{i}t)^{-1} \widehat{x}) \mathrm{d}t \right| \\
&\quad \leq a(t_0, x)^{1/2} \left(\int_{t_0}^{+\infty} (A(A - \mathrm{i}t)^{-1} B (D - \mathrm{i}t)^{-1} \widehat{x}, \right. \\
&\qquad \left. (A - \mathrm{i}t)^{-1} B (D - \mathrm{i}t)^{-1} \widehat{x})\, \mathrm{d}t \right)^{1/2} \\
&\quad \leq a(t_0, x)^{1/2} \left(\frac{\|B\|^2}{2} \int_{t_0}^{\infty} \frac{1}{t} \|(D - \mathrm{i}t)^{-1} \widehat{x}\|^2 \mathrm{d}t \right)^{1/2}
\end{aligned} \tag{3.9}
$$

$$\leq a(t_0, x)^{1/2} \frac{\|B\|}{\sqrt{2}} \frac{1}{\delta} \left(\ln \frac{\sqrt{t_0^2 + \delta^2}}{t_0} \right)^{1/2} \|\widehat{x}\|$$

$$=: a(t_0, x)^{1/2} \gamma_2^\delta(t_0) \|\widehat{x}\|.$$

Here we have used the Cauchy-Schwarz inequality, the estimate

$$\|(A + \mathrm{i}t)^{-1} A (A - \mathrm{i}t)^{-1}\| \leq \frac{1}{2t}, \qquad t > 0,$$

and, with $(E_\lambda)_{\lambda \in \mathbb{R}}$ denoting the spectral function of the self-adjoint operator D, the estimate

$$\begin{aligned}
\int_{t_0}^{\infty} \frac{1}{t} \|(D - \mathrm{i}t)^{-1} \widehat{x}\|^2 \mathrm{d}t &= \int_{t_0}^{\infty} \frac{1}{t} \int_{-\infty}^{-\delta} \frac{1}{\lambda^2 + t^2} \mathrm{d}(E_\lambda \widehat{x}, \widehat{x}) \, \mathrm{d}t \\
&= \int_{-\infty}^{-\delta} \int_{t_0}^{\infty} \frac{1}{t} \frac{1}{\lambda^2 + t^2} \mathrm{d}t \mathrm{d}(E_\lambda \widehat{x}, \widehat{x}) \\
&= \int_{-\infty}^{-\delta} \frac{1}{\lambda^2} \ln \frac{\sqrt{t_0^2 + \lambda^2}}{t_0} \mathrm{d}(E_\lambda \widehat{x}, \widehat{x}) \\
&\leq \frac{1}{\delta^2} \ln \frac{\sqrt{t_0^2 + \delta^2}}{t_0} \|\widehat{x}\|^2.
\end{aligned}$$

In an analogous way, for the first mixed terms from the last expressions in (3.6) one can show that

$$\left| \int_{t_0}^{+\infty} (D(D - \mathrm{i}t)^{-1} \widehat{x}, (D - \mathrm{i}t)^{-1} B^* (A - \mathrm{i}t)^{-1} x) \mathrm{d}t \right|$$

$$\leq d(t_0, \widehat{x})^{1/2} \frac{\|B\|}{\sqrt{2}} \frac{1}{\alpha} \left(\ln \frac{\sqrt{t_0^2 + \alpha^2}}{t_0} \right)^{1/2} \|x\| \tag{3.10}$$

$$=: d(t_0, \widehat{x})^{1/2} \gamma_2^\alpha(t_0) \|x\|.$$

We now consider, with an arbitrary positive $c_1 < 1$, the difference of $c_1(a(t_0, x) + d(t_0, \widehat{x}))$ and the two mixed terms estimated in (3.9) and in (3.10). It is greater than or equal to

$$c_1 a(t_0, x) - a(t_0, x)^{1/2} \gamma_2^\delta(t_0) \|\widehat{x}\| + c_1 d(t_0, \widehat{x}) - d(t_0, \widehat{x})^{1/2} \gamma_2^\alpha(t_0) \|x\|,$$

which, in turn, using e.g. the inequality

$$(a(t_0, x))^{1/2} \gamma_2^\delta(t_0) \|\widehat{x}\| \leq \frac{1}{2} \left(c_1 a(t_0, x) + \frac{1}{c_1} (\gamma_2^\delta(t_0))^2 \|\widehat{x}\|^2 \right),$$

can be estimated from below by

$$\frac{1}{2}(c_1(a(t_0,x)+d(t_0,\widehat{x}))-c_1^{-1}\gamma_2^{\delta}(t_0)^2\|\widehat{x}\|^2-c_1^{-1}\gamma_2^{\alpha}(t_0)^2\|x\|^2).$$

If we now use the inequalities (3.7) and (3.8), we find that a lower bound for this is given by

$$\begin{aligned}&\frac{1}{2}(c_1\gamma_1^{\alpha}(t_0)-c_1^{-1}\gamma_2^{\alpha}(t_0)^2)\|x\|^2+\frac{1}{2}(c_1\gamma_1^{\delta}(t_0)-c_1^{-1}\gamma_2^{\delta}(t_0)^2)\|\widehat{x}\|^2\\&\quad=:\gamma(t_0)\,\|x\|^2+\widehat{\gamma}(t_0)\,\|\widehat{x}\|^2.\end{aligned}$$

It is not difficult to see that $\gamma(t_0),\widehat{\gamma}(t_0)\searrow 0$ if $t_0\to+\infty$. Indeed, e.g.

$$\begin{aligned}\gamma(t_0)&=c_1\left(\frac{\pi}{2}-\operatorname{Arc\,tan}\frac{t_0}{\alpha}\right)-\frac{\|B\|^2}{2\alpha^2c_1}\ln\frac{\sqrt{t_0^2+\alpha^2}}{t_0}\\&=c_1\left(\frac{\pi}{2}-\operatorname{Arc\,tan}\frac{t_0}{\alpha}\right)\left(1-\frac{\|B\|^2\ln\frac{\sqrt{t_0^2+\alpha^2}}{t_0}}{2\alpha^2c_1^2\left(\frac{\pi}{2}-\operatorname{Arc\,tan}\frac{t_0}{\alpha}\right)}\right),\end{aligned}$$

and it is easy to check that the first factor tends to 0 from above, while the second factor tends to 1 for $t_0\to\infty$. Hence for all sufficiently large $t_0\in[0,\infty)$ we find $\gamma(t_0),\widehat{\gamma}(t_0)>0$.

If we use the estimates $\|A(A-\mathrm{i}t)^{-1}\|\le 1$, $\|(A-\mathrm{i}t)^{-1}\|$, $\|(D-\mathrm{i}t)^{-1}\|\le|t|^{-1}$, and $\|S_A(\mathrm{i}t)^{-1}\|\le\kappa\,|t|^{-1}$ for some number κ, it follows that the other integrals are of the size $\mathrm{O}(t_0^{-2})$, multiplied by $\|x\|^2$, $\|x\|\,\|\widehat{x}\|$ or $\|\widehat{x}\|^2$. Now, we can proceed in a similar way as above and estimate these terms against a multiple ($\le 1-c_1$) of the leading terms. Altogether, this yields that there exist a $t_0>0$ and a $\gamma>0$ such that

$$[\widetilde{x},\widetilde{x}]\ge\frac{1}{\pi}\int_{t_0}^{\infty}\Re[\widetilde{A}(\widetilde{A}-\mathrm{i}t)^{-1}\widetilde{x},(\widetilde{A}-\mathrm{i}t)^{-1}\widetilde{x}]\,dt\ge\gamma\,\|\widetilde{x}\|^2.$$ □

Under the assumptions of Theorem 3.1 we denote $\widetilde{\mathcal{D}}_\pm:=\mathcal{D}(\widetilde{A}_0)\cap\widetilde{\mathcal{L}}_\pm$ and

$$\begin{aligned}\mathcal{D}_+&:=\left\{x\in\mathcal{H}:\begin{pmatrix}x\\Kx\end{pmatrix}\in\widetilde{\mathcal{D}}_+\right\}\\&=\{x\in\mathcal{H}:x\in\mathcal{D}(A)\cap\mathcal{D}(B^*),\ Kx\in\mathcal{D}(B)\cap\mathcal{D}(D)\},\end{aligned}\tag{3.11}$$

$$\begin{aligned}\mathcal{D}_-&:=\left\{\widehat{x}\in\widehat{\mathcal{H}}:\begin{pmatrix}-K^*\widehat{x}\\\widehat{x}\end{pmatrix}\in\widetilde{\mathcal{D}}_-\right\}\\&=\{\widehat{x}\in\widehat{\mathcal{H}}:\widehat{x}\in\mathcal{D}(D)\cap\mathcal{D}(B),\ K^*\widehat{x}\in\mathcal{D}(A)\cap\mathcal{D}(B^*)\}.\end{aligned}\tag{3.12}$$

In general, apparently, it is not clear if these sets are sufficiently large.

Corollary 3.2 *The angular operator K satisfies the Riccati type equation*

$$KBKx + KAx - DKx - B^*x = 0, \qquad x \in \mathcal{D}_+.$$

The operator $\widetilde{A}|_{\widetilde{\mathcal{D}}_+}$ in $(\widetilde{\mathcal{L}}_+, (\cdot,\cdot))$ is isomorphic to the operator $(A+BK)|_{\mathcal{D}_+}$ in the space $(\mathcal{H}, (G\cdot,\cdot))$ with $G := I + K^*K$. The operator $\widetilde{A}|_{\widetilde{\mathcal{D}}_-}$ in $(\widetilde{\mathcal{L}}_-, (\cdot,\cdot))$ is isomorphic to the operator $(D - B^*K^*)|_{\mathcal{D}_-}$ in the space $(\widehat{\mathcal{H}}, (\widehat{G}\cdot,\cdot))$ with $\widehat{G} := I + KK^*$. On the set $\mathcal{D}_+ \oplus \mathcal{D}_-$ the block operator matrix $\widetilde{A}$ admits a 'diagonalization'

$$\begin{pmatrix} I & -K^* \\ K & I \end{pmatrix}^{-1} \begin{pmatrix} A & B \\ B^* & D \end{pmatrix} \begin{pmatrix} I & -K^* \\ K & I \end{pmatrix} \begin{pmatrix} x \\ \widehat{x} \end{pmatrix}$$
$$= \begin{pmatrix} A+BK & 0 \\ 0 & D - B^*K^* \end{pmatrix} \begin{pmatrix} x \\ \widehat{x} \end{pmatrix}, \quad x \in \mathcal{D}_+, \, \widehat{x} \in \mathcal{D}_-.$$

Clearly, e.g. the isomorphism between $(\widetilde{\mathcal{L}}_+, (\cdot,\cdot))$ and $(\mathcal{H}, (G\cdot,\cdot))$ is given by

$$\widetilde{\mathcal{L}}_+ \ni \begin{pmatrix} x \\ Kx \end{pmatrix} \longleftrightarrow x \in \mathcal{H}.$$

Remark 3.3 If in Theorem 3.1 the assumptions (3.3) are satisfied, then

$$\{z \in \mathbb{C} : -\delta < \Re(z) < \alpha\} \subset \rho(\widetilde{A}).$$

Remark 3.4 Under the stronger conditions (3.2) or (3.3) the assumption $0 \notin \sigma_p(\widetilde{A})$ in Theorem 3.1 is trivially satisfied. This assumption can be avoided if $\ker \widetilde{A}$ has the splitting property (see (2.11)).

Indeed, then the subspace $\widetilde{\mathcal{L}}_+$ is chosen to be the spectral subspace of $\widetilde{A}$ corresponding to $(0, \infty)$ and augmented by the subspace $\left\{\begin{pmatrix} x_0 \\ 0 \end{pmatrix} : \begin{pmatrix} x_0 \\ \widehat{x}_0 \end{pmatrix} \in \ker \widetilde{A}_0\right\}$, and the subspace $\widetilde{\mathcal{L}}_-$ is chosen to be the spectral subspace of $\widetilde{A}$ corresponding to $(-\infty, 0)$ and augmented by the subspace $\left\{\begin{pmatrix} 0 \\ \widehat{x}_0 \end{pmatrix} : \begin{pmatrix} x_0 \\ \widehat{x}_0 \end{pmatrix} \in \ker \widetilde{A}_0\right\}$. Thus, according to Theorem 2.3, the assumption $0 \notin \sigma_p(\widetilde{A})$ can be avoided if at least one of the operators $A,\ B,\ D$ is bounded.

Remark 3.5 We conjecture, that if in Theorem 3.1 the operators A and D are bounded and the assumptions (3.3) are satisfied, one has

$$\mathcal{D}(\widetilde{A}|_{\widetilde{\mathcal{L}}_+}) = \left\{\begin{pmatrix} x \\ Kx \end{pmatrix} : x \in \mathcal{D}(B^*)\right\},$$

that is, K maps $\mathcal{D}(B^*)$ into $\mathcal{D}(B)$. This holds true, e.g., in the example of the abstract Dirac operator, which will be considered in the next subsection.

3.2 As an example for the self-adjoint case we consider the abstract Dirac operator with supersymmetry (see [T]). In this case the operator K can be calculated explicitly. We shall do this under the assumptions that in (2.2) the operators A and $-D$ are positive, bounded and boundedly invertible, and that B is a closed operator. Then also the operator $\widetilde{A}_0$, defined on $\mathcal{D}(B^*) \oplus \mathcal{D}(B)$, is closed: $\widetilde{A}_0 = \widetilde{A}$. Recall that the supersymmetry assumption means that

$$(3.13) \qquad D(\mathcal{D}(B)) \subset \mathcal{D}(B), \qquad B\,Dx = -A\,Bx, \qquad x \in \mathcal{D}(B),$$

$$(3.14) \qquad A(\mathcal{D}(B^*)) \subset \mathcal{D}(B^*), \qquad B^*Ax = -D\,B^*x, \qquad x \in \mathcal{D}(B^*).$$

As a first consequence of these assumptions, the (closed) subspace $\ker B^*$ is invariant under A and hence the subspace $\widetilde{\mathcal{L}}_0 := (\ker B^*) \oplus \{0\}$ of $\widetilde{\mathcal{H}} = \mathcal{H} \oplus \widehat{\mathcal{H}}$ is invariant under $\widetilde{A}$. If $y \in \ker B^*$ then $\left(\widetilde{A}\begin{pmatrix} y \\ 0 \end{pmatrix}, \begin{pmatrix} y \\ 0 \end{pmatrix}\right) = (Ay, y) \geq 0$, therefore $\widetilde{\mathcal{L}}_0$ is contained in the spectral subspace of $\widetilde{A}$ corresponding to the positive half axis. In the following description of this spectral subspace we can split it off, that is, we replace $\mathcal{H}$ by $\mathcal{H} \ominus \ker B^*$, or in other words, we suppose that $\ker B^* = (\operatorname{ran} B)^{\perp} = \{0\}$. A similar reasoning applies for $\ker B$, hence we can suppose that the operators B and B^* are injective or, equivalently, that $\operatorname{ran} B$ and $\operatorname{ran} B^*$ are dense.

The assumptions (3.13) and (3.14) further imply that

$$(3.15) \quad A\mathcal{D}(BB^*) \subset \mathcal{D}(BB^*), \quad BB^*Ax = ABB^*x, \quad x \in \mathcal{D}(BB^*).$$

Indeed, if $x \in \mathcal{D}(BB^*)$, then (3.14) yields $B^*Ax = -DB^*x$. Since $B^*x \in \mathcal{D}(B)$ it now follows from (3.13) that $BB^*Ax = -BDB^*x = ABB^*x$.

We introduce the operator

$$(3.16) \qquad K := B^*(A + (A^2 + BB^*)^{1/2})^{-1}.$$

Lemma 3.6 *The operator K has the properties*

$$\mathcal{D}(K) = \mathcal{H}, \quad K\mathcal{D}(B^*) \subset \mathcal{D}(B) \quad \text{and} \quad \|Kx\| < \|x\| \text{ for } x \in \mathcal{H},\ x \neq 0.$$

Proof: Since the operator A is strictly positive, the inverse in (3.16) is bounded. With the theory of quadratic forms (see, e.g., [K], Chapter VI) it is easy to see that

$$(3.17) \qquad \mathcal{D}(B^*) = \mathcal{D}((BB^*)^{1/2}) = \mathcal{D}((A^2 + BB^*)^{1/2}).$$

If $x \in \mathcal{D}(B^*)$, the relation (3.17) implies $x \in \mathcal{D}(A + (A^2 + BB^*)^{1/2})$ and hence

$$(A + (A^2 + BB^*)^{1/2})^{-1}x \in \mathcal{D}((A + (A^2 + BB^*)^{1/2})^2) = \mathcal{D}(BB^*),$$

which yields $Kx \in \mathcal{D}(B)$. Finally, for $y \in \mathcal{D}(B^*)$, $y \neq 0$, we have

$$\|B^*y\|^2 < \|(A + (A^2 + BB^*)^{1/2})y\|^2,$$

and with $x = (A + (A^2 + BB^*)^{1/2})y$ it follows that K is a strict contraction. $\square$

We introduce the subspace

$$\widetilde{\mathcal{L}}_+ := \left\{ \begin{pmatrix} x \\ Kx \end{pmatrix} : x \in \mathcal{H} \right\}, \tag{3.18}$$

and we shall show that it has the following properties:

(i) $\widetilde{\mathcal{L}}_+ \cap \mathcal{D}(\widetilde{A}) = \left\{ \begin{pmatrix} x \\ Kx \end{pmatrix} : x \in \mathcal{D}(B^*) \right\}.$

(ii) $\widetilde{A}(\widetilde{\mathcal{L}}_+ \cap \mathcal{D}(\widetilde{A})) \subset \widetilde{\mathcal{L}}_+$.

(iii) $\widetilde{\mathcal{L}}_+$ is the spectral subspace of $\widetilde{A}$ corresponding to $(0, \infty)$.

The property (i) is an immediate consequence of Lemma 3.6. The property (ii) will be proved if we show that the relation

$$K(A + BK)x = (B^* + DK)x, \quad x \in \mathcal{D}(B^*), \tag{3.19}$$

holds. To this end we introduce the operator $F := (A^2 + BB^*)^{1/2}$ and start from the identity

$$(A - F)(A + F)y + BB^*y = 0, \quad y \in \mathcal{D}(BB^*).$$

If $x \in \mathcal{D}(B^*)$ and $y := (A + F)^{-1}x$, it follows that

$$(A - F)x + BB^*(A + F)^{-1}x = 0,$$

and hence

$$\begin{aligned} (A + BB^*(A + F)^{-1})x &= (A + F)x - Ax, \\ (A + F)^{-1}(A + BB^*(A + F)^{-1})x &= x - A(A + F)^{-1}x, \\ B^*(A + F)^{-1}(A + BK)x &= B^*x - B^*A(A + F)^{-1}x, \end{aligned}$$

which is, due to (3.14), the relation (3.19). In order to prove (iii), we show that

$$(\widetilde{A}\widetilde{x}, \widetilde{x}) \geq 0, \quad \widetilde{x} \in \mathcal{D}(\widetilde{A}) \cap \widetilde{\mathcal{L}}_+, \tag{3.20}$$

and that with the orthogonal complement

$$\widetilde{\mathcal{L}}_- := \widetilde{\mathcal{L}}_+^{\perp} = \left\{ \begin{pmatrix} -K^*\widehat{x} \\ \widehat{x} \end{pmatrix} : \widehat{x} \in \widehat{\mathcal{H}} \right\} \tag{3.21}$$

we have

$$(\widetilde{A}\widetilde{x}, \widetilde{x}) \leq 0 \quad \widetilde{x} \in \mathcal{D}(\widetilde{A}) \cap \widetilde{\mathcal{L}}_-.$$

If we observe (3.19), then the relation (3.20) can be written as

$$\begin{aligned} &\left(\begin{pmatrix} A & B \\ B^* & D \end{pmatrix} \begin{pmatrix} x \\ Kx \end{pmatrix}, \begin{pmatrix} x \\ Kx \end{pmatrix} \right) \\ &= ((I + KK^*)(A + BK)x, x) \geq 0, \quad x \in \mathcal{D}(B^*), \end{aligned} \tag{3.22}$$

and it is not hard to see that for $x \in \mathcal{D}(B^*)$ all the numbers (Ax, x), (K^*KAx, x), (BKx, x), and (K^*KBKx, x) are nonnegative. For the first term this is trivial since A is positive, for the second term it follows from the fact that $K^*K = (A+F)^{-1}BB^*(A+F)^{-1}$ and that this operator commutes with A and therefore also with $A^{1/2}$. Further, $BKx = (-A + (A^2 + BB^*)^{1/2})x$ for $x \in \mathcal{D}(B^*)$, and $(BKx, x) \geq 0$ follows. Finally, $K^*KBK = K^*B^*(A + (A^2 + BB^*)^{1/2})^{-1}BK$ and this is, evidently, also a nonnegative operator.

The relation (3.13) implies $BD^2 = A^2B$ and $B(D^2 + B^*B)^{1/2} = (A^2 + BB^*)^{1/2}B$, hence

$$-B(D - (D^2 + B^*B)^{1/2}) = (A + (A^2 + BB^*)^{1/2})B,$$

and for K^* we obtain

$$K^* = (A + (A^2 + BB^*)^{1/2})^{-1}B = B(-D + (D^2 + B^*B)^{1/2})^{-1}.$$

This operator is completely analogous to K, with A replaced by $-D$ and B replaced by B^*. Therefore it follows in the same way that the subspace $\widetilde{\mathcal{L}}_-$ is also $\widetilde{A}$-invariant and that $\widetilde{A}$ is nonpositive there. But then $\widetilde{\mathcal{L}}_+$ must be the spectral subspace of $\widetilde{A}$ corresponding to the positive half axis and $\widetilde{\mathcal{L}}_-$ the spectral subspace of $\widetilde{A}$ corresponding to the negative half axis, and the statements (i)–(iii) are proved.

The relations (3.19) or (3.22) imply that the restriction $\widetilde{A}|_{\widetilde{\mathcal{L}}_+}$ is isomorphic to the operator $A + BK$ in the space $\mathcal{H}$ if the latter is equipped with the Hilbert space inner product $((I + K^*K)\cdot, \cdot)$, and, analogously, the restriction $\widetilde{A}|_{\widetilde{\mathcal{L}}_-}$ is isomorphic to the operator $D - B^*K^*$ in the space $\widehat{\mathcal{H}}$ if the latter is equipped with the Hilbert space inner product $((I + KK^*)\cdot, \cdot)$. In other words, the block operator matrix $\widetilde{A}$ can be diagonalized as follows:

$$\begin{pmatrix} A & B \\ B^* & D \end{pmatrix}\begin{pmatrix} I & -K^* \\ K & I \end{pmatrix} = \begin{pmatrix} I & -K^* \\ K & I \end{pmatrix}\begin{pmatrix} A + BK & 0 \\ 0 & D - B^*K^* \end{pmatrix}.$$

The transforming matrix can also be chosen unitary. With $\Gamma := (I + K^*K)^{-1/2}$, $\Delta := (I + KK^*)^{-1/2}$ this relation becomes:

$$\begin{aligned}&\begin{pmatrix} A & B \\ B^* & D \end{pmatrix}\begin{pmatrix} \Gamma & -K^*\Delta \\ K\Gamma & \Delta \end{pmatrix}\\ &= \begin{pmatrix} \Gamma & -K^*\Delta \\ K\Gamma & \Delta \end{pmatrix}\begin{pmatrix} \Gamma^{-1}(A + BK)\Gamma & 0 \\ 0 & \Delta^{-1}(D - B^*K^*)\Delta \end{pmatrix}.\end{aligned}$$

This coincides with the so-called Foldy-Wouthuysen transformation well-known in quantum mechanics (see e.g. [T]).

The case that $A = D = 0$ was excluded in this subsection. In this case it can easily be checked directly that the spectral subspaces $\widetilde{\mathcal{L}}_+$ and $\widetilde{\mathcal{L}}_-$ of

$$\widetilde{A} = \begin{pmatrix} 0 & B \\ B^* & 0 \end{pmatrix},$$

corresponding to the positive and the negative half axis, have representations as in (3.18) and (3.21) with $K = V$ where V is the partial isometry in the polar decomposition $B^* = V(BB^*)^{1/2}$ of B^*.

4 The Non-self-adjoint Case

4.1 We introduce the following assumptions about the operators A and D in (2.2) which combine the assumptions (2.7) and (2.8), (2.9):

$$W(A) \subset \{z \in \mathbb{C} : |\arg z| \le \Theta_A,\ \Re(z) \ge \alpha\}, \tag{4.1}$$

$$W(D) \subset \{z \in \mathbb{C} : |\arg z| \ge \pi - \Theta_D,\ \Re(z) \le -\delta\} \tag{4.2}$$

for some $\Theta_A,\ \Theta_D \in [0, \frac{\pi}{2})$ and $\alpha,\ \delta > 0$. Then e.g. the additional assumption that A and D are boundedly invertible implies that A and D are m-sectorial.

Theorem 4.1 *Suppose that the operators A and D satisfy the assumptions* (4.1) *and* (4.2), *respectively, that they are boundedly invertible and that the operator B satisfies the following conditions:*

b$_1$) $\mathcal{D}(D) \subset \mathcal{D}(B),\ \mathcal{D}(A) \subset \mathcal{D}(B^*)$,

b$_2$) *there exist positive numbers* $\gamma,\ \eta$ *such that for* $t \in \mathbb{R}$

$$\|B(D - \mathrm{i}t)^{-1}\| \le \frac{\gamma}{1 + |t|^\eta}, \qquad \|B^*(A - \mathrm{i}t)^{-1}\| \le \frac{\gamma}{1 + |t|^\eta}.$$

Then $\mathrm{i}\mathbb{R} \subset \rho(\widetilde{\mathrm{A}})$,

$$\frac{1}{\pi \mathrm{i}} \int_{\mathrm{i}\mathbb{R}}' (\widetilde{A} - z)^{-1} \mathrm{d}z = \widetilde{P}_+ - \widetilde{P}_- \tag{4.3}$$

with two complementary projections $\widetilde{P}_\pm$ *such that the corresponding subspaces* $\widetilde{\mathcal{L}}_+ := \widetilde{P}_+\widetilde{\mathcal{K}}$ *and* $\widetilde{\mathcal{L}}_- := \widetilde{P}_-\widetilde{\mathcal{K}}$, *respectively, can be represented as*

$$\widetilde{\mathcal{L}}_+ = \left\{ \begin{pmatrix} x \\ K_+ x \end{pmatrix} : x \in \mathcal{H} \right\}, \qquad \widetilde{\mathcal{L}}_- = \left\{ \begin{pmatrix} K_- \widehat{x} \\ \widehat{x} \end{pmatrix} : \widehat{x} \in \widehat{\mathcal{H}} \right\}$$

where $K_+ \in L(\mathcal{H}, \widehat{\mathcal{H}})$ *and* $K_- \in L(\widehat{\mathcal{H}}, \mathcal{H})$ *are strict contractions. In particular,* $\widetilde{\mathcal{H}} = \widetilde{\mathcal{L}}_+ \dot{+} \widetilde{\mathcal{L}}_-$, *the relations* $\widetilde{\mathcal{L}}_\pm \cap \mathcal{D}(\widetilde{A}) = \widetilde{P}_\pm \mathcal{D}(\widetilde{A})$, $\widetilde{A}(\widetilde{\mathcal{L}}_\pm \cap \mathcal{D}(\widetilde{A})) \subset \widetilde{\mathcal{L}}_\pm$ *hold, and*

$$\begin{aligned} \widetilde{\mathcal{L}}_+ \cap \mathcal{D}(\widetilde{A}) &= \left\{ \begin{pmatrix} x \\ K_+ x \end{pmatrix} : x \in \mathcal{D}(A) \right\}, \\ \widetilde{\mathcal{L}}_- \cap \mathcal{D}(\widetilde{A}) &= \left\{ \begin{pmatrix} K_- \widehat{x} \\ \widehat{x} \end{pmatrix} : \widehat{x} \in \mathcal{D}(D) \right\}. \end{aligned} \tag{4.4}$$

Proof: We apply Theorem 1.3 to the block operator matrices

$$\widetilde{S} = \begin{pmatrix} A & 0 \\ 0 & D \end{pmatrix}, \qquad \widetilde{T} = \begin{pmatrix} 0 & B \\ B^* & 0 \end{pmatrix}.$$

Since the inclusions (4.1), (4.2) hold and the operators A and D are boundedly invertible, it follows that A and D satisfy the assumptions (i) and (ii) of Theorem 1.1, in particular $\mathrm{i}\mathbb{R} \subset \rho(\mathrm{A}) \cap \rho(\mathrm{D})$, and it is easy to see that

$$\frac{1}{\pi i} \int_{\mathrm{i}\mathbb{R}}' (A - z)^{-1} \mathrm{d}z = I, \qquad \frac{1}{\pi i} \int_{\mathrm{i}\mathbb{R}}' (D - z)^{-1} \mathrm{d}z = -I.$$

Therefore also the operator $\widetilde{S}$ fulfills the assumptions (i) and (ii) of Theorem 1.1, the integral $\frac{1}{\pi \mathrm{i}} \int_{\mathrm{i}\mathbb{R}}' (\widetilde{S} - z)^{-1} \mathrm{d}z$ exists and

$$\frac{1}{\pi \mathrm{i}} \int_{\mathrm{i}\mathbb{R}}' (\widetilde{S} - z)^{-1} \mathrm{d}z = \begin{pmatrix} I & 0 \\ 0 & -I \end{pmatrix}.$$

According to the assumptions b_1) and b_2) the operator $\widetilde{A} = \widetilde{S} + \widetilde{T}$ is a closed operator on its domain $\mathcal{D}(A) \oplus \mathcal{D}(D)$, and sufficiently large points on the imaginary axis belong to $\rho(\widetilde{A})$, hence Theorem 2.1 implies that $\mathrm{i}\mathbb{R} \subset \rho(\widetilde{\mathrm{A}})$, and all assumptions of Theorem 1.3 are satisfied, which yields the existence of the invariant subspaces $\widetilde{\mathcal{L}}_{\pm}$ and their properties.

In order to show e.g. the first equality in (4.4), we observe that the relation $(\widetilde{A} - z)^{-1} \widetilde{\mathcal{L}}_+ = \widetilde{\mathcal{L}}_+ \cap \mathcal{D}(\widetilde{A})$ holds for $z \in \rho(\widetilde{A})$ (see (1.2)). Therefore it is sufficient to prove that

$$Q_+ (\widetilde{A} - z)^{-1} \widetilde{\mathcal{L}}_+ = \mathcal{D}(A) \tag{4.5}$$

for some $z \in \rho(\widetilde{A})$, where Q_+ denotes the orthogonal projection in $\widetilde{\mathcal{H}}$ onto $\mathcal{H}$. On the left hand side of (4.5) there is the set of all elements $(A - z - B(D - z)^{-1} B^*)^{-1} (I - B(D - z)^{-1} K_+) x$ with $x \in \mathcal{H}$. If we choose z on the imaginary axis such that $\| B(D - z)^{-1} \| < 1$, then the operator $I - B(D - z)^{-1} K_+$ maps $\mathcal{H}$ bijectively onto itself. Further, since $z \in \rho(\widetilde{A})$, the operator $(A - z - B(D - z)^{-1} B^*)^{-1}$ maps $\mathcal{H}$ bijectively onto $\mathcal{D}(A)$. □

We mention that all the assumptions about B in Theorem 4.1 are satisfied if B is a bounded operator. Therefore Theorem 4.1 generalizes the main result of [LT], where the operators B and D were supposed to be bounded. In addition to [LT] where it has been proved that $\| K_+ K_- \|$, $\| K_- K_+ |$ < 1, Theorem 4.1 shows that in fact K_+ and K_- are strict contractions.

Let G be a bounded selfadjoint operator in a Hilbert space $(\mathcal{H}, (\cdot, \cdot))$. The operator V in $\mathcal{H}$ is called *G-accretive* if

$$\Re\,(GVx, x) \geq 0 \quad x \in \mathcal{D}(V).$$

Corollary 4.2 *The Riccati equations*

$$K_+BK_+x + K_+Ax - DK_+x - B^*x = 0, \quad x \in \mathcal{D}(A), \tag{4.6}$$

$$K_-B^*K_-\widehat{x} + K_-D\widehat{x} - AK_-\widehat{x} - B\widehat{x} = 0, \quad \widehat{x} \in \mathcal{D}(D), \tag{4.7}$$

hold. Moreover, the operator $\widetilde{A}|_{\widetilde{\mathcal{L}}_+\cap\mathcal{D}(\widetilde{A})}$ *in* $(\widetilde{\mathcal{L}}_+, [\cdot,\cdot])$ *is isomorphic to the operator* $A + BK_+|_{\mathcal{D}(A)}$ *in* $(\mathcal{H}, (G\cdot,\cdot))$ *with* $G := I - K_+^*K_+$*, in particular,* $A + BK$ *is G-accretive in* $\mathcal{H}$*. The operator* $\widetilde{A}|_{\widetilde{\mathcal{L}}_-\cap\mathcal{D}(\widetilde{A})}$ *in* $(\widetilde{\mathcal{L}}_-, [\cdot,\cdot])$ *is isomorphic to the operator* $D + B^*K_-|_{\mathcal{D}(D)}$ *in* $(\widehat{\mathcal{H}}, (\widehat{G}\cdot,\cdot))$ *with* $\widehat{G} := I - K_-^*K_-$*, in particular,* $D + B^*K_-$ *is* $\widehat{G}$*-accretive in* $\widehat{\mathcal{H}}$*.*

Proof: The Riccati equations (4.6) and (4.7) follow from the invariance of the subspaces $\widetilde{\mathcal{L}}_\pm$ under the operator $\widetilde{A}$. Further, using (4.6) we find

$$\begin{aligned}\left[\widetilde{A}\begin{pmatrix} x \\ K_+x\end{pmatrix}, \begin{pmatrix} x \\ K_+x\end{pmatrix}\right] &= ((A + BK_+ - K_+^*(B^* + DK_+))x, x)\\ &= ((I - K_+^*K_+)(A + BK_+)x, x), \quad x \in \mathcal{D}(A).\end{aligned}$$

The real part of this expression is given by $\Re\,(Ax, x) - \Re\,(DK_+x, K_+x)$ which is nonnegative by the assumptions (4.1) and (4.2). □

4.2 In Theorem 4.1 the off-diagonal operators B, B^* were supposed to be 'small' with respect to the diagonal operators. In other examples, connected e.g. with the Dirac operator, the domain of $\widetilde{A}$ is determined by the operators B and B^*. For this situation we prove the following theorem.

Theorem 4.3 *Suppose that the operator B is boundedly invertible, that A and D satisfy the conditions* (4.1) *and* (4.2), *respectively, that*

a₁) $\mathcal{D}(B^*) \subset \mathcal{D}(A)$,

d₁) $\mathcal{D}(B) \subset \mathcal{D}(D)$,

and that there exist positive numbers γ, η *such that for* $t > 0$

a₂)
$$\|A(BB^* + t^2)^{-1}\| \le \frac{\gamma}{1 + |t|^{1+\eta}},$$
$$\|AB(B^*B + t^2)^{-1}\| \le \frac{\gamma}{1 + |t|^{\eta}},$$

d₂)
$$\|D(B^*B + t^2)^{-1}\| \le \frac{\gamma}{1 + |t|^{1+\eta}},$$
$$\|DB^*(BB^* + t^2)^{-1}\| \le \frac{\gamma}{1 + |t|^{\eta}}.$$

Then $i\mathbb{R} \subset \rho(\widetilde{A})$,

$$\frac{1}{\pi i}\int_{i\mathbb{R}}'(\widetilde{A}-z)^{-1}dz = \widetilde{P}_+ - \widetilde{P}_- \tag{4.8}$$

with two complementary projections $\widetilde{P}_\pm$ *such that the corresponding subspaces* $\widetilde{\mathcal{L}}_+ := \widetilde{P}_+\widetilde{\mathcal{K}}$ *and* $\widetilde{\mathcal{L}}_- := \widetilde{P}_-\widetilde{\mathcal{K}}$, *can be represented as*

$$\widetilde{\mathcal{L}}_+ = \left\{\begin{pmatrix} x \\ K_+x \end{pmatrix} : x \in \mathcal{H}\right\}, \quad \widetilde{\mathcal{L}}_- = \left\{\begin{pmatrix} K_-\widehat{x} \\ \widehat{x} \end{pmatrix} : \widehat{x} \in \widehat{\mathcal{H}}\right\}$$

where $K_+ \in L(\mathcal{H}, \widehat{\mathcal{H}})$ *and* $K_- \in L(\widehat{\mathcal{H}}, \mathcal{H})$ *are strict contractions,* $\widetilde{\mathcal{H}} = \widetilde{\mathcal{L}}_+ \dot{+} \widetilde{\mathcal{L}}_-$, *and the relations* $\widetilde{\mathcal{L}}_\pm \cap \mathcal{D}(\widetilde{A}) = \widetilde{P}_\pm\mathcal{D}(\widetilde{A})$, $\widetilde{A}(\widetilde{\mathcal{L}}_\pm \cap \mathcal{D}(\widetilde{A})) \subset \widetilde{\mathcal{L}}_\pm$ *hold.*

Proof: We apply Theorem 1.3 with the block operator matrices

$$\widetilde{S} := \begin{pmatrix} 0 & B \\ B^* & 0 \end{pmatrix}, \quad \widetilde{T} := \begin{pmatrix} A & 0 \\ 0 & D \end{pmatrix}.$$

According to the assumptions (4.1), (4.2), Theorem 2.1 yields that the strip

$$\{z \in \mathbb{C} : -\delta < \Re(z) < \alpha\} \tag{4.9}$$

consists of points of regular type of $\widetilde{A}$. It will follow that the whole strip (4.9) belongs to $\rho(\widetilde{A})$ if at least one point of this strip belongs to $\rho(\widetilde{A})$. In order to show the latter, we observe the relation

$$(\widetilde{A}-z)^{-1} = (\widetilde{S}+\widetilde{T}-z)^{-1} = (\widetilde{S}-z)^{-1}(\widetilde{T}(\widetilde{S}-z)^{-1}+I)^{-1} \tag{4.10}$$

which holds if these inverses exist. Further,

$$\widetilde{T}(\widetilde{S}-z)^{-1} = \begin{pmatrix} zA(BB^*-z^2)^{-1} & AB(B^*B-z^2)^{-1} \\ DB^*(BB^*-z^2)^{-1} & zD(B^*B-z^2)^{-1} \end{pmatrix}$$

If we choose $z = it$, then the assumptions a$_2$), d$_2$) imply that for sufficiently large t the norm $\|\widetilde{T}(\widetilde{S}-z)^{-1}\|$ becomes arbitrarily small. Since $\widetilde{S}$ is self-adjoint, all the inverses in (4.10) exist.

According to the assumptions a$_1$), d$_1$), a$_2$), d$_2$) and the condition that B is boundedly invertible, the assumptions of Theorem 1.3 are fulfilled, and all assertions of the theorem follow. □

Note that the assumptions a$_1$) and a$_2$) (d$_1$) and d$_2$), respectively) of Theorem 4.3 are fulfilled if the operator A (D, respectively) is bounded.

The following corollary can be proved in the same way as Corollary 4.2. In accordance with (3.11) and (3.12) we now denote

$$\mathcal{D}_+ := \{x \in \mathcal{H} : x \in \mathcal{D}(B^*),\ K_+x \in \mathcal{D}(B)\},$$

$$\mathcal{D}_- := \{\widehat{x} \in \widehat{\mathcal{H}} : \widehat{x} \in \mathcal{D}(B),\ K_-\widehat{x} \in \mathcal{D}(B^*)\}.$$

Corollary 4.4 *The Riccati equations*

$$K_+BK_+x + K_+Ax - DK_+x - B^*x = 0, \quad x \in \mathcal{D}_+, \tag{4.11}$$

$$K_-B^*K_-\widehat{x} + K_-D\widehat{x} - AK_-\widehat{x} - B\widehat{x} = 0, \quad \widehat{x} \in \mathcal{D}_-, \tag{4.12}$$

hold. Moreover, the operator $\widetilde{A}|_{\widetilde{\mathcal{L}}_+\cap\mathcal{D}(\widetilde{A})}$ *in* $(\widetilde{\mathcal{L}}_+,\ [\cdot,\cdot])$ *is isomorphic to the operator* $A + BK_+|_{\mathcal{D}_+}$ *in* $(\mathcal{H},\ (G\cdot,\cdot))$ *with* $G := I - K_+^*K_+$, *in particular,* $A + BK$ *is* G*-accretive in* $\mathcal{H}$. *The operator* $\widetilde{A}|_{\widetilde{\mathcal{L}}_-\cap\mathcal{D}(\widetilde{A})}$ *in* $(\widetilde{\mathcal{L}}_-,\ [\cdot,\cdot])$ *is isomorphic to the operator* $D + B^*K_-|_{\mathcal{D}_-}$ *in* $(\widehat{\mathcal{H}},\ (\widehat{G}\cdot,\cdot))$ *with* $\widehat{G} := I - K_-^*K_-$, *in particular,* $D + B^*K_-$ *is* $\widehat{G}$*-accretive in* $\widehat{\mathcal{H}}$.

5 Dirac Operators with Potential

5.1 In this section we apply the results of Section 3.1 to the Dirac operator with a potential in the space $\widetilde{\mathcal{H}} = L_2(\mathbb{R}^3)^4$ with respect to the natural decomposition

$$L_2(\mathbb{R}^3)^4 = L_2(\mathbb{R}^3)^2 \oplus L_2(\mathbb{R}^3)^2.$$

We consider the operator

$$H = H_0 + H_1$$

on the domain $\mathcal{D}(H) = H^1(\mathbb{R}^3)^4$ ($H^1(\mathbb{R}^3)$ denotes the first order Sobolev space) where H_0 is the free Dirac operator,

$$H_0 = \begin{pmatrix} mc^2 & -ic\sigma\cdot\nabla \\ -ic\sigma\cdot\nabla & -mc^2 \end{pmatrix},$$

and H_1 is a potential of the form

$$H_1 = \begin{pmatrix} V_1 & 0 \\ 0 & V_2 \end{pmatrix}.$$

Here m is the mass of the particle, c the velocity of light, $\sigma = (\sigma_1\,\sigma_2\,\sigma_3)$ with the Pauli matrices $\sigma_1, \sigma_2, \sigma_3$, and V_i, $i = 1, 2$, are 2×2 matrices with entries V_i^{jk} which are functions on $\mathbb{R}^3$ satisfying the estimates

$$|V_i^{jk}(x)| \le a\frac{c}{2|x|} + b, \qquad x \in \mathbb{R}^3\backslash\{0\},\ i, j, k = 1, 2,$$

with some constants $b > 0$ and $a < 1$. Then the operator H is self-adjoint on the domain $\mathcal{D}(H) = \mathcal{D}(H_0) = H^1(\mathbb{R}^3)^4$ of the free Dirac operator (see e.g. [T], Chapter 4.3).

The subsequent theorem follows immediately from Theorem 3.1, Remark 3.4 and Theorem 2.3.

Theorem 5.1 *Assume that the multiplication in $L_2(\mathbb{R}^3)^2$ with V_i satisfies the norm estimate*

$$\|V_i\| \le mc^2, \qquad i = 1, 2. \tag{5.1}$$

Then the decomposition $L_2(\mathbb{R}^3)^4 = \widetilde{\mathcal{L}}_+ \dot{+} \widetilde{\mathcal{L}}_-$ holds where $\widetilde{\mathcal{L}}_+$ and $\widetilde{\mathcal{L}}_-$ are the spectral subspaces of H corresponding to $[0, \infty)$ and $(-\infty, 0]$, respectively. The subspaces $\widetilde{\mathcal{L}}_\pm$ can be represented as

$$\begin{aligned} \widetilde{\mathcal{L}}_+ &= \left\{ \begin{pmatrix} x \\ Kx \end{pmatrix} : x \in L_2(\mathbb{R}^3)^2 \right\}, \\ \widetilde{\mathcal{L}}_- &= \left\{ \begin{pmatrix} -K^*\widehat{x} \\ \widehat{x} \end{pmatrix} : \widehat{x} \in L_2(\mathbb{R}^3)^2 \right\} \end{aligned}$$

with a contraction K in $L_2(\mathbb{R}^3)^2$, which satisfies a Riccati equation. The operator $H|_{\widetilde{\mathcal{L}}_+}$ is isomorphic to the operator $mc^2 + V_1 - \mathrm{ic}\sigma\cdot\nabla\mathrm{K}$ on the domain $\mathcal{D}_+ := \{x \in H^1(\mathbb{R}^3)^2 : Kx \in H^1(\mathbb{R}^3)^2\}$ which is self-adjoint in $L_2(\mathbb{R}^3)^2$ with respect to the inner product $[\cdot, \cdot]$ given by

$$[x, y] := ((I + K^*K)x, y), \qquad x, y \in L_2(\mathbb{R}^3)^2,$$

the operator $H|_{\widetilde{\mathcal{L}}_-}$ is isomorphic to the operator $-mc^2 + V_2 - \mathrm{ic}\sigma\cdot\nabla\mathrm{K}^*$ on the domain $\mathcal{D}_- := \{\widehat{x} \in H^1(\mathbb{R}^3)^2 : K^*\widehat{x} \in H^1(\mathbb{R}^3)^2\}$ which is self-adjoint in $L_2(\mathbb{R}^3)^2$ with respect to the inner product $[\cdot, \cdot]_\wedge$ given by

$$[\widehat{x}, \widehat{y}]_\wedge := ((I + KK^*)\widehat{x}, \widehat{y}), \qquad \widehat{x}, \widehat{y} \in L_2(\mathbb{R}^3)^2.$$

On $\mathcal{D}_+ \oplus \mathcal{D}_-$ the operator H admits a diagonalization

$$\begin{aligned} &U^{-1} \begin{pmatrix} mc^2 + V_1 & -\mathrm{ic}\sigma\cdot\nabla \\ -\mathrm{ic}\sigma\cdot\nabla & -mc^2 + V_2 \end{pmatrix} \\ &U = \begin{pmatrix} mc^2 + V_1 - \mathrm{ic}\sigma\cdot\nabla\mathrm{K} & 0 \\ 0 & -mc^2 + V_2 - \mathrm{ic}\sigma\cdot\nabla\mathrm{K}^* \end{pmatrix}, \end{aligned}$$

where the transformation matrix is given by

$$U = \begin{pmatrix} I & -K^* \\ K & I \end{pmatrix}.$$

If the norm inequality (5.1) *is strengthened to*

$$\|V_i\| < mc^2, \qquad i = 1, 2,$$

then the contraction K is uniform.

5.2 The perturbed Dirac operator H can also be considered with respect to the decomposition

$$L_2(\mathbb{R}^3)^4 = \widetilde{\mathcal{L}}_+^0 \oplus \widetilde{\mathcal{L}}_-^0$$

where $\widetilde{\mathcal{L}}_+^0$ and $\widetilde{\mathcal{L}}_-^0$ are the spectral subspaces of the free Dirac operator H_0 corresponding to $[mc^2, \infty)$ and $(-\infty, -mc^2]$, respectively. If we denote by $\mathcal{Q}_+$ and $\mathcal{Q}_-$ the projections of $L_2(\mathbb{R}^3)^4$ onto $\widetilde{\mathcal{L}}_+^0$ and $\widetilde{\mathcal{L}}_-^0$, respectively, then H has a block operator matrix representation

$$H = \begin{pmatrix} \mathcal{Q}_+(H_0 + H_1)\mathcal{Q}_+ & \mathcal{Q}_+ H_1 \mathcal{Q}_- \\ \mathcal{Q}_- H_1 \mathcal{Q}_+ & \mathcal{Q}_-(H_0 + H_1)\mathcal{Q}_- \end{pmatrix}. \tag{5.2}$$

It is not difficult to see (see e.g. [GS]) that for the domains of the entries of H in this representation the relations

$$\mathcal{D}(\mathcal{Q}_+(H_0 + H_1)\mathcal{Q}_+) = \mathcal{D}(\mathcal{Q}_- H_1 \mathcal{Q}_+) = \mathcal{Q}_+\mathcal{D}(H_0),$$
$$\mathcal{D}(\mathcal{Q}_-(H_0 + H_1)\mathcal{Q}_-) = \mathcal{D}(\mathcal{Q}_+ H_1 \mathcal{Q}_-) = \mathcal{Q}_-\mathcal{D}(H_0),$$

hold and hence

$$\mathcal{D}(H) = \mathcal{Q}_+\mathcal{D}(H_0) \oplus \mathcal{Q}_-\mathcal{D}(H_0).$$

Also here, under some assumptions on the potential H_1, Theorem 3.1 applies and yields:

Theorem 5.2 *Assume that the potential H_1 satisfies the estimates*

$$\|\mathcal{Q}_+ H_1 \mathcal{Q}_+\| \le mc^2, \qquad \|\mathcal{Q}_- H_1 \mathcal{Q}_-\| \le mc^2. \tag{5.3}$$

If $0 \notin \sigma_p(H)$, or if $0 \in \sigma_p(H)$ and $\ker H$ has the splitting property (see (2.11)), then the decomposition $L_2(\mathbb{R}^3)^4 = \widetilde{\mathcal{L}}_+ \dot{+} \widetilde{\mathcal{L}}_-$ holds where $\widetilde{\mathcal{L}}_+$ and $\widetilde{\mathcal{L}}_-$ are the spectral subspaces of H corresponding to $[0, \infty)$ and $(-\infty, 0]$, respectively. The subspaces $\widetilde{\mathcal{L}}_\pm$ can be represented as

$$\widetilde{\mathcal{L}}_+ = \left\{ \begin{pmatrix} x \\ K'x \end{pmatrix} : x \in \widetilde{\mathcal{L}}_+^0 \right\}, \qquad \widetilde{\mathcal{L}}_- = \left\{ \begin{pmatrix} -K'^*\widehat{x} \\ \widehat{x} \end{pmatrix} : \widehat{x} \in \widetilde{\mathcal{L}}_-^0 \right\}$$

with a contraction K' from $\widetilde{\mathcal{L}}_+^0$ to $\widetilde{\mathcal{L}}_-^0$, which satisfies a Riccati equation. The operator $H|_{\widetilde{\mathcal{L}}_+}$ is isomorphic to the operator $\mathcal{Q}_+(H_0 + H_1)\mathcal{Q}_+ + \mathcal{Q}_- H_1 \mathcal{Q}_+ K'$ on the domain $\mathcal{D}_+ := \{x \in \mathcal{Q}_+\mathcal{D}(H_0) : K'x \in \mathcal{Q}_-\mathcal{D}(H_0)\}$ which is self-adjoint in $\widetilde{\mathcal{L}}_+^0$ with respect to the inner product $[\cdot,\cdot]$ given by

$$[x, y] := ((I + K'^*K')x, y), \qquad x, y \in \widetilde{\mathcal{L}}_+^0,$$

the operator $H|_{\widetilde{\mathcal{L}}_-}$ is isomorphic to the operator $\mathcal{Q}_-(H_0+H_1)\mathcal{Q}_- + \mathcal{Q}_+ H_1 \mathcal{Q}_- K'^*$ on the domain $\mathcal{D}_- := \{\widehat{x} \in \mathcal{Q}_-\mathcal{D}(H_0) : K'^*\widehat{x} \in \mathcal{Q}_+\mathcal{D}(H_0)\}$ which is self-adjoint in $\widetilde{\mathcal{L}}_-^0$ with respect to the inner product $[\cdot,\cdot]_\wedge$ given by

$$[\widehat{x}, \widehat{y}]_\wedge := ((I + K'K'^*)\widehat{x}, \widehat{y}), \qquad \widehat{x}, \widehat{y} \in \widetilde{\mathcal{L}}_-^0.$$

On $\mathcal{D}_+ \oplus \mathcal{D}_-$ the operator H admits a diagonalization

$$V^{-1}\begin{pmatrix} \mathcal{Q}_+(H_0+H_1)\mathcal{Q}_+ & \mathcal{Q}_+H_1\mathcal{Q}_- \\ \mathcal{Q}_-H_1\mathcal{Q}_+ & \mathcal{Q}_-(H_0+H_1)\mathcal{Q}_- \end{pmatrix} V$$

$$= \begin{pmatrix} \mathcal{Q}_+(H_0+H_1)\mathcal{Q}_+ + \mathcal{Q}_-H_1\mathcal{Q}_+K' & 0 \\ 0 & \begin{matrix}\mathcal{Q}_-(H_0+H_1)\mathcal{Q}_- \\ +\mathcal{Q}_+H_1\mathcal{Q}_-K'^*\end{matrix} \end{pmatrix},$$

where the transformation matrix is given by

$$V = \begin{pmatrix} I & -K'^* \\ K' & I \end{pmatrix}.$$

References

[AL] V.M. Adamjan and H. Langer, Spectral properties of a class of rational operator valued functions; *J. Operator Theory* **33** (1995), 259–277.

[ALMS] V.M. Adamjan, H. Langer, R. Mennicken and J. Saurer, Spectral components of selfadjoint block operator matrices with unbounded entries; *Math. Nachr.* **178** (1996), 43–80.

[AI] T.Ya. Azizov and I.S. Iohvidov, Linear Operators in Spaces with an Indefinite Metric; John Wiley & Sons, Chichester New York Brisbane Toronto Singapore, 1989.

[BGK] H. Bart, I.C. Gohberg and M.A. Kaashoek, Wiener-Hopf factorization, inverse Fourier transforms and exponentially dichotomous operators; *J. Funct. Anal.* **68** (1986), 1–42.

[E] K.-J. Engel, Operator Matrices and Systems of Evolution Equations; Manuscript, Tuebingen, 1996.

[GGK] I.C. Gohberg, S. Goldberg and M.A. Kaashoek, Classes of Linear Operators vol. I; Operator Theory: *Adv. Appl.* **49**, Birkhäuser, Basel, 1990.

[GS] M. Griesemer and H. Siedentop, A minimax principle for the eigenvalues in gaps; *J. London Math. Soc.*, to appear.

[IK] I.S. Iohvidov and M.G. Krein, Spectral theory of operators in spaces with an indefinite metric, *I. Amer. Math. Soc. Transl.* (2) **13** (1960), 105–175.

[IKL] I.S. Iohvidov, M.G. Krein and H. Langer, Introduction to the Spectral Theory of Operators in Spaces with an Indefinite Metric; *Math. Research* vol. 9, Akademie-Verlag, Berlin, 1982.

[KVL] M.A. Kaashoek and S.M. Verduyn-Lunel, An Integrability Condition on the Resolvent for Hyperbolicity of the Semigroup; *J. Differential Equations* **112:2** (1994), 374–406.

[K] T. Kato, Perturbation Theory for Linear Operators; Springer Verlag, Berlin Heidelberg New York, 1966.

[LT] H. Langer and C. Tretter, Spectral decomposition of some nonselfadjoint block operator matrices; *J. Operator Theory* **39** (1998), 339–359.

[LMMT] H. Langer, A. Markus, V. Matsaev and C. Tretter, A new concept for block operator matrices: The quadratic numerical range; *Math. Ann.*, submitted.

[M] A.K. Motovilov, Removal of the energy dependence from the resolvent-like energy-dependent interactions, preprint JINR E5-94-259, Dubna, 1994.

[MS] R. Mennicken and A.A. Shkalikov, Spectral decomposition of symmetric operator matrices; *Math. Nachr.* **179** (1996), 259–273.

[T] B. Thaller, The Dirac Equation; Springer-Verlag, Berlin Heidelberg New York, 1992.

[V] K. Veselič, On spectral properties of a class of J-selfadjoint operators, I; *Glas. Mat. Ser. III*, **7** (1972), 229–248.

Inst. for Analysis and Technical Mathematics
University of Technology Vienna
Wiedner Hauptstr. 8–10
A-1040 Vienna
Austria
hlanger@email.tuwien.ac.at

Dept. of Mathematics
University of Regensburg
Universitätsstr. 31
D-93053 Regensburg
Germany
christiane.tretter@mathematik.uni-regensburg.de

Operator Theory:
Advances and Applications, Vol. 122
© 2001 Birkhäuser Verlag Basel/Switzerland

Stability of Pseudospectral Factorizations

André Ran[1], *Leiba Rodman*[2] *and Dirk Temme*

Dedicated to Professor M.A. Kaashoek on the occasion of his 60th birthday with gratitude to our friend and mentor

For rational matrix functions that are of the form identity plus a contractive function, or have positive semidefinite real part, a class of pseudospectral factorizations is considered. Within the class, stable factorizations, i.e., those that persist after a small perturbation of a minimal realization of the original function, are fully described. The description is based on stability results concerning semidefinite invariant subspaces of dissipative matrices with respect to an indefinite inner product.

1 Introduction

The state space approach (based on realization theory) to study minimal factorizations of rational matrix functions originated with Bart, Gohberg, Kaashoek, Van Dooren [5], [4]. In essence this approach reduces factorization to a pair of mutually complemented subspaces, one invariant for the main realization operator of the function to be factored, and the other invariant for the main operator of the inverse of the factored function. Since then, this approach has been successfully used to study many types of factorizations of various classes of rational matrix functions, as well as of more general operator valued functions, including Wiener-Hopf, canonical (or spectral), inner-outer, coprime, non-minimal, etc. factorizations, in classes such as matrix polynomials, matrix functions with symmetries (Hermitian, unitary, etc.), also matrix functions with symmetries with respect to an indefinite inner product. Many advances in this area have been motivated by applications, including applications in linear control systems. In particular, stable minimal factorizations, i.e., those for which nearby rational matrix functions admit nearby minimal factorizations, have been described in [3] (the polynomial case) and [4], and see also [7]. In this approach, the distance between rational matrix functions is measured in terms of minimal realizations. Stable minimal factorizations for various classes of rational matrix functions with symmetries have been studied in [14], [15], [16], [17], [18].

[1]The work of first two authors partially supported by the NATO grant CRG 960700.

[2]The work of the second author is partially supported by NSF grant DMS 9800794 and by the Summer Research Grant from the College of William and Mary.

In the present paper we consider a particular class of minimal factorizations, namely, pseudospectral factorizations, and study stability properties within this class. A factorization $W(\lambda) = W_1(\lambda)W_2(\lambda)$ is called *pseudospectral*, or more precisely, *left pseudospectral*, with respect to the real axis, if W_1 (resp., W_2) has all its poles in the open lower (resp. upper) halfplane and all its zeros in the closed lower (resp., upper) halfplane, including infinity. Thus, pseudospectral factorizations are in between canonical Wiener-Hopf (or spectral) and minimal factorizations. Pseudospectral factorizations have been introduced and studied in [20], [21], where applications of this notion to Wiener-Hopf integral equations are also given. It turns out that for rational matrix valued functions that are of the form identity plus a contractive function, or have positive semidefinite real part, the existence of pseudospectral factorizations is guaranteed (see [9], [10] for the identity plus a contraction functions, and [15] and [19] for the positive semidefinite real part functions). These classes of pseudospectral factorizations we consider are closely linked with matrices that are dissipative in an indefinite inner product, and their invariant maximal semidefinite subspaces. Such matrices and their invariant subspaces were studied in [19] and in more detail in [23] (see also [22]).

Our attention will be focused on the problem of stability of pseudospectral factorizations of identity plus contractive functions, and of functions with positive semidefinite real part, under perturbations of the matrices that appear in a minimal realization of the function in question. We completely describe stable pseudospectral factorizations in the situations when they are connected to corresponding stability properties of the invariant maximal semidefinite subspaces.

The paper consists of three sections besides this introduction. In the second section we study stability of invariant maximal nonnegative subspaces of matrices that are dissipative in an indefinite inner product. The third and fourth section are devoted to the study of stability of pseudospectral factorizations for the cases of identity plus a contraction and positive real part, respectively.

All matrices in this paper are assumed to have entries in the field of complex numbers $\mathbb{C}$, and $\sigma(A)$ stands for the set of eigenvalues of a square size matrix A. For a square size matrix A, and a subset Λ of $\mathbb{C}$, the notation $\mathcal{R}(A, \Lambda)$ is used for the A-invariant spectral subspace corresponding to $\sigma(A) \cap \Lambda$. The open upper, open lower halfplanes, and the real axis are denoted $\mathbb{C}_{upp}$, $\mathbb{C}_{low}$, and $\mathbb{R}$, respectively. For Hermitian $n \times n$ matrices X and Y, $X \geq Y$ (resp. $X > Y$) means that $X - Y$ is positive semidefinite (resp., positive definite). Scalar matrices λI, $\lambda \in \mathbb{C}$, are often written as simply λ. We denote by $\langle \cdot, \cdot \rangle$ the standard inner product in $\mathbb{C}^n$. The operator norm (the largest singular value) is used throughout.

2 Dissipative Matrices and their Invariant Subspaces

We fix an invertible Hermitian $n \times n$ matrix H. Basic notions pertaining to the indefinite scalar product in $\mathbb{C}^n$ induced by H will be used, for instance, H-nonnegative subspaces. We refer to [8] for definitions and properties of these notions.

Let A be an H-dissipative $n \times n$ matrix. This means that

$$\frac{1}{2i}(HA - A^*H) \geq 0. \tag{2.1}$$

Such a matrix always has an A-invariant maximal H-nonnegative subspace, as well as an A-invariant maximal H-nonpositive subspace. (See, e.g., [1], [11].) A constructive proof of this fact was given in [19], see also [22]. We recall here the construction without going into details.

For dissipative matrices in an indefinite inner product space there does not exist a canonical form; however, by applying similarity to A and congruence to H (with the same invertible matrix) we can transform the pair (A, H) into a simple form, from which it is easy to read off that invariant maximal semidefinite subspaces always exist. In fact, the simple form allows one to construct an A-invariant maximal H-nonnegative subspace in essentially the same way as was done for matrices that are selfadjoint in an indefinite inner product in [8]. The part of the construction pertaining to the real eigenvalues of A stays the same, yielding a subspace $\mathcal{M}_0$; an A-invariant maximal H-nonnegative subspace is then constructed by taking the direct sum of $\mathcal{M}_0$ with the spectral subspace of A corresponding to the open upper half plane. We refer the reader to [19], where the main results concerning the simple form are proved. We remark that if $x_1, \ldots, x_k$ is a Jordan chain of A corresponding to $\lambda \in \mathbb{R}$ such that it cannot be extended to a longer Jordan chain then $\mathcal{M}_0$ contains span $\{x_1, \ldots, x_{\frac{k}{2}}\}$ if k is even and span $\{x_1, \ldots, x_{[\frac{k}{2}]-1}\}$ if k is odd. Whether or not the vector $x_{[\frac{k}{2}]+1}$ (for k odd) is in $\mathcal{M}_0$ depends on the sign of $\langle Hx_{[\frac{k}{2}]+1}, x_{[\frac{k}{2}]+1}\rangle$.

Let $H = H^*$ be invertible, and let A be H-dissipative. An A-invariant maximal H-nonnegative subspace $\mathcal{M}$ is called *stable* if for every $\varepsilon > 0$ there is a $\delta > 0$ such that every H-dissipative matrix B with $\|A - B\| < \delta$ has an invariant maximal H-nonnegative subspace $\mathcal{N}$ with $\text{gap}(\mathcal{M}, \mathcal{N}) < \varepsilon$. Here the well-known notion of a gap between two subspaces $\mathcal{M}, \mathcal{N}$ of $\mathbb{C}^n$ is used:

$$\text{gap}(\mathcal{M}, \mathcal{N}) = \|P_{\mathcal{M}} - P_{\mathcal{N}}\|,$$

where $P_{\mathcal{M}}$, $P_{\mathcal{N}}$ is the orthogonal projection on $\mathcal{M}$, $\mathcal{N}$, respectively. (See, e.g., [7] for more details.)

Before we can state the main result on stability of invariant maximal nonnegative subspaces for an H-dissipative matrix A we need to introduce the notion of the numerical range condition for the pair (A, H) (see [19], also [23], Chapter 3).

Let us denote the number of negative eigenvalues of H by κ. It is well-known that the maximal length of a Jordan chain of A corresponding to a real eigenvalue is then $2\kappa + 1$.

Let λ be a real eigenvalue of A. Consider a Jordan basis for $R(A, \{\lambda\})$ and split it into the sets $J(\lambda, j)$, where $J(\lambda, j)$ consists of the basis vectors belonging to Jordan chains of length j. Denote by $n_{\lambda,j}$ the number of chains of length j. Now denote the basis vectors in $J(\lambda, j)$ that are not in the set Ker $(A - \lambda)^{j-1}$ by $x_{j;1}, \ldots, x_{j;n_{\lambda,j}}$. Thus, these vectors are vectors at the end of a Jordan chain of lenght j. Let m_j be $\frac{j+1}{2}$ in case j is odd, and $\frac{j}{2}$ in case j is even. Define $y_{j;k} = (A-\lambda)^{m_j-1}x_{j;k}$.

From here on there is a difference in the situation for Jordan chains of odd length and Jordan chains of even length. We first deal with the case of Jordan chains of odd length.

For $j = 1, 3, \dots, 2\kappa+1$ let

$$CM_j = \begin{pmatrix} \langle Hy_{j;1}, y_{j;1}\rangle & \dots & \langle Hy_{j;n_{\lambda,j}}, y_{j;1}\rangle \\ \vdots & & \vdots \\ \langle Hy_{j;1}, y_{j;n_{\lambda,j}}\rangle & \dots & \langle Hy_{j;n_{\lambda,j}}, y_{j;n_{\lambda,j}}\rangle \end{pmatrix}$$

(here CM stands for *characteristic matrix*). Define

$$CM_{odd}(A,\lambda) = \text{diag}\,(CM_1, CM_3, \dots, CM_{2\kappa+1}),$$

and, finally, put

$$NR_{odd}(A,\lambda) = \{\langle CM_{odd}(A,\lambda)x, x\rangle \mid x \neq 0\}.$$

(here NR stands for numerical range).

It turns out that $CM_{odd}(A,\lambda)$ is Hermitian and invertible and that $NR_{odd}(A,\lambda)$ is independent of the choice of the Jordan basis one starts with.

The pair (A, H) is said to satisfy the *odd numerical range condition* if $0 \notin NR_{odd}(A,\lambda)$ for all real eigenvalues λ of A.

For even length chains, let $j, k \in \{2, 4, \dots, 2\kappa\}$ and put

$$CM_{j,k} = \begin{pmatrix} \langle H(A-\lambda)y_{j;1}, y_{k;1}\rangle & \dots & \langle H(A-\lambda)y_{j;n_{\lambda,j}}, y_{k;1}\rangle \\ \vdots & & \vdots \\ \langle H(A-\lambda)y_{j;1}, y_{k;n_{\lambda,k}}\rangle & \dots & \langle H(A-\lambda)y_{j;n_{\lambda,j}}, y_{k;n_{\lambda,k}}\rangle \end{pmatrix}.$$

One can show that for $j < k$ one has $CM_{j,k} = 0$, and that $CM_{j,j}$ is invertible. Put

$$CM_{even}(A,\lambda) = \begin{pmatrix} CM_{2,2} & CM_{4,2} & \dots & CM_{2\kappa,2} \\ \vdots & & & \vdots \\ CM_{2,2\kappa} & CM_{4,2\kappa} & \dots & CM_{2\kappa,2\kappa} \end{pmatrix}.$$

Then $CM_{even}(A,\lambda)$ is an invertible block upper triangular matrix. Define

$$NR_{even}(A,\lambda) = \{\langle CM_{even}(A,\lambda)x, x\rangle \mid x \neq 0\}.$$

It turns out that $NR_{even}(A,\lambda)$ is independent of the choice of the Jordan basis one starts with.

The pair (A, H) is said to satisfy the *even numerical range condition* if $0 \notin NR_{even}(A,\lambda)$ for all real eigenvalues of A, and we say that (A, H) satisfies the *numerical range condition* if it satisfies both the odd numerical range condition and the even numerical range condition.

Even though the odd and even numerical ranges do not depend on the choice of the Jordan basis, and hence neither does the numerical range condition, a basis free definition of the numerical range condition does not exist so far.

It can be shown [23, 19] that if A is H-selfadjoint then the pair (A, H) satisfies the numerical range condition if and only if it satisfies the sign condition introduced for selfadjoint matrices in an indefinite inner product in [13].

The following theorem describes uniqueness and stability of A-invariant maximal H-nonnegative subspaces.

Theorem 2.1 *The following statements are equivalent for given H-dissipative matrix A:*

(i) *there exists a stable A-invariant maximal H-nonnegative subspace,*

(ii) *there exists a stable A-invariant maximal H-nonpositive subspace,*

(iii) *the numerical range condition holds for the pair (A, H),*

(iv) *there is a unique A-invariant maximal H-nonnegative subspace $\mathcal{M}$ with $\sigma(A|_{\mathcal{M}})$ contained in the closed upper half plane,*

(v) *there is a unique A-invariant maximal H-nonpositive subspace $\mathcal{M}$ with $\sigma(A|_{\mathcal{M}})$ contained in the closed lower half plane.*

In that case, there is a unique stable A-invariant maximal H-nonnegative subspace, being the one with $\sigma(A|_{\mathcal{M}})$ contained in the closed upper half plane, and there is a unique stable A-invariant maximal H-nonpositive subspace, being the one with $\sigma(A|_{\mathcal{M}})$ contained in the closed lower half plane.

Proof: The equivalence between (iii), (iv) and (v) was established in [19], see also [23], Chapter 3.

We shall show the equivalence between (i), (iii) and (iv), the equivalence between (ii), (iii) and (v) is proved in the same manner. First assume that (iii) holds and (i) does not hold. In particular the unique A-invariant maximal H-nonnegative subspace $\mathcal{M}_+$ such that $\sigma(A|_{\mathcal{M}_+}) \subset \mathbb{C}_{upp} \cup \mathbb{R}$ is not stable. Then there exists an $\varepsilon > 0$ and a sequence $A_m \to A$ of H-dissipative matrices such that for any A_m- invariant maximal H-nonnegative subspace $\mathcal{M}_m$ we have gap$(\mathcal{M}_m, \mathcal{M}_+) \geq \varepsilon$. However, take for $\mathcal{M}_m$ any A_m-invariant maximal H-nonnegative subspace with $\sigma(A|_{\mathcal{M}_m}) \subset \mathbb{C}_{upp} \cup \mathbb{R}$. As the set of subspaces of a certain fixed dimension is compact in the gap topology, there is a converging subsequence $\mathcal{M}_{m_k}$, say with limit $\mathcal{M}$. Then by continuity it follows that $\mathcal{M}$ is A-invariant maximal H-nonnegative and $\sigma(A|_{\mathcal{M}}) \subset \mathbb{C}_{upp} \cup \mathbb{R}$. Thus $\mathcal{M} = \mathcal{M}_+$ by uniqueness of such a subspace. This contradicts gap$(\mathcal{M}_m, \mathcal{M}_+) \geq \varepsilon$, so it follows that $\mathcal{M}_+$ must be stable.

Suppose the numerical range condition is not satisfied. Let $\mathcal{M}_1$ and $\mathcal{M}_2$ be different A-invariant maximal H-nonnegative subspaces with $\sigma(A|\mathcal{M}_i) \subset \mathbb{C}_{upp} \cup \mathbb{R}$.

Let $\mathcal{N}$ be a maximal H-negative subspace such that $\mathbb{C}^n = \mathcal{M}_1 \dot{+} \mathcal{N}$. Write A and H with respect to this decomposition as

$$\tilde{A} = S^{-1}AS = \begin{pmatrix} A_{11} & A_{12} \\ 0 & A_{22} \end{pmatrix}, \quad \tilde{H} = S^*HS = \begin{pmatrix} H_{11} & H_{12} \\ H_{12}^* & H_{22} \end{pmatrix}. \tag{2.2}$$

Note that H_{11} is positive semidefinite and H_{22} is negative definite. Then $\tilde{A}$ is $\tilde{H}$-dissipative and $\tilde{\mathcal{M}}_i = S^{-1}\mathcal{M}_i$ $(i = 1, 2)$ are $\tilde{A}$-invariant maximal $\tilde{H}$-nonnegative subspaces. These are stable if and only if the original subspaces $\mathcal{M}_i$ are stable. So, from this point on we shall assume that A and H are already in the form (2.2), and drop the tildes.

For $\varepsilon > 0$ consider

$$A_\varepsilon = \begin{pmatrix} A_{11} + \varepsilon i I & A_{12} \\ 0 & A_{22} - \varepsilon i I \end{pmatrix}.$$

Then

$$\frac{1}{2i}(HA_\varepsilon - A_\varepsilon^* H) = \frac{1}{2i}(HA - A^*H) + \varepsilon \begin{pmatrix} H_{11} & 0 \\ 0 & -H_{22} \end{pmatrix}.$$

Since $H_{11} \geq 0$ and $H_{22} < 0$, we see that A_ε is H-dissipative. Moreover, $\sigma(A_\varepsilon) \cap \mathbb{R} = \emptyset$ for ε close to zero. We shall show that there is a unique A_ε-invariant maximal H-nonnegative subspace, which must then be $\mathcal{R}(A_\varepsilon, \mathbb{C}_{upp}) = \mathcal{M}_1$.

Indeed, suppose this is not the case. Then, according to Theorem 4.9 in [19] there exists an A_ε-invariant H-neutral subspace $\mathcal{N}$ with $\sigma(A_\varepsilon|_{\mathcal{N}}) \subset \mathbb{C}_{low}$. In particular, let $\begin{pmatrix} x \\ y \end{pmatrix}$ be an eigenvector of A_ε in $\mathcal{N}$, and let λ be the corresponding eigenvalue. Then $\operatorname{Im}\lambda < 0$. Consider

$$\frac{1}{2i}\left\langle (HA_\varepsilon - A_\varepsilon^* H)\begin{pmatrix} x \\ y \end{pmatrix}, \begin{pmatrix} x \\ y \end{pmatrix}\right\rangle = \frac{1}{2i}(\lambda - \bar{\lambda})\left\langle H\begin{pmatrix} x \\ y \end{pmatrix}, \begin{pmatrix} x \\ y \end{pmatrix}\right\rangle = 0.$$

On the other hand, we have that this is equal to

$$\frac{1}{2i}\left\langle (HA - A^*H)\begin{pmatrix} x \\ y \end{pmatrix}, \begin{pmatrix} x \\ y \end{pmatrix}\right\rangle + \varepsilon(\langle H_{11}x, x\rangle - \langle H_{22}y, y\rangle) \geq 0.$$

As either term in this expression is larger than or equal to zero we derive that in particular $\langle H_{22}y, y\rangle = 0$. But as $H_{22} < 0$ it follows that $y = 0$. Then $\begin{pmatrix} x \\ y \end{pmatrix} = \begin{pmatrix} x \\ 0 \end{pmatrix}$ which cannot be an eigenvalue of A_ε corresponding to an eigenvalue in the lower half plane. Thus such an A_ε-invariant H-neutral subspace cannot exist.

Now let $\varepsilon \downarrow 0$. We see that $\mathcal{R}(A_\varepsilon, \mathbb{C}_{upp})$ converges to $\mathcal{M}_1$ (in fact, it is $\mathcal{M}_1$ for all ε). So $\mathcal{M}_1$ is the only possibility for a stable invariant maximal nonnegative subspace. But replacing $\mathcal{M}_1$ by $\mathcal{M}_2$ throughout the previous argument we see that there is no stable invariant maximal nonnegative subspace if the numerical range condition is not satisfied.

The argument above also proves that if the numerical range condition is satisfied only the unique invariant maximal nonnegative subspace $\mathcal{M}$ for which $\sigma(A|_{\mathcal{M}}) \subset \mathbb{C}_{upp} \cup \mathbb{R}$ can be stable. As it was already shown that it is in fact stable, the second part of the theorem is also proved. □

The concept of stability can also be defined by allowing H, as well as A, to change. Thus, for an H-dissipative matrix A, an A-invariant maximal H-nonnegative subspace $\mathcal{M}$ will be called *strongly stable* if for every $\varepsilon > 0$ there is a $\delta > 0$ such that every G-dissipative matrix B with $G = G^*$ and $\|A - B\| + \|H - G\| < \delta$ has an invariant maximal G-nonnegative subspace $\mathcal{N}$ with $\text{gap}(\mathcal{M}, \mathcal{N}) < \varepsilon$. It turns out that strong stability is equivalent to stability; one can prove this by following the Proof of Theorem 2.1 and making the obvious adjustments.

In connection with Theorem 2.1 observe that the numerical range condition is not stable, i.e., if it holds for (A, H), it still may fail for H-dissipative matrices arbitrarily close to A, as shown in the following example:

Example: Let

$$A = \begin{pmatrix} 0 & 1 & 0 \\ 0 & 0 & 0 \\ 0 & 0 & 0 \end{pmatrix}, \qquad H = \begin{pmatrix} 0 & 1 & 0 \\ 1 & 0 & 0 \\ 0 & 0 & -1 \end{pmatrix}.$$

Then A is H-selfadjoint, and a computation shows that $CM_{odd}(A, 0) = -1$, $CM_{even}(A, 0) = 1$. So the numerical range condition is satisfied for (A, H). For positive ε, let

$$A_\varepsilon = \begin{pmatrix} 0 & 1 & 0 \\ \varepsilon^2 & 0 & 0 \\ 0 & 0 & \varepsilon \end{pmatrix}.$$

The matrix A_ε is H-selfadjoint. One computes

$$CM_{odd}(A, \varepsilon) = \begin{pmatrix} 2\varepsilon & 0 \\ 0 & -1 \end{pmatrix},$$

and so the numerical range condition is not satisfied for (A_ε, H).

3 Functions of the Form Identity Plus a Contraction

Let $W(\lambda)$ be an $m \times m$ rational matrix function of the form $W(\lambda) = I + U(\lambda)$, where $U(\lambda)$ is a strict contraction, i.e., $\|U(\lambda)\| < 1$, for all λ on the extended real line. Observe that such a function $W(\lambda)$ can have neither poles nor zeros on the real line.

A factorization

$$W(\lambda) = W_+(\lambda)W_-(\lambda) \tag{3.1}$$

where W_+ and W_- are rational matrix functions is called a *left canonical Wiener-Hopf factorization* if W_+ has no poles and zeros in the closed upper half plane and W_- has no poles and zeros in the closed lower half plane. A factorization

$$W(\lambda) = V_-(\lambda)V_+(\lambda) \tag{3.2}$$

where V_+ and V_- are rational matrix functions is called a *right canonical Wiener-Hopf factorization* if V_+ (resp., V_-) has no poles and zeros in the closed upper (resp., lower) half plane. Note that if either left or right Wiener-Hopf factorization exist $W(\lambda)$ has neither poles nor zeros on the real line. It is well-known, see e.g. [6], Section 3.1, that a function of the form $W(\lambda) = I + U(\lambda)$, where $\|U(\lambda)\| < 1$ for all λ on the extended real line, admits both left and right canonical Wiener-Hopf factorization.

In the particular case when $U(\lambda)$ is strictly proper, i.e., $U(\infty) = 0$ formulas for the factors in the left and right Wiener-Hopf factorization may be found following [4]. These formulas are given in terms of a realization of $W(\lambda)$. Any rational matrix function $W(\lambda)$ with $W(\infty) = I$ admits a *realization*, that is, it can be written as $W(\lambda) = I_m + C(\lambda I_n - A)^{-1}B$. In fact, $W(\lambda)$ admits a so-called minimal realization, i.e., a realization where the number n, the order of A, is as small as possible.

In [9] this result was generalized to the case when the function $W(\lambda)$ has zeros on the real line. More precisely, we consider functions of the form $W(\lambda) = I + U(\lambda)$, where $U(\lambda)$ is merely contractive for all real λ, instead of being strictly contractive. Note that such a function $W(\lambda)$ has no poles on the real line, but may have real zeros. We shall however assume throughout that $U(\infty)$ is a strict contraction. In fact, we shall assume for simplicity that $U(\infty) = 0$, i.e., U is strictly proper. A factorization (3.1) of such a function $W(\lambda)$ is called a *left pseudospectral factorization* if W_+ has no poles and zeros in the open upper half plane, W_- has no poles and zeros in the open lower half plane, and neither W_+ nor W_- has poles on the real line. Also, a factorization (3.2) of such a function $W(\lambda)$ is called a *right pseudospectral factorization* if V_+ has no poles and zeros in the open upper half plane, V_- has no poles and zeros in the open lower half plane, and neither V_+ nor V_- has poles on the real line. Compare [20, 21], where this notion was first introduced in more generality. Note that the definition in [20, 21] involves the concept of minimal factorization. In case $W(\lambda)$ has no poles on the real line our definition implies the minimality of the factorizations.

The factorizations introduced above all play an important role in the study of invertibility properties of integral operators of Wiener-Hopf type (see [20]), as well as singular integral operators ([6]) and Toeplitz operators.

Left and right pseudospectral factorizations may be characterized in terms of a realization. This was remarked in [20, 21] as a corollary from [4].

Theorem 3.1 *Let $W(\lambda) = I_m + C(\lambda I_n - A)^{-1}B$ be a minimal realization. Then $W(\lambda)$ admits a left pseudospectral factorization if and only if there exist subspaces $\mathcal{M}_-$ and $\mathcal{M}_+^\times$, such that with $A^\times = A - BC$:*

(i) $A\mathcal{M}_- \subseteq \mathcal{M}_-, \quad A^\times \mathcal{M}_+^\times \subseteq \mathcal{M}_+^\times$,

(ii) $\mathbb{C}^n = \mathcal{M}_- \dot{+} \mathcal{M}_+^\times$ *(direct sum of subspaces).*

(iii) $\sigma(A|_{\mathcal{M}_-}) \subset \{\lambda | \text{Im } \lambda \le 0\}$, $\sigma(A^\times|_{\mathcal{M}_+^\times}) \subset \{\lambda | \text{Im } \lambda \ge 0\}$,

(iv) *$\mathcal{M}_-$ contains the A-invariant spectral subspace corresponding to eigenvalues λ of A with* Im $\lambda < 0$, *$\mathcal{M}_+^\times$ contains the $A^\times$-invariant spectral subspace corresponding to eigenvalues λ of $A^\times$ with* Im $\lambda > 0$.

Let π_l be the projection along $\mathcal{M}_-$ onto $\mathcal{M}_+^\times$. Put

$$\begin{aligned} W_+(\lambda) &= I + C(\lambda - A)^{-1}(I - \pi_l)B, \\ W_-(\lambda) &= I + C\pi_l(\lambda - A)^{-1}B. \end{aligned} \tag{3.3}$$

Then

$$W(\lambda) = W_+(\lambda)W_-(\lambda) \tag{3.4}$$

is a left pseudospectral factorization.

In case $W(\lambda)$ has no zeros or poles on the real line, and hence A and $A^\times$ have no real eigenvalues, then $\mathcal{M}_-$ is just equal to $\mathcal{R}(A, \mathbb{C}_{low})$, and $\mathcal{M}_+^\times$ is equal to $\mathcal{R}(A^\times, \mathbb{C}_{upp})$. The pseudospectral factorization then becomes a canonical Wiener-Hopf factorization. We note that for minimal realizations of W as in the theorem there is a one-one correspondence between left pseudospectral factorizations and pairs of subspaces $\mathcal{M}_-$, $\mathcal{M}_+^\times$ satisfying (i), (ii), (iii) and (iv). This one-one correspondence is given by (3.3). For right pseudospectral factorizations an analogous statement holds:

Theorem 3.2 *Let $W(\lambda) = I_m + C(\lambda I_n - A)^{-1}B$ be a minimal realization. Then $W(\lambda)$ admits a right pseudospectral factorization if and only if there exist subspaces $\mathcal{M}_+$ and $\mathcal{M}_-^\times$, such that*

(i) $A\mathcal{M}_+ \subseteq \mathcal{M}_+, \quad A^\times \mathcal{M}_-^\times \subseteq \mathcal{M}_-^\times$,

(ii) $\mathbb{C}^n = \mathcal{M}_+ \dot{+} \mathcal{M}_-^\times$.

(iii) $\sigma(A|_{\mathcal{M}_+}) \subset \{\lambda | \text{Im } \lambda \ge 0\}$, $\sigma(A^\times|_{\mathcal{M}_-^\times}) \subset \{\lambda | \text{Im } \lambda \le 0\}$,

(iv) *$\mathcal{M}_+$ contains the A-invariant spectral subspace corresponding to eigenvalues of A with* Im $\lambda > 0$, *$\mathcal{M}_-^\times$ contains the $A^\times$-invariant spectral subspace corresponding to eigenvalues of $A^\times$ with* Im $\lambda < 0$.

Let π_r be the projection along $\mathcal{M}_+$ onto $\mathcal{M}_-^\times$. Put

$$(3.5)\qquad \begin{aligned} V_-(\lambda) &= I + C(\lambda - A)^{-1}(I - \pi_r)B, \\ V_+(\lambda) &= I + C\pi_r(\lambda - A)^{-1}B. \end{aligned}$$

Then

$$(3.6)\qquad W(\lambda) = V_-(\lambda)V_+(\lambda)$$

is a right pseudospectral factorization.

Our aim is to study pseudospectral factorizations for rational matrix functions of the form $W(\lambda) = I + U(\lambda)$, where $U(\infty) = 0$, and $U(\lambda)$ is a contraction for all real λ. In case we have a minimal realization $U(\lambda) = C(\lambda I_n - A)^{-1}B$, it is a well-known fact that $U(\lambda)$ is contractive for all real λ if and only if there is a Hermitian matrix X solving the Riccati equation

$$(3.7)\qquad XC^*CX - iXA^* + iAX + BB^* = 0$$

(see, e.g. [10], Theorem 3.2). In the special case of stable A this is known as the *bounded real lemma* in systems theory. By Theorem 3.4 in [10] every Hermitian solution X of (3.7) is invertible; in fact, the controllability of (A, B) alone guarantees invertibility of every Hermitian solution of (3.7) (see also the Proof of Theorem 21.2.1 in [12]). From (3.7) it is seen that

$$\frac{1}{2i}(XA^* - AX) = -\frac{1}{2}(XC^*CX + BB^*).$$

Thus for all $x \in \mathbb{C}^n$

$$\operatorname{Im}\langle X^{-1}Ax, x\rangle \geq 0.$$

So A is an X^{-1}-dissipative matrix. Moreover, with $A^\times = A - BC$ it is straightforward to see from (3.7) that

$$(3.8)\qquad \frac{1}{2i}(X(A^\times)^* - A^\times X) = -\frac{1}{2}(iXC^* - B)(iXC^* - B)^*.$$

Thus for all $x \in \mathbb{C}^n$

$$(3.9)\qquad \operatorname{Im}\langle X^{-1}A^\times x, x\rangle \geq 0.$$

So, also $A^\times$ is X^{-1}-dissipative.

Recall that a rational matrix function of the form $W(\lambda) = I + U(\lambda)$, where $U(\lambda)$ is contractive for all real λ cannot have poles on the real line. Thus, if U is strictly proper and $C(\lambda I_n - A)^{-1}B$ is a minimal realization for U, the matrix A has no real eigenvalues. We quote from [9] the following theorem.

Theorem 3.3 *Let $W(\lambda) = I_m + U(\lambda)$ be a rational matrix valued function such that $U(\lambda)$ has contractive values for all real λ, and assume that $U(\lambda_0)$ is a strict contraction for some $\lambda_0 \in \mathbb{R}$. Then $W(\lambda)$ admits both left and right pseudospectral factorization.*

In fact, assuming $U(\infty) = 0$, let $W(\lambda) = I_m + C(\lambda I_n - A)^{-1}B$ be a minimal realization. Let $\mathcal{M}_+$, resp. $\mathcal{M}_-$, be the spectral subspace of A corresponding to the open upper, resp. lower, half plane. Let $A^\times = A - BC$. Let X be an invertible Hermitian solution of (3.7). *Let $\mathcal{M}_-^\times$, resp. $\mathcal{M}_+^\times$ be $A^\times$-invariant subspaces such that $\mathcal{M}_+^\times$ is maximal X^{-1}-nonnegative and $\mathcal{M}_-^\times$ is maximal X^{-1}-nonpositive, and in addition*

$$\sigma(A^\times|_{M_-^\times}) \subset \{\lambda | \operatorname{Im} \lambda \leq 0\}, \quad \mathcal{M}_-^\times \supseteq \mathcal{R}(A^\times, \mathbb{C}_{low});$$

$$\sigma(A^\times|_{\mathcal{M}_+^\times}) \subset \{\lambda | \operatorname{Im} \lambda \geq 0\}, \quad \mathcal{M}_+^\times \supseteq \mathcal{R}(A^\times, \mathbb{C}_{upp}).$$

Then

$$\mathbb{C}^n = \mathcal{M}_+ \dot{+} \mathcal{M}_-^\times = \mathcal{M}_- \dot{+} \mathcal{M}_+^\times. \tag{3.10}$$

The factors in a left pseudospectral factorization (3.1) *are given by* (3.3), *the factors in a right pseudospectral factorization* (3.2) *are given by* (3.5).

The formulas for the factors, in case U is not strictly proper, as well as the Riccati equation (3.7), become less transparent, therefore we have chosen to present only the strictly proper case in full detail here. Compare [10]. We also note that the results of Theorem 3.3 will hold as well for non-minimal realizations, provided we assume that there exists an invertible Hermitian solution of the Riccati equation (3.7).

Observe that the choices for $\mathcal{M}_+^\times$ and $\mathcal{M}_-^\times$ can depend on which solution X of (3.7) we take. If we assume that the realization $W(\lambda) = I + C(\lambda - A)^{-1}B$ is minimal, then the algebraic Riccati equation

$$C^*C - iA^*X^{-1} + iX^{-1}A + X^{-1}BB^*X^{-1} = 0 \tag{3.11}$$

has the *maximal* Hermitian solution X_{max}^{-1} and the *minimal* Hermitian solution X_{min}^{-1} (see, e.g., Section 7.5 in [12]). They are characterized by the property that $X_{min}^{-1} \leq X^{-1} \leq X_{max}^{-1}$ for every Hermitian solution X^{-1} of (3.11). Again, controllability of (A, B) is sufficient to guarantee existence of $X_{min}^{-1}, X_{max}^{-1}$.

Since only Hermitian solutions of (3.7) and (3.11) (as well as of other Riccati equations in the text) will be considered, we shall use "solution" to mean "Hermitian solution". Note that X_{min}^{-1} and X_{max}^{-1} need not be the inverses of the maximal and minimal solution, respectively, of (3.7). Observe also that any solution of (3.7) has the same inertia, and the same is true for (3.11) (see [10]). So, maximal X^{-1}-nonnegative subspaces have the same dimension for any solution X^{-1} of (3.11). Moreover, if $\mathcal{M}$ is a maximal X^{-1}-nonnegative subspace for some solution X^{-1} of (3.11) then $\mathcal{M}$ is also maximal X_{max}^{-1}-nonnegative, since $X^{-1} \le X_{max}^{-1}$. On the other hand, if $\mathcal{M}$ is maximal X_{min}^{-1}-nonnegative then it is maximal X^{-1}-nonnegative for any solution X^{-1} of (3.11), as $X_{min}^{-1} \le X^{-1}$.

We have seen before that stability of invariant maximal nonnegative subspaces has to do with the numerical range condition. In this respect the following result is of interest.

Proposition 3.4 *Let X and Y be two solutions of the algebraic Riccati equation (3.7). Assume that the pair is (A, B) controllable. Then $(A^\times, X^{-1})$ satisfies the numerical range condition if and only if $(A^\times, Y^{-1})$ does.*

In that case the unique $A^\times$-invariant maximal X^{-1}-nonnegative subspace $\mathcal{M}$ with

$$\sigma(A^\times|_{\mathcal{M}}) \subset \{\lambda \in \mathbb{C} : \operatorname{Im} \lambda \ge 0\}$$

does not depend on X.

Proof: We shall show that for any real eigenvalue λ of $A^\times$ and for every Jordan chain $x_1, \dots, x_k$ of $A^\times$ corresponding to λ we have that

$$X^{-1}x_j = Y^{-1}x_j \tag{3.12}$$

whenever $j \le [\frac{k}{2}]+1$ if k is odd, and whenever $j \le \frac{k}{2}$ if k is even. As the numerical range condition holding or not is determined by the last of these vectors, the result of the proposition will follow once this is established.

We shall proceed by induction. Suppose that k is odd. Then, as $A^\times x_1 = \lambda x_1$ and λ is real, we have

$$\frac{1}{2i}x_1^*((A^\times)^*X^{-1} - X^{-1}A^\times)x_1 = 0,$$

while on the other hand this equals $-\frac{1}{2}\|(iC^* - X^{-1}B)^*x_1\|^2$ by (3.8). So $iCx_1 = B^*X^{-1}x_1$. As the same holds with Y instead of X we see that $B^*(X^{-1} - Y^{-1})x_1 = 0$. Moreover, again from the equation (3.8) we see that

$$(A^\times)^*X^{-1}x_1 - \lambda X^{-1}x_1 = -\frac{1}{2}(iC^* - X^{-1}B)(iC^* - X^{-1}B)^*x_1 = 0.$$

Again, the same holds with Y in place of X. Thus we have

$$((A^\times)^* - \lambda)(X^{-1} - Y^{-1})x_1 = 0, \quad B^*(X^{-1} - Y^{-1})x_1 = 0.$$

Controllability of (A, B) now implies that $X^{-1}x_1 = Y^{-1}x_1$.

Now suppose that we have shown that $X^{-1}x_j = Y^{-1}x_j$ for $j = 1, \dots, l$, and suppose that $l < [\frac{k}{2}] + 1$. Then

$$\frac{1}{2i}x_{l+1}^*\{((A^\times)^* - \lambda)X^{-1} - X^{-1}(A^\times - \lambda)\}x_{l+1}$$

$$= \frac{1}{2i}(x_l^* X^{-1} x_{l+1} - x_{l+1}^* X^{-1} x_l) = -\text{Im } \langle X^{-1} x_l, x_{l+1} \rangle$$

$$= -\text{Im } \langle Y^{-1} x_l, x_{l+1} \rangle = 0,$$

where the last equation is true as $l < [\frac{k}{2}] + 1$. This follows from the simple form for dissipative matrices in an indefinite inner product space (see [23], Section 2.2, also [19]). On the other hand, this equals $-\frac{1}{2}\|(iC - B^* X^{-1})x_{l+1}\|^2$, so $iCx_{l+1} = B^* X^{-1} x_{l+1}$. Repeating the argument for Y instead of X we see that $Cx_{l+1} = B^* Y^{-1} x_{l+1}$, and we obtain

$$B^*(X^{-1} - Y^{-1})x_{l+1} = 0.$$

Also,

$$((A^\times)^* X^{-1} - X^{-1} A^\times) x_{l+1} = ((A^\times)^* - \lambda) X^{-1} x_{l+1} - X^{-1} x_l.$$

Again repeating the argument with Y in place of X and recalling that $X^{-1} x_l = Y^{-1} y_l$ we get

$$((A^\times)^* - \lambda)(X^{-1} - Y^{-1})x_{l+1} = 0.$$

Controllability implies that also $X^{-1} x_{l+1} = Y^{-1} x_{l+1}$.

The case where k is even is dealt with in the same way.

Suppose that $\mathcal{M}_{min}$ is the unique $A^\times$-invariant maximal X_{min}^{-1}-nonnegative subspace with $\sigma(A^\times|_{\mathcal{M}_{min}}) \subset \mathbb{C}_{upp}$. Let $\mathcal{M}$ be the unique $A^\times$-invariant maximal X^{-1}-nonnegative subspace with $\sigma(A^\times|_{\mathcal{M}}) \subset \mathbb{C}_{upp}$. As $X_{min}^{-1} \le X^{-1}$ we have that $\mathcal{M}_{min}$ is also X^{-1}-nonnegative. It follows that $\mathcal{M} = \mathcal{M}_{min}$. The proposition is proved. □

It is natural to ask whether a subspace that is maximal X^{-1}-nonnegative for some solution X of the algebraic Riccati equation (3.7) is necessarily also Y^{-1}-nonnegative for any other solution Y of (3.7). The next example shows that this is not always the case.

Example: Take

$$A^\times = \begin{pmatrix} 0 & 1 & 0 & 0 \\ 0 & 0 & 0 & 0 \\ 0 & 0 & 0 & 1 \\ 0 & 0 & 0 & 0 \end{pmatrix}, \quad B = \begin{pmatrix} 0 & 0 \\ 1 & 0 \\ 0 & 0 \\ 0 & 1 \end{pmatrix}, \quad C = \begin{pmatrix} -1 & 0 & -2\sqrt{2} & -2i \\ 0 & 0 & -2 & 0 \end{pmatrix}.$$

Note that $(A^\times, B)$ is controllable and $(C, A^\times)$ is observable. One easily checks that

$$X^{-1} = \begin{pmatrix} 0 & -i & 0 & 0 \\ i & -\sqrt{2} & 2\sqrt{2}i & 0 \\ 0 & -2\sqrt{2}i & 0 & -2i \\ 0 & 0 & 2i & 0 \end{pmatrix} \qquad (3.13)$$

solves the Riccati equation

$$\frac{1}{2i}\{(A^\times)^* X^{-1} - X^{-1}A^\times\} = -\frac{1}{2}(iC^* - X^{-1}B)(iC^* - X^{-1}B)^*, \tag{3.14}$$

which is equivalent to (3.7). Both sides of this equality are equal to

$$\begin{pmatrix} 0 & 0 & 0 & 0 \\ 0 & -1 & 0 & -\sqrt{2} \\ 0 & 0 & 0 & 0 \\ 0 & -\sqrt{2} & 0 & -2 \end{pmatrix}$$

if (3.13) is substituted for X^{-1}. Observe that the pair $(A^\times, X^{-1})$ does not satisfy the numerical range condition. Indeed, we claim that $0 \in W\left(\begin{pmatrix} i & 2\sqrt{2}i \\ 0 & 2i \end{pmatrix}\right)$, which is easily seen by $\left\langle \begin{pmatrix} i & 2\sqrt{2}i \\ 0 & 2i \end{pmatrix}\begin{pmatrix} 1 \\ -\frac{1}{2}\sqrt{2} \end{pmatrix}, \begin{pmatrix} 1 \\ -\frac{1}{2}\sqrt{2} \end{pmatrix}\right\rangle = 0$. Thus there are many $A^\times$-invariant maximal X^{-1}-nonnegative subspaces. From what we already know, (see the proof of the previous proposition) any other solution of (3.14) has the form

$$Y^{-1} = \begin{pmatrix} 0 & -i & 0 & 0 \\ i & \alpha_1 & 2\sqrt{2}i & \alpha_2 \\ 0 & -2\sqrt{2}i & 0 & -2i \\ 0 & \overline{\alpha_2} & 2i & \alpha_3 \end{pmatrix} \tag{3.15}$$

for some real α_1, α_3 and complex α_2. Then $iC^* - Y^{-1}B$ is of the form

$$\begin{pmatrix} 0 & 0 \\ -\alpha_1 & -\alpha_2 \\ 0 & 0 \\ -2-\overline{\alpha_2} & -\alpha_3 \end{pmatrix},$$

and it follows from (3.14) that we must have

$$\begin{pmatrix} \alpha_1 & \alpha_2 \\ 2+\overline{\alpha_2} & \alpha_3 \end{pmatrix}\begin{pmatrix} \alpha_1 & \alpha_2 \\ 2+\overline{\alpha_2} & \alpha_3 \end{pmatrix}^* = \begin{pmatrix} 2 & -2\sqrt{2} \\ -2\sqrt{2} & 4 \end{pmatrix} = \begin{pmatrix} \sqrt{2} & 0 \\ -2 & 0 \end{pmatrix}\begin{pmatrix} \sqrt{2} & -2 \\ 0 & 0 \end{pmatrix}. \tag{3.16}$$

It is easily seen that taking $\alpha_1 = -\frac{\sqrt{2}}{3}$, $\alpha_2 = -\frac{4}{3}$ and $\alpha_3 = \frac{4\sqrt{2}}{3}$ satisfies (3.16). So

$$Y_0^{-1} = \begin{pmatrix} 0 & -i & 0 & 0 \\ i & -\frac{\sqrt{2}}{3} & 2\sqrt{2}i & \frac{-4}{3} \\ 0 & -2\sqrt{2}i & 0 & -2i \\ 0 & \frac{-4}{3} & 2i & \frac{4\sqrt{2}}{3} \end{pmatrix}$$

satisfies the Riccati equation (3.14) as well. Again, there are many $A^\times$-invariant Y_0^{-1}-maximal nonnegative subspaces, and indeed, $\mathcal{M}_0 = \operatorname{span}\left(\begin{pmatrix} 1 \\ 0 \\ -\frac{\sqrt{2}}{2} \\ 0 \end{pmatrix}, \begin{pmatrix} 0 \\ 1 \\ 0 \\ -\frac{\sqrt{2}}{2} \end{pmatrix}\right)$ is Y_0^{-1}-nonnegative, as is easily checked. The Gram matrix of any solution as in (3.15) for the subspace $\mathcal{M}_0$ is $\begin{pmatrix} 0 & 0 \\ 0 & \beta \end{pmatrix}$ where

$$\beta = \left\langle \begin{pmatrix} \alpha_1 & \alpha_2 \\ \overline{\alpha_2} & \alpha_3 \end{pmatrix} \begin{pmatrix} 1 \\ -\frac{\sqrt{2}}{2} \end{pmatrix}, \begin{pmatrix} 1 \\ -\frac{\sqrt{2}}{2} \end{pmatrix} \right\rangle = \alpha_1 - \sqrt{2}\operatorname{Re}\alpha_2 + \frac{1}{2}\alpha_3.$$

So $\mathcal{M}_0$ is Y_0^{-1}-nonnegative but not X^{-1}-nonnegative.

We now turn to stability of the special pseudospectral factorizations that are obtained in the way described in Theorem 3.3. Note that in general not all pseudospectral factorizations are obtained from Theorem 3.3; for this reason we shall call the left pseudospectral factorization $W(\lambda) = W_1(\lambda)W_2(\lambda)$ X^{-1}-*special* (or *special*, if X^{-1} is not specified) if it corresponds to a projection π for which the image $\operatorname{Im}\pi$ is maximal X^{-1}-nonnegative and the inclusions

$$\sigma(A^\times|_{\operatorname{Im}\pi}) \subset \{\lambda|\operatorname{Im}\lambda \geq 0\}, \quad \operatorname{Im}\pi \supseteq \mathcal{R}(A^\times, \mathbb{C}_{upp}) \tag{3.17}$$

hold. Here, X^{-1} is a solution of (3.11), or equivalently X is a solution of (3.7). (Results similar to the ones to be developed below are valid also for the case where only right pseudospectral factorizations are considered for which $\operatorname{Im}\pi$ is maximal X^{-1}-nonpositive and the inclusions analogous to (3.17) hold true.) We shall be especially concerned with X_{max}^{-1}-special and X_{min}^{-1}-special pseudospectral factorizations. Note that if a factorization is X_{min}^{-1}-special then it is X^{-1}-special for any solution X of (3.7), while if it is X^{-1}-special for some solution X of (3.7), then it is X_{max}^{-1}-special.

Proposition 3.5 *If the pair* $(A^\times, X^{-1})$ *satisfies the numerical range condition for some solution* X *of* (3.7), *then there is a unique* Y^{-1}*-special pseudospectral factorization for any* Y *which solves the algebraic Riccati equation* (3.7), *and moreover, this special pseudospectral factorization is independent of the particular choice of the solution of the algebraic Riccati equation.*

Proof: The proposition follows immediately from Proposition 3.4 and the remarks preceding the proposition, keeping in mind that A has no real eigenvalues. □

Let $W(\lambda) = I + U(\lambda) = I + C(\lambda - A)^{-1}B$ be a minimal realization. A special pseudospectral factorization

$$W(\lambda) = W_1(\lambda)W_2(\lambda)$$

of $W(\lambda)$, where $W_i(\lambda) = I + C_i(\lambda - A_i)^{-1}B_i$ $(i = 1, 2)$ are minimal realizations, is called *unconditionally stable* if for every $\varepsilon > 0$ there is a $\delta > 0$ such that for every triple of matrices A', B', C' with

$$\|A - A'\| + \|B - B'\| + \|C - C'\| < \delta$$

the function $U'(\lambda) = C'(\lambda - A')^{-1}B'$ is contractive and $W'(\lambda) = I + U'(\lambda)$ has a pseudospectral factorization $W'(\lambda) = W_1'(\lambda)W_2'(\lambda)$ where the factors $W_i'(\lambda) = I + C_i'(\lambda - A_i')^{-1}B_i'$ satisfy

$$\|A_i - A_i'\| + \|B_i - B_i'\| + \|C_i - C_i'\| < \varepsilon \qquad (i = 1, 2). \tag{3.17}$$

The X_{max}^{-1}-special pseudospectral factorization of $W(\lambda)$, is called *conditionally stable with respect to* X_{max}^{-1} if for every $\varepsilon > 0$ there is a $\delta > 0$ such that for every triple of matrices A', B', C' with

$$\|A - A'\| + \|B - B'\| + \|C - C'\| < \delta$$

for which the function $U'(\lambda) = C'(\lambda - A')^{-1}B'$ is contractive, the function $W'(\lambda) = I + U'(\lambda)$ has a $(X_{max}')^{-1}$-special pseudospectral factorization $W'(\lambda) = W_1'(\lambda)W_2'(\lambda)$ where the factors $W_i'(\lambda) = I + C_i'(\lambda - A_i')^{-1}B_i'$ satisfy (3.17). Here $(X_{max}')^{-1}$ is the maximal solution of the analogue of the Riccati equation (3.11) for the function $W'(\lambda)$.

Likewise, replacing X_{max}^{-1} by X_{min}^{-1}, we obtain the definition of *conditionally stable with respect to* X_{min}^{-1}.

Theorem 3.6 *There exists an unconditionally stable special pseudospectral factorization of* $W(\lambda) = I + U(\lambda)$, *where* $\|U(\lambda)\| \leq 1$ *for* $\lambda \in \mathbb{R}$, *if and only if* $U(\lambda)$ *is a strict contraction.*

Let $U(\lambda) = C(\lambda - A)^{-1}B$ *be a minimal realization, where* $\|U(\lambda)\| \leq 1$ *for* $\lambda \in \mathbb{R}$.

If the pair $(A^\times, X_{max}^{-1})$ *satisfies the numerical range condition then there is a unique* X_{max}^{-1}*-special pseudospectral factorization of* $W(\lambda) = I + U(\lambda)$. *This pseudospectral factorization is conditionally stable with respect to* X_{max}^{-1}.

There exists a X_{min}^{-1}*-special pseudospectral factorization of* $W(\lambda) = I + U(\lambda)$ *that is conditionally stable with respect to* X_{min}^{-1} *if and only if the pair* $(A^\times, X_{min}^{-1})$ *satisfies the numerical range condition. In that case, there is a unique* X_{min}^{-1}*-special pseudospectral factorization of* W. *This factorization is also conditionally stable with respect to* X_{max}^{-1}.

For the Proof of Theorem 3.6 it is convenient to formulate a lemma which is perhaps of independent interest.

Lemma 3.7 *Let* Q *be a positive semidefinite* $n \times n$ *matrix written in the form* $Q = ZZ^*$, *where* Z *is also* $n \times n$. *Then there exists a positive constant* C, *depending on Q and Z only, such that every $n \times n$ positive semidefinite $n \times n$ matrix Q' has a representation $Q' = Z'(Z')^*$ in which*

$$\|Z' - Z\| \leq C\,\|Q' - Q\|.$$

Proof: Using the transformation $Q \to SQS^*$, $Z \to SZ$ for a suitable invertible matrix S, we may (and do) assume, without loss of generality, that

$$Q = \begin{pmatrix} I_k & 0 \\ 0 & 0 \end{pmatrix},$$

where I_k is the $k \times k$ identity matrix ($0 \leq k \leq n$). Then Z must be of the form

$$Z = \begin{pmatrix} Z_1 & Z_2 \\ 0 & 0 \end{pmatrix},$$

with Z_1 of size $k \times k$. Replacing Z by ZU, for a suitable unitary matrix U, we may further assume that $Z_1 = I_k$, $Z_2 = 0$.

Next, observe that we have to prove Lemma 3.7 only for Q' sufficiently close to Q. Indeed, suppose Q' is positive semidefinite, and $\|Q' - Q\| \geq \varepsilon_0$. Then for any Z' such that $Q' = Z'(Z')^*$ we have (provided $\|Q'\| \geq 1$)

$$\begin{aligned} \|Z' - Z\| &\leq \|Z\| + \|Z'\| = \|Q\|^{1/2} + \|Q'\|^{1/2} \\ &\leq \|Q\|^{1/2} + \|Q'\| \leq \|Q\|^{1/2} + \|Q' - Q\| + \|Q\| \\ &\leq C_1\varepsilon_0 + \|Q' - Q\| \leq (C_1 + 1)(\|Q' - Q\|), \end{aligned} \tag{3.18}$$

where $C = \varepsilon_0^{-1}(\|Q\|^{1/2} + \|Q\|)$. If $\|Q'\| \leq 1$, then analogously to (3.18)

$$\|Z' - Z\| \leq \|Q\|^{1/2} + 1 \leq C_2\|Q' - Q\|,$$

where $C_2 = \varepsilon^{-1}(\|Q\|^{1/2} + 1)$. Thus, if the statement of Lemma 3.7 is proved for $\|Q' - Q\| < \varepsilon_0$ with a certain constant C, replace C by $\max(C, C_1 + 1, C_2)$ to prove Lemma 3.7 for all positive semidefinite Q'.

Now, with

$$Q = Z = \begin{pmatrix} I_k & 0 \\ 0 & 0 \end{pmatrix},$$

write a positive semidefinite Q' in the form

$$Q' = \begin{pmatrix} I_k + V_{11} & V_{12} \\ V_{12}^* & V_{22} \end{pmatrix}$$

for some blocks V_{11}, V_{12}, and V_{22}. As explained in the preceding paragraph, we can assume that V_{11} is small. Now let

$$Z' = \begin{pmatrix} (I_k + V_{11})^{1/2} & 0 \\ V_{12}^*\,(I_k + V_{11})^{-1/2} & (V_{22} - V_{12}^*\,(I_k + V_{11})^{-1}\,V_{12})^{1/2} \end{pmatrix}.$$

It is easy to see that Z' satisfies the required properties (if V_{11} is small). □

Proof of Theorem 3.6: Suppose first that $W(\lambda) = I + U(\lambda)$, where $U(\lambda)$ is contractive but not strictly contractive. Then $(1+\varepsilon)U(\lambda)$ converges to $U(\lambda)$ as $\varepsilon \downarrow 0$, but is not contractive. Because of the fact that we required the perturbed function to be automatically contractive, no special pseudospectral factorization can be unconditionally stable in this case.

Conversely, assume that $U(\lambda)$ is a strict contraction. Then for δ small enough also $U'(\lambda)$ will be strictly contractive, and hence $W'(\lambda) = I + U'(\lambda)$ will allow canonical Wiener-Hopf factorization with the factors close to those of $W(\lambda)$ in the sense of the definition above (see [4]).

Turning to conditional stability, assume $(A^\times, X_{max}^{-1})$ satisfies the numerical range condition. Let A', B' and C' be perturbations of A, B and C such that $C'(\lambda - A')^{-1}B'$ is contractive. So, the perturbed Riccati equation (3.11) has a solution by assumption. It is well-known, see e.g. [12], Chapter 11, also [14], that the maximal solution X_{max}^{-1} is stable under small perturbations of A, B and C (as is the minimal solution). Denote by X'^{-1}_{max} the maximal solution of (3.11), where we replace A, B, C by A', B' and C', respectively. By Theorem 2.1 for δ small enough there exists an $A'^\times$-invariant maximal X'^{-1}_{max}-nonnegative subspace $\mathcal{M}'^\times_+$ which is such that all eigenvalues of $A'^\times$ restricted to this subspace are in the closed upper half plane, and with $\operatorname{gap}(\mathcal{M}^\times_+, \mathcal{M}'^\times_+)$ as small as we wish. It then follows that the special pseudospectral factorization, which is unique (by Proposition 3.4), is conditionally stable with respect to X_{max}^{-1}.

The same argument, replacing X_{max}^{-1} by X_{min}^{-1} also proves the "if" part of the third statement of the theorem.

Conversely, assume there is a conditionally stable X_{min}^{-1}-special factorization $W(\lambda) = W_1(\lambda)W_2(\lambda)$ of $W(\lambda) = I + U(\lambda)$ with respect to X_{min}^{-1}.

Let $A'^\times$ be any X_{min}^{-1}-dissipative matrix close to $A^\times$. Write

$$\frac{1}{2i}(X_{min}(A'^\times)^* - A'^\times X_{min}) = -\frac{1}{2}DD^*,$$

then by Lemma 3.7 D can be chosen in such a way that $\|D - (iX_{min}C^* - B)\|$ is small. Put $C' = \{-iX_{min}^{-1}(D+B)\}^*$, then $\|C - C'\|$ is small as well, and put $A' = A'^\times + BC'$, then also $\|A - A'\|$ is small. Now let

$$U'(\lambda) = C'(\lambda - A')^{-1}B,$$

and $W'(\lambda) = I + U'(\lambda)$. We show that $U'(\lambda)$ is contractive. Indeed, X_{min} satisfies

$$\begin{aligned} &\frac{1}{2i}(X_{min}(A'^\times)^* - A'^\times X_{min}) \\ &= -\frac{1}{2}(iX_{min}C'^* - B)(iX_{min}C'^* - B)^*, \end{aligned} \tag{3.19}$$

which is equivalent to (3.7) for the realization for $U'(\lambda)$. Now let the X_{min}^{-1}-special pseudospectral factorization $W(\lambda) = W_1(\lambda)W_2(\lambda)$ which is conditionally stable with respect to X_{min}^{-1} be the one corresponding to the $A^\times$-invariant maximal

X_{min}^{-1}-nonnegative subspace $\mathcal{M}_+^\times$. From the stability it follows that there exists an $A'^\times$-invariant maximal $X_{min}'^{-1}$-nonnegative subspace $\mathcal{M}_+'^\times$ such that $\operatorname{gap}(\mathcal{M}_+^\times, \mathcal{M}_+'^\times)$ is as small as we wish. As X_{min}^{-1} satisfies the equation (3.19) for the perturbed function $U'(\lambda)$, we have $X_{min}'^{-1} \leq X_{min}^{-1}$, and so $\mathcal{M}_+'^\times$ is also maximal X_{min}^{-1}-nonnegative. It then follows that $\mathcal{M}_+^\times$ is stable as an $A^\times$-invariant maximal X_{min}^{-1}-nonnegative subspace. Applying Theorem 2.1 then shows that $(A^\times, X_{min}^{-1})$ satisfies the numerical range condition. From Proposition 3.4 it follows that $(A^\times, X_{max}^{-1})$ satisfies the numerical range condition as well. Applying the already proved part of the theorem then finishes the proof.

4 Positive Real Functions

In this section rational matrix functions $W(\lambda)$ which have a positive semidefinite real part will be studied. Such functions allow pseudospectral factorizations (under the non-degeneracy condition that the real part of $W(\lambda_0)$ is positive definite for some real λ_0). We start with reviewing several results from [19] concerning formulas for such factorizations (their existence was established already in [15].

In the following $W(\lambda)$ will denote an $m \times m$ rational matrix function. The next three propositions are taken from [19].

Proposition 4.1 *Let $W(\lambda) = D + C(\lambda I - A)^{-1}B$ be a controllable realization, and assume $D + D^* > 0$. Then the following are equivalent.*

(i) $W(\lambda) + W(\lambda)^* \geq 0$ *for* $\lambda \in \mathbb{R} \setminus \{poles\ of\ W\}$;

(ii) *there exists a Hermitian matrix X such that*

$$X(i(A - BG^{-1}C)) + (i(A - BG^{-1}C))^*$$

(4.1)
$$X - XBG^{-1}B^*X - C^*G^{-1}C = 0,$$

where $G = D + D^*$.

As in the previous section, only Hermitian solutions of (4.1) will be considered. For the case when A is stable this result is known as the positive real lemma (see, e.g., [2]). Also we remark that the implication (ii) $\Rightarrow$ (i) is true without any assumption on controllability of the realization, provided there exists an invertible solution X of (4.1), and $G = D + D^* > 0$. Indeed, it is a matter of straightforward checking to see that in that case, we have the equality

$$W(\lambda) + W(\lambda)^* = V(\lambda)GV(\lambda)^*, \quad \lambda \in \mathbb{R} \setminus \{\text{poles of } W(\lambda)\},$$

where

$$V(\lambda) = I + C(\lambda - A)^{-1}(B - iX^{-1}C^*)G^{-1}.$$

It turns out that under an additional minimality assumption every solution X of (4.1) is invertible:

Proposition 4.2 *Let $W(\lambda) = D + C(\lambda I - A)^{-1}B$ be an observable realization. Assume $D + D^* > 0$ and $W(\lambda) + W(\lambda)^* \geq 0$ for all $\lambda \in \mathbb{R} \setminus \sigma(A)$. Then every Hermitian solution X of* (4.1) *is invertible.*

An important consequence of (4.1) is the following observation.

Proposition 4.3 *Let $W(\lambda) = D + C(\lambda I - A)^{-1}B$ be a minimal realization. Assume $D + D^* > 0$ and $W(\lambda) + W(\lambda)^* \geq 0$ for all $\lambda \in \mathbb{R} \setminus \sigma(A)$. Then for any invertible solution X of* (4.1) *the matrices A and $A^\times = A - BD^{-1}C$ are both $(-X)$-dissipative. More precisely,*

$$\frac{1}{2i}(XA - A^*X) = -\frac{1}{2}(XB - iC^*)G^{-1}(B^*X + iC); \tag{4.2}$$

$$\begin{aligned}\frac{1}{2i}(XA^\times - (A^\times)^*X) = -\frac{1}{2}&(XBD^{-1}D^* + iC^*)G^{-1}\\ &(D(D^*)^{-1}B^*X - iC).\end{aligned} \tag{4.3}$$

The following theorem (proved in [19]) describes a way to construct a pseudospectral factorization of $W(\lambda)$.

Theorem 4.4 *Let $W(\lambda) = D + C(\lambda I - A)^{-1}B$ be a minimal realization. Assume $D + D^* > 0$ and $W(\lambda) + W(\lambda)^* \geq 0$ for all $\lambda \in \mathbb{R} \setminus \sigma(A)$. Let X be a Hermitian invertible solution of* (4.1) *for which the matrices A and $A^\times = A - BD^{-1}C$ are both $(-X)$-dissipative. Let $\mathcal{M}$ and $\mathcal{M}^\times$ be subspaces with the following properties:*

(i) *$\mathcal{M}$ is A-invariant maximal $(-X)$-nonnegative, $\mathcal{M}^\times$ is $A^\times$-invariant maximal $(-X)$-nonpositive;*

(ii) *$A|_{\mathcal{M}}$ has all its eigenvalues in the closed upper half plane, $A^\times|_{\mathcal{M}^\times}$ has all its eigenvalues in the closed lower half plane;*

(iii) *$\mathcal{M} \supseteq \mathcal{R}(A, \mathbb{C}_{upp})$; $\mathcal{M}^\times \supseteq \mathcal{R}(A^\times, \mathbb{C}_{low})$.*

Then $\mathcal{M} \dot{+} \mathcal{M}^\times = \mathbb{C}^n$.

Let π be the projection onto $\mathcal{M}^\times$ along $\mathcal{M}$. Put

$$\begin{aligned} W_-(\lambda) &= D + C(\lambda I - A)^{-1}(I - \pi)B;\\ W_+(\lambda) &= I + D^{-1}C\pi(\lambda I - A)^{-1}B. \end{aligned} \tag{4.4}$$

Then

$$W(\lambda) = W_-(\lambda)W_+(\lambda)$$

is a right pseudospectral factorization of $W(\lambda)$.

In a similar way it can be shown that $W(\lambda)$ admits a left pseudospectral factorization, and formulas for the factors may be given.

First we state the analogue of Proposition 3.4 for this case.

Proposition 4.5 *Let* $W(\lambda) = D + C(\lambda I - A)^{-1}B$ *be a minimal realization, and assume that* $G = D + D^* > 0$. *Let* X *be an invertible solution of* (4.1).

(i) *If* $(A, -X)$ *satisfies the numerical range condition, so does* $(A, -Y)$ *for any other solution* Y *of* (4.1).

(ii) *If* $(A^\times, -X)$ *satisfies the numerical range condition, so does* $(A^\times, -Y)$ *for any other solution* Y *of* (4.1).

Proof: The proof is the same as the Proof of Proposition 3.4, using (4.2) and (4.3) instead of (3.8). □

Let $W(\lambda)$ have positive semidefinite real part, and let $W(\lambda) = D + C(\lambda - A)^{-1}B$ be a minimal realization. A pseudospectral factorization of W will be called X-*special* (or *special* if X is not specified) if the image and kernel of the corresponding supporting projection π are maximal $(-X)$-nonnegative and maximal $(-X)$-nonpositive, respectively, where X is a solution of (4.1).

Assume, in addition, that $G = D + D^* > 0$, and let X be a solution of (4.1). An X-special pseudospectral factorization

$$W(\lambda) = W_1(\lambda)W_2(\lambda)$$

of $W(\lambda)$, where $W_1(\lambda) = D + C_1(\lambda - A_1)^{-1}B_1$ and $W_2(\lambda) = I + C_2(\lambda - A_2)^{-1}B_2$, is called *unconditionally stable* if for every $\varepsilon > 0$ there is a $\delta > 0$ such that for every quadruple of matrices A', B', C', D' with

$$\|A - A'\| + \|B - B'\| + \|C - C'\| + \|D - D'\| < \delta$$

the function $W'(\lambda) = D' + C'(\lambda - A')^{-1}B'$ satisfies $W'(\lambda) + W'(\lambda)^* \geq 0$ for all real λ not poles of W', and $W'(\lambda)$ has a pseudospectral factorization $W'(\lambda) = W_1'(\lambda)W_2'(\lambda)$ where the factors $W_1'(\lambda) = D' + C_1'(\lambda - A_1')^{-1}B_1'$ and $W_2'(\lambda) = I + C_2'(\lambda - A_2')^{-1}B_2'$ satisfy

$$\|A_i - A_i'\| + \|B_i - B_i'\| + \|C_i - C_i'\| < \varepsilon \qquad (i = 1, 2). \tag{4.5}$$

Theorem 4.6 *Assume that* $W(\lambda) = D + C(\lambda I - A)^{-1}B$ *is a minimal realization and that* W *satisfies* $W(\lambda) + W(\lambda)^* \geq 0$ *for all real* λ *which are not poles of* W. *There exists an unconditionally stable special pseudospectral factorization of* W *if and only if* W *has no real poles and zeros, i.e., if and only if* $W(\lambda) + W(\lambda)^* > 0$ *for all real* λ.

Proof: Put $W'(\lambda) = (1-\varepsilon)D + C(\lambda I - A)^{-1}B$. If α is a real zero of W, then there is a nonzero vector x such that $\lim_{\lambda\to\alpha} W(\lambda)x = 0$. But then also $\lim_{\lambda\to\alpha}(x^*(W(\lambda)^* + W(\lambda))x) = 0$. Then

$$\lim_{\lambda\to\alpha}(x^*(W'(\lambda) + W'(\lambda)^*)x) = -\varepsilon x^*(D + D^*)x < 0$$

so $W' + (W')^*$ is not positive semidefinite on the real line.

Observe that if $W + W^*$ is positive semidefinite on the real line, then so is $W^{-1} + W^{-1*}$, as $0 \leq W(\lambda)^{-1}(W(\lambda) + W(\lambda)^*)W(\lambda)^{-1*} = W(\lambda)^{-1} + W(\lambda)^{-1*}$, for any real λ which is neither a zero nor a pole of W. So, if α is a real pole of W one applies the reasoning of the previous paragraph to W^{-1} to see that there is a perturbation W' for which $W' + W'^*$ is not positive semidefinite on the real line. □

Let X_{max} (resp., X_{min}) be the maximal (resp., minimal) solution of (4.1). The X_{max}-special pseudospectral factorization of $W(\lambda)$, is called *conditionally stable with respect to* X_{max} if for every $\varepsilon > 0$ there is a $\delta > 0$ such that for every triple of matrices A', B', C' with

$$\|A - A'\| + \|B - B'\| + \|C - C'\| < \delta$$

for which the function $W'(\lambda) = D' + C'(\lambda - A')^{-1}B'$ satisfies $W'(\lambda) + W'(\lambda)^* \geq 0$ for all real λ not poles of W', the function $W'(\lambda)$ has a X'_{max}-special pseudospectral factorization $W'(\lambda) = W'_1(\lambda)W'_2(\lambda)$ where the factors $W'_1(\lambda) = D' + C'_1(\lambda - A'_1)^{-1}B'_1$ and $W'_2(\lambda) = I + C'_2(\lambda - A'_2)^{-1}B'_2$ satisfy (4.5).

Likewise, replacing X_{max} by X_{min}, we obtain the definition of *conditionally stable with respect to* X_{min}.

Theorem 4.7 *Assume that* $W(\lambda) = D + C(\lambda I - A)^{-1}B$ *is a minimal realization and that* W *satisfies* $W(\lambda) + W(\lambda)^* \geq 0$ *for all real* λ *which are not poles of* W.

If $(A^\times, -X_{max})$ *and* $(A, -X_{max})$ *both satisfy the numerical range condition, then the following hold true:*

(i) *there is a unique* X*-special pseudospectral factorization for any* X *that solves* (4.1),

(ii) *this special pseudospectral factorization does not depend on* X,

(iii) *this special pseudospectral factorization is conditionally stable with respect to* X_{max} *as well as with respect to* X_{min}.

Conversely, suppose there is a pseudospectral factorization that is both X_{max}*-special and* X_{min}*-special, and is conditionally stable with respect to both* X_{max} *and* X_{min}. *Then* $(A^\times, -X)$ *and* $(A, -X)$ *both satisfy the numerical range condition for every solution* X *of* (4.1).

Proof: First observe that if a pseudospectral factorization is X-special then the image $\mathcal{M}^\times$ of the corresponding supporting projection is $(-X_{min})$-nonnegative, since it is $(-X)$-nonnegative and $-X \leq -X_{min}$. Likewise, the kernel $\mathcal{M}$ of the supporting projection is $(-X_{max})$-nonpositive since it is $(-X)$-nonpositive and $-X_{max} \leq -X$.

Assume that both $(A, -X_{max})$ and $(A^\times, -X_{max})$ satisfy the numerical range condition. Then, according to Proposition 4.5 there is for any solution X of (4.1) a unique A-invariant maximal $(-X)$-nonpositive subspace $\mathcal{M}$ and a unique $A^\times$-invariant maximal $(-X)$-nonnegative subspace $\mathcal{M}^\times$ such that the eigenvalues of $A|_{\mathcal{M}}$ are in the closed lower half plane and the eigenvalues of $A^\times|_{\mathcal{M}^\times}$ are in the closed upper half plane. A priori these subspaces depend on X. However, using the remarks of the preceding paragraph we see that they are the same for all X. Indeed, let $\mathcal{M}_{min}$ be the unique A-invariant maximal $(-X_{min})$-nonpositive subspace such that the eigenvalues of $A|_{\mathcal{M}_{min}}$ are in the closed lower half plane. Then, as $-X \leq -X_{min}$ for every solution X of (4.1), we see that $\mathcal{M}_{min}$ is also $(-X)$-nonpositive. Thus it is the unique A-invariant maximal $(-X)$-nonpositive subspace for which the eigenvalues of the restriction of A to it are all in the closed lower half plane. This holds for all solutions X, so this subspace is independent of X. Likewise, one shows that the unique $A^\times$-invariant maximal $(-X)$-nonnegative subspace does not depend on X. This proves (i) and (ii).

To prove (iii), let A', B', C' and D' be perturbations of A, B, C and D for which the perturbed Riccati equation of the form (4.1) has a Hermitian solution. Again by [14] or [12], Chapter 11, the maximal and minimal solutions of (4.1) are stable under perturbations of the coefficients. By Theorem 2.1 the subspaces $\mathcal{M}$ and $\mathcal{M}^\times$ above are stable under small perturbations. So the perturbed function $W'(\lambda)$ admits a pseudospectral factorization with respect to X'_{max} as well with respect to X'_{min}, with factors close to the factors of W (as it follows from the formulas (4.4)) in its unique pseudospectral factorization with respect to X_{max} and X_{min}. This proves part (iii).

Conversely, assume that there is pseudospectral factorization that is special with respect to both X_{max} and X_{min} and also conditionally stable with respect to both. Let $A'^\times$ be any $(-X_{max})$-dissipative matrix close to $A^\times$. Write (compare (4.3))

$$\frac{1}{2i}(X_{max}A'^\times - (A'^\times)^* X_{max}) = -\frac{1}{2}E^*G^{-1}E. \tag{4.6}$$

By Lemma 3.7, we may assume that E is close to $DD^{*-1}B^*X_{max} - iC$. Put $C' = i(E - DD^{*-1}B^*X_{max})$, then $\|C - C'\|$ is small as well. Put $A' = A'^\times + BD^{-1}C'$. Then $W'(\lambda) = D + C'(\lambda - A')^{-1}B$ satisfies $W' + W'^* \geq 0$ on the real line, as X_{max} solves the corresponding algebraic Riccati equation (4.6) (by Proposition 4.1). Let the special pseudospectral factorization that is conditionally stable with respect to both X_{max} and X_{min} be the one corresponding to the $A^\times$ invariant subspace $\mathcal{M}^\times$ and the A-invariant subspace $\mathcal{M}$. Thus $\mathcal{M}$ is maximal $(-X_{max})$-nonpositive and maximal $(-X_{min})$-nonpositive, while $\mathcal{M}^\times$ is maximal $(-X_{max})$-nonnegative as well as maximal $(-X_{min})$-nonnegative. From the stability it follows that there exists an $A'^\times$-invariant maximal $(-X'_{max})$-nonnegative subspace $\mathcal{M}'$ close to $\mathcal{M}$.

Here X'_{max} is the maximal solution of the Riccati equation (4.6) corresponding to W'. As X_{max} also satisfies the same Riccati equation, we have $-X'_{max} \leq -X_{max}$, and hence $\mathcal{M}'$ is $(-X_{max})$-nonnegative as well. It follows from Theorem 2.1 that $(A^{\times}, -X_{max})$ satisfies the numerical range condition, and Proposition 4.5 implies that the same holds for $(A^{\times}, -X)$, for any other solution X of (4.1).

To show that $(A, -X_{min})$ satisfies the numerical range condition, let A' be an arbitrary $(-X_{min})$-dissipative matrix sufficiently close to A, and write (cf. (4.2))

$$\frac{1}{2i}(X_{min}A - A^*X_{min}) = -\frac{1}{2}E^*G^{-1}E.$$

Here, again by Lemma 3.7, E is close to $B^*X - iC$. Put $C' = i(E - B^*X_{min})$. Defining W' as in the previous paragraph we follow the same line of reasoning to see that there is an A'-invariant maximal $(-X'_{min})$-nonpositive subspace close to $\mathcal{M}$. Here X'_{min} is the minimal solution of the perturbed Riccati equation corresponding to W'. As X_{min} also satisfies the same Riccati equation $-X_{min} \leq -X'_{min}$, and thus $\mathcal{M}'$ is $(-X_{min})$-nonpositive as well. Again applying Theorem 2.1 and Proposition 4.5 implies that $(A, -X)$ satisfies the numerical range condition for any solution X of (4.1). □

References

[1] T.Ya. Azizov and I.S. Iokhvidov, *Linear Operators in Spaces with an Indefinite Metric*. Wiley-Interscience, 1989.

[2] B.D.O. Anderson and S. Vongpanitlerd, *Network Analysis and Synthesis*. Prentice Hall, Englewood Cliffs, 1973.

[3] H. Bart, I. Gohberg and M.A. Kaashoek, Stable factorization of monic matrix polynomials and stable invariant subspaces. *Integral Equations and Operator Theory* **1** (1978), 496–517.

[4] H. Bart, I. Gohberg and M.A. Kaashoek, *Minimal Factorization of Matrix and Operator Functions*. OT 1, Birkhäuser, Basel, 1979.

[5] H. Bart, I. Gohberg, M.A. Kaashoek and P. Van Dooren, Factorizations of transfer functions *SIAM J. Control and Optimization*, **18** (1980), 675–696.

[6] K. Clancey and I. Gohberg, *Factorization of Matrix Functions and Singular Integral Operators*. OT 3, Birkhäuser Verlag, Basel, 1981.

[7] I. Gohberg, P. Lancaster and L. Rodman, *Invariant Subspaces of Matrices with Applications*. J.Wiley & Sons, 1986.

[8] I. Gohberg, P. Lancaster and L. Rodman, *Matrices and Indefinite Scalar Products*. OT8, Birkhäuser Verlag, Basel, 1983.

[9] I. Gohberg and A.C.M. Ran, On pseudo-canonical factorization of rational matrix functions. *Indagationes Mathematicae N.S.* **4** (1993), 51–63.

[10] I. Gohberg and S. Rubinstein, Proper contractions and their unitary minimal completions, in: *Topics in Interpolation Theory of Rational Matrix-valued Functions*, OT 33, Ed. I. Gohberg, Birkhäuser Verlag, Basel, 1988, 109–123.

[11] I.S. Iohvidov, M.G. Krein and H. Langer, *Introduction to Spectral Theory of Operators in Spaces with Indefinite Metric*. Reihe "Math. Forschung", Akademie Verlag, Berlin, 1981.

[12] P. Lancaster and L. Rodman, *Algebraic Riccati Equations*. Oxford University Press, 1995.

[13] A.C.M. Ran and L. Rodman, Stability of invariant maximal semidefinite subspaces I. *Linear Algebra and its Applications* **62** (1984), 51–86.

[14] A.C.M. Ran and L. Rodman, Stability of invariant maximal semidefinite subspaces II. Applications: self-adjoint rational matrix functions, algebraic Riccati equations. *Linear Algebra and its Applications* **63** (1984), 133–173.

[15] A.C.M. Ran and L. Rodman, Generalized canonical factorization of matrix and operator functions with definite Hermitian part. *Integral Equations and Operator Theory* **15** (1992), 673–696.

[16] A.C.M. Ran and L. Rodman, Stability of neutral invariant subspaces in indefinite inner products and stable symmetric factorizations. *Integral Equations and Operator Theory* **6** (1983), 536–571.

[17] A.C.M. Ran and L. Rodman, Stable real invariant semidefinite subspaces and stable factorizations of symmetric rational matrix functions, *Linear and Multilinear Algebra* **22** (1987),25–55.

[18] A.C.M. Ran and L. Rodman, Stable invariant Lagrangian subspaces: factorization of symmetric rational matrix functions and other applications, *Linear Algebra and Applications* **137–138** (1990), 575–620.

[19] A.C.M. Ran and D. Temme, Dissipative Matrices and Invariant Maximal Semidefinite Subspaces, Linear Algebra and its Applications **212/213** (1994), 169–214.

[20] L. Roozemond, *Systems of Non-Normal and First Kind Wiener-Hopf Equations*, Ph. D. Thesis, Vrije Universiteit, Amsterdam, 1987.

[21] L. Roozemond, Canonical pseudo-spectral factorization and Wiener-Hopf integral equations, in: *Constructive methods of Wiener-Hopf factorization*, Eds. I. Gohberg, M.A. Kaashoek, OT 21, Birkhäuser Verlag, Basel, 1986, 127–156.

[22] A.A. Shkalikov, Selection principles and properties of some proper and adjoined elements of operator pencils. *Moscow Univ. Math. Bull.* **43** (1988), 16–25.

[23] Dirk Temme, *Dissipative Operators in Indefinite Scalar Product Spaces*. Ph. D. Thesis, Vrije Universiteit, 1996.

A.C.M. Ran
Divisie Wiskunde en Informatica
Faculteit Exacte Wetenschappen
Vrije Universiteit
De Boelelaan 1081a
1081 HV Amsterdam
The Netherlands
ran@cs.vu.nl

D. Temme
Pieter Nieuwlandstraat 77c
Amsterdam
The Netherlands
dtemme@dds.nl

L. Rodman
Department of Mathematics
The College of William and Mary
Williamsburg, VA 23187-8795 USA
lxrodm@math.wm.edu

AMS Classification numbers: 47B50, 47A68

Operator Theory:
Advances and Applications, Vol. 122
© 2001 Birkhäuser Verlag Basel/Switzerland

Factorization of Almost Periodic Matrix Functions of Several Variables and Toeplitz Operators

Leiba Rodman, Ilya M. Spitkovsky and Hugo J. Woerdeman

Dedicated with respect and affection to Professor M.A. Kaashoek on the occasion of his 60-th birthday

We study connections between operator theoretic properties of Toeplitz operators acting on suitable Besikovitch spaces and factorizations of their symbols which are matrix valued almost periodic functions of several real variables. Among other things, we establish the existence of a twisted canonical factorization for locally sectorial symbols, and characterize one-sided invertibility of Toeplitz operators in terms of their symbols. In addition, we study stability of factorizations, and factorizations of hermitian valued almost periodic matrix functions of several variables.

1 Introduction

Operator theoretic properties of Toeplitz operators (invertibility, Fredholmness, description of the kernel and of the range), and closely related Wiener-Hopf equations and Riemann boundary value problems, have been studied using factorization techniques for a long time. Starting with the ground breaking papers by Krein [38] and Gohberg-Krein [22], factorization properties of the Wiener-Hopf type of the symbol function, which may be scalar, matrix, or operator valued, and may have symmetries, e.g., Hermitian valued, or unitary valued, played a major role in these studies. The literature on this subject is extensive; we mention here only a few relevant books. Singular integral equations on a (more or less) general curve are studied using generalized factorizations (in the L_p classes) in [23] (scalar symbols) and [11, 41] (matrix symbols). The book [9] provides a comprehensive treatment of Toeplitz operators with matrix symbols (including various classes of discontinuous symbols) on the unit circle or the real line. A graduate level textbook exposition of basic theory of Toeplitz operators and their symbols is given in [19]. For rational matrix (and operator) valued symbols, an approach using realization (a concept borrowed from control and systems theory), leads to explicit descriptions of the corresponding Toeplitz operators and solutions of systems of Wiener-Hopf equations, see [20], [21].

Recently, in connection with convolution type equations on finite intervals, factorization of almost periodic matrix functions became a subject of considerable attention, see [32, 33]. Factorization properties of 2×2 triangular almost periodic matrix functions (motivated by applications to Wiener-Hopf equations on finite intervals) were considered in [35, 34, 8, 45]. As it happens, the factorization problem for such matrices is closely related with the almost periodic version of the Corona problem, see [6, 7, 46]. General relations between Fredholm properties of Toeplitz operators with almost periodic matrix symbols and factorization of the symbols were discussed in [31, 36, 4]. Very recently, the almost periodic factorization found still another area of application to positive and contractive extension problems in the context of the abstract band method: see [53] (scalar functions), [47] (matrix functions), [2, 3] (matrix periodic functions of two, and in certain cases more than two, variables). So far, factorization of almost periodic symbols of Toeplitz operators was studied exclusively for functions of one variable.

In the present paper the main theme is connections between Toeplitz operators and factorizations of their symbols, which are matrix valued almost periodic functions of several real variables. In our approach, it is advantegeous to allow the consideration of functions with Fourier spectrum in a given additive subgroup of $\mathbb{R}^k$. By doing so, we are able to treat simultaneously many particular cases, for example, periodic functions, or functions of mixed type (e.g., periodic in one variable and non-periodic in other variables). One of our main results states that for locally sectorial matrix valued functions a factorization exists that is canonical up to a scalar multiple (a so-called twisted canonical factorization). Other results pertain to Hermitian matrix functions and to continuity properties of canonical factorizations.

We conclude the introduction with a brief description of the contents section by section. Basic properties of algebras of almost periodic functions are addressed in Section 2, where also the notion of canonical factorizations with respect to a halfspace are introduced. Section 3 contains the invertibility criteria for Toeplitz operators whose matrix symbols belong to Wiener algebras of almost periodic functions of several variables, in terms of their (canonical) factorization. In Section 4, the existence of such factorization is established for the *sectorial* matrix functions. For *locally sectorial* matrix functions we prove existence of canonical factorizations up to multiplication by a scalar elementary exponential. Specifics of the factorization of Hermitian matrix functions, both definite and indefinite, are discussed in Section 5. Results of Sections 4 and 5 are then used in Section 6, where one-sided invertibility of Toeplitz operators with almost periodic matrix symbols of several variables is characterized, again in terms of factorizations of the symbols. The factorizations here are generally non-canonical, and may involve non-diagonalizable middle factors. In the last Section 7 we deal with continuous families of almost periodic matrix functions, and discuss the continuity of their canonical factorizations.

2 Algebras of Almost Periodic Functions and Factorizations

In this section we present some background results on algebras of almost periodic functions, and introduce the notion of a factorization with respect to a halfspace, which is central to this paper.

We let (AP^k) denote the algebra of complex valued almost periodic functions of k real variables, i.e., the closed subalgebra of $L^\infty(\mathbb{R}^k)$ (with respect to the standard Lebesgue measure) generated by all the functions $e_\lambda(t) = e^{i\langle\lambda,t\rangle}$, where $\lambda = (\lambda_1, \ldots, \lambda_k) \in \mathbb{R}^k$. Here the variable $t = (t_1, \ldots, t_k) \in \mathbb{R}^k$, and

$$\langle\lambda, t\rangle = \sum_{j=1}^{k} \lambda_j t_j$$

is the standard inner product of λ and t. The norm in (AP^k) will be denoted by $\|\cdot\|_\infty$.

The next proposition is standard (see Section 1.1 in [43]).

Proposition 2.1 *(AP^k) is a commutative unital C^*-algebra, and therefore can be identified with the algebra $C(\mathcal{B})$ of complex valued continuous functions on a compact Hausdorff topological space $\mathcal{B}$. Moreover, $\mathbb{R}^k$ is dense in $\mathcal{B}$.*

The space $\mathcal{B}$ is called the *Bohr compactification* of $\mathbb{R}^k$.

Recall that for any $f(t) \in (AP^k)$ its *Fourier series* is defined by the formal sum

$$\sum_{\lambda} f_\lambda e^{i\langle\lambda,t\rangle}, \tag{2.1}$$

where

$$f_\lambda = \lim_{T\to\infty} \frac{1}{(2T)^k} \int_{[-T,T]^k} e^{-i\langle\lambda,t\rangle} f(t)dt, \ \lambda \in \mathbb{R}^k, \tag{2.2}$$

and the sum in (2.1) is taken over the set $\sigma(f) = \{\lambda \in \mathbb{R}^k : f_\lambda \neq 0\}$, called the *spectrum* of f. The spectrum of every $f \in (AP^k)$ is at most a countable set. The *mean* $M\{f\}$ of $f \in (AP^k)$ is defined by $M\{f\} = f_0 = \lim_{T\to\infty} \frac{1}{(2T)^k} \int_{[-T,T]^k} f(t)dt$. The *Wiener algebra* (APW^k) is defined as the set of all $f \in (AP^k)$ such that the Fourier series of f converges absolutely. The Wiener algebra is a Banach $*$-algebra with respect to the *Wiener norm* $\|f\|_W = \sum_{\lambda\in\mathbb{R}^k} |f_\lambda|$ (the multiplication in (APW^k) is pointwise). Note that (APW^k) is dense in (AP^k). For the general theory of almost periodic functions of one and several variables we refer the reader to the books [12, 39, 40] and to Chapter 1 in [43].

Let Δ be a non-empty subset of $\mathbb{R}^k$. Denote

$$(AP^k)_\Delta = \{f \in (AP^k) : \sigma(f) \subseteq \Delta\},$$

$$(APW^k)_\Delta = \{f \in (APW^k) : \sigma(f) \subseteq \Delta\}.$$

If Δ is an additive subgroup of $\mathbb{R}^k$, then $(AP^k)_\Delta$ (resp. $(APW^k)_\Delta$) is a unital subalgebra of (AP^k) (resp. (APW^k)).

Proposition 2.2 *Let Δ be an additive subgroup of $\mathbb{R}^k$. Then:*

(a) *$f \in (AP^k)_\Delta$ is invertible in $(AP^k)_\Delta$ if and only if f is invertible in (AP^k) if and only if there exists $\epsilon > 0$ such that $|f(t)| \geq \epsilon$ for every $t \in \mathbb{R}^k$.*

(b) *$f \in (APW^k)_\Delta$ is invertible in $(APW^k)_\Delta$ if and only if f is invertible in (APW^k) if and only if there exists $\epsilon > 0$ such that $|f(t)| \geq \epsilon$ for every $t \in \mathbb{R}^k$.*

Proof: The second "if and only if" statement of (a) is easily obtained using the identification of (AP^k) with $C(\mathcal{B})$ and the denseness of $\mathbb{R}^k$ in $\mathcal{B}$. The first "if and only if" statement of (a) follows form a general result on C^*-algebras: If $\mathcal{A}_2$ is a unital C^*-subalgebra of a unital C^*-algebra $\mathcal{A}_1$ then $a \in \mathcal{A}_2$ is invertible in $\mathcal{A}_2$ if and only if a is invertible in $\mathcal{A}_1$ (see, e.g., Theorem 2.1.11 in [42]).

The part (b) for one variable $k = 1$ is a classical result which can be found, for example, in [17] (Corollary 2, p. 175). For several variables the result follows using the approach of Section 29 of [17]. We briefly outline this approach. First of all, the algebra (APW^k) is identified with $\ell^1(\mathbb{R}^k_d)$, where by $\mathbb{R}^k_d$ we denote the Abelian group $\mathbb{R}^k$ with the discrete topology. Next, the maximal ideal space of $\ell^1(\mathbb{R}^k_d)$ is identified with the group Γ_k of characters on $\mathbb{R}^k_d$, when Γ_k is given the weak* topology (cf. Theorem 5.1 in [17]); recall that a character of $\mathbb{R}^k_d$ is a group homomorphism from $\mathbb{R}^k_d$ into the multiplicative group of complex numbers of absolute value one. Namely, the value of $x \in \ell^1(\mathbb{R}^k_d)$ on the maximal ideal $\mathcal{M}_\kappa$ of $\ell^1(\mathbb{R}^k_d)$ that corresponds to the character $\kappa \in \Gamma_k$ is

$$x(\mathcal{M}_\kappa) = \sum_{\lambda \in \mathbb{R}^k_d} x(\lambda)\kappa(\lambda).$$

Next, one proves that the set of characters

$$\{\kappa_t\}_{t \in \mathbb{R}^k} \tag{2.3}$$

defined by $\kappa_t(\lambda) = e^{i\langle \lambda, t\rangle}$, $t \in \mathbb{R}^k$ is dense in Γ_k. For $k = 1$, this is Theorem 6.2 in [17]; for $k > 1$ this follows from Theorem 23.18 in [29] that asserts that the character group of a direct product of finitely many locally compact Abelian groups coincides with the direct product of the corresponding character groups. Therefore, if $f \in (APW^k)$ is such that

$$|f(t)| \geq \epsilon, \qquad t \in \mathbb{R}^k \tag{2.4}$$

where $\epsilon > 0$ is independent of t, in other words,

$$|f(\mathcal{M}_{\kappa_t})| \geq \epsilon, \qquad t \in \mathbb{R}^k,$$

then the density of the set (2.3) in Γ_k implies that f takes nonzero values on the set of maximal ideals of (APW^k), and hence is invertible in (APW^k). This proves the second "if and only if" statement in (b). If (2.4) holds, and moreover $f \in (APW^k)_\Delta$, then by the already proved parts of Proposition 2.2, the inverse of f belongs to $(APW^k) \cap (AP^k)_\Delta = (APW^k)_\Delta$, which proves the first "if and only if" statement in (b). □

If (X) is a set (typically a Banach space or an algebra), we denote by $(X)^{m\times n}$ the set of $m \times n$ matrices with entries in (X).

Many properties of the algebras $(AP^k)_\Delta^{n\times n}$ and $(APW^k)_\Delta^{n\times n}$ can be extended from the one-variable almost periodic functions without difficulties. We state one such property, which is especially useful.

Proposition 2.3 *Let Δ be an additive subgroup of $\mathbb{R}^k$. Let $f \in (AP^k)_\Delta^{n\times n}$ (resp. $f \in (APW^k)_\Delta^{n\times n}$), and let*

$$\Omega = \{z \in \mathbb{C} : \; z \textit{ is an eigenvalue of } f(t) \textit{ for some } t \in \mathbb{R}^k\}.$$

If Ψ is an analytic function in an open neighborhood of the closure of Ω, then $\Psi(f) \in (AP^k)_\Delta^{n\times n}$ (resp. $\Psi(f) \in (APW^k)_\Delta^{n\times n}$).

Here, for every fixed $t \in \mathbb{R}^k$, $\Psi(f(t))$ is understood as the $n \times n$ matrix defined by the standard functional calculus.

Proof: We give the proof for the case $f \in (APW^k)_\Delta^{n\times n}$; if $f \in (AP^k)_\Delta^{n\times n}$, the proof is analogous. The Proof is modeled after the Proof of Proposition 2.3 in [47]. Since $f(t)$ is a bounded function, the set Ω is also bounded. Let $z_0 \notin \overline{\Omega}$ (the closure of Ω). Then $z_0 I - f(t)$ has eigenvalues $z_0 - \lambda_1(t), \ldots, z_0 - \lambda_n(t)$, where $\lambda_1(t), \ldots, \lambda_n(t)$ are eigenvalues of $f(t)$, so

$$|\det(z_0 I - f(t))| = \prod_{j=1}^{n} |z_0 - \lambda_j(t)| \geq \alpha^n,$$

where α is the (positive) distance from z_0 to $\overline{\Omega}$. By Proposition 2.2,

$$(\det(z_0 I - f(t)))^{-1} \in (APW^k)_\Delta,$$

and therefore the matrix function $z_0 I - f$ is invertible in $(APW^k)_\Delta^{n\times n}$. Thus, z_0 belongs to the resolvent set of f as an element of the Banach algebra $(APW^k)_\Delta^{n\times n}$. Now we can define $\Psi(f) \in (APW^k)_\Delta^{n\times n}$ using the functional calculus:

$$\Psi(f) = \frac{1}{2\pi i} \int_\Gamma \Psi(z)(zI - f)^{-1} dz$$

for a suitable contour Γ that contains $\overline{\Omega}$ in its interior, where the integral converges in the norm of $(APW^k)^{n\times n}$. Since the convergence in $(APW^k)^{n\times n}$ implies pointwise convergence, for every $t \in \mathbb{R}^k$ we have

$$(\Psi(f))(t) = \frac{1}{2\pi i}\int_\Gamma \Psi(z)(zI - f(t))^{-1}dz. \tag{2.5}$$

But the right-hand side of (2.5) is just the definition of $\Psi(f(t))$. It follows that $\Psi(f(t))$ (defined pointwise for every $t \in \mathbb{R}^k$) is the value of $\Psi(f)$ at t; since $\Psi(f) \in (APW^k)^{n\times n}_\Delta$, we are done. □

Next, we present a theorem that introduces the notion of mean motion of a multivariable almost periodic function.

Theorem 2.4 *Let $f \in (AP^k)$ and let Λ be the minimal additive subgroup of $\mathbb{R}^k$ that contains $\sigma(f)$. Assume that $|f(t)| \geq \epsilon > 0$ for all $t \in \mathbb{R}^k$, where ϵ is independent of t. Define the continuous real valued function $y(t)$, $t \in \mathbb{R}^k$ by the conditions $f(t) = |f(t)|e^{iy(t)}$, $-\pi < y(0) \leq \pi$. Then there exists a unique $c \in \mathbb{R}^k$ such that $y(t) = \langle c, t\rangle + u(t)$, where $u \in (AP^k)$. Moreover, $c \in \Lambda$, $\sigma(u) \subseteq \Lambda$ and, if $f \in (APW^k)$, then also $u \in (APW^k)$.*

As in the one variable ($k = 1$) case, we shall call c the *mean motion* of f, and denote it by $w(f)$. In the one variable case, Theorem 2.4, is the celebrated Bohr's theorem (see, e.g., [39]).

Theorem 2.4 (without the (APW^k) part) was proved in [44]. The proof in [44] involves the consideration of $e^{iy(st)}$, $s \in \mathbb{R}$, for a fixed t as an almost periodic function of one variable s, applying the Bohr's theorem, and proving that the mean motion of $e^{iy(st)}$ is a linear function of t. The uniqueness is easy: If

$$y(t) = \langle c_1, t\rangle + u_1(t) = \langle c_2, t\rangle + u_2(t),$$

where c_j and u_j satisfy the requirements of Theorem 2.4, then $\langle c_1 - c_2, t\rangle \in (AP^k)$, and since the functions in (AP^k) are bounded, we must have $c_1 = c_2$.

To obtain the (APW^k) part of this theorem, argue as follows. Represent the function u as the sum $u = u_0 + u_1$, where $u_0 \in (APW^k)$ and $\|u_1\|_{AP^k} < \frac{\pi}{2}$. Then

$$e^{iu_1(t)} = f(t)\ |f(t)|^{-1}\, e^{-i\langle c,t\rangle} e^{-iu_0(t)}. \tag{2.6}$$

Assume $f \in (APW^k)$; then $|f|^2 = f\overline{f} \in (APW^k)$ as well. Since the function $\phi(z) = z^{-1/2}$ is analytic in a (complex) neighborhood of $[\epsilon^2, \infty)$ and $|f|^{-1} = \phi(|f|^2)$, by Proposition 2.3 the function $|f|^{-1}$ belongs to (APW^k). The other two multiples in the right hand side of (2.6) obviously belong to (APW^k); hence, so does the function $\xi(t) = e^{iu_1(t)}$. On the other hand, the values of ξ lie in the right open half-plane. Using Proposition 2.3 again, we may define $u_2 \in (APW^k)$

so that $e^{iu_2} = \xi$. Since u_1 and u_2 are both continuous on $\mathbb{R}^k$, this means that they differ by a constant summand. But then u_1 belongs to (APW^k) simultaneously with u_2. Finally, the function u $(= u_0 + u_1)$ belongs to (APW^k) as well.

We now introduce the notion of canonical factorization with respect to a halfspace. A subset S of $\mathbb{R}^k$ is called a *halfspace* if it has the following properties:

(i) $\mathbb{R}^k = S \cup (-S)$;

(ii) $S \cap (-S) = \{0\}$;

(iii) if $x, y \in S$ then $x + y \in S$;

(iv) if $x \in S$ and α is a nonnegative real number, then $\alpha x \in S$.

A standard example of a halfspace is given by

$$\begin{aligned} E_k &= \{(x_1, \ldots, x_k)^T \in \mathbb{R}^k \setminus \{0\} \ : \ x_1 = x_2 = \cdots = x_{j-1} = 0, \\ &\qquad x_j \neq 0 \Rightarrow x_j > 0\} \cup \{0\}. \end{aligned}$$

(The vectors in $\mathbb{R}^k$ are understood as column vectors; the superscript ^{T} denotes the transpose.) Clearly, when $k = 1$ the only halfspaces are $[0, \infty)(= E_1)$ and $(-\infty, 0]$. In general, we have the following statement.

Proposition 2.5 *A set $S \subseteq \mathbb{R}^k$ is a halfspace if and only if there exists a real invertible $k \times k$ matrix A (which is unique up to multiplication with a positive real scalar) such that*

$$S = AE_k \stackrel{\text{def}}{=} \{Ax \ : \ x \in E_k\}. \tag{2.7}$$

This proposition is a special case of basic results on linearly ordered real vector spaces, see [14] or Section IV.5 in [15]. Indeed, a halfspace S induces a linear order $x \succeq_S y$ in $\mathbb{R}^k$ by the rule $x \succeq_S y$ if and only if $x - y \in S$. Conversely, every linear order in $\mathbb{R}^k$ which is compatible with the addition and multiplication by positive scalars determines a halfspace, consisting of vectors that are greater than or equal to zero with respect to the linear order. By the results of [14], for every linear order $x \succeq y$ in $\mathbb{R}^k$ (compatible with addition and multiplication by positive scalars) there exists a basis $b_1, \ldots, b_k$ of $\mathbb{R}^k$ such that $x \succeq y$ if and only if the vectors of coefficients $\alpha = (\alpha_1, \ldots, \alpha_k)^T$ and $\beta = (\beta_1, \ldots, \beta_k)^T$ taken from the linear combinations $x = \sum_{j=1}^k \alpha_j b_j$ and $y = \sum_{j=1}^k \beta_j b_j$ satisfy $\alpha - \beta \in E_k$. Now $A = [b_1 b_2 \ldots b_k]$ satisfies Proposition 2.5.

Let $G \in (AP^k)^{n \times n}$. A representation

$$G(t) = G_+(t)G_-(t), \quad t \in \mathbb{R}^k, \tag{2.8}$$

where

$$G_+^{\pm 1} \in (AP^k)_S^{n \times n}, \quad G_-^{\pm 1} \in (AP^k)_{-S}^{n \times n} \tag{2.9}$$

is called a *left AP_S canonical factorization* of G. Using instead of (2.8) the equality $G(t) = G_-(t)G_+(t)$ (with the same properties of $G_\pm(t)$), we obtain a *right AP_S canonical factorization* of G. Note that the right AP_S canonical factorization coincides with the left AP_{-S} canonical factorization. Therefore it is in principle not necessary to introduce both left and right AP_S canonical factorizations. However, to allow easy comparison with the classical results we will use both notions. Canonical factorizations, their generalizations and applications for one variable ($k = 1$ and $S = [0, \infty)$) have been studied in [32, 52, 33, 36], for example. We are not aware of previous studies of AP_S canonical factorizations for functions of several variables.

We say that (2.8) is a left APW_S canonical factorization of G if $G_\pm$ satisfy the stronger than (2.9) conditions $G_+^{\pm 1} \in (APW^k)_S^{n\times n}$, $G_-^{\pm 1} \in (APW^k)_{-S}^{n\times n}$. If Λ is an additive subgroup of $\mathbb{R}^k$, then a representation (2.8) is called a left canonical $(AP_S)_\Lambda$ factorization if $G_+^{\pm 1} \in (AP^k)_{S\cap\Lambda}^{n\times n}$, $G_-^{\pm 1} \in (AP^k)_{(-S)\cap\Lambda}^{n\times n}$, and a left canonical $(APW_S)_\Lambda$ factorization if $G_+^{\pm 1} \in (APW^k)_{S\cap\Lambda}^{n\times n}$, $G_-^{\pm 1} \in (APW^k)_{(-S)\cap\Lambda}^{n\times n}$. Of course, G must belong to $(APW^k)^{n\times n}$ (respectively, $(AP^k)_\Lambda^{n\times n}$, or $(APW^k)_\Lambda^{n\times n}$) in order to potentially admit a left canonical APW_S (respectively, $(AP_S)_\Lambda$, or $(APW_S)_\Lambda$) factorization. The notions of right APW_S canonical factorizations etc. are introduced analogously.

3 Toeplitz Operators

We start with general and well-known remarks concerning Toeplitz operators. The exposition in the first part of this section follows that in [4].

Let $\mathcal{H}$ be a Hilbert space, and let $\mathfrak{B}(\mathcal{H})$ stand for the C^* algebra of (bounded linear) operators acting on $\mathcal{H}$. For any orthoprojection $P \in \mathfrak{B}(\mathcal{H})$ and an arbitrary $A \in \mathfrak{B}(\mathcal{H})$ an abstract *Toeplitz operator* $T_P(A)$ ($\in \mathfrak{B}(\mathcal{L})$ where $\mathcal{L} = \operatorname{Im} P$) is defined by the formula

$$T_P(A)x = PAx, \quad x \in \mathcal{L}.$$

The next lemma is a basic result connecting invertibility of abstract Toeplitz operators and operator factorizations.

Lemma 3.1 *Let $A \in \mathfrak{B}(\mathcal{H})$ be invertible, and let P and Q $(= I - P)$ be a pair of complementary orthoprojections on $\mathcal{H}$. Then the following statements are equivalent:*

i) *the operator $T_P(A)$ is invertible;*

ii) *the operator $T_Q(A^{-1})$ is invertible;*

iii) *$A = A_+A_-$ where $A_\pm$ are invertible elements of $\mathfrak{B}(\mathcal{H})$ and $A_+^{\pm 1}\operatorname{Im} Q \subseteq \operatorname{Im} Q$, $A_-^{\pm 1}\operatorname{Im} P \subseteq \operatorname{Im} P$.*

Lemma 3.1 is a combination of Theorem 5 and Corollary 3 from [13]; see also [9], [5] or Section III.4 in [18] for a simpler exposition from the point of view of matricially coupled operators.

In general (as was also observed in [13]), the invertibility of $T_P(A)$ does not imply the invertibility of A. However, this is the case for *classical Toeplitz operators*, in other words, when $\mathcal{H}$ is the Lebesgue space $(L^2)^{n\times 1}$ on the unit circle $\mathbb{T}$, P is the Riesz orthoprojection onto the Hardy space $(H^2)^{n\times 1}$ and A is the multiplication operator M_f defined by $(M_f x)(t) = f(t)x(t)$, with $f \in (L^\infty(\mathbb{T}))^{n\times n}$; see the original paper [50] or the monographs [9, 11, 41]. Moreover, for certain classes of functions f it can be guaranteed that $A_\pm$ in statement iii) of Lemma 3.1 also are multiplication operators: $A_\pm = M_{f_\pm}$. This is true, in particular, when f has an absolutely convergent Fourier series; the same property is then inherited by $f_\pm$ (see the original paper [22] or later expositions in [11, 41]). Thus, a basic approach to express invertibility of Toeplitz operators in terms of factorization of symbol functions consists of using Lemma 3.1 and interpreting, if possible, the operators A, A_+ and A_- as multiplication operators. As we shall see in Theorem 3.2 below, all the properties mentioned above (and much more) are valid in the context of the algebras (AP^k) and (APW^k).

Introduce an inner product on (AP^k) by the formula

$$\langle f, g\rangle = M\{fg^*\}, \quad f, g \in (AP^k). \tag{3.1}$$

The completion of (AP^k) with respect to this inner product is called the *Besikovitch space* and is denoted by (B^k). Thus (B^k) is a Hilbert space. For a nonempty set $\Lambda \subseteq \mathbb{R}^k$, define the projection

$$\Pi_\Lambda\left(\sum_{\lambda\in\sigma(f)} f_\lambda e^{i\langle\lambda,t\rangle}\right) = \sum_{\lambda\in\sigma(f)\cap\Lambda} f_\lambda e^{i\langle\lambda,t\rangle},$$

where $f \in (APW^k)$. The projection Π_Λ extends by continuity to the orthogonal projection (also denoted Π_Λ) on (B^k). We denote by $(B^k)_\Lambda$ the range of Π_Λ, or, equivalently, the completion of $(AP^k)_\Lambda$ with respect to the inner product (3.1). The vector valued Besikovitch space $(B^k)^{n\times 1}$ consists of $n \times 1$ columns with components in (B^k), with the standard Hilbert space structure. Similarly, $(B^k)^{n\times 1}_\Lambda$ is the Hilbert space of $n \times 1$ columns with components in $(B^k)_\Lambda$. In the periodic case ($\Lambda = \mathbb{Z}^k$) we may identify $(B^k)_\Lambda$ with $L_2(\mathbb{T}^k)$.

We let A be a multiplication operator M_f with $f \in (APW^k)^{n\times n}$. Denote by Λ the smallest additive subgroup of $\mathbb{R}^k$ which contains $\sigma(f)$. Then A can be considered as an operator on $\mathcal{H} = (B^k)^{n\times 1}_{\Lambda'}$ for any additive subgroup Λ' of $\mathbb{R}^k$ which contains Λ (the cases $\Lambda' = \Lambda$ and $\Lambda' = \mathbb{R}^k$ are not excluded). For a fixed half-space S, the Toeplitz operators $T_P(M_f)$ corresponding to $P = \Pi_{S\cap\Lambda'}, \Pi_{(S\setminus\{0\})\cap\Lambda'}$, $\Pi_{(-S)\cap\Lambda'}$ and $\Pi_{((-S)\setminus\{0\})\cap\Lambda'}$ will be denoted by $T(f)^{[S}_{\Lambda'}$, $T(f)^{(S}_{\Lambda'}$, $T(f)^{-S]}_{\Lambda'}$ and $T(f)^{-S)}_{\Lambda'}$, respectively. Thus, for example, $T(f)^{-S]}_{\Lambda'}$ is defined on $(B^k)^{n\times 1}_{(-S)\cap\Lambda'}$.

We denote by G^T the transpose of a matrix function G.

Theorem 3.2 *Assume that $G \in (APW^k)^{n\times n}$ and let $\Lambda = \Lambda(G)$ be the minimal additive subgroup of $\mathbb{R}^k$ that contains $\sigma(G)$. Then the following statements are equivalent:*

i) *the operator $T(G)_{\Lambda'}^{-S]}$ is invertible for some additive subgroup $\Lambda' \supseteq \Lambda(G)$;*

ii) *the operator $T(G)_{\Lambda'}^{-S]}$ is invertible for every additive subgroup $\Lambda' \supseteq \Lambda(G)$;*

iii) *the operator $T(G)_{\Lambda'}^{-S)}$ is invertible for some additive subgroup $\Lambda' \supseteq \Lambda(G)$;*

iv) *the operator $T(G)_{\Lambda'}^{-S)}$ is invertible for every additive subgroup $\Lambda' \supseteq \Lambda(G)$;*

v) *the operator $T(G^T)_{\Lambda'}^{[S}$ is invertible for some additive subgroup $\Lambda' \supseteq \Lambda(G)$;*

vi) *the operator $T(G^T)_{\Lambda'}^{[S}$ is invertible for every additive subgroup $\Lambda' \supseteq \Lambda(G)$;*

vii) *the operator $T(G^T)_{\Lambda'}^{(S}$ is invertible for some additive subgroup $\Lambda' \supseteq \Lambda(G)$;*

viii) *the operator $T(G^T)_{\Lambda'}^{(S}$ is invertible for every additive subgroup $\Lambda' \supseteq \Lambda(G)$;*

ix) *G is invertible (in $(L^\infty(\mathbb{R}^k))^{n\times n}$ and therefore in $(APW^k)_\Lambda^{n\times n}$) and the operator $T(G^{-1})_{\Lambda'}^{[S}$ is invertible for some additive subgroup $\Lambda' \supseteq \Lambda(G)$;*

x) *G is invertible (in $(L^\infty(\mathbb{R}^k))^{n\times n}$ and therefore in $(APW^k)_\Lambda^{n\times n}$) and the operator $T(G^{-1})_{\Lambda'}^{[S}$ is invertible for every additive subgroup $\Lambda' \supseteq \Lambda(G)$;*

xi) *G is invertible (in $(L^\infty(\mathbb{R}^k))^{n\times n}$ and therefore in $(APW^k)_\Lambda^{n\times n}$) and the operator $T(G^{-1})_{\Lambda'}^{(S}$ is invertible for some additive subgroup $\Lambda' \supseteq \Lambda(G)$;*

xii) *G is invertible (in $(L^\infty(\mathbb{R}^k))^{n\times n}$ and therefore in $(APW^k)_\Lambda^{n\times n}$) and the operator $T(G^{-1})_{\Lambda'}^{(S}$ is invertible for every additive subgroup $\Lambda' \supseteq \Lambda(G)$;*

xiii) *G admits a left canonical AP_S factorization;*

xiv) *G admits a left canonical APW_S factorization;*

xv) *G admits a left canonical $(AP_S)_\Lambda$ factorization;*

xvi) *G admits a left canonical $(APW_S)_\Lambda$ factorization.*

Remark: A left canonical AP_S factorization, if it exists, is unique up to the transformation

$$G_+ \mapsto G_+C,\ G_- \mapsto C^{-1}G_-, \tag{3.2}$$

where C is an arbitrary constant invertible matrix. Hence, the equivalence of statements xiii)–xvi) in Theorem 3.2 means that a left canonical AP_S factorization of the matrix function $G \in (APW^k)_\Lambda^{n\times n}$ is in fact its left canonical $(APW_S)_\Lambda$ factorization.

Observe also that the product $M\{G_+\}M\{G_-\}$ is invariant under the transformation (3.2), and is therefore defined uniquely by the matrix function G itself. We denote it by $d_S(G)$ and call it the (S-) *geometric mean* of G.

The particular case $k = 1$ and $S = [0, \infty)$ was considered in [4, Theorem 2.3]. Most of the Proof of Theorem 2.3 of [4] goes through in the more general situation of Theorem 3.2 without essential changes. Several equivalences in Theorem 3.2 are simply particular instances of i) $\Leftrightarrow$ ii) of Lemma 3.1.

We give below only some details of the proof of the implication i) $\Rightarrow$ xvi), which is a difficult part of Theorem 3.2. As in the Proof of Theorem 2.3 of [4], we can restrict the proof to the case $\Lambda' = \Lambda$. Analogously, we show that the generalized Riemann operators $R_1 = \Pi_{(S\setminus\{0\})\cap\Lambda} + M_G\Pi_{(-S)\cap\Lambda}$ and $R_2 = \Pi_{(-S\setminus\{0\})\cap\Lambda} + M_{G^T}\Pi_{S\cap\Lambda}$ are invertible on $(B^k)^{n\times 1}_\Lambda$, and letting $\phi_j = R_1^{-1}E_j$, $\psi_j = R_2^{-1}E_j$, where E_j is the j-th column of I, the functions ϕ_j and ψ_j belong to $(APW^k)^{n\times 1}_\Lambda$. Denote by Φ^+ (Φ^-, Ψ^+, Ψ^-) the $n \times n$ matrix function the j-th column of which is $\Pi_{(S\setminus\{0\})}\phi_j$ (respectively, $\Pi_{(-S)}\phi_j$, $\Pi_S\psi_j$, $\Pi_{(-S)\setminus\{0\}}\psi_j$). Then

$$\Phi^+ \in (APW^k)^{n\times n}_{\Lambda\cap(S\setminus\{0\})}, \qquad \Phi^- \in (APW^k)^{n\times n}_{\Lambda\cap(-S)},$$

$$\Psi^+ \in (APW^k)^{n\times n}_{\Lambda\cap S}, \qquad \Psi^- \in (APW^k)^{n\times n}_{\Lambda\cap(-S)\setminus\{0\}}.$$

Using the definition of ϕ_j, ψ_j, it also follows that

$$\Phi^+ + G\Phi^- = I \tag{3.3}$$

and

$$\Psi^- + G^T\Psi^+ = I. \tag{3.4}$$

Introduce now a matrix function $C = (\Psi^+)^T G\Phi^-$. On the one hand, (3.3) implies that $C = (\Psi^+)^T(I - \Phi^+) \in (APW^k)^{n\times n}_{\Lambda\cap S}$. On the other hand, from (3.4):

$$C = (G^T\Psi^+)^T\Phi^- = (I - \Psi^-)^T\Phi^- \in (APW^k)^{n\times n}_{\Lambda\cap(-S)}.$$

Hence, C is a constant matrix: $C \in \mathbb{C}^{n\times n}$. We will show that C is non-singular. To this end, introduce matrix functions $\widehat{\Phi}^\pm$, $\widehat{\Psi}^\pm$ of one complex variable z according to the formulas

$$\widehat{\Phi}^\pm(z) = \Phi^\pm(A^{*-1}[z, \underbrace{0, \dots, 0}_{k-1 \text{ times}}]^T),$$

$$\widehat{\Psi}^\pm(z) = \Psi^\pm(A^{*-1}[z, \underbrace{0, \dots, 0}_{k-1 \text{ times}}]^T),$$

where A is defined by (2.7). Then $\widehat{\Phi}^+$, $\widehat{\Psi}^+$ ($\widehat{\Phi}^-$, $\widehat{\Psi}^-$) can be extended analytically into the upper half plane Π^+ (respectively, the lower half plane Π^-), and

$\lim_{y\to+\infty}\widehat{\Phi}^+(iy)(=\lim_{y\to-\infty}\widehat{\Psi}^-(iy)) = 0$. If $\det C = 0$, then $\det\widehat{\Psi}^+ \cdot \det(I-\widehat{\Phi}^+) = 0$, and therefore either $\det\widehat{\Psi}^+$ or $\det(I-\widehat{\Phi}^+)$ vanishes in Π^+ identically. Since $\lim_{y\to+\infty}(I-\widehat{\Phi}^+(iy)) = I$, $\det(I-\widehat{\Phi}^+(iy))$ differs from zero for y large enough. Hence, $\det\widehat{\Psi}^+(z) = 0$ for all $z \in \Pi^+$. From here and (3.4) we conclude that $\det(I-\widehat{\Psi}^-) = 0$ identically (on $\mathbb{R}$, and therefore in Π^-). This, however, is in contradiction with the property $\lim_{y\to-\infty}(I-\widehat{\Psi}^-(iy)) = I$. After the non-singularity of C is established, the proof of i) $\Rightarrow$ xvi) is completed as in the proof of [4, Theorem 2.3].

4 Factorization of Sectorial Matrix Functions

In this section we prove several results on sectorial (in particular, positive definite) Toeplitz operators. Recall that an operator $A \in \mathfrak{B}(\mathcal{H})$ is called *α-sectorial* for a given $\alpha \in [0, \frac{\pi}{2})$ if

$$|\mathrm{Im}\langle Af, f\rangle| \le (\tan\alpha)\,\mathrm{Re}\langle Af, f\rangle \tag{4.1}$$

and there exists an $\epsilon > 0$ such that

$$\mathrm{Re}\langle Af, f\rangle \ge \epsilon\|f\|^2 \tag{4.2}$$

for all $f \in \mathcal{H}$. More precisely, we say that an operator $A \in \mathfrak{B}(\mathcal{H})$ is *(α,ϵ)-sectorial* if (4.1) and (4.2) hold for every $f \in \mathcal{H}$. Condition (4.2) clearly implies

$$|\langle Af, f\rangle| \ge \epsilon\|f\|^2, \qquad f \in \mathcal{H}; \tag{4.3}$$

conversely, (4.3) and (4.1) imply $\mathrm{Re}\langle Af, f\rangle \ge \epsilon(\cos(\alpha))^{-1}\|f\|^2$.

For $\alpha = 0$, condition (4.1) means that $\mathrm{Im}\langle Af, f\rangle = 0$, $\mathrm{Re}\langle Af, f\rangle \ge 0$, and condition (4.2) can therefore be rewritten as

$$\langle Af, f\rangle \ge \epsilon\|f\|^2 \text{ for all } f \in \mathcal{H}. \tag{4.4}$$

Hence, the operator A is 0-sectorial if and only if it is *positive definite* (notation: $A > 0$).

Proposition 4.1 *Let Λ be an additive subgroup of $\mathbb{R}^k$, and let $G \in (APW^k)_\Lambda^{n\times n}$. Then the following statements are equivalent for any $\alpha \in [0, \frac{\pi}{2})$, $\epsilon > 0$, and for a given halfspace S:*

(i) *$T(G)_\Lambda^{[S}$ is (α,ϵ)-sectorial;*

(ii) *$T(G)_{\Lambda'}^{[S}$ is (α,ϵ)-sectorial for some additive subgroup $\Lambda' \supseteq \Lambda$;*

(iii) *$T(G)_{\Lambda'}^{[S}$ is (α,ϵ)-sectorial for every additive subgroup $\Lambda' \supseteq \Lambda$.*

(iv) *the matrix function $G(t)$ is (α,ϵ)-sectorial for every $t \in \mathbb{R}^k$.*

Proof: We first consider the case $\alpha = 0$. In that case, condition (iv) means that $M_G \geq \epsilon I$ on $(B^k)^{n\times 1}$. Since $T(G)^{[S}_{\Lambda'}$ is a compression of M_G, the implication (iv) $\Rightarrow$ (iii) holds. Implications (iii) $\Rightarrow$ (ii) and (iii) $\Rightarrow$ (i) are obvious. It remains to show that (i) $\Rightarrow$ (iv); the implication (ii) $\Rightarrow$ (iv) would follow from there by considering Λ' in place of Λ.

The following proof of the implication (i) $\Rightarrow$ (iv) is a simple generalization of the respective reasoning from [27] for the case of classical Toeplitz operators on the Hardy space $H^2(\mathbb{T})$.

Suppose that $T(G)^{[S}_{\Lambda} \geq \epsilon I$, that is,

$$\langle T(G)^{[S}_{\Lambda}\phi, \phi\rangle \geq \epsilon\|\phi\|^2$$

for all $\phi \in (B^k)^{n\times 1}_{S\cap\Lambda}$. Since

$$\langle \Pi_{S\cap\Lambda} G\phi, \phi\rangle = \langle G\phi, \Pi_{S\cap\Lambda}\phi\rangle = \langle G\phi, \phi\rangle,$$

this implies that

$$\langle G\phi, \phi\rangle \geq \epsilon\|\phi\|^2 \text{ for all } \phi \in (B^k)^{n\times 1}_{S\cap\Lambda}.$$

Denote by W the multiplication operator

$$(W\phi)(t) = e^{i\langle t,a\rangle}\phi(t), \quad \phi \in (B^k)^{n\times 1},$$

where a is the first column of the matrix A in (2.7). Then $W^* = W^{-1}$, $WM_G = M_G W$, and

$$\langle M_G W^m\phi, W^m\phi\rangle = \langle W^{-m} M_G W^m\phi, \phi\rangle = \langle M_G\phi, \phi\rangle, \quad m \in \mathbb{Z}.$$

Therefore, the inequality

$$\langle M_G\phi, \phi\rangle \geq \epsilon\|\phi\|^2$$

does not only hold for $\phi \in (B^k)^{n\times 1}_{S\cap\Lambda}$, but also for $\phi \in W^m(B^k)^{n\times 1}_{S\cap\Lambda}$, $m \in \mathbb{Z}$. Since the lineal $\{W^m\chi : \chi \in (B^k)^{n\times 1}_{S\cap\Lambda},\ m \in \mathbb{Z}\}$ is dense in $(B^k)^{n\times 1}_{\Lambda}$, this proves that $M_G \geq \epsilon I$ on $(B^k)^{n\times 1}_{\Lambda}$. That $M_G \geq \epsilon I$ holds on the space $(B^k)^{n\times 1}$ now follows as in the Proof of Lemma 8.2 of [47]. Indeed, let $\{\Lambda_j\}_{j\in J}$ be the complete collection of the cosets of Λ in $\mathbb{R}^k$; then we have the orthogonal decomposition $(B^k)^{n\times 1} = \oplus_{j\in J}(B^k)^{n\times 1}_{\Lambda_j}$, and

$$M_G = \oplus_{j\in J}(M_G)_j,$$

where $(M_G)_j$ is an operator on $(B^k)^{n\times 1}_{\Lambda_j}$ which is unitarily similar to M_G.

Recall that (B^k) and $(AP)^k$ can be thought of as $L^2(\mathcal{B})$ and $C(\mathcal{B})$, respectively, where $\mathcal{B}$ is, as before, the Bohr compactification of $\mathbb{R}^k$. Therefore, M_G is a multiplication operator on the space $(L^2(\mathcal{B}))^{n\times 1}$ with a continuous matrix function $G \in (C(\mathcal{B}))^{n\times n}$. The positivity of the operator $M_G - \epsilon I$ implies that $G - \epsilon I$ itself is positive semi-definite (on $\mathcal{B}$, and therefore on $\mathbb{R}^k$). This finishes the proof in the case that $\alpha = 0$.

For general α, observe that condition (4.1) can be rewritten in terms of the real part $A_R = \frac{1}{2}(A + A^*)$ and imaginary part $A_J = \frac{1}{2i}(A - A^*)$ of the operator A as follows:

$$\langle (A_R \tan\alpha \pm A_J) f, f\rangle \geq 0. \tag{4.5}$$

Therefore, an operator A is α-sectorial if and only if the operators $cA_R + A_J$ and $cA_R - A_J$ are both positive definite for all $c > \tan\alpha$. Moreover, A is (α, ϵ)-sectorial if and only if

$$cA_R \pm A_J \geq \epsilon(c - \tan\alpha)I, \qquad c \geq \tan\alpha. \tag{4.6}$$

for every $c \geq \tan\alpha$. Indeed, if (4.5) and (4.2) hold, then (4.6) is immediate. Conversely, (4.6) clearly implies (4.5); dividing both sides of (4.6) by c and passing to the limit when $c \to \infty$, we obtain (4.2).

Applying the $\alpha = 0$ case to the matrix functions $cG_R \pm G_J$ and the respective Toeplitz operators $T(cG_R \pm G_J)^{[S}_{\Lambda'}$ $(= c(T(G)^{[S}_{\Lambda'})_R \pm (T(G)^{[S}_{\Lambda'})_J)$, we arrive at the general case. □

Corollary 4.2 *Let $\Lambda \subseteq \Lambda'$ be additive subgroups of $\mathbb{R}^k$. If $G \in (APW^k)^{n\times n}_{\Lambda'}$ is such that $G(t)$ is (α, ϵ)-sectorial for every $t \in \mathbb{R}^k$, then $\Pi_\Lambda G \in (APW^k)^{n\times n}_{\Lambda}$ is also (α, ϵ)-sectorial for every $t \in \mathbb{R}^k$.*

Proof: Note that $T(G)^{[S}_{\Lambda'}$ is (α, ϵ)-sectorial by Proposition 4.1. Now for every $g \in (B^k)^{n\times 1}_{\Lambda}$ we have

$$\begin{aligned} \langle T(G)^{[S}_{\Lambda'} g, g\rangle &= \langle \Pi_{S\cap\Lambda'}(Gg), \Pi_\Lambda g\rangle \\ &= \langle \Pi_\Lambda \Pi_{S\cap\Lambda'}(Gg), g\rangle \\ &= \langle \Pi_{S\cap\Lambda}((\Pi_\Lambda G)g + (\Pi_{\Lambda'\setminus\Lambda} G)g), g\rangle. \end{aligned} \tag{4.7}$$

But because the Fourier spectrum of g is contained in Λ, the Fourier spectrum of $(\Pi_{\Lambda'\setminus\Lambda} G)g$ does not intersect Λ. The expression (4.7) is therefore equal to

$$\langle \Pi_{S\cap\Lambda}((\Pi_\Lambda G)g), g\rangle = \langle \Pi_{S\cap\Lambda}(Gg), g\rangle = \langle T(\Pi_\Lambda G)^{[S}_{\Lambda} g, g\rangle.$$

Comparing with the left hand side of (4.7) we see that the operator $T(\Pi_\Lambda G)^{[S}_{\Lambda}$ is (α, ϵ)-sectorial, and by the same Proposition 4.1 we are done. □

Using the sectoriality of the Toeplitz operators, we obtain the following factorization result.

Theorem 4.3 *Let $G \in (APW^k)^{n\times n}$, and let Λ be the minimal additive subgroup of $\mathbb{R}^k$ that contains $\sigma(G)$. If $G(t)$ is (α, ϵ)-sectorial for every $t \in \mathbb{R}^k$, where $\alpha \in [0, \frac{\pi}{2})$, $\epsilon > 0$ are independent of t, then G admits left and right canonical $(APW_S)_\Lambda$ factorization.*

Proof: According to Proposition 4.1, the operator $T(G)^{[S}_{\Lambda}$ is (α, ϵ)-sectorial, and therefore invertible, simultaneously with G. Theorem 3.2 implies then that G admits the left canonical $(APW_S)_\Lambda$ factorization.

The matrix functions G and G^T are (α, ϵ)-sectorial only simultaneously. Thus, the reasoning above, applied to G^T in place of G, yields the left canonical $(APW_S)_\Lambda$ factorization $G^T = X_+X_-$ of G^T. But then $G = X_-^T X_+^T$ is a right canonical $(APW_S)_\Lambda$ factorization of G. □

Theorem 4.3 can be extended to locally sectorial matrix functions, provided that one allows for the following modification in the definition of the canonical AP_S factorization. Let us say that G admits a *left twisted* canonical AP_S factorization if there exist $c \in \mathbb{R}^k$ and $G_\pm$ satisfying (2.9) such that, in place of (2.8), the following representation holds:

$$G(t) = e^{i\langle c,t\rangle} G_+(t)G_-(t), \quad t \in \mathbb{R}^k. \tag{4.8}$$

The *right twisted* canonical AP_S factorization, as well as $(AP_S)_\Lambda$, (APW_S), $(APW_S)_\Lambda$ variations of both left and right twisted canonical factorizations, are defined in the obvious way. For the left twisted canonical $(AP_S)_\Lambda$ and $(APW_S)_\Lambda$ factorizations we require, in particular, that c in the formula (4.8) belongs to Λ, and analogously for their right counterparts.

Theorem 4.4 *Let $G \in (APW^k)^{n\times n}$, and let Λ be the minimal additive subgroup of $\mathbb{R}^k$ that contains $\sigma(G)$. Assume that the numerical range of $G(t)$ is bounded away from zero, that is, there exists $\epsilon > 0$ such that*

$$|\langle G(t)\xi, \xi\rangle| \geq \epsilon \|\xi\|^2 \tag{4.9}$$

for all $\xi \in \mathbb{C}^n$ and $t \in \mathbb{R}^k$. Then G admits left and right twisted canonical $(APW_S)_\Lambda$ factorization.

For the one variable case Theorem 4.4 was proved in [1].

Taking the determinants of both sides in (4.8), we observe that $c = w(\det G)/n$. Hence, the vector c in the left/right twisted canonical factorization of G is defined by G uniquely and is the same for both factorizations (provided, of course, that they exist).

Proof of Theorem 4.4: Consider G as a continuous matrix function on $\mathcal{B}$. Condition (4.9) can be extended from $t \in \mathbb{R}^k$ to all $t \in \mathcal{B}$. Denote by $S(t)$ the smallest sector with the vertex at the origin containing the numerical range of $G(t)$, by $\omega(t)$ the unit vector in the direction of its bisector, and by $l(t)$ the angle between the sides of $S(t)$. The numerical range is a continuous function of the matrix (see [27]); hence, $\omega: \mathcal{B} \to \mathbb{T}$ and $l: \mathcal{B} \to [0, \pi)$ also are continuous functions. But then $\sup_{t \in \mathbb{R}^k} l(t) = 2\alpha < \pi$, $\omega^{\pm 1}|_{\mathbb{R}^k} \in (AP^k)$, and the matrix function $F = \omega^{-1} G \in (AP^k)$ is (α, ϵ)-sectorial. Choose a function $\omega_0 \in (APW^k)$ so close to ω that $F_0 = \omega_0^{-1} G$ is (α_0, ϵ_0)-sectorial (perhaps, with α_0 a little bigger than α and ϵ_0 a little smaller than ϵ). On top of that, F_0 belongs to (APW^k) together with ω_0^{-1} and G, and therefore (Theorem 4.3) admits left and right canonical APW_S factorizations: $F_0 = X_+ X_- = Y_- Y_+$.

There also exists a canonical APW_S factorization of the scalar positive definite function $|\omega_0|$ in the form

$$|\omega_0| = \omega_+ \overline{\omega_+}. \tag{4.10}$$

(Indeed, the scalar case of Theorem 4.3 gives existence of a canonical APW_S factorization of $|\omega_0|$; passing to complex conjugates and comparing with the original factorization produces the form (4.10)). From Theorem 2.4 we then obtain:

$$\omega_0(t) = \omega_+(t)\overline{\omega_+}(t) e^{i\langle c,t\rangle} e^{i(\Pi_S u)(t)} e^{i(\Pi_{-S\setminus\{0\}} u)(t)}.$$

Hence,

$$\begin{aligned} G &= \omega_0 F_0 = e^{i\langle c,t\rangle} (e^{i(\Pi_S u)(t)} \omega_+(t) X_+(t)) \\ &\quad (e^{i(\Pi_{-S\setminus\{0\}} u)(t)} \overline{\omega_+}(t) X_-(t)) \\ &= e^{i\langle c,t\rangle} (e^{i(\Pi_{-S\setminus\{0\}} u)(t)} \overline{\omega_+}(t) Y_-(t)) \\ &\quad (e^{i(\Pi_S u)(t)} \omega_+(t) Y_+(t)) \end{aligned} \tag{4.11}$$

are, respectively, left and right twisted canonical APW_S factorizations of G.

To finish the Proof of Theorem 4.4, it suffices to show that every left twisted canonical APW_S factorization (4.8) of a locally sectorial matrix function G is automatically a $(APW_S)_\Lambda$ factorization; the case of right factorizations can then be tackled by taking transposes.

Since the diagonal entries of any matrix are contained in its numerical range, the left upper entry g_{11} of the locally sectorial matrix function G is invertible. The matrix function $g_{11}^{-1} G$ is locally sectorial simultaneously with G. In addition, its numerical range for all $t \in \mathbb{R}^k$ contains the point 1, and therefore skips the non-positive half-axis $\mathbb{R}_-$. Hence, the continuous branch of $\arg \det(\frac{1}{g_{11}} G)$ is bounded, and the mean motion of $\det(\frac{1}{g_{11}} G)$ equals zero. But then $w(\det G) = n\, w(g_{11})$, and therefore $c = w(g_{11})$. Since $g_{11} \in (APW^k)_\Lambda$, Theorem 2.4 implies that $c \in \Lambda$.

Observe now that (4.8) leads to a left canonical APW_S factorization of the matrix function $e^{-i\langle c,t\rangle} G(t)$. Since $c \in \Lambda$, the latter matrix function belongs to $(APW^k)_\Lambda$ simultaneously with G. But then $G_+^{\pm 1} \in (APW_S^k)_\Lambda^{n\times n}$, $G_-^{\pm 1} \in (APW_{-S}^k)_\Lambda^{n\times n}$ (see the remark after Theorem 3.2).

Corollary 4.5 *Let G be as in Theorem 4.4. Then G admits a canonical AP_S factorization if and only if the mean motion of* $\det G$ *(or, equivalently, of any diagonal entry of G) equals zero. If this condition is satisfied, then the numerical range of $d_S(G)$ does not contain zero.*

Proof: The first statement follows from the existence of the representation (4.8) with $c = \frac{1}{n}w(\det G) = w(g_{jj})$. To prove the second statement, we turn to equality (4.11).

According to (4.11), a factorization of G can be obtained by multiplying out factorizations of a scalar function ω_0 and a (α_0, ϵ_0)-sectorial matrix function F_0. But then $d_S(G)$ differs from $d_S(F_0)$ by a scalar (non-zero) multiple $d_S(\omega_0)$. Thus, it suffices to show that the numerical range of $d_S(F_0)$ does not contain zero. To this end, observe that the matrix function

$$X_-^{*-1}X_+ = X_-^{*-1}(X_+X_-)X_-^{-1}$$

is α_0-sectorial simultaneously with $F_0 = X_+X_-$. Hence, its mean value $M\{X_-^{*-1}X_+\}$ is α_0-sectorial as well. Next we note the equality

$$M\{X_-^{*-1}X_+\} = M\{X_-\}^{*-1}M\{X_+\},$$

which follows from the fact that the mean is an additive and multiplicative functional on $(APW^k)_S^{n\times n}$ (the multiplicativity is a consequence of S being a half-space). So $M\{X_-\}^{*-1}M\{X_+\}$ is α_0-sectorial. This, in its turn, implies the α_0-sectoriality of

$$M\{X_-\}^*(M\{X_-\}^{*-1}M\{X_+\})M\{X_-\} = M\{X_+\}M\{X_-\} = d_S(F_0).$$

□

For certain classes of matrices, an explicit description of the numerical range is known, see [26, 37]. Whenever values of G belong to such classes, the condition (4.9) can be recast accordingly. For instance, for 2×2 matrix functions the following result holds.

Corollary 4.6 *Let $G \in (APW^k)_\Lambda^{2\times 2}$. Suppose that the eigenvalues $\lambda_1(t)$, $\lambda_2(t)$ of $G(t)$, the Frobenius norm $\|G_0(t)\|_F$ of $G_0(t) = G(t) - \frac{1}{2}(\lambda_1(t) + \lambda_2(t))I$, and its determinant* $\det G_0(t)$ *uniformly on $\mathbb{R}^k$ satisfy the inequality*

$$|\lambda_1(t)| + |\lambda_2(t)| > \sqrt{|\lambda_1(t) - \lambda_2(t)|^2 + (\|G_0(t)\|_F - 2\,|\det G_0(t)|)^2}.$$

Then G admits left and right twisted canonical $(APW_S)_\Lambda$ factorization.

The one variable ($k = 1$) version of Corollary 4.6 for $\Lambda = \mathbb{R}$ was stated in [52]; its discrete ($\Lambda = \mathbb{Z}$) prototype goes back to [51].

5 Factorization of Hermitian Matrix Functions

Another class of almost periodic matrix functions for which factorization results are often useful, is the class of Hermitian valued functions. Before we formulate the theorem, recall that $J \in \mathbb{C}^{n\times n}$ is a *signature* matrix if $J = J^* = J^{-1}$, that is, if J is simultaneously Hermitian and unitary. An example of signature matrices is delivered by

$$J = \begin{bmatrix} I_{p_+} & 0 \\ 0 & -I_{p_-} \end{bmatrix}, \quad p_+ + p_- = n. \tag{5.1}$$

If J is a signature matrix, $U \in \mathbb{C}^{n\times n}$ is called J*-unitary* provided that $U^*JU = J$. For an invertible Hermtian matrix G we define its signature (notation: sign G) by its number of positive eigenvalues minus its number of negative eigenvalues (counting multiplicities).

Theorem 5.1 *Let $G \in (APW^k)^{n\times n}$ and assume that the matrix $G(t)$ is Hermitian for every $t \in \mathbb{R}^k$. Let also Λ be the minimal additive subgroup of $\mathbb{R}^k$ which contains $\sigma(G)$. If the Toeplitz operator $T(G)_{\Lambda'}^{-S]}$ is invertible for some additive subgroup $\Lambda' \supseteq \Lambda$, then G admits a factorization*

$$G(t) = A_+(t)J(A_+(t))^*, \tag{5.2}$$

where $A_+^{\pm 1} \in (APW^k)^{n\times n}_{S\cap\Lambda}$, J is given by (5.1), and the sizes p_+ and p_- are uniquely determined by the additional condition $p_+ - p_- = \text{sign}\, G$ (the signature of $G(t)$, which is constant for $t \in \mathbb{R}^k$). The matrix function A_+ is unique up to a constant right J-unitary multiple.

Proof: Using Theorem 3.2, the proof is not much different from that for the canonical Wiener-Hopf factorization of matrix functions continuous on the unit circle [49]. Namely, from the invertibility of $T(G)_{\Lambda'}^{-S]}$ it follows that G has a canonical $(APW_S)_\Lambda$ factorization (2.8). Since G is Hermitian, we have that at the same time

$$G(t) = G_-^*(t)G_+^*(t), \quad t \in \mathbb{R}^k,$$

so that $(G_-^*)^{-1}G_+ = G_+^*G_-^{-1}$.

According to the latter equality, the matrix function $H = (G_-^*)^{-1}G_+$ is Hermitian, invertible, and belongs to $(AP^k)_S^{n\times n} \cap (AP^k)_{-S}^{n\times n} = \mathbb{C}^{n\times n}$. As such, H admits a representation $H = C^*JC$. From here, $G_- = C^{-1}J(C^*)^{-1}G_+^*$, so that

$$G = (G_+C^{-1})J(G_+C^{-1})^*,$$

and (5.2) holds with $A_+ = G_+C^{-1}$. In particular, $p_+ - p_-$ coincides with the signature of $G(t)$ for all t.

Finally, let also $G = B_+JB_+^*$, where $B_+ \in (AP^k)_S^{n\times n}$. Then $A_+^{-1}B_+J = JB_+^*A_+^{*-1}$. In other words, the matrix function $U = A_+^{-1}B_+$ satisfies the condition

$UJ = JU^{*-1}$, and is therefore J-unitary. It is also constant, because $U(= JU^{*-1}J)$ belongs simultaneously to $(AP^k)_S^{n\times n}$ and $(AP^k)_{-S}^{n\times n}$. It remains to observe that $B_+ = A_+U$. □

Corollary 5.2 *Let $G \in (APW^k)^{n\times n}$ be Hermitian for every $t \in \mathbb{R}^k$. If G admits a canonical AP_S factorization, then $d_S(G)$ is a Hermitian matrix with the same signature as $G(t)$.*

Indeed, (5.2) implies that $d_S(G) = M\{A_+\}JM\{A_+\}^*$.
Combining Theorem 5.1 with Theorem 4.3, the following result follows.

Corollary 5.3 *Let $G \in (APW^k)^{n\times n}$ and assume that the matrix $G(t)$ is positive definite for every $t \in \mathbb{R}^k$, and $|\det G(t)| \geq \epsilon$ for every $t \in \mathbb{R}^k$, where $\epsilon > 0$ is independent of t. Let also Λ be the minimal additive subgroup of $\mathbb{R}^k$ which contains $\sigma(G)$. Then $G(t)$ admits left and right canonical factorizations of the forms*

$$G(t) = A_+(t) \cdot (A_+(t))^* = (\tilde{A}_+(t))^* \cdot \tilde{A}_+(t), \tag{5.3}$$

where $A_+^{\pm 1}, \tilde{A}_+^{\pm 1} \in (AP^k)_S^{n\times n}$. The factors $A_+^{\pm 1}, \tilde{A}_+^{\pm 1}$ are defined up to the right/left constant unitary multiple, and in fact satisfy the condition $A_+^{\pm 1}, \tilde{A}_+^{\pm 1} \in (APW^k)_{S\cap\Lambda}^{n\times n}$.

For the case $\Lambda = \mathbb{Z}^k$, the result of Corollary 5.3 is well-known, see [28]. For the one-variable case (with $S = [0, \infty)$ and $\Lambda = \mathbb{R}$) Corollary 5.3 was proved in [52]. The one variable periodic ($\Lambda = \mathbb{Z}$) case was considered much earlier in [48].

As soon as Corollary 5.3 is established, the factorization of any (invertible) matrix function in $(APW^k)^{n\times n}$ can be reduced to the factorization of unitary matrix functions. The reasoning is exactly the same as in the case of classical Wiener-Hopf factorization (see [41, Chapter 7]) or when $k = 1$, $S = [0, \infty)$ [52]. Namely, for a fixed subgroup Λ of $\mathbb{R}^k$, if $G \in (APW^k)_\Lambda^{n\times n}$ is invertible, then GG^*, $(GG^*)^{1/2}$ and $(G^*G)^{1/2}$ are invertible in $(APW^k)_\Lambda^{n\times n}$ and are positive definite (this follows by applying Proposition 2.3 first with $\Psi(z) = \sqrt{z}$, and then with $\Psi(z) = z^{-1}$). Then, according to Corollary 5.3, $GG^* = XX^*$, $(GG^*)^{1/2} = X_1X_1^*$ and $(G^*G)^{1/2} = X_2X_2^*$, where X, X_1 and X_2 are invertible in $(APW^k)_{S\cap\Lambda}^{n\times n}$. The matrix functions $U = X^{-1}G$ and $V = X_1^{-1}GX_2^{*-1}$ are unitary, and they both admit left (twisted) canonical factorization only simultaneously with G. If G itself is Hermitian, then $(GG^*)^{1/2} = (G^*G)^{1/2}$ and it is possible to choose $X_2 = X_1$. But then V is Hermitian as well (observe that U does not necessarily have this property; this is our main reason for introducing V along with a more straightforwardly defined U). In other words, the factorization problem for any Hermitian matrix function reduces to such a problem for a signature matrix function, the values of which are simultaneously Hermitian and unitary. A necessary and sufficient condition for the factorability of such matrix functions is presently not known. We will give here one sufficient condition of geometrical nature.

Theorem 5.4 *Let $V \in (APW^k)_\Lambda^{n\times n}$ be a signature matrix function. Suppose that there exists a maximal V-positive subspace $\mathfrak{L}$ of $\mathbb{C}^n$, i.e.:*

$$(5.4) \qquad \langle V(t)\xi, \xi\rangle \geq \epsilon \|\xi\|^2$$

for all $\xi \in \mathfrak{L}$ and $t \in \mathbb{R}^k$ (with $\mathfrak{L}$ and $\epsilon > 0$ not depending on t), and the dimension of $\mathfrak{L}$ is equal to the multiplicity of the eigenvalue 1 of $V(t)$. Then V admits left and right canonical $(APW_S)_\Lambda$ factorizations.

Proof: Introduce an orthonormal basis of $\mathbb{C}^n$ in such a way that the first $p_+ = \dim \mathfrak{L}$ vectors of it constitute a basis of $\mathfrak{L}$, and denote the corresponding change of coordinates matrix by U_0. Since U_0 is a constant matrix, the factorability properties of V and $U_0^* V U_0$ are the same. Consider the partition

$$U_0^* V U_0 = \begin{bmatrix} X & Y \\ Y^* & Z \end{bmatrix}$$

with a $p_+ \times p_+$ upper left block X. Due to (5.4), X is a uniformly positive definite matrix function. From the unitarity of V (and therefore of $U_0^* V U_0$) it follows that

$$(5.5) \quad X = (I - YY^*)^{1/2}, \;\; Z^2 = I - Y^*Y, \text{ and } Y((I - Y^*Y)^{1/2} + Z) = 0.$$

The latter two equalities of (5.5), combined with the maximal positivity of $\mathfrak{L}$, imply that $Z = -(I - Y^*Y)^{1/2}$. (Indeed, if Z had a positive eigenvalue μ with a normalized eigenvector x, then we would have $Yx = 0$, and therefore $\mathfrak{L} \dot{+} \operatorname{span}\left\{\begin{bmatrix} 0 \\ x \end{bmatrix}\right\}$ would be an $U_0^* V U_0$-positive subspace of dimension larger than p_+, a contradiction with the hypothesis that $\mathfrak{L}$ is a maximal positive subspace.) Then, due to the first equality of (5.5), $-Z$ and X are unitarily equivalent on the orthogonal complements to the subspaces of their fixed vectors. The matrix function

$$W = \begin{bmatrix} I_{p_+} & 0 \\ 0 & -I_{n-p_+} \end{bmatrix} U_0^* V U_0 = \begin{bmatrix} X & Y \\ -Y^* & -Z \end{bmatrix}$$

then has a uniformly positive real part $W_R = \begin{bmatrix} X & 0 \\ 0 & -Z \end{bmatrix}$ and a contractive imaginary part $W_J = \begin{bmatrix} 0 & -iY \\ iY^* & 0 \end{bmatrix}$. The existence of left and right canonical $(APW_S)_\Lambda$ factorization of W follows now from Theorem 4.3. Indeed, W is $(\arctan(1/\epsilon), \epsilon)$-sectorial where $\epsilon > 0$ is such that $W_R \geq \epsilon I$. Since W and V differ by constant multiples only, the same is true for V. □

6 One-sided Invertibility of Toeplitz Operators

In this section, we use Theorem 5.1 to establish one-sided invertibility conditions for Toeplitz operators in terms of factorizations. As in [36], where the case $k = 1$, $\Lambda = \mathbb{R}$ was considered, we take advantage of a simple but useful observation that T is right invertible if and only if $\begin{bmatrix} I & T^* \\ T & 0 \end{bmatrix}$ is two-sided invertible. The main result here is the following extension of Theorem 3.2.

Theorem 6.1 *Assume that $G \in (APW^k)^{n\times n}$ and let $\Lambda = \Lambda(G)$ be the minimal additive subgroup of $\mathbb{R}^k$ that contains $\sigma(G)$. Then the following statements are equivalent:*

i) *the operator $T(G)_{\Lambda'}^{-S]}$ is right invertible for some additive subgroup $\Lambda' \supseteq \Lambda(G)$;*

ii) *the operator $T(G)_{\Lambda'}^{-S]}$ is right invertible for every additive subgroup $\Lambda' \supseteq \Lambda(G)$;*

iii) *the operator $T(G)_{\Lambda'}^{-S)}$ is right invertible for some additive subgroup $\Lambda' \supseteq \Lambda(G)$;*

iv) *the operator $T(G)_{\Lambda'}^{-S)}$ is right invertible for every additive subgroup $\Lambda' \supseteq \Lambda(G)$;*

v) *the operator $T(G^T)_{\Lambda'}^{[S}$ is left invertible for some additive subgroup $\Lambda' \supseteq \Lambda(G)$;*

vi) *the operator $T(G^T)_{\Lambda'}^{[S}$ is left invertible for every additive subgroup $\Lambda' \supseteq \Lambda(G)$;*

vii) *the operator $T(G^T)_{\Lambda'}^{(S}$ is left invertible for some additive subgroup $\Lambda' \supseteq \Lambda(G)$;*

viii) *the operator $T(G^T)_{\Lambda'}^{(S}$ is left invertible for every additive subgroup $\Lambda' \supseteq \Lambda(G)$;*

ix) *G is invertible (in $(L^\infty(\mathbb{R}^k))^{n\times n}$ and therefore in $(APW^k)_\Lambda^{n\times n}$) and the operator $T(G^{-1})_{\Lambda'}^{[S}$ is right invertible for some additive subgroup $\Lambda' \supseteq \Lambda(G)$;*

x) *G is invertible (in $(L^\infty(\mathbb{R}^k))^{n\times n}$ and therefore in $(APW^k)_\Lambda^{n\times n}$) and the operator $T(G^{-1})_{\Lambda'}^{[S}$ is right invertible for every additive subgroup $\Lambda' \supseteq \Lambda(G)$;*

xi) *G is invertible (in $(L^\infty(\mathbb{R}^k))^{n\times n}$ and therefore in $(APW^k)_\Lambda^{n\times n}$) and the operator $T(G^{-1})_{\Lambda'}^{(S}$ is right invertible for some additive subgroup $\Lambda' \supseteq \Lambda(G)$;*

xii) *G is invertible (in $(L^\infty(\mathbb{R}^k))^{n\times n}$ and therefore in $(APW^k)_\Lambda^{n\times n}$) and the operator $T(G^{-1})_{\Lambda'}^{(S}$ is right invertible for every additive subgroup $\Lambda' \supseteq \Lambda(G)$;*

xiii) *G admits a representation* $G(t) = G_+(t)G_-(t)$, $t \in \mathbb{R}^k$, *where*

$$G_+ \in (AP^k)_S^{n\times n},\ G_-^{-1} \in (AP^k)_{-S}^{n\times n};\ G_+^{-1}, G_- \in (AP^k)^{n\times n};$$

xiv) *G admits a representation* $G(t) = G_+(t)G_-(t)$, $t \in \mathbb{R}^k$, *with*

$$\begin{aligned} &G_+ \in (APW^k)_{S\cap\Lambda}^{n\times n},\ G_-^{-1} \in (APW^k)_{(-S)\cap\Lambda}^{n\times n}; \\ &G_+^{-1}, G_- \in (APW^k)_{\Lambda}^{n\times n}; \end{aligned} \tag{6.1}$$

xv) *G admits a representation*

$$G = \Phi_+ U \Phi_-, \tag{6.2}$$

where $\Phi_+^{\pm 1} \in (AP^k)_S^{n\times n}$, $\Phi_-^{\pm 1} \in (AP^k)_{-S}^{n\times n}$, *and where* $U \in (AP^k)_S^{n\times n}$ *is unitary valued,*

xvi) *G admits a representation* (6.2), *where* $\Phi_+^{\pm 1} \in (APW^k)_{S\cap\Lambda}^{n\times n}$, $\Phi_-^{\pm 1} \in (APW^k)_{(-S)\cap\Lambda}^{n\times n}$, *and* $U = \begin{bmatrix} V & 0 \\ 0 & I \end{bmatrix}$ *with unitary valued* $V \in (APW^k)_{S\setminus\{0\}}^{q\times q}$, *for some* $0 \le q \le n$.

Proof: Obviously, xvi)$\Rightarrow$xv)$\Rightarrow$xiii) and xvi)$\Rightarrow$xiv)$\Rightarrow$xiii). If xiii) holds, then

$$\begin{aligned} T(G)^{-S]}G_-^{-1}\Pi_{-S}G_+^{-1}\Pi_{-S} &= \Pi_{-S}G_+G_-\Pi_{-S}G_-^{-1}\Pi_{-S}G_+^{-1}\Pi_{-S} \\ &= \Pi_{-S}G_+G_-G_-^{-1}\Pi_{-S}G_+^{-1}\Pi_{-S} = \Pi_{-S}G_+\Pi_{-S}G_+^{-1}\Pi_{-S}. \end{aligned} \tag{6.3}$$

On the other hand, using the equality $G_+\Pi_{S\setminus\{0\}} = \Pi_{S\setminus\{0\}}G_+\Pi_{S\setminus\{0\}}$, we have

$$\begin{aligned} \Pi_{-S} &= \Pi_{-S}G_+G_+^{-1}\Pi_{-S} = \Pi_{-S}G_+\Pi_{-S}G_+^{-1}\Pi_{-S} \\ &+\Pi_{-S}G_+\Pi_{S\setminus\{0\}}G_+^{-1}\Pi_{-S} = \Pi_{-S}G_+\Pi_{-S}G_+^{-1}\Pi_{-S}. \end{aligned}$$

Comparing with (6.3), we obtain that $G_-^{-1}\Pi_{-S}G_+^{-1}\Pi_{-S}$ is a right inverse of the operator $T(G)^{-S]}$. In other words, xiii)$\Rightarrow$i) for $\Lambda' = \mathbb{R}$.

We will show now that i)$\Rightarrow$xiv). To this end, introduce the matrix function $\hat{G} = \begin{bmatrix} I & G^* \\ G & 0 \end{bmatrix}$ and consider the Toeplitz operator $T(\hat{G})_{\Lambda'}^{-S]}$ on the direct sum of two copies of the space $(B^k)_{\Lambda'}^{n\times 1}$. Since

$$T(\hat{G})_{\Lambda'}^{-S]} = \begin{bmatrix} I & (T(G)_{\Lambda'}^{-S]})^* \\ T(G)_{\Lambda'}^{-S]} & 0 \end{bmatrix},$$

the right invertibility of $T(G)_{\Lambda'}^{-S]}$ implies the two sided invertibility of $T(\hat{G})_{\Lambda'}^{-S]}$ (see [36, Lemma 1.4]). According to implication i)$\Rightarrow$xvi) of Theorem 3.2, $\hat{G}$ admits a left canonical $(APW_S)_\Lambda$ factorization. But $\hat{G}$ is an Hermitian matrix function with zero signature. Due to Theorem 5.1, its left canonical factorization can be written in the form $\hat{G} = XJX^*$ with $J = \begin{bmatrix} I_n & 0 \\ 0 & -I_n \end{bmatrix}$ and $X^{\pm 1} \in (APW^k)^{2n\times 2n}_{S\cap\Lambda}$. We now partition the matrix function X into $n \times n$ blocks: $X = \begin{bmatrix} X_{11} & X_{12} \\ X_{21} & X_{22} \end{bmatrix}$, and let

$$G_+ = X_{22}, \quad G_- = (X_{11}X_{21}^{-1}X_{22} - X_{12})^*.$$

The proof of the invertibility of X_{21} (in $(APW^k)_\Lambda$), the equality $G = G_+G_-$, and of the properties (6.1) goes through in exactly the same way as in the implication 5)⇒2) of Theorem 1.5 in [36], where the case $k = 1$, $S = [0, \infty)$ was considered.

The proof of xiv)⇒xvi) is the same as for the implication 2)⇒3)⇒4) in the same theorem of [36], the only difference being that, when factoring positive definite matrix functions in $(APW^k)^{n\times n}_\Lambda$, one should refer to Corollary 5.3 instead of its one-variable version from [52].

Hence, statements xiii)–xvi) and i) (for $\Lambda' = \mathbb{R}$) are pairwise equivalent. To prove the equivalence of statements i) (for general subgroups $\Lambda' \supseteq \Lambda$) through xii), one simply should apply the respective equivalences of Theorem 3.2 to the matrix function $\hat{G}$. □

Note that the left twisted canonical factorization (4.8) is of the form (6.2) if $c \in S$. Comparing with Theorem 4.4, the following corollary results.

Corollary 6.2 *Let $G \in (APW^k)^{n\times n}$, and let Λ be the minimal additive subgroup of $\mathbb{R}^k$ that contains $\sigma(G)$. Assume that the numerical range of $G(t)$ is bounded away from zero. Then $T(G)^{-S]}_{\Lambda'}$ is right (resp. left) invertible for every additive supergroup Λ' of Λ if the mean motion of the determinant of G belongs to S (resp. $w(\det G) \in -S$).*

7 Robustness and Continuity of Factorizations

In this section we consider continuity properties of canonical and twisted canonical, factorizations. It turns out that these factorizations persist under small perturbations of the original function, and the factors are well-behaved. A basic result in this direction is the following theorem.

Theorem 7.1 *If $G \in (APW^k)^{n\times n}_\Lambda$ admits a left, resp. right, canonical AP_S factorization, then there exists $\epsilon > 0$ such that every $\tilde{G} \in (APW^k)^{n\times n}_\Lambda$ with $\|\tilde{G} - G\|_\infty < \epsilon$ admits a left, resp. right, canonical $(APW_S)_\Lambda$ factorization. Moreover, there exists $\epsilon_1 \le \epsilon$, $\epsilon_1 > 0$, such that for every $\tilde{G} \in (APW^k)^{n\times n}_\Lambda$ satisfying $\|\tilde{G} - G\|_W < \epsilon_1$ the left, resp. right, canonical $(APW_S)_\Lambda$ factorization of $\tilde{G}$ can be chosen so that its factors are continuous functions of $\tilde{G}$ (in the Wiener norm $\|\cdot\|_W$).*

Note that we have used the (AP^k) norm in the first statement and the Wiener norm in the second statement. Note also that Proposition 2.2 guarantees invertibility (in $(APW^k)_\Lambda^{n\times n}$) of every $\tilde{G} \in (APW^k)_\Lambda^{n\times n}$ which is sufficiently close to G in the (AP^k) norm. This is a necessary condition for existence of a canonical $(APW_S)_\Lambda$ factorization.

Proof: First, we verify the equality

$$\|T(G)_{\Lambda'}^{-S]}\| = \|G\|_\infty \tag{7.1}$$

for any supergroup Λ' of the minimal additive subgroup that contains $\sigma(G)$. Let $M_G : (B)_{\Lambda'}^{n\times 1} \to (B)_{\Lambda'}^{n\times 1}$ be the operator of multiplication by G. Then a standard argument (see, for example, the Proof of Lemma 4.11 in [4]) shows that

$$\|M_G\| = \|G\|_\infty. \tag{7.2}$$

Since $T(G)_{\Lambda'}^{-S]}$ is a compression of M_G, we have $\|T(G)_{\Lambda'}^{-S]}\| \leq \|G\|_\infty$. Conversely, fix $\delta > 0$, and let $x \in (B)_{\Lambda'}^{n\times 1}$, $\|x\| = 1$, be such that $\|y\| \geq \|M_G\| - \delta$, where $y = M_G x$. We may also assume that the Fourier spectrum of x is a finite set. Let $\{\lambda_1, \ldots, \lambda_q\} = \sigma(x) \cap S$. Then for every $\lambda' \in \Lambda' \cap (-S)$ such that $\lambda' + \sum_{j=1}^q \lambda_j \in -S$ we have that the spectrum of $e^{i\langle\lambda',t\rangle}x$ belongs to $-S$. Now, denoting by $\sum_{\lambda\in\sigma(y)} y_\lambda e^{i\langle\lambda,t\rangle}$ the Fourier series of y:

$$\begin{aligned}\|T(G)_{\Lambda'}^{-S]}(e^{i\langle\lambda',t\rangle}x)\|^2 &= \|\Pi_{\Lambda'\cap(-S)} M_G(e^{i\langle\lambda',t\rangle}x)\|^2 \\ &= \|\Pi_{\Lambda'\cap(-S)}(e^{i\langle\lambda',t\rangle}M_G x)\|^2 = \sum_{\lambda\in\sigma(y),\ \lambda+\lambda'\in -S} \|y_\lambda\|^2.\end{aligned} \tag{7.3}$$

Choosing a suitable λ', we can make $\sum_{\lambda\in\sigma(y),\ \lambda+\lambda'\in -S} \|y_\lambda\|^2$ as close as we wish to $\|y\|^2$. Thus, the expression (7.3) can be made greater than or equal to $(\|M_G\| - \delta)^2 - \delta$. Since $\delta > 0$ was arbitrary, we obtain $\|T(G)_{\Lambda'}^{-S]}\| \geq \|M_G\|$, and in view of (7.2), the equality (7.1) follows.

Once (7.1) is established, the proof of the first part of Theorem 7.1 for the left canonical factorizations follows by applying Theorem 3.2. For the right canonical factorizations, use the already obtained result for the transposed matrix function G^T.

We now prove continuity of the factors (in the Wiener norm), for the left canonical factorization (the proof for the right one is completely analogous). Let $G = G_+G_-$ be a left canonical $(APW_S)_\Lambda$ factorization. Replacing G by $G_+^{-1}GG_-^{-1}$, we reduce the proof to the case $G = I$. Now the continuity follows from general results on factorization of elements close to unity in abstract Banach algebras (see [10]). Alternatively, use the explicit construction of the factors in the proof of i) $\Rightarrow$ xvi) of Theorem 3.2.

Theorem 7.1 leads to a global continuity result for canonical factorizations:

Corollary 7.2 *Let $\mathcal{G}_\ell$ (resp. $\mathcal{G}_r$) be the set of all functions $G \in (APW^k)^{n\times n}_\Lambda$ that admit a left, resp. right, canonical $(APW_S)_\Lambda$ factorization $G = G_+G_-$, resp. $G = G_-G_+$. Then the factors $G_\pm$ in this factorization can be chosen continuous functions of $G \in \mathcal{G}_\ell$, resp. $G \in \mathcal{G}_r$ (in the Wiener norm).*

Proof: Consider the case of the left factorization. By Theorem 7.1, $G_\pm$ are continuous functions of $G \in \mathcal{G}_\ell$ locally, in a neighborhood of every given G. Multiplying G_+ on the right by $(M\{G_+\})^{-1}$ (which is a locally continuous function of G), we may assume without loss of generality that $M\{G_+\} = I$. But a left canonical $(APW_S)_\Lambda$ factorization with this additional property is unique, so in fact G_+ is a (globally) continuous function of $G \in \mathcal{G}_\ell$. □

Recall that $M\{G_+\}M\{G_-\}$ is the geometric mean $d_S(G)$ of the matrix function G. Hence, for the factorizations constructed in the Proof of Corollary 7.2, $M\{G_-\} = d_S(G)$. This proves that $d_S(G)$ also is a continuous function of G on $\mathcal{G}_l$. Splitting this factor out of G_-, we can rewrite the (left) canonical factorization of $G \in \mathcal{G}_l$ as

$$G = G_+ d_S(G) G_-, \tag{7.4}$$

where all three factors $G_+ \in (APW^k)^{n\times n}_{S\cap\Lambda}$, $d_S \in \mathbb{C}^{n\times n}$, $G_+ \in (APW^k)^{n\times n}_{(-S)\cap\Lambda}$ are defined by G uniquely and depend on it continuously.

We note that with obvious changes Theorem 7.1 and Corollary 7.2, and their proofs, extend to twisted canonical factorizations. In particular, the mean motion of $\tilde{G} \in (APW^k)^{n\times n}_\Lambda$ remains constant in a $\|\cdot\|_\infty$-neighborhood of $G \in (APW^k)^{n\times n}_\Lambda$, provided G admits a twisted canonical $(AP_S)_\Lambda$ factorization.

The results of Theorem 7.1 and Corollary 7.2 apply to all matrix functions having canonical APW_S factorization, independent of their additional algebraic properties. However, it is a nontrivial question whether or not the symmetric factorizations of Hermitian matrix functions considered in Theorem 5.1 and Corollary 5.3 are also continuous, locally or globally. The answer is affirmative for positive definite matrix functions.

Corollary 7.3 *Let $\mathcal{P}$ be the set of all uniformly positive definite matrix functions in $(APW^k)^{n\times n}$. Then the factors A_+, $\tilde{A}_+$ in the factorizations* (5.3) *can be chosen continuous functions of $G \in \mathcal{P}$ (in the Wiener norm).*

Proof: As usual, it suffices to consider the case of left factorization. According to Corollary 7.2, G can be represented in the form (7.4). Since G is positive definite, the matrix $d_S(G)$ is positive definite as well (Corollary 5.2), and the representation $G_-^* d_S(G) G_+^*$ also delivers a left canonical factorization of $G (= G^*)$. But then G_+ and G_-^* differ only by a constant right multiple. Since both these matrices have the mean value I, it follows from here that in fact they are equal: $G_+ = G_-^*$, so that

$$G = G_+ d_S(G) G_+^*. \tag{7.5}$$

Let H be the positive square root of $d_S(G)$. Then H, and therefore $A_+ = G_+H$, depend continuously on G in the Wiener norm. It remains to observe that (7.5) can be rewritten as $G = A_+A_+^*$. □

Local continuity persists also for symmetric factorizations of arbitrary (not necessarily definite) Hermitian matrix functions.

Theorem 7.4 *Let $G \in (APW^k)_\Lambda^{n\times n}$, and assume that G admits a left canonical AP_S factorization of the form* (5.2) *(in particular, G is Hermitian valued). Then there exists $\epsilon > 0$, such that every $\tilde{G} \in (APW^k)_\Lambda^{n\times n}$ having Hermitian values and satisfying $\|\tilde{G} - G\|_W < \epsilon$ admits a left canonical $(APW_S)_\Lambda$ factorization*

$$\tilde{G}(t) = \tilde{A}_+(t)J(\tilde{A}_+(t))^*, \tag{7.6}$$

in which $\tilde{A}$ is a continuous function of $\tilde{G}$ (in the Wiener norm).

An analogous result holds for right canonical factorizations.

Proof: Using Theorem 7.1, and arguing as in the Proof of Theorem 5.1, we obtain a left canonical factorization of the form

$$\tilde{G}(t) = \tilde{A}_+(t)H(\tilde{A}_+(t))^*, \tag{7.7}$$

where $\tilde{A}_+ \in (APW^k)_{S\cap\Lambda}^{n\times n}$ and the constant (that is, independent of $t \in \mathbb{R}^k$) invertible matrix $H = H^*$ are continuous functions of $\tilde{G}$. To make explicit the dependence of H on $\tilde{G}$, write $H = H(\tilde{G})$. Without loss of generality, we assume that $H(G) = J$. Taking a smaller ϵ, if necessary, we may also assume that the determinants of the left upper $k \times k$ blocks of $H(\tilde{G})$ are nonzero, for $k = 1, \ldots, n$, and for every $\tilde{G} \in (APW^k)_\Lambda^{n\times n}$ having Hermitian values and satisfying $\|\tilde{G} - G\|_W < \epsilon$. One can then apply to $H(\tilde{G})$ the method of Lagrange (which is essentially based on the Gaussian algorithm; see [16], for example) for reduction of a Hermitian form to a sum of signed squares. As a result, we obtain continuous (as a function of H in a neighborhood of $H(G)$) invertible matrix S such that $H = S^*JS$. Substituting this expression for H in (7.7), the claim of the theorem follows. □

It turns out, however, that for Hermitian non-positive definite valued functions the global continuity of canonical symmetric factorizations fails. This failure occurs already for 2×2 constant matrices:

Lemma 7.5 *There does not exist a continuous function f on the set $\mathcal{H}_-$ of 2×2 Hermitian matrices with negative determinants, with f taking values in the group $\mathcal{GL}_2$ of all invertible (complex) 2×2 matrices, and such that*

$$(f(X))^*Xf(X) = \begin{bmatrix} 1 & 0 \\ 0 & -1 \end{bmatrix} \tag{7.8}$$

for every $X \in \mathcal{H}_-$.

Proof: We show that a function f with the required properties does not exist even on a subset $\mathcal{H}_{hu}$ of $\mathcal{H}_-$ consisting of all 2×2 matrices that are simultaneously Hermitian and unitary and have determinant -1. Arguing by contradiction, suppose that $f : \mathcal{H}_{hu} \to \mathcal{GL}_2$ is a continuous function such that (7.8) holds for every $X \in \mathcal{H}_{hu}$. Taking determinants, we see that $|\det(f(X))| = 1$. We claim that in fact $f(X)$ can be chosen unitary for every $X \in \mathcal{H}_{hu}$. To prove the claim, consider the singular value decomposition $f(X) = U_1 D U_2$, where U_1, U_2 are unitaries, and $D = \begin{bmatrix} d_1 & 0 \\ 0 & d_2 \end{bmatrix}$ with $d_1, d_2 > 0$. In fact, $d_1 d_2 = 1$. Substituting the singular value decomposition in (7.8), and denoting the Hermitian unitary matrix $U_1^* X U_1$ by Y, we obtain that DYD is unitary. Writing out $(DYD)^2 = I$, a simple algebra yields $YD^2 = D^{-2}Y$, which in turn implies that either $d_1 = d_2 = 1$ (and then our claim is proved), or $Y = \begin{bmatrix} 0 & \omega \\ \overline{\omega} & 0 \end{bmatrix}$ for some $|\omega| = 1$. In the latter case a calculation shows that

$$(7.9) \qquad \left(U_1 \begin{bmatrix} x_1 & 0 \\ 0 & x_2 \end{bmatrix} U_2 \right)^* X \left(U_1 \begin{bmatrix} x_1 & 0 \\ 0 & x_2 \end{bmatrix} U_2 \right) = \begin{bmatrix} 1 & 0 \\ 0 & -1 \end{bmatrix}$$

for any positive number x_1 and $x_2 = x_1^{-1}$.

At this point we note that if $Z = V_1 D V_2$ is a singular value decomposition of an *invertible* matrix Z (V_1, V_2 unitaries, D positive diagonal), then the product $V_1 V_2$ is uniquely defined by Z and is a continuous function of Z. We indicate a proof. Let $Z = \sqrt{ZZ^*} \cdot V$ be the polar decomposition of Z, and let $\sqrt{ZZ^*} = WDW^*$, where W is unitary and D is positive diagonal. Then $Z = WD(W^*V)$ is a singular value decomposition, and $W \cdot W^*V = V = (ZZ^*)^{-\frac{1}{2}} Z$ is a continuous function of Z. The uniqueness of $V_1 V_2$ follows from the uniqueness of the polar decomposition of Z.

Returning to the Proof of Lemma 7.5, we let $\tilde{f}(X) = U_1 U_2$ which in view of the preceding paragraph is a continuous function of $f(X)$, and therefore also of $X \in \mathcal{H}_{hu}$. By (7.9), the equality $\tilde{f}(X)^* X \tilde{f}(X) = \begin{bmatrix} 1 & 0 \\ 0 & -1 \end{bmatrix}$ holds, and replacing f by $\tilde{f}$, we prove the claim.

We assume therefore that $f(X)$ is unitary. The equation (7.8) is now a similarity as well. Hence the first column of $f(X)$ is an eigenvector of X corresponding to the eigenvalue 1. There is a homeomorphism between the set $\mathcal{H}_{hu}$ and the set $\mathcal{GR}_1$ of one-dimensional subspaces in $\mathbb{C}^2$, with the standard gap topology in $\mathcal{GR}_1$. The homeomorphism matches a matrix $X \in \mathcal{H}_{hu}$ with the (one-dimensional) eigenspace of X corresponding to the eigenvalue 1. Using this homeomorphism, we now have a continuous nowhere zero function g from $\mathcal{GR}_1$ into $\mathbb{C}^2$ such that $g(\mathcal{M}) \in \mathcal{M}$ for every $\mathcal{M} \in \mathcal{GR}_1$. The value $g(\mathcal{M})$ is simply the first column of $f(X)$, when X is identified with its eigenspace $\mathcal{M}$ corresponding to the eigenvalue 1 via the homeomorphism. Write $\mathcal{M} = \operatorname{Span} \begin{bmatrix} 1 \\ z \end{bmatrix}$, where $z \in \mathbb{C}$; then

$g(\mathcal{M}) = \begin{bmatrix} r(z) \\ zr(z) \end{bmatrix}$, for some continuous nowhere zero function $r : \mathbb{C} \to \mathbb{C}$. As $z \to \infty$, the subspace $\mathcal{M}$ converges to Span $\begin{bmatrix} 0 \\ 1 \end{bmatrix}$. Therefore, $zr(z) \to z_0$ for some nonzero $z_0 \in \mathbb{C}$ as $z \to \infty$. We now obtain that the index $\frac{1}{2\pi}\{\arg[r(Re^{i\theta})]_{\theta=0}^{\theta=2\pi}\}$ is equal to -1 for $R > 0$ sufficiently large, and (because $r(0) \neq 0$ and $r(z)$ is continuous at 0) is equal to 0 for $R > 0$ sufficiently small. Since the index (as a function of the radius R) is locally constant for continuous nowhere zero complex valued functions, a continuous function $r(z)$ with the indices as above cannot exist. □

If we restrict consideration to a Hermitian function parameterized by an interval, then global continuity of canonical factorizations still holds:

Theorem 7.6 *Let $G \in (APW^k)_\Lambda^{n\times n}$ be a continuous function (with respect to the Wiener norm) of a parameter $\alpha \in [0, 1]$: $G = G_\alpha$. Assume that for every $\alpha \in [0, 1]$ the function G_α is Hermitian valued and admits a left canonical AP_S factorization. Then G_α admits a left canonical $(APW_S)_\Lambda$ factorization*

(7.10) $$G_\alpha(t) = A_{\alpha+}(t)J(A_{\alpha+}(t))^*,$$

where J is given by (5.1) *and $A_{\alpha+} \in (APW^k)_{\Lambda\cap S}^{n\times n}$ is a continuous function of $\alpha \in [0, 1]$ (in the Wiener norm).*

Again, an analogous result holds for the right canonical factorizations.

Proof: All matrix functions G_α are invertible. Therefore, their signatures do not depend on α. Theorem 5.1 then implies that for each $\alpha \in [0, 1]$ the factorizations (7.7) exist and J there does not depend on α. According to the same theorem, for a fixed α, if (7.10) and

$$G_\alpha(t) = B_{\alpha+}(t)J(B_{\alpha+}(t))^*$$

are left canonical $(APW_S)_\Lambda$ factorizations, then $B_{\alpha+}(t) = A_{\alpha+}(t)S$, where S is a constant J-unitary matrix (which depends on α, of course). Denote by $\mathcal{U}$ the multiplicative group of all J-unitary matrices $S \in \mathbb{C}^{n\times n}$, with J given by (5.1).

Using Theorem 7.4, find points $0 = \alpha_0 < \alpha_1 < \cdots < \alpha_q < \alpha_{q+1} = 1$ so that for each closed interval $[\alpha_{j-1}, \alpha_j]$ ($j = 1, \ldots, q+1$), the function G_α admits a left canonical $(APW_S)_\Lambda$ factorization

(7.11) $$G_\alpha(t) = A_{\alpha+}^{(j)}(t)J(A_{\alpha+}^{(j)}(t))^*,$$

where the factor $A_{\alpha+}^{(j)}$ is continuous on $[\alpha_{j-1}, \alpha_j]$. Because of the uniqueness of factors in a canonical symmetric factorization, up to multiplication $S \in \mathcal{U}$, the matrices $S_j = (A_{\alpha_j+}^{(j)})^{-1}A_{\alpha_j+}^{(j+1)}$, ($j = 1, \ldots, q$), are constant (i.e., independent of $t \in \mathbb{R}^k$) and belong to $\mathcal{U}$. Using the fact that $\mathcal{U}$ is connected (see, e.g., Theorem IV.3.1 in [24], where a topological description of this group is given), for each $j = 1, \ldots, q$, select a continuous path of matrices $S_\alpha^{(j)} \in \mathcal{U}$, $\alpha \in [\alpha_{j-1}, \alpha_j]$, such that $S_{\alpha_{j-1}}^{(j)} = I$ and $S_{\alpha_j}^{(j)} = S_j$. Now define $A_{\alpha+} = A_{\alpha+}^{(j)} S_\alpha^{(j)}$ for $\alpha \in [\alpha_{j-1}, \alpha_j]$, $(j = 1, \ldots, q)$, and $A_{\alpha+} = A_{\alpha+}^{(q+1)}$ for $\alpha \in [\alpha_q, 1]$. The construction of $S_\alpha^{(j)}$ shows that $A_{\alpha+}$ is a well-defined continuous function of α on the whole interval $[0, 1]$, and the factorization (7.10) holds. □

Theorem 7.6 can be extended to the case when a paracompact Hausdorff contractible topological space X is used in place of the interval $[0, 1]$. Recall that a topological space is called *contractible* if all its homotopy groups are trivial. This extension of Theorem 7.6 is based on the triviality of the set of principal $\mathcal{U}$-bundles on X, where $\mathcal{U}$ is the Lie group of J-unitary matrices. In turn, this set can be identified with the set of $\mathcal{U}$-valued cocycles on X modulo equivalence, after passing to the limit with respect to refinements of open coverings of X. See the book [30], for example, for more details. An elementary proof of triviality of cocycles, for the case when X is a contractible compact of a finite dimensional Euclidean space, is found in [25] (Anhang).

Acknowledgements

The research of all three authors is partially supported by NSF Grant DMS 9800704. The research of LR is also partially supported by a Faculty Research Assignment Grant of the College of William and Mary. We thank Prof. C. Schneiderer of Universität Regensburg for elucidation of some material concerning principal bundles and cocycles.

References

[1] R.G. Babadzhanyan and V.S. Rabinovich, On factorization of almost periodic matrix functions, *Differential and Integral Equations, and Complex Analysis. University Press, Elista* 1986, 13–22 (in Russian).

[2] M. Bakonyi, L. Rodman, I.M. Spitkovsky and H.J. Woerdeman, Positive extensions of matrix functions of two variables with support in an infinite band, *C. R. Acad. Sci. Paris Sér. I Math.* **323(8)** (1996), 859–863.

[3] ———, Positive matrix functions on the bitorus with prescribed coefficients in a band, *J. Fourier Analysis and Applications* **5** (1999), 789–812.

[4] J.A. Ball, Yu.I. Karlovich, L. Rodman and I.M. Spitkovsky, Sarason interpolation and Toeplitz corona theorem for almost periodic matrix functions, *Integral Equations and Operator Theory* **32** (1998), 243–281.

[5] H. Bart, I. Gohberg and M.A. Kaashoek, The coupling method for solving integral equations, *Operator Theory: Advances and Applications* **12** (1984), 39–73.

[6] M.A. Bastos, Yu.I. Karlovich, F.A. dos Santos and P.M. Tishin, The Corona theorem and the existence of canonical factorization of triangular AP-matrix functions, *J. Math. Anal. Appl.* **223** (1998), 494–522.

[7] ———, The Corona theorem and the canonical factorization of triangular AP-matrix functions – Effective criteria and explicit formulas, *J. Math. Anal. Appl.* **223** (1998), 523–550.

[8] M.A. Bastos, Yu.I. Karlovich, I.M. Spitkovsky and P.M. Tishin, On a new algorithm for almost periodic factorization, *Operator Theory: Advances and Applications* **103** (1998), 53–74.

[9] A. Böttcher and B. Silbermann, Analysis of Toeplitz Operators, Springer-Verlag, Berlin, Heidelberg, New York, 1990.

[10] M.S. Budjanu and I.C. Gohberg, The factorization problem in abstract Banach algebras. I. Splitting algebras, *Amer. Math. Soc. Transl.* **110** (1977), 107–123.

[11] K.F. Clancey and I.Gohberg, Factorization of Matrix Functions and Singular Integral Operators, Birkhäuser, Basel and Boston, 1981.

[12] C. Corduneanu, Almost Periodic Functions, J. Wiley & Sons, 1968.

[13] A. Devinatz and M. Shinbrot, General Wiener-Hopf operators, *Trans. Amer. Math. Soc.* **145** (1969), 467–494.

[14] J. Erdös, On the structure of ordered real vector spaces, *Publ. Math. Debrecen* **4** (1956), 334–343.

[15] L. Fuchs, Partially Ordered Algebraic Systems, Pergamon Press, Oxford, 1963.

[16] F.R. Gantmacher, The Theory of Matrices, volume 1, Chelsea Publishing Company, New York, N.Y. 1959.

[17] I. Gelfand, D. Raikov and G. Shilov, Commutative Normed Rings, Chelsea, Bronx, N.Y. 1964.

[18] I. Gohberg, S. Goldberg and M.A. Kaashoek, Classes of Linear Operators. I, Birkhäuser, Basel and Boston, 1990.

[19] ———, Classes of Linear Operators. II, Birkhäuser, Basel and Boston, 1993.

[20] I. Gohberg and M.A. Kaashoek, (Eds.) Constructive Methods of Wiener-Hopf Interpolation Birkhäuser, Basel and Boston, 1986.

[21] I. Gohberg and M.A. Kaashoek, Block Toeplitz operators with rational symbols, *Operator Theory: Advances and Applications* **35** (1988), 385–440.

[22] I. Gohberg and M.G. Krein, Systems of integral equations on a half-line with kernel depending upon the difference of the arguments, *Amer. Math. Soc. Transl.* **14** (1960), 217–287.

[23] I. Gohberg and N. Krupnik, One-Dimensional Linear Singular Integral Equations. Introduction, volume 1 and 2, Birkhäuser, Basel and Boston, 1992.

[24] I. Gohberg, P. Lancaster and L. Rodman, Matrices and Indefinite Scalar Products, Birkhäuser, Basel and Boston, 1983.

[25] I. Gohberg and J. Leiterer, Über Algebren stetiger Operatorfunktionen, *Studia Mathematica* **17** (1976), 1–26.

[26] K.E. Gustafson and D.K.M. Rao, Numerical Range. The Field of Values of Linear Operators and Matrices, Springer, New York, 1997.

[27] P. Halmos, A Hilbert Space Problem Book, Van Nostrand, Princeton, N.J. 1967.

[28] H. Helson and D. Lowdenslager, Prediction theory and Fourier series in several variables. I., *Acta Math.* **99** (1958), 165–202.

[29] E. Hewitt and K.A. Ross, Abstract Harmonic Analysis, volume 1, Springer-Verlag, Berlin-Göttingen-Heidelberg, 1963.

[30] D. Husemoller, Fibre Bundles, Springer-Verlag, New York-Berlin-Heidelberg, 1994.

[31] Yu.I. Karlovich, On the Haseman problem, *Demonstratio Math.* **26** (1993), 581–595.

[32] Yu.I. Karlovich and I.M. Spitkovsky, Factorization of almost periodic matrix-valued functions and the Noether theory for certain classes of equations of convolution type, *Mathematics of the USSR, Izvestiya* **34** (1990), 281–316.

[33] ———, (Semi)-Fredholmness of convolution operators on the spaces of Bessel potentials, *Operator Theory: Advances and Applications* **71** (1994), 122–152.

[34] ———, Almost periodic factorization: An analogue of Chebotarev's algorithm, *Contemporary Math.* **189** (1995), 327–352.

[35] ———, Factorization of almost periodic matrix functions, *J. Math. Anal. Appl.* **193** (1995), 209–232.

[36] ———, Semi-Fredholm properties of certain singular integral operators, *Operator Theory: Advances and Applications* **90** (1996), 264–287.

[37] D. Keeler, L. Rodman and I.M. Spitkovsky, The numerical range of 3×3 matrices, *Linear Algebra Appl.* **252** (1997), 115–139.

[38] M.G. Krein, Integral equations on a half-line with kernel depending upon the difference of the arguments, *Amer. Math. Soc. Transl., Series 2* **22** (1962), 163–288.

[39] B.M. Levitan, Almost Periodic Functions, GITTL, Moscow, 1953 (in Russian).

[40] B.M. Levitan and V.V. Zhikov, Almost Periodic Functions and Differential Equations, Cambridge University Press, 1982.

[41] L.S. Litvinchuk and I.M. Spitkovsky, Factorization of Measurable Matrix Functions, Birkhäuser Verlag, Basel and Boston, 1987.

[42] G.J. Murphy, C^*-Algebras and Operator theory, Academic Press 1990.

[43] A.A. Pankov, Bounded and Almost Periodic Solutions of Nonlinear Differential Operator Equations, Kluwer, Dordrecht/Boston/London, 1990.

[44] A.I. Perov and A.V. Kibenko, A theorem on the argument of an almost periodic function of several variables, *Litovskii Matematicheskii Sbornik* **7** (1967), 505–508 (in Russian).

[45] D. Quint, L. Rodman and I.M. Spitkovsky, New cases of almost periodic factorization of triangular matrix functions, *Michigan Math. J.* **45(1)** (1998), 73–102.

[46] L. Rodman and I.M. Spitkovsky, Almost periodic factorization and corona theorem, *Indiana Univ. Math. J.* 1998, to appear.

[47] L. Rodman, I.M. Spitkovsky and H.J. Woerdeman, Carathéodory-Toeplitz and Nehari problems for matrix valued almost periodic functions, *Trans. Amer. Math. Soc.* **350** (1998), 2185–2227.

[48] Yu.L. Shmulyan, The Riemann problem with a positive definite matrix, *Uspekhi Matem. Nauk*, **8(2)** (1953), 143–145 (in Russian).

[49] ———, The Riemann problem with a Hermitian matrix, *Uspekhi Matem. Nauk*, **9(4)** (1954), 243–248 (in Russian).

[50] I.B. Simonenko, The Riemann boundary value problem for n pairs of functions with measurable coefficients and its application to the investigation of singular integrals in the spaces L^p with weight, *Izv. Akad. Nauk SSSR. Ser. Mat.* **28(2)** (1964), 277–306 (in Russian).

[51] I.M. Spitkovsky, Stability of partial indices of the Riemann boundary value problem with a strictly nondegenerate matrix, *Soviet Math. Dokl.* **15** (1974), 1267–1271.

[52] I.M. Spitkovsky, On the factorization of almost periodic matrix functions, *Math. Notes* **45(5–6)** (1989), 482–488.

[53] I.M. Spitkovsky and H.J. Woerdeman, The Carathéodory-Toeplitz problem for almost periodic functions, *J. Functional Analysis* **115(2)** (1993), 281–293.

Department of Mathematics
College of William and Mary
Williamsburg VA 23187-8795
USA

1991 Mathematics Subject Classification. Primary 47A68; Secondary 26B99, 43A60, 47A53, 47B35

Received Date inserted by the Editor

Operator Theory:
Advances and Applications, Vol. 122
© 2001 Birkhäuser Verlag Basel/Switzerland

Simultaneous Similarity of Pairs of Companions to their Transposes

R.A. Zuidwijk

Dedicated to professor M.A. Kaashoek on occasion of his sixtieth birthday

This paper considers simultaneous similarity of pairs of Companion matrices to their transposes. It turns out that the similarity transformation in question always is a (non-zero) multiple of the Bezoutian associated with the characteristic polynomials of the Companion matrices.

1 Introduction

One of the research directions inspired by the book [1] by Bart, Gohberg and Kaashoek on minimal factorization of rational matrix functions is the study of simultaneous similarity of pairs of square matrices to pairs in complementary triangular forms; see for example Bart & Hoogland [2] and Bart & Zuidwijk [6].

Another notion of simultaneous similarity, which again has a connection to the subject matter of [1], is the following notion which is the topic of this paper.

For a pair of complex $m \times m$ matrices A and Z, we shall discuss the existence of an invertible $m \times m$ matrix S, such that

$$S^{-1}AS = A^T, \qquad S^{-1}ZS = Z^T.$$

In words, we study simultaneous similarity of a pair of square matrices to their transposes. This notion of simultaneous similarity is connected to the theory of rational matrix functions as follows: If the rational $n \times n$ matrix function $W(\lambda)$ with minimal realization (see [1])

$$W(\lambda) = I_n + C(\lambda I_m - A)^{-1}B$$

is symmetrical, i.e., if $W(\lambda) = W(\lambda)^T$, then there exists an invertible $m \times m$ matrix S, such that $S^{-1}AS = A^T$ and $S^{-1}(A - BC)S = (A - BC)^T$.

In the case when A and Z are first Companion matrices, or in the case that A is a first companion and Z is a third companion matrix, necessary and sufficient conditions are known for the existence of an invertible S, which reduces A and Z to their transposes (c.f. Bart & Thijsse [4] and Bart & Wimmer [5]). In this paper, it is shown that the invertible matrix S is uniquely determined up to a scalar multiple.

2 Companion and Bezoutian Matrices

A *first companion* $m \times m$ matrix is of the form (we assume that $m > 1$)

$$C_d = \begin{pmatrix} 0 & 1 & & & \\ & 0 & 1 & & \\ & & \ddots & \ddots & \\ & & & 0 & 1 \\ -d_0 & -d_1 & & -d_{m-2} & -d_{m-1} \end{pmatrix}.$$

It holds that $\det(\lambda - C_d) = d_0 + \cdots + d_{m-1}\lambda^{m-1} + \lambda^m = d(\lambda)$. In this manner, each monic polynomial $d(\lambda)$ of degree m can be associated with an $m \times m$ first companion matrix C_d. Let R denote the *reversed identity* or *rotation matrix*, given by $Re_k = e_{m-k+1}$ $(k = 1, \ldots, m)$. The *third companion* matrix $\hat{C}_d$ is defined as $\hat{C}_d = RC_dR$. For a polynomial $d(\lambda)$ of degree m, define the *reversed* polynomial $\hat{d}(\lambda) = \lambda^m d(\frac{1}{\lambda})$. Note that $\hat{d}(\lambda)$ is a polynomial of degree m if and only if $d(0) \neq 0$. If $d(0) \neq 0$, then

$$\widehat{\hat{d}(\lambda)} = \lambda^m \widehat{d}\left(\frac{1}{\lambda}\right) = \lambda^m \left(\frac{1}{\lambda}\right)^m d(\lambda) = d(\lambda).$$

In the case when $d(0) \neq 0$, the first companion matrix C_d is invertible, with inverse $C_d^{-1} = \hat{C}_{\hat{d}/d(0)}$, a third companion matrix. The following two propositions are special cases of Theorem 1.4 in [4] and Theorem 4.3 in [5], respectively.

Proposition 2.1 *Let $a(\lambda)$ and $b(\lambda)$ be distinct monic polynomials of degree m. Then the following are equivalent:*

1. *There exists an invertible matrix X, such that $X^{-1}C_aX = C_a^T$ and $X^{-1}C_bX = C_b^T$*
2. $\alpha - \beta \neq 0$ *for* $\alpha \in \sigma(C_a)$ *and* $\beta \in \sigma(C_b)$
3. $\gcd(a, b) = 1$

Proposition 2.2 *Let $a(\lambda)$ and $b(\lambda)$ be monic polynomials of degree m, such that $\hat{b}(\lambda) \neq a(\lambda)b(0)$. Then the following are equivalent:*

1. *There exists an invertible matrix X, such that $X^{-1}C_aX = C_a^T$ and $X^{-1}\hat{C}_bX = \hat{C}_b^T$*
2. $1 - \alpha\beta \neq 0$ *for* $\alpha \in \sigma(C_a)$ *and* $\beta \in \sigma(\hat{C}_b)$
3. $\gcd(a, \hat{b}) = 1$

In Proposition 2.2, we require that $\hat{b}(\lambda) \neq a(\lambda)b(0)$. This is equivalent to $C_a\hat{C}_b \neq I$. Indeed, $\hat{b}(\lambda) = a(\lambda)b(0)$, with $b(\lambda)$ not identically zero, implies that $b(0) \neq 0$, so that $\hat{C}_b$ is invertible. In fact,

$$\hat{C}_b^{-1} = C_{\hat{b}(\lambda)/b(0)} = C_a.$$

By reversing the argument given above, one proves the converse. The Propositions 2.1 and 2.2 give necessary and sufficient conditions for the existence of an invertible matrix, which reduces the relevant pair of companion matrices to their transposes. We now introduce the Bezoutian matrix and some of its elementary properties. Let $d(\lambda)$ be a polynomial of degree m and let $r(\lambda)$ be a polynomial of degree at most m. The expression

$$\frac{d(\lambda)r(\mu) - r(\lambda)d(\mu)}{\lambda - \mu}$$

is a polynomial in the variables λ and μ. Write

$$\frac{d(\lambda)r(\mu) - r(\lambda)d(\mu)}{\lambda - \mu} = \sum_{i,j=1}^{m} B_{ij}\lambda^{i-1}\mu^{j-1},$$

then the $m \times m$ matrix $B = (B_{ij})_{i,j=0}^{m-1}$ is called the *Bezout matrix* or *Bezoutian* associated with the polynomials $d(\lambda)$ and $r(\lambda)$. Write $B = \mathrm{Bez}(d, r)$. For proofs of the elementary properties of the Bezoutian stated below, we refer to [10], Section 13.3. To avoid trivial ambiguities, the order of the matrix $\mathrm{Bez}(d, r)$ (here m) is defined as the maximum of the degrees of the polynomials $d(\lambda)$ and $r(\lambda)$. In this manner, we could also define the $m \times m$ matrix $\mathrm{Bez}(r, d)$. It then easily follows that $\mathrm{Bez}(r, d) = -\mathrm{Bez}(d, r)$. If $d(\lambda)$ is monic of degree m and $r(\lambda)$ is of degree at most m, then the *Barnett factorization* of $\mathrm{Bez}(d, r)$ reads

(1) $$\mathrm{Bez}(d, r) = S(d)r(C_d),$$

where $S(d)$ denotes the *symmetrizer* of $d(\lambda)$, given by

$$S(d) = \begin{pmatrix} d_1 & d_2 & \cdots & 1 \\ d_2 & \cdots & 1 & 0 \\ \vdots & \iddots & \iddots & \vdots \\ 1 & 0 & \cdots & 0 \end{pmatrix}.$$

If $r(\lambda)$ is a polynomial of degree at most m, then $\mathrm{Bez}(d, r)C_d = C_d^T \mathrm{Bez}(d, r)$. In particular, this equation is satisfied for $r(\lambda) = 1(\lambda) = 1$, and we get (use $S(d) = B(d, 1)$)

(2) $$S(d)C_d = C_d^T S(d).$$

This explains the term “symmetrizer”. If $a(\lambda)$ and $b(\lambda)$ are (monic) polynomials of degree m, then

$$R\mathrm{Bez}(a, \hat{b})R = -\mathrm{Bez}(\hat{a}, b).$$

The Bezout matrix $\mathrm{Bez}(d, r)$ is invertible if and only if $\gcd(d, r) = 1$. Therefore, if the polynomials $d(\lambda)$ and $r(\lambda)$ are relatively prime, then $X = \mathrm{Bez}(d, r)$ is an invertible solution to

$$XC_d = C_d^T X. \tag{3}$$

The content of the following lemma is, that *all* (invertible) solutions to the Lyapunov equation (3) arise from (invertible) Bezoutians. This result is related to results in Heinig & Rost [9] on solutions to other Lyapunov and Stein equations with companion coefficients.

Lemma 2.3 *Let $d(\lambda)$ be a monic polynomial of degree m. Then all solution matrices X to* (3) *are of the form $X = \mathrm{Bez}(d, p)$, where $p(\lambda)$ is a polynomial of degree less than m, the zero polynomial included.*

Proof: The matrix $X = \mathrm{Bez}(d, 1) = S(d)$ is an invertible solution to (3). Let Y be another solution. Then

$$X^{-1}YC_d = X^{-1}C_d^T Y = C_d X^{-1}Y,$$

so $X^{-1}Y$ commutes with C_d. Since C_d is nonderogatory, there exists a polynomial $p(\lambda)$ of degree less than m, such that $X^{-1}Y = p(C_d)$. It follows that $Y = Xp(C_d) = S(d)p(C_d) = \mathrm{Bez}(d, p)$. The lemma is proved. □

The following two technical lemmas describe how identities between (rotated) Bezoutians impose restrictions on the underlying polynomials.

Lemma 2.4 *Let $a(\lambda)$ and $b(\lambda)$ be monic polynomials of degree m, and let $p(\lambda)$ and $q(\lambda)$ be non-zero polynomials of degree less than m. Then*

$$\mathrm{Bez}(a, p) = \mathrm{Bez}(b, q),$$

if and only if $p(\lambda) = q(\lambda)$, and $a(\lambda) - b(\lambda) = \gamma p(\lambda)$ for some complex number γ.

Proof: If $\mathrm{Bez}(a, p) = \mathrm{Bez}(b, q)$, then by definition,

$$a(\lambda)p(\mu) - a(\mu)p(\lambda) = b(\lambda)q(\mu) - b(\mu)q(\lambda).$$

Divide this expression by λ^m and let $\lambda \to \infty$ to obtain $p(\mu) = q(\mu)$. Then

$$[a(\lambda) - b(\lambda)]p(\mu) = [a(\mu) - b(\mu)]p(\lambda).$$

Let $\mu_0 \in \mathbf{C}$, such that $p(\mu_0) \neq 0$. Then

$$a(\lambda) - b(\lambda) = \frac{a(\mu_0) - b(\mu_0)}{p(\mu_0)} p(\lambda) = \gamma p(\lambda).$$

Moreover, if $a(\lambda) \neq b(\lambda)$, then $\gamma \neq 0$, and

$$\mathrm{Bez}(a, p) = \frac{1}{\gamma}\mathrm{Bez}(a, a - b) = \frac{-1}{\gamma}\mathrm{Bez}(a, b).$$

The lemma is proved. □

Lemma 2.5 *Let $a(\lambda)$ and $b(\lambda)$ be monic polynomials of degree m, and $p(\lambda)$ and $q(\lambda)$ non-zero polynomials of degree less than m. Then*

$$\mathrm{Bez}(a, p) = R\mathrm{Bez}(b, q)R$$

is equivalent to the following:

1. *If $b(0) \neq 0$, then $a(\lambda) - \hat{b}(\lambda)/b(0) = \gamma p(\lambda)$ for some complex number γ*
2. *If $b(0) = 0$, then $r(\lambda) = \lambda^m q(1/\lambda)$ is of degree m, with non-zero leading coefficient $q(0)$, and $p(\lambda) = q(0)\hat{b}(\lambda)$, and $a(\lambda) - r(\lambda)/q(0) = \gamma \hat{b}(\lambda)$ for a complex number γ*

Proof: Write $\mathrm{Bez}(b, q) = (B_{kl})_{k,l=1}^m$. Then, by definition,

$$\frac{b(\lambda)q(\mu) - q(\lambda)b(\mu)}{\lambda - \mu} = \sum_{k,l=1}^{m} B_{kl}\lambda^{k-1}\mu^{l-1}.$$

Since $\mathrm{Bez}(a, p) = R\mathrm{Bez}(b, q)R = (B_{m-k+1,m-l+1})_{k,l=1}^m$, one may write

$$\frac{a(\lambda)p(\mu) - p(\lambda)a(\mu)}{\lambda - \mu} = \sum_{k,l=1}^{m} B_{m-k+1,m-l+1}\lambda^{k-1}\mu^{l-1}$$

$$= \sum_{k,l=1}^{m} B_{kl}\lambda^{m-k}\mu^{m-l} = \frac{\lambda^m\mu^m}{\lambda\mu}\sum_{k,l=1}^{m} B_{kl}(1/\lambda)^{k-1}(1/\mu)^{l-1}$$

$$= -\frac{\lambda^m b(1/\lambda)\mu^m q(1/\mu) - \lambda^m q(1/\lambda)\mu^m b(1/\mu)}{\lambda - \mu}$$

$$= -\frac{\hat{b}(\lambda)r(\mu) - r(\lambda)\hat{b}(\mu)}{\lambda - \mu}.$$

Consequently,

$$a(\lambda)p(\mu) - p(\lambda)a(\mu) = r(\lambda)\hat{b}(\mu) - \hat{b}(\lambda)r(\mu). \tag{4}$$

We will now distinguish two cases.

Case 1 $b(0) \neq 0$, so $\hat{b}(\lambda)$ is of degree m.

The polynomial $s(\lambda) = b(0)r(\lambda) - q(0)\hat{b}(\lambda)$ is of degree less than m. Lemma 2.4 now gives that $p(\lambda) = -s(\lambda)$, and that $a(\lambda) - \hat{b}(\lambda)/b(0) = \gamma p(\lambda)$ for a complex number γ. The first case is proved.

Case 2 $b(0) = 0$, so $\hat{b}(\lambda)$ is of degree less than m.

If we divide both sides of equation (4) by λ^m and let $\lambda \to \infty$, then $p(\mu) = q(0)\hat{b}(\mu)$. Since $p(\mu)$ is non-zero, $q(0) \neq 0$, and $r(\lambda)$ must be of degree m. In addition, as in the Proof of Lemma 2.4, there exists a constant γ, such that $a(\lambda) - r(\lambda)/q(0) = \gamma\hat{b}(\lambda)$.

The lemma is proved. □

3 Simultaneous Similarity

We first discuss simultaneous similarity to transposes of two first companion matrices. The proposition has been proved in Bart & Kroon [3], using arguments inspired by systems theory. A different proof, using Lemma 2.4, will be given here.

Proposition 3.1 *Let $a(\lambda)$ and $b(\lambda)$ be monic polynomials of degree m, with* gcd $(a, b) = 1$. *The (invertible) solution matrices X to the pair of equations*

$$XC_a = C_a^T X, \qquad XC_b = C_b^T X \tag{5}$$

are all of the form $X = c\mathrm{Bez}(a, b)$, where c is a (non-zero) constant.

Proof: Assume that $X \neq O$ satisfies (5). By Lemma 2.3, $XC_a = C_a^T X$ implies that $X = \mathrm{Bez}(a, p)$ for a non-zero polynomial $p(\lambda)$ of degree less than m. Further, $XC_b = C_b^T X$ implies that $X = \mathrm{Bez}(b, q)$ with $q(\lambda)$ a non-zero polynomial of degree less than m. Conclusively, $X = \mathrm{Bez}(a, p) = \mathrm{Bez}(b, q)$, and Lemma 2.4 yields $p(\lambda) = q(\lambda)$and $a(\lambda) - b(\lambda) = \gamma p(\lambda)$ for a complex number γ. Since $a(\lambda) \neq b(\lambda)$ we get, as discussed at the end of the Proof of Lemma 2.4, that $X = c\mathrm{Bez}(a, b)$ for a non-zero constant c. □

The next proposition deals with simultaneous similarity to transposes of a first companion matrix and a third companion matrix.

Proposition 3.2 *Let $a(\lambda)$ and $b(\lambda)$ be monic polynomials of degree m, with* gcd $(a, \hat{b}) = 1$. *The (invertible) solution matrices X to the pair of equations*

$$XC_a = C_a^T X, \qquad X\hat{C}_b = \hat{C}_b^T X \tag{6}$$

are all of the form $X = c\mathrm{Bez}(a, \hat{b})$, where c is a (non-zero) constant.

Proof: Assume that $X \neq O$ satisfies equation (6). First, $XC_a = C_a^T X$ implies, by Lemma 2.3, that $X = \text{Bez}(a, p)$ for some polynomial $p(\lambda)$ of degree less than m. Second, $X\hat{C}_b = \hat{C}_b X$ is equivalent to $RXRC_b = C_b RXR$ (use $\hat{C}_b = RC_b R$). Now apply Lemma 2.3 to obtain $X = R\text{Bez}(b, q)R$ for a polynomial $q(\lambda)$ of degree less than m. Conclusively, $X = \text{Bez}(a, p) = R\text{Bez}(b, q)R$. We now apply Lemma 2.5. If $b(0) \neq 0$, then $a(\lambda) - \hat{b}(\lambda) = \gamma p(\lambda)$, where $\gamma \neq 0$. It follows that

$$X = \text{Bez}(a, p) = \frac{-1}{\gamma}\text{Bez}(a, \hat{b} - a) = c\text{Bez}(a, \hat{b}),$$

where c is a non-zero complex number. If $b(0) = 0$, then $p(\lambda) = q(0)\hat{b}(\lambda)$, so

$$X = \text{Bez}(a, p) = q(0)\text{Bez}(a, \hat{b}).$$

The proposition is proved. □

A set of $m \times m$ matrices $\mathcal{A}$ can be *reduced simultaneously to transposes*, if there exists an invertible $m \times m$ matrix X, such that $X^{-1}AX = A^T$, for all $A \in \mathcal{A}$. In particular, if a set of first companion $m \times m$ matrices $\mathcal{A}$ has this property, each pair $C_a, C_b \in \mathcal{A}$ can be simultaneously reduced to transposes. By Proposition 2.1, distinct $C_a, C_b \in \mathcal{A}$ have no common eigenvalues. We give an example of such a set $\mathcal{A}$. Let C_a and C_b be first companion $m \times m$ matrices, such that $\sigma(C_a) \cap \sigma(C_b) = \emptyset$, and define the one-dimensional affine space

$$\mathcal{A}(a, b) = \{\rho C_a + (1 - \rho)C_b \mid \rho \in \mathbf{C}\}.$$

All elements in $\mathcal{A}(a, b)$ are first companion $m \times m$ matrices, and have mutually disjoint spectrum. Since

$$\begin{aligned} X^{-1}(\rho C_a + (1 - \rho)C_b)X &= \rho X^{-1}C_a X + (1 - \rho)X^{-1}C_b X \\ = \rho C_a^T + (1 - \rho)C_b^T &= (\rho C_a + (1 - \rho)C_b)^T, \end{aligned}$$

the set $\mathcal{A}(a, b)$ can be reduced simultaneously to transposes. The content of Proposition 3.3 below is, that each set of first companion $m \times m$ matrices (with more than one element) that can be reduced simultaneously to transposes, is contained in $\mathcal{A}(a, b)$, where $C_a, C_b \in \mathcal{A}$ are distinct elements.

Proposition 3.3 *Let $\mathcal{A}$ be a set of first companion $m \times m$ matrices, consisting of more than one element, that can be reduced simultaneously to transposes. Then*

$$\mathcal{A} \subseteq \mathcal{A}(a, b) = \{\rho C_a + (1 - \rho)C_b \mid \rho \in \mathbf{C}\},$$

if $C_a, C_b \in \mathcal{A}$ are distinct elements.

Proof: Let C_a, C_b be two distinct elements in $\mathcal{A}$, and let C_d be any other companion matrix in $\mathcal{A}$. There exists an invertible matrix X, such that

$$X^{-1}C_aX = C_a^T, \qquad X^{-1}C_bX = C_b^T, \qquad X^{-1}C_dX = C_d^T.$$

Proposition 3.1 applied to the pairs C_a, C_d and C_b, C_d, gives that there exist $c_1, c_2 \in \mathbf{C}$, such that

$$X = c_1\text{Bez}(a, d) = c_2\text{Bez}(b, d),$$

and consequently,

$$\text{Bez}(a, c_1(d - a)) = \text{Bez}(b, c_2(d - b)).$$

Note that the polynomials $c_1(d(\lambda) - a(\lambda))$ and $c_2(d(\lambda) - b(\lambda))$ are of degree less than m. Therefore, by Lemma 2.4, we get

$$c_1(d(\lambda) - a(\lambda)) = c_2(d(\lambda) - b(\lambda)).$$

Since $a(\lambda) \neq b(\lambda)$, we get $c_1 \neq c_2$. Write $\rho = c_1/(c_1 - c_2)$ to obtain $d(\lambda) = \rho a(\lambda) + (1 - \rho)b(\lambda)$. This proves that $C_d \in \mathcal{A}(a, b)$, and the proposition is proved. □

A result in the same direction for first and third companion matrices reads as follows: If $\mathcal{A}$ is a non-empty set of first companions, and $\mathcal{Z}$ is a non-empty set of third companions, such that $\mathcal{A} \cup \mathcal{Z}$ is simultaneously reduced to transposes, then both $\mathcal{A}$ and $\mathcal{Z}$ consist of only one element.

This paper dealt with companion and Bezoutian matrices associated with scalar polynomials. An important result used here was that for a pair of monic polynomials a and b, $\gcd(a, b) = 1$ implies that $\text{Bez}(a, b)$ is invertible and solves certain Lyapunov equations. Such notions and results have been extended to the case of matrix polynomials; see Lerer & Tismenetsky [11, 12, 13] and Gohberg, Kaashoek, Lerer & Rodman [7, 8]. It is an aim of further research to extend the results in this paper to the case of matrix polynomials.

Acknowledgement

The author would like to thank H. Bart and L. Lerer for interesting discussions which have contributed to the contents of this paper.

References

[1] H. Bart, I. Gohberg and M.A. Kaashoek, Minimal Factorization of Matrix and Operator Functions, *Operator Theory: Adv. Appl.* 1, Birkhäuser Verlag, Basel, 1979.

[2] H. Bart and H. Hoogland, Complementary Triangular Forms of Matrices, Realizations with Prescribed Main Matrices and Complete Factorization of Rational Matrix Functions, *Linear Algebra Appl.* **103** (1988), 193–228.

[3] H. Bart and L.G. Kroon, Companion Based Matrix Functions: Description and Factorization, *Linear Algebra Appl.* **248** (1996), 1–46.

[4] H. Bart and G.Ph.A. Thijsse, Simultaneous Reduction to Companion and Triangular Forms of Sets of Matrices, *Linear and Multilinear Algebra* **26** (1990), 231–241.

[5] H. Bart and H.K. Wimmer, Simultaneous Reduction to Triangular and Companion Forms of Pairs of Matrices: the Case $rank(I - AZ) \leq 1$, *Linear Algebra Appl.* **150** (1991), 443–461.

[6] H. Bart and R.A. Zuidwijk, Simultaneous Reduction to Triangular Forms after Extension with Zeroes, *Linear Algebra Appl.* **281** (1998), 105–135.

[7] I. Gohberg, M.A. Kaashoek, L. Lerer and L. Rodman, Common Multiples and Common Divisors of Matrix Polynomials, I. Spectral Method, *Indiana Univ. Math. J.* **30** (3) (1981), 321–356.

[8] I. Gohberg, M.A. Kaashoek, L. Lerer and L. Rodman, Common Multiples and Common Divisors of Matrix Polynomials, II. Vandermonde and Resultant Matrices, *Linear and Multilinear Algebra* **12** (1982), 159–203.

[9] G. Heinig and K. Rost, *Algebraic Methods for Toeplitz-like Matrices and Operators*, Akademie-Verlag, Berlin, 1984.

[10] P. Lancaster and M. Tismenetsky, *The Theory of Matrices with Applications*, Academic Press, Orlando, 1985.

[11] L. Lerer and M. Tismenetsky, The Bezoutian and the Eigenvalue-Separation Problem for Marix Polynomials, *Integral Eq. Oper. Th.* **5** (1982), 386–445.

[12] L. Lerer and M. Tismenetsky, Generalized Bezoutian and Matrix Equations, *Linear Algebra Appl.* **99** (1988), 123–160.

[13] L. Lerer and M. Tismenetsky, Generalized Bezoutian and the Inversion Problem for Block Matrices, I. General Scheme, *Integral Eq. Oper. Th.* **9** (1986), 790–819.

Rotterdam School of Management
Erasmus University Rotterdam
P.O. Box 1738
3000 DR Rotterdam
R.Zuidwijk@fac.fbk.eur.nl

Programme
Workshop on Operator Theory and Analysis

Amsterdam, 12–14 November 1997

On the occasion of the sixtieth birthday of prof. dr. M.A. Kaashoek

The Workshop is funded by

- Vakgroep Wiskunde VU
- Stichting VU Computer Science & Math. Research Centre
- Thomas Stieltjes Institute for Mathematics
- Faculteit Wiskunde en Informatica VU
- Faculteit Economische Wetenschappen EUR

Wednesday, November 12

Coffee and tea

chairman **André Ran** (Vrije Universiteit Amsterdam)

13.30–14.00 opening **Israel Gohberg** (Tel-Aviv University, VU Amsterdam), presented by **Seymour Goldberg** (University of Maryland, College Park)

14.10–15.00 **Seymour Goldberg** (University of Maryland, College Park)
Traces and determinants of linear operators

15.05–15.40 **Bernhard Gramsch** (Johannes Gutenberg Universität, Mainz)
Lifting of idempotents in operator algebras

Coffee and tea

chairman **Freek van Schagen** (Vrije Universiteit Amsterdam)

16.00–16.35 **Cor van der Mee** (Università di Cagliari)
Inverse scattering of Schrödinger equations with rational reflection

16.40–17.15 **Paul Van Dooren** (Université Catholique de Louvain, Louvain la Neuve)
About stability radii

Thursday, November 13

Coffee and tea

chairman **Winfried Kaballo** (Universität Dortmund)

09.15–10.05 **Heinz Langer** (Technische Universität Wien)
Linearization and compact perturbations of symmetric holomorphic operator functions (joint work with V. Matsaev and A. Markus)

10.10–10.45 **Sjoerd Verduyn Lunel** (VU Amsterdam)
Effects of small time delays on the dynamics of evolutionary systems

Coffee and tea

chairman **Gilbert Groenewald** (University of Western Cape)

11.00–11.35 **Peter Lancaster** (University of Calgary)
Numerical solution of algebraic Riccati equations in the singular case (joint work with Guo Chun-Hua)

11.40–12.15 **Robert Ellis** (University of Maryland, College Park)
Inversion Formulas For Infinite Generalized Toeplitz Matrices

Lunch break

chairman **Karl-Heinz Förster** (Technische Universität Berlin)

14.00–14.35 **Reinhard Mennicken** (Universität Regensburg)
On the essential spectrum of a class of 2×2 block operator matrices

14.40–15.15 **Trevor West** (Trinity College, Dublin)
Perron-Frobenius & the Minimal Diagonal Element

Coffee and tea

chairman **Ben de Pagter** (Technische Universiteit Delft)

15.30–16.05 **Erhard Meister** (Technische Hochschule Darmstadt)
On some mathematical problems of diffraction theory for canonical obstacles

16.10–17.00 **Joe Ball** (Virginia Tech, Blacksburg)
Interpolation problems for Schur-class functions: The confluence of operator model theory, linear system theory and scattering

Friday, November 14

Coffee and tea

chairman **Leen Roozemond** (KSLA, Rijswijk)

09.15–10.05 **Hugo Woerdeman** (College of William and Mary, Williamsburg)
Multivariable Extensions and Nonlinear Matrix Completions

10.10–10.45 **Harm Bart** (Erasmus Universiteit Rotterdam)
Sums of idempotents and analytic vector valued functions

Coffee and tea

chairman **Derk Pik** (Vrije Universiteit Amsterdam)

11.00–11.35 **Art Frazho** (Purdue University, West Lafayette)
Constructing Interpolants with Commutant Lifting

11.40–12.15 **Leiba Rodman** (College of William and Mary, Williamsburg)
Factorization and Corona

Lunch break

chairman **Henk de Snoo** (Rijksuniversiteit Groningen)

14.00–14.35 **Ruth Curtain** (Rijksuniversiteit Groningen)
The Positive-real lemma

14.40–15.15 **Freek van Schagen** (VU Amsterdam)
Operator blocks and quadruples of subspaces: classification and the eigenvalue completion problem

Coffee and tea

chairman **Jan van Schuppen** (CWI Amsterdam)

15.30–16.20 **Leonid Lerer** (Technion, Israel Institute of Technology, Haifa)
Variants of Gohberg-Heinig theorems

16.35–17.25 **Rien Kaashoek** (VU Amsterdam)
The band method and generalized Carathéodory-Toeplitz interpolation

Saturday, November 15

Trip to Haarlem and dinner, organized by the Kaashoeks.

Rooms

All lectures will be held in room S 201 of the "Beta" building, De Boelelaan 1081.

For discussions and preparation of talks we have a small room available: on Wednesday afternoon this is room R 224, on Thursday room R 240, and on Friday R 224 again. These are located just around the corner of S 201.

List of Participants
Workshop Operator Theory and Analysis

Amsterdam, 12–14 November 1997

On the occasion of the sixtieth birthday of prof. dr. M.A. Kaashoek

J.A. Ball	Virginia Tech, Blacksburg, USA
H. Bart	Erasmus Universiteit, Rotterdam
P. Beneker	Universiteit van Amsterdam
H. den Boer	Itsec, Bonn, Germany
R.F. Curtain	Rijksuniversiteit Groningen
P. Dewilde	Technische Universiteit Delft
R.L. Ellis	University of Maryland, College Park, USA
A.E. Frazho	Purdue University, USA
K.-H. Förster	Technische Universität Berlin, Germany
S. Goldberg	University of Maryland, College Park, USA
B. Gramsch	Johannes Gutenberg Universität, Mainz, Germany
G. Groenewald	University of Western Cape, South Africa
Th. Hintermann	Birkhäuser Verlag AG, Basel
M.A. Kaashoek	Vrije Universiteit, Amsterdam
W. Kaballo	Universität Dortmund, Germany
J. Korevaar	Universiteit van Amsterdam
J. Kos	NLR
P. Lancaster	University of Calgary, Canada
H. Langer	Technische Universität Wien, Austria
L. Lerer	Technion, Haifa, Israel
C.V.M. van der Mee	Università di Cagliari, Italy
R. Mennicken	Universität Regensburg, Germany
E. Meister	Technishce Hochschule Darmstad, Germany
B. de Pagter	Technische Universiteit Delft
M. Petersen	University of Western Cape, South Africa
D. Pik	Vrije Universiteit, Amsterdam
A.C.M. Ran	Vrije Universiteit, Amsterdam
L. Rodman	College of William and Mary, USA
L. Roozemond	KSLA, Rijswijk
F. van Schagen	Vrije Universiteit, Amsterdam
Jan Schermer	Castricum
J.H. van Schuppen	CWI, Amsterdam
H.S.V. de Snoo	Rijksuniversiteit Groningen
D. Temme	IP-groep
G.Ph.A. Thijsse	Erasmus Universiteit, Rotterdam
P. Van Dooren	Université Catholique Louvain, Belgium
S.M. Verduyn Lunel	Vrije Universiteit, Amsterdam
R. Vreugdenhil	Hogeschool Utrecht
T.T. West	Trinity College, Dublin, Ireland
J.J.O.O. Wiegerinck	Universiteit van Amsterdam
H.J. Woerdeman	College of William and Mary, USA
A.C. Zaanen	Wassenaar
Rob Zuidwijk	CWI and Vrije Universiteit, Amsterdam

MIX
Papier aus verantwortungsvollen Quellen
Paper from responsible sources
FSC® C105338

If you have any concerns about our products,
you can contact us on
ProductSafety@springernature.com

In case Publisher is established outside the EU,
the EU authorized representative is:
Springer Nature Customer Service Center GmbH
Europaplatz 3, 69115 Heidelberg, Germany

Printed by Libri Plureos GmbH
in Hamburg, Germany